Lecture Notes in Earth System Sciences

Lev Eppelbaum • Boris Khesin†

Geophysical Studies in the Caucasus

Springer

Lev Eppelbaum
Dept. Geophysics and Planetary Sciences,
Faculty of Exact Sciences
Tel Aviv University
Tel Aviv 69978
Israel

Boris Khesin
Dept. of Geological and Environmental
Sciences
Ben-Gurion University of the Negev
Beer Sheva 84110
Israel

ISBN 978-3-540-76618-6 ISBN 978-3-540-76619-3 (eBook)
DOI 10.1007/978-3-540-76619-3
Springer Heidelberg New York Dordrecht London

Library of Congress Control Number: 2012938641

Printed on acid-free paper

Springer is part of Springer Science+Business Media (www.springer.com)

Preface

This book is dedicated to the memory of my teachers, colleagues and internationally renowned geophysicists, Prof. Boris E. Khesin (Azerbaijan – Israel) and Dr. Vyacheslav V. Alexeyev (Azerbaijan – Russia). I began to prepare this book together with Boris, but he died peacefully on November 2010 (see Eppelbaum (2011b)). Vyacheslav passed away tragically earlier in December 1998 in Russia.

The Caucasian Mountains occupy an area of about 440,000 km^2. A number of important mineral resources are concentrated there. Geophysical data on the geological structure of Caucasus can shed light on the basic principles of evolution of the Earth, the distribution of minerals and seismic activity. However, geophysical surveys under complex conditions are generally riddled by poor accessibility to certain mountainous regions, the unevenness of observation surfaces, as well as by the great variety and frequent changes of tectonic structures and geological bodies with variable physical properties. These factors either restrict geophysical surveys in difficult environments or confine the scope of useful information that can be obtained from the results. This has led to the development of special techniques in geophysical surveys, data processing and interpretation that draw heavily on the experience accumulated in the specific conditions of these mountainous regions.

The Caucasian mountain-fold belt is generally considered to have been a result of the interaction of several microplates – fragments of the Afro-Arabian and Eurasian lithospheric plates. Endogenic activity during the Hercynian and Alpine cycles, especially along the boundaries of interacting microplates, caused the formation of a number of large polymetallic, copper, gold-bearing, iron-ore, and other deposits. Within the sedimentary basins there are raw materials for the chemical and building industries. Some of sedimentary basins in the Caucasus and around it contain rich hydrocarbon deposits.

The following conditions are typical for the Caucasus: rugged terrain relief, a highly variable geological medium, oblique magnetization (polarization) and often an unknown level of the normal field. Besides these mountainous conditions that are difficult to traverse, there are also woods, swamps, rivers and lakes. In these areas, the influence of oblique magnetization is of paramount importance since these regions are studied primarily by magnetic survey. Some geophysical investigations have been conducted in small isolated areas (for instance, archaeogeophysical studies), where it is difficult to determine the level of the normal field.

This paints a composite picture of the complexities of Caucasian region. Nevertheless, almost all the methods and geophysical field interpretation procedures described in this book can be efficiently used in simpler geological situations.

The interpretation of geophysical fields in Caucasian conditions remains a challenging process. Successful interpretation calls for (a) knowledge of the geological regularities and the geological situation, (b) availability of petrophysical support, (c) mathematical methods to solve direct and inverse problems of geophysics, and (d) application of statistical, logical-informational, and wavelet procedures to the analysis and synthesis of observational data to reveal specific objects and the peculiarities of geological structures.

This book illustrates the advantages and uses of potential and quasi-potential field interpretation in the complex situation of the Caucasus. The peculiarities of the media and geophysical surveys are discussed. Results show that a common approach to the analysis of geophysical fields that cover magnetic, gravitational and thermal resistivity; self-potential (SP); electromagnetic field of very low frequency (VLF) transmitters and the field of induced polarization (IP) is not only possible but fruitful. This book introduces the reader to the informational content and the structure of interpretation processes developed by the authors. Our methods of reducing various noise effects, especially rugged relief, are exemplified. In particular, several ways of identifying the precursors of earthquakes from the noise background (some of which were developed by the authors) are discussed.

This book not only discusses the techniques recommended for geophysical field transformation and rapid methods of interpretation of anomalies in inclined surfaces and arbitrary polarization of objects. It also describes a process of 3D physical-geological modelling of gravity and magnetic fields in complex environments. It introduces the reader to different variants of integrated interpretation based on probabilistic or deterministic approaches, their reliability estimation and data presentation. The authors of this book (first of all, Boris E. Khesin) worked in the former USSR for many years. They recognized that geophysics in this country had been developing in isolation for many decades. This may help explain why different methods and technologies were simultaneously developed in the East and in the West. Therefore, the references mainly cover the works of Soviet authors. I hope that an acquaintance with this literature will interest Western readers who are conversant with Western investigators' contribution to geophysics.

The authors have attempted to present common overview of Caucasian geophysics that also includes their own work. At the same time I stress that this book should in no way be considered a detailed reference book on Caucasian geophysics since it was not intended to be so.

My goal in this book is interest not only geophysicists, geologists and other environmental specialists working in the Caucasus and students of the corresponding specialties, but also experts investigating other mountainous systems of the world.

Finally, I would like to thank Esther Singer for her detailed careful editing.

Tel Aviv, Israel
December 2011

Lev V. Eppelbaum

Contents

Biographies of the Authors

Lev V. Eppelbaum

Lev V. Eppelbaum has been involved in research and investigations of the Caucasus since 1977 when he began to work as a student-technician on a geophysical survey. His M.Sc. and Ph.D. theses both deal with integrated geophysical prospecting of polymetallic ores in the Greater and Lesser Caucasus. He has also played an active role in the development of regional physical-geological models of this tectonically very complex region and did not sever his ties with the Caucasus after immigrating to Israel in 1990. Currently Lev V. Eppelbaum is a Professor at the Department of Geophysical, Atmospheric and Planetary Sciences at Tel Aviv University. He is the author of more than 290 publications including 4 books, over 100 articles and 40 proceedings. His scientific interests range from potential and quasi-potential geophysical field interpretation to the combined analysis of different geophysical fields in the presence of noise, thus spanning the entire spectrum from highly detailed work in archaeological prospection to the much broader issues involved in regional constructions.

Boris E. Khesin

Professor Boris E. Khesin (1932–2010) worked in the Caucasus for more than 30 years and continued to study and write on geophysical-geological materials from this region after his immigration to Israel in 1990. In Israel he was a Professor at the Department of Geological and Environmental Sciences at Ben-Gurion University of the Negev. Professor Khesin was the founder of the scientific school of integrated geophysical studies in complex mountainous environments. He authored more than 260 publications including 10 books and monographs. Professor Khesin truly made an enormous contribution to geophysical studies of the Caucasian region in the fields of regional and ore geophysics, hydro-geophysics and geophysical-geological mapping. He had an extraordinary breadth of scientific views and a remarkable capacity for work. His professional activity, accomplishments, and seminal discoveries are acknowledged worldwide.

Chapter 1
Introduction

The Caucasus is a large mountainous area covering about 440,000 km^2 between the Black, Azov and Caspian Seas (Fig. 1.1). The mountains are more than 1,000 km long, in a WNW-ESE striking range between the Caspian and Black Seas and reach elevations of over 5 km above sea level.

The Caucasus, one of the most attractive mountain regions in the world, is the corridor between two continents. In addition to cities and villages, the region has numerous lakes, rivers, forests, abundant economic minerals, enigmatic archaeological remains, and many other natural and artificial sites that attract resort-goers as well as geological and environmental scientists and engineers.

This book describes the methodology and results of numerous exploratory geophysical studies in the Caucasus focusing on mineral prospecting, petroleum and water and describes the specific features of the region (mud volcanoes, engineering geology, archaeology, and seismic hazards).

The Caucasian isthmus between the Black and Caspian seas within the Mediterranean mobile belt has a luxuriant geological history and numerous mineral resources.

There are large oil and gas deposits (Azerbaijan and North-Caucasian regions of the Russian Federation), copper and molybdenum (Armenia), manganese (Georgia), magnetite, alunite and polymetallic ores (Azerbaijan), tungsten and molybdenum ores (North Caucasus) (Andruschuk 1968; Aslanyan 1970; Azizbekov 1972; Smirnov 1978; Tvalchrelidze 1976, 1978; Borodaevskaya et al. 1977; Sokolov and Khain 1982; Gasanov 2001; Baba-Zadeh et al. 2002; Markus (2002); Khain and Alizadeh 2005). Many other ores and non-metallic resources are well known such as Caucasian mineral waters (Shempelev et al. 2005), Caucasian gold deposits (Kholodov and Kiknadze 1989; Bogush et al. 2004; Kekelia et al. 2004, 2008), coal and copper-pyrite ores in Georgia (Gamkrelidze 1964; Tvalchrelidze 2002), lead-zinc and copper ores in the North Caucasus, and uranium mineralization in Armenia and the North Caucasus (Smirnov 1978).

Geophysical methods in the former USSR were mainly designed for mineral prospecting in the Caucasus which is characterized by the complex topographical, geological and geophysical features typical of this region (Eppelbaum 2012).

L.V. Eppelbaum and B.E. Khesin, *Geophysical Studies in the Caucasus*,
Lecture Notes in Earth System Sciences, DOI 10.1007/978-3-540-76619-3_1,

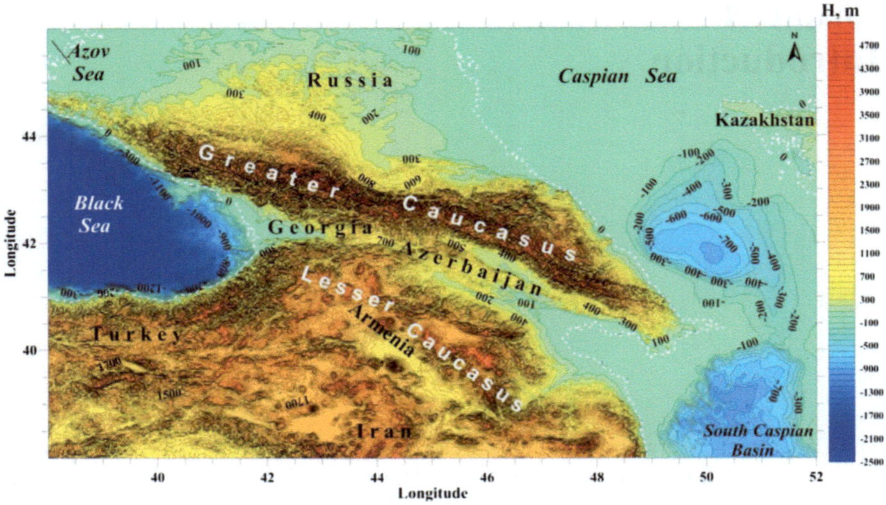

Fig. 1.1 Caucasus region map with topography and bathymetry data (constructed on the basis of the World Topography Database (Smith and Sandwell 1997))

The Caucasus form a wide variety of environments and the range of geophysical experiences is enormous.

The authors worked in Azerbaijan for many years and participated in geophysical studies of other Caucasian countries. They carried out field surveys, research and development of effective methodologies for complex environments such as uneven observation surfaces (including mountain/sea transition zones with disturbed patterns of gravity fields), inclined (oblique) magnetization (which is characteristic of temperate and tropical latitudes of the Earth), and variable media with a wide spectrum of physical properties of rocks (e.g., Khesin 1969, 1976, 1981; Khesin et al. 1983, 1988, 1996; Eppelbaum 1989; Eppelbaum and Khesin 2002, 2004, 2011). The principles of optimal methodology were developed for: (1) fieldwork in mountainous regions based on a combination of natural geophysical field mapping with electric soundings and seismic prospecting along reference profiles; (2) analysis and utilization of uneven topography in geophysical study zones to obtain new geological information; (3) statistical modeling of complex geophysical fields based on combined data on targets such as node intersections, concentric structures and other typical targets (Khesin 1978b, 1981, 2005; Khesin et al. 1993b, 1996; Khesin and Eppelbaum 1997; Eppelbaum 2007a).

This integrated approach and the modeling of geophysical data greatly facilitate the detection of important geological features. For example, hidden intersections of linear structures can identify the location of large commercial ore and oil and gas deposits and can reveal dangerous geodynamic events at a depth. For instance, advanced inverse problem solution techniques and 3D modeling of geophysical fields under conditions of inclined polarization and uneven topography with the utilization of a huge petrophysical database successfully captured the features of the

Caucasus deep structure and its near-surface peculiarities. By taking the idiosyncrasies of the Caucasus into account when developing methodologies and techniques important problems have been solved, including the identification of new mineral resources and the prevention of dangerous geodynamical phenomena. On the basis of geophysical studies (mainly mobile and low cost magnetic and electric methods), a new copper-polymetallic province in the southern slope of the Greater Caucasus was discovered. These studies have also facilitated the discovery of new deposits and reserves of iron, gold, kaolin and other ores in the Lesser Caucasus and new sources of underground water in sub-mountainous and mountainous areas, including freshwater to supply Baku (the capital of Azerbaijan). Examination of magnetic and other geophysical data by deep and super-deep well drilling has indicated that the magmatic rocks of the Lesser Caucasus in fact extend northward under the thick sedimentary cover of the Kura Depression up to the Greater Caucasus. These rocks form hidden petroleum-bearing traps of a new type. In addition to its prospecting value, this result is important for our understanding of the geological structure of one of the key regions of the Earth where different tectonic hypotheses are tested.

The first oil well in the world was drilled in Azerbaijan in 1847 in the Bibi-Heybat area. More than half of the world's oil production was located in Baku at the turn of the nineteenth century(e.g., Balat 2006). Marine electric prospecting started in Azerbaijan as early as in the 1930s (Litvinov 1941). After the discovery of the first offshore oil field "Neft Dashlary" ("Petroleum Stones") in 1949, marine seismic prospecting began midcentury (e.g., Shapirovsky and Ganbarov 1961; Rapoport et al. 1964). The Caspian Sea region contains oil and gas reserves that are comparable to those of other fossil-fuel-producing regions in the world, excluding the Middle East (e.g., Rabinowitz et al. 2004). Azerbaijan's oil production increased considerably once again as of 1997 (Balat 2006). The exploitation of hydrocarbons requires huge international collaboration to build a pipeline system (e.g., Tanircan et al. 2011). Its design needs to incorporate knowledge of seismic hazards including deep geophysical data on seismic-active zones. Regional mapping of active faults has successfully pinpointed vast seismic hazards for pipelines within the Mediterranean belt (Fig. 1.2).

For instance, Fig. 1.2 shows the first version of the planned Baku (Azerbaijan) – Ceyhan (Turkey) pipeline that was located within the active East Anatolian Fault (EAF in Fig. 1.2) zone. Subsequently the pipeline was relocated to a longer route that skirts the fault: in Turkey, the route runs south to Erzurum and then west to Erzincan and Sivas before turning south to the Ceyhan terminal (Balat 2006). This illustrates the role of the Caucasus as an east-west energy corridor, in addition to its value and interest for researchers studying the Earth's crust. Thus, the findings and methodological experience have implications for a broad swath of specialists in the field of Earth sciences (e.g., Kerimov 1996; Alizadeh 2005). New geophysical developments reported here will naturally appeal the most to geologists and scientists in related areas, whereas the methodological advances should primarily attract geophysicists.

The authors' goal in writing this book is to combine their own personal experience with the modern achievements of the wonderful community of Caucasian

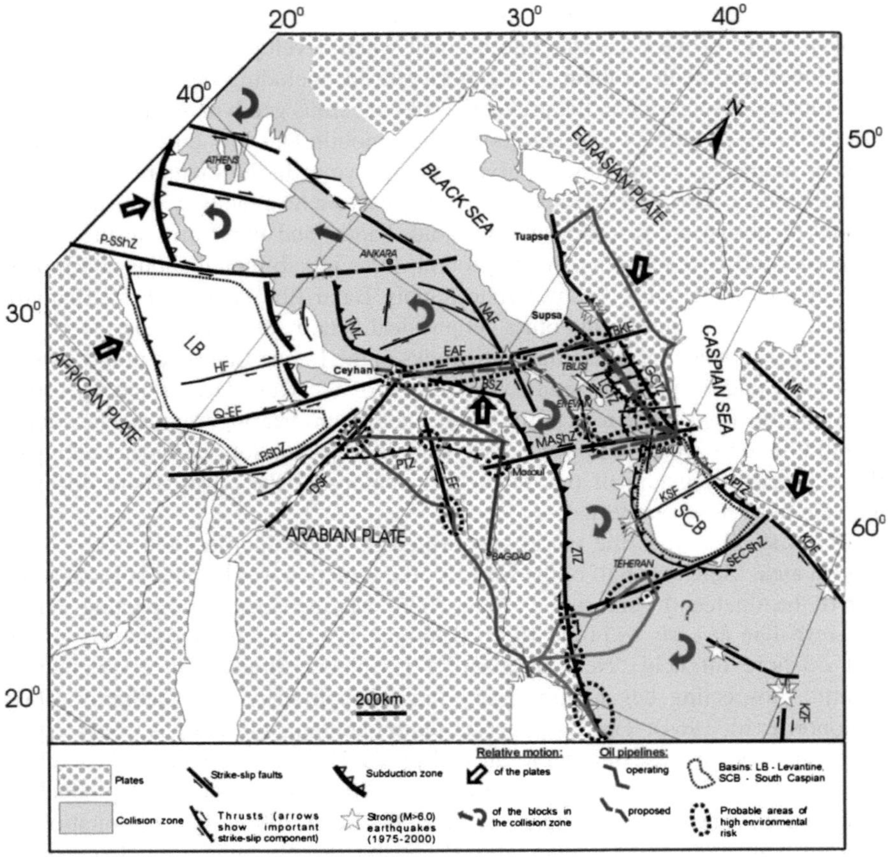

Fig. 1.2 Major fault zones within the Mediterranean belt and adjacent regions (After Khesin and Metaxas 2000)

exploration-geophysicists. The authors studied under some of the pioneers of Soviet geophysics. We have cooperated with many others and have initiated younger colleagues. We are grateful to geophysicists such as V. V. Fedynsky and A. I. Zaborousky, who started in Azerbaijan, where our friends V. V. Alexeyev, R. M. Gadjiev, T. A. Ismail-Zadeh, A. Sh. Mamed-Zadeh, S. A. Muradkhanov, A. L. Shakhnazaryan worked later, including M. A. Alexidze, B. K. Balavadze, G. E. Gugunava, T. L. Chelidze, in Georgia, S. V. Badalyan, G. B. Petrosyan, G. M. Vanyan, in Armenia, A. G. Shempelev, A. P. Vinogradov in the North Caucasus region of the Russian Federation (i.e. in southern part of the Giscaucasus, including the northern immersion of the Greater Caucasus). We apologize to all those not mentioned in person because of the size limitations of this book, however but all our colleagues have had an invisible hand in this volume and, for the most part, are cited in the bibliography.

Chapter 2
Tectonical-Geophysical Setting of the Caucasus

2.1 The Origin of the Caucasus, Geological Evolution and Main Features

Many outstanding geologists such as G. V. Abich, I. M. Gubkin, V. E. Khain, K. N. Paffenholtz studied the Caucasus. The Caucasus comprises four main morphological and tectonic units (Khain and Koronovsky 1997): (1) the Ciscaucasian plain (Scythian platform), including the foredeeps of the Greater Caucasus; (2) the Greater Caucasus itself, stretching in a WNW-ESE direction; (3) the Transcaucasian system of intermontane basins, and (4) the Lesser Caucasus with its an arcuate N-convex shape and the most heterogeneous structure. This geological division of the Caucasus is traditional, although it is sometimes modified (Fig. 2.1).

One of the main characteristics of the Caucasus region is the complexity of its active tectonics, which exhibit both compressive structures such as reverse and strike-slip faults, and extensional features, such as normal faults (Rebai et al. 1993).

Today, the origin and evolution of the Caucasus are mainly explained in terms of plate tectonics (e.g., Khain and Ryabukhin 2002). In a brief review, Kopf et al. (2003) noted that the present-day Caucasus is dominated by thrust faulting due to continental collision. From the Jurassic to the Paleogenic eras, subduction of the Tethian seafloor occurred along the southern margin of the Turkish and Iranian blocks, resulting in calc-alkaline arc volcanism and a wide backarc basin system. The spread of the Red Sea began during the Early Miocene, and the Arabian Plate migrated northward, accompanied by a reduction in width of the Tethys. After its closure (~20 Ma), subduction shifted to the north. As a result of the indentation of the Arabian block, the continuous backarc basin was separated, and the oceanic crust only remained in the Black Sea and the southern Caspian Sea. The continuous northward drift of the Anatolian plate led to initial continental collision expressed by the formation of the Lesser Caucasus, and the subsequent resurrection of the Greater Caucasus during the Middle Pliocene. Currently, continental convergence continues at a rate of up to ~30 mm per year along strike slip faults, where most of the modern tectonic activity is localized. For example, the intersection of deep

L.V. Eppelbaum and B.E. Khesin, *Geophysical Studies in the Caucasus*, Lecture Notes in Earth System Sciences, DOI 10.1007/978-3-540-76619-3_2,

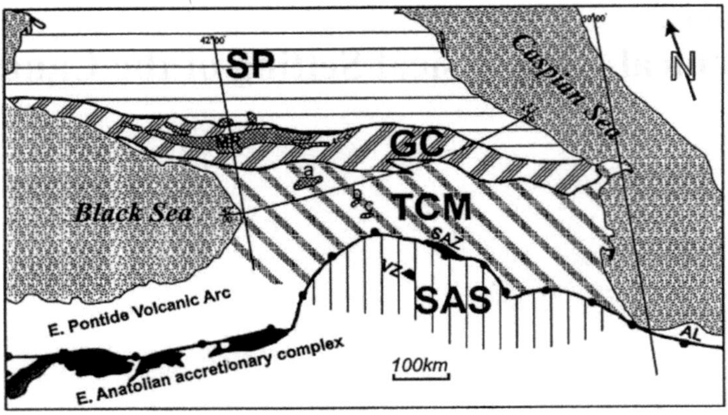

Fig. 2.1 Major crustal units of the Caucasus (After Zakariadze et al. 2007). *SP* Scythian Platform, *GC* Greater Caucasus (*MR* Main Range), *TCM* Transcaucasian Massif ((*a*) Dzirula, (*b*) Khrami, (*c*) Loki salients), *SAS* South Armenian Subplatform, *SAZ* Sevan-Akera ophiolite zone, *VZ* Vedi ophiolite zone, *AL* Albortz Mountains. The line with *black dots* indicates the Eastern Pontide – Lesser Caucasus paleo-oceanic suture zone and corresponds to the southern border of the Eastern Pontide and Caucasus Hercynides

faults was detected at the northern point of sharp tectonic wedge where the destructive Spitak earthquake of 1988 (coordinates: 40°49′ N, 44°15′ E) took place (Fig. 2.2).

Gadjiev et al. (1989) identified this feature from this map of the Caucasus (Fig. 2.2) by calculating the lithospheric heterogeneity at different depths. The sum total length of the lineaments within square cells with different sides (corresponding to cubic blocks of different depths) served as a measure of heterogeneity – one that had previously been suggested (Khesin and Metaxas 1974; Khesin 1981).

Note that this map of the deep structure of the Caucasian lithosphere (Fig. 2.2) is consistent with the depiction of horizontal gradients of isostatic and lithospheric gravity anomalies shown in Fig. 2.3.

The presence of a mosaic of microplates in the complex structure of the Caucasus led to additional modifications of plate tectonic theories such as two-stepped plate tectonics (L. I. Lobkowsky) and "shoal tectonics" (I. I. Abramovich), which argues that crust fragments move along inter-crust astenolenses. Shoal tectonics is exemplified by the differences in the Paleogene evolution of the Lesser-Caucasian "andesite belt" between the areas of sluggish crust subduction (Megri-Ordubad Pluton) and the energetic lateral heat/mass transition (Tezhsar Complex in the Pambak ridge).

V. E. Khain (e.g., 1984, 1991–1993, 1995, 2000, 2007) described the geology of the Caucasus extensively. He defined three stages in the evolution of this region during the last 1 Ga: (1) the Baikalian (mainly, the Neoproterozoic) which produced the consolidated basement of the Transcaucasian Massif, (2) the Hercynian (Paleozoic), and the Alpine (Mesozoic-Cenozoic). According to Khain (2000), the most

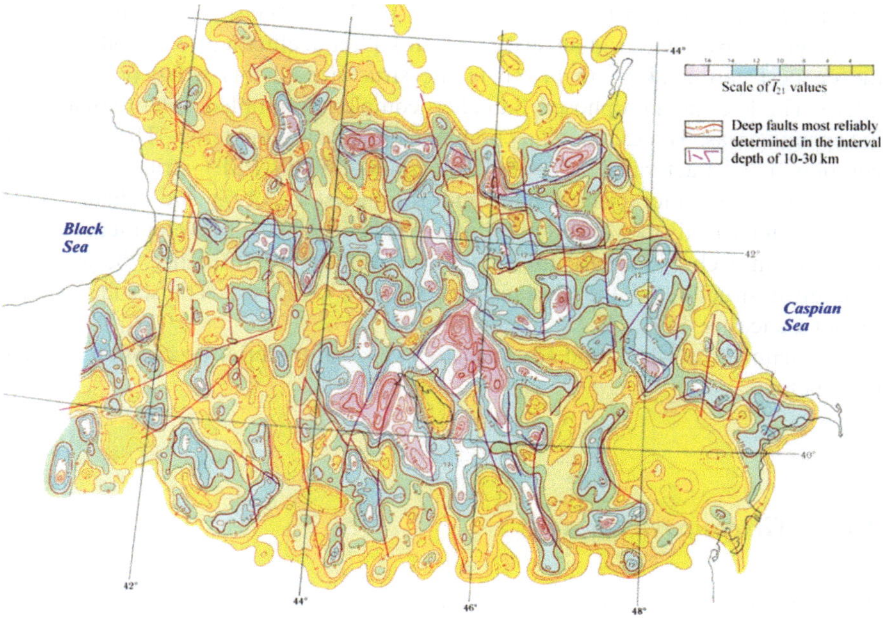

Fig. 2.2 Deep structure of the Caucasian lithosphere (the cut at ≈10 km in depth) according to satellite data (After Gadjiev et al. 1989)

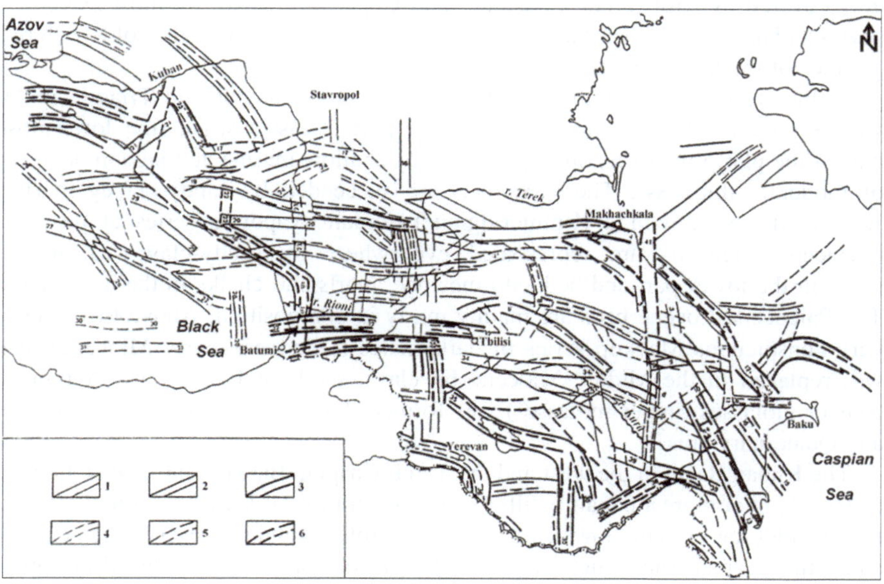

Fig. 2.3 Diagram of horizontal gradients of isostatic and lithospheric gravity anomalies (After Gorshkov and Niauri 1984). Zones of horizontal gradients of isostatic anomalies in mGal/km: (*1*) 0.5–1; (*2*) 1–2; (*3*) 3, zones of horizontal gradients of lithosperic anomalies in mGal/km; (*4*) 0.5–1; (*5*) 1–2; (*6*) 2

ancient Pre-Baikalian structural complex (mainly, gneisses and marbles) is characterized by sub-meridian strikes, often with a declination to the NE. In fact, the deepest earthquakes are linked to the structures of this strike (e.g., Khesin 1976). The less metamorphosed Baikalian complex is rumpled to latitudinal folds in separate areas. The Caledonian complex is practically unknown. The Hercynian complex is characterized by the "Caucasian" (WNW-ESE) strike as are the overlaying Mesozoic rocks. Khain and Koronovsky (1997) and Khain (2000, 2007) emphasized the differences in tectonic-magmatic development of the Caucasus in each of the last three stages, as well as the differences between the geological structure of the Ciscaucasus, Greater Caucasus, Transcaucasus, and Lesser Caucasus (Fig. 2.4).

We mainly cite Khain and Koronovsky (1997) in the brief description of these units below.

2.1.1 Giscaucasus

In the southern part of the Ciscaucasian platform, directly adjoining the Greater Caucasus but separated from the latter by a deep fault, is its basement which dates to the Late Proterozoic age and is represented by greenshists and crystalline schists of amphibolite facies. The Upper Proterozoic metamorphic complex is unconformably covered by Ordovician sandstones and Upper Silurian-Lowermost Devonian shales and limestones. This basement is pierced with numerous bodies of K-rich red granites of the Late Paleozoic age.

In the western part of the platform, terrigenous and volcanic Triassic rocks were discovered by drilling. In its eastern part, Triassic strata (the lowermost part of the platform cover) are represented by shallow marine and lagoon deposits of moderate thickness. These deposits are succeeded by sandy-clayey marine Jurassic deposits, with a Kimmeridgian-Turonian evaporate series at the top. Cretaceous deposits are represented everywhere by sandy-clayey-calcareous rocks in the lower part and bedded limestone, marls and chalks in the upper part. The Paleocene-Eocene beds are mainly marly in composition. Above them lies a very conspicuous Maykop series of dark shales (Oligocene-Early Miocene) that was replaced in the Middle Miocene by clays, sands and limestone. A typical coarser molasses (upper molasses) was formed during Upper Miocene, Pliocene and Quaternary eras.

The Kuban Basin in the west and the Terek-Caspian Basin in the east at a depth of 10–12 km or more are mainly filled by the upper molasses. In the Terek foredeep, two parallel lines of anticlines contain large petroleum and gas reserves and extend along the Caspian shore into Dagestan. The Terek foredeep is replaced in north-eastern Azerbaijan by the narrow Kusar-Divitchi foredeep, which is shifted to the south with respect to the former.

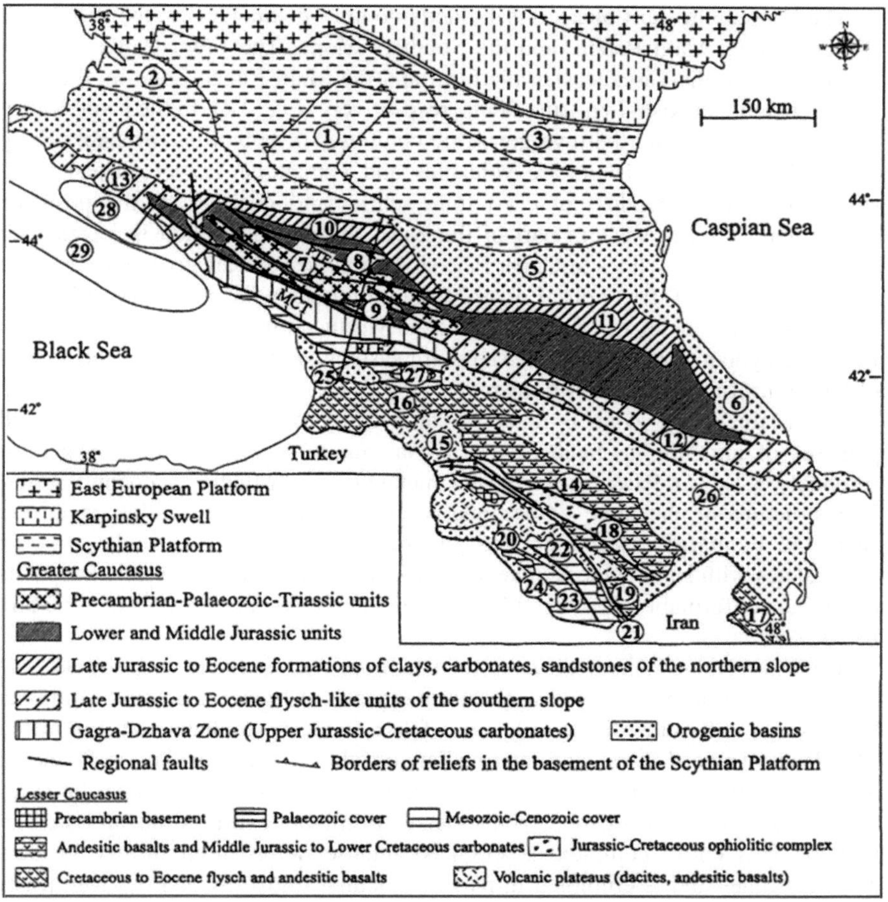

Fig. 2.4 Simplified geological map of the Caucasus (After Milanovsky and Khain 1963; Saintot et al. 2006; with slight modification of the captions). *PTF* Pshekish-Tyrnauz Fault, *MCT* Main Caucasian Thrust, *RLFZ* Racha-Lechkhumy Fault Zone. Circled numbers: *1–6* are zones of the Scythian Platform: (*1*) Stavropol High; (*2*) Azov-Berezan High; (*3*) Manych Basin; (*4*) Kuban Basin; (*5*) Terek-Caspian Basin; (*6*) Kusar-Divitchi Basin; *7–13* are zones of the Greater Caucasus (GS): (*7*) Peredovoy Zone; (*8*) Betcha Anticline; (*9*) Svanetia Anticline; (*10*) Laba-Malka Monocline; (*11*) Dagestan Folded Zone; (*12*) Flysch Zone of southeastern GC; (*13*) Flysch Zone of north-western GC; *14–24* are zones of the Lesser Caucasus: (*14*) Somkhet-Karabakh Zone; (*15*) Artvin-Bolnisi Zone; (*16*) Adzharo-Trialet; (*17*) Talysh; (*18*) Sevan-Akera; (*19*) Kafan; (*20*) Vedin; (*21*) Zangezur; (*22*) Mishkhan-Zangezur Massif; (*23*) Ararat-Djulfa Massif; (*24*) Araks Basin; *25–29* are intramontane zones of the Transcaucasus and Black Sea: (*25*) Rioni Basin; (*26*) Kura Basin; (*27*) Dzirula Massif; (*28*) Tuapse Basin; (*29*) Shatsky Ridge

2.1.2 The Greater Caucasus

This mountainous edifice stretches over a distance ~1,300 km and attains a width of 150–200 km in its central part. Its highest summits reach 5,642 m (the comparatively recently (~2000 years) extinct volcano Elbrus) and 5,033 m (another volcano, Kazbek).

The Central Greater Caucasus is divided into three distinct zones: the Peredovoy (Front) Range, Main Range, and the Southern Slope. The Front Range Zone is separated from the Ciscaucasian platform and the Main Range Zone by the deep Northern and Pshekish-Tyrnyauz faults, respectively (Fig. 2.4). In the Front Range Zone, the nappes of northern vergence are composed of ophiolites (probably Early Paleozoic), Devonian island-arc volcanics and Famennian-Lower Visean limestone. A tectonic window of crystalline schists and gneisses of the probable autochthon exists in western part of the zone. Its neo-autochthon is represented by superimposed synclines of Upper Paleozoic molasses that are coal-bearing in the lower part (C_2). Terrigenous Lower-Middle Jurassic with some volcanics lies unconformably on all the older rocks. It is worth noting that the fragments of Early-Hercynian copper-sulphide island-arc system can be seen within the nappes of the Front Range and Main Range.

The Main Range Zone is made up of a complex of crystalline schists and gneisses of amphibolite facies and a K-Na granite series: the metaterrigenous (probably a Lower Paleozoic age) and metavolcanic (Middle Paleozoic) series. The structure of the zone domes and swells formed by crystalline rocks are separated by narrow, steep graben-synclines filled by Lower-Middle Jurassic black slates with some intermediate and felsic volcanics. Marine Permian deposits overlie unconformably the crystalline complex at the southern border of the Main Ridge Zone.

The Main Ridge Zone is thrust to the south onto the Southern Slope Zone. The zone was built up mainly by a very thick sequence of Lower and Middle Jurassic black slates with subordinate volcanics: felsic at the base, diabase-spilite in the Middle and Upper Liassic, and intermediate or felsic in the Aalenian. In Svanetia, Triassic and Paleozoic rocks appear under the Jurassic ones. A thick Upper Jurassic-Cretaceous-Lower Paleogene flysch series conformably overlays the Jurassic formation. This entire complex experienced intense deformation mostly in Late Miocene times and is crumpled in narrow isoclinal folds, divided by overthrusts with a general southern vergence. The zone terminates in the south by a large overthrust where it bounds with the Gagra-Dzhava Zone, which is transitional to the Northern Transcaucasus.

The Gagra-Dzhava Zone is characterized by an ancient crystalline basement and the development of a so-called porphyritic, island-arc volcanic Bajocian series, Bathonian coal measures, Kimmeridgian-Tithonian red beds that are replaced with Upper Jurassic reef limestones in the north, and marly-calcareous Cretaceous and Lower Paleogene series.

In the axial part of the eastern segment of the Greater Caucasus, the Lower-Middle Jurassic slate formation is very tightly folded into two narrow anticlinoria – the Main Range and the Side Range, divided by an even narrower synclinorium. To the north of the Side Range the Upper Jurassic reef formation expressed in the relief by the Rocky Range, is succeeded by Lower-Cretaceous-Cenomanian terrigenous-carbonate, Upper-Cretaceous carbonate and Lower Paleogene marly series. This whole sequence occupies a large strip of Dagestan, where it is folded in a series of box-shaped anticlines divided by narrower synclines.

In Azerbaijan, the Greater Caucasus edifice plunge under the Kusar-Divitchi foredeep. The synclinorium between the Side Range and Main Range is filled by the Cretaceous wildflysh. The main flysch zone of the southern slope enlarges, and the eastern end of the Gagra-Dzhava (here termed Kakheti-Vandam) zone plunges under a thick Oligocene-Miocene unconformable series overlain by Pontian (uppermost Miocene) and Ahchagyl (Upper Pliocene) beds. This is the Shamakha-Qobystan Zone, separated from the Lower Kura molasses basin in the south by the overthrust. Pliocene deposits comprising the petroleum-rich Productive (or Balakhan) Series at their base occupy most of the surface of the Absheron Peninsula and Caspian shore to the south of it. Older rocks appear here in the cores of diapiric anticlines and mud volcanoes are abundant.

Similar changes took place along the strike of the Greater Caucasus edifice in the western direction, but over a shorter distance and accordingly more abruptly.

In the southern slope of the Greater Caucasus, copper-pyrrhotite stratiform and vein ores within Jurassic slates (e.g., Filizchay or Djikhikh deposits in Azerbaijan) were not initially considered to be a significant source of mineral resources. However, geological-geophysical studies, beginning in the middle of twentieth century revised this opinion (see Sect. 2.2.2).

2.1.3 The Transcaucasus

The Dzirula salient of the pre-Mesozoic basement within the Northern Transcaucasian intermontane region represents a zone of transverse Okriba-Dzirula uplift (salient), which divides the main molasses basins into the Rioni Depression in the west and the Kura Depression in the east (see Figs. 2.1 and 2.4). The Upper Precambrian crystalline schists, paragneisses and amphibolites of the salient are the oldest. On them are thrust rocks composed of a Lower or Middle Paleozoic meta-ophiolitic mélange. Both complexes are intruded by Late Paleozoic granites and overlain by Late Visean – Early Serpukhovian acid tuffs. A Mesozoic-Cenozoic cover overlays these Paleozoic rocks with a strong unconformity, beginning with Liassic-Aalenian terrigenous and carbonate deposits, Bajocian island-arc intermediate volcanics ("porphyritic series"), Bathonian limnic coal measures and Kimmeridgian-Tithonian red beds including alkaline basalts. In the west there are also Upper Jurassic evaporites, Cretaceous-Lower Paleogene limestones and marl strata, including Cenomanian volcanics. All these folded strata plunger under the younger deposits of the Rioni and Kura Basins. Oligocene-Miocene deposits fill two narrow troughs on the northern and southern periphery of the Okriba-Dzirula uplift, corresponding to seaways which connected the Rioni and Kura molasses basins up to the Late Miocene.

The Rioni Basin is not as deep (up to 9 km according to geophysical data) and is structurally simpler than the Kura Basin. The main phase of deformation took place in the Late Miocene, and younger deposits lie almost horizontally, except at the southern border of the basin, where the Adzharo-Trialet folded system of the NW framing of the Lesser Caucasus is thrust over the Pliocene deposits.

The Kura Basin extends from the Dzirula Massif to the Caspian shore and continues offshore in the South Caspian deep basin. This basin can be divided into two sub-basins (depressions) – the Middle Kura and Lower Kura, which differ in their width and inner structure. The Middle Kura Depression is bordered on the north by the Gagra-Dzhava marginal zone of the Greater Caucasus and its semi-buried eastern extension and on the south by the Adzharo-Trialet folded system and the northwestern slope of the Lesser Caucasus. The depth of the premolasse basement increases to the east, reaching 15 km according to geophysical data. In the west, the molasses only comprises Oligocene-Miocene marine beds that were folded by the end of the Miocene or the beginning of the Pliocene, whereas in the east they are supplemented by a very thick series of predominantly coarse clastic (partly continental, partly very shallow marine) strata of the uppermost Miocene-Lower Pleistocene, folded by the time of the Middle Pleistocene.

The Lower Kura Depression is bordered on the north by the Shamakha-Qobystan Zone of the southeastern Greater Caucasus. The main part of this depression, between the lower course of the Kura river and the foothills of Greater Caucasus, is filled with a very thick series of Oligocene-Quaternary deposits (including the Lower Pliocene Productive Series), that was deformed in the Middle Pleistocene into several strings of brachyanticline bearing large mud volcanoes. These strings continue in the Caspian offshore, where the mud volcanoes are expressed by islands or submarine rises. The strike of the anticline zones comes progressively closer to meridional offshore. Only the northernmost folds of the Absheron offshore preserve their "Caucasian" WNW-ESE strike and continue with it to the east, forming the Absheron Sill in the sea bottom relief between the Middle Caspian and the South Caspian Basins. The thickness of the sedimentary cover in the South Caspian Basin is very great. The Productive Series contains large petroleum and gas deposits here as well as on shore.

Southwest of the Lower Kura Valley the thickness of the Oligocene-Quaternary deposits diminishes, partly due to the wedging out of the Productive Series. Near the confluence of Kura and Araks a large buried salient of deep rocks was formed by the Talysh-Vandam gravity maximum (Fedynsky 1937), where later the first Soviet super-deep (SD-1) borehole was planned.

2.1.4 The Lesser Caucasus

The Lesser Caucasus comprises parts of the former North Transcaucasian and Iranian microcontinents and remnants of the oceanic basin separating them. The Somkhet-Karabakh zone of the Transcaucasian microcontinent (the Transcaucasian Massif) was uplifted in the post-Eocene and is characterized by the echeloned Shakhdag, Mrovdag and Karabakh ridges that are more than 3 km in height. Metamorphic rocks of the Pre-Mesozoic basement constitute outcroppings in the

Khrami and Loki, and other minor salients and are, in general, similar to the rocks of the Dzirula salient. A thin Lias-Aalenian terrigenous sequence at the base of the Mesozoic section is covered by a thick Bajocian-Bathonian arc volcanic series. The Middle Jurassic volcanism was followed by Bathonian-Neocomian granite intrusions. In the Late Jurassic, reef carbonates accumulated along the NE slope of the volcanic arc and evaporates formed behind them. Then, a new volcanic arc appeared, superimposed on the Middle Jurassic arc. After the cessation of island-arc volcanism in the middle of the Senonian, a cover of Upper Senonian limestone and Lower Paleogene marls accumulated in the Somkhet-Karabakh zone.

The Middle Jurassic volcanics of the Somkhet-Karabakh zone are bordered in the south and south-west by overthrusts of the Sevan-Akera paleo-oceanic zone (Fig. 2.4) which is remarkable for the development of ophiolites. These ophiolites date to the Late Paleozoic and Mesozoic up to Cenomanian. They form tectonic nappes and were obducted on the Transcaucasian and Iranian continental blocks during the Turonian-Coniacian. Then, Senonian general subsidence caused the accumulation of the limestone formations mentioned earlier.

The southernmost zone of the Lesser Caucasus belongs to the former Iranian microcontinent or, according to another opinion, to a separate continental block. Its basement is exposed in the Mishkhan and Western Zangezur massifs and is represented by different metamorphic schists, amphibolites and marbles (Late Proterozoic or earliest Paleozoic?) with intrusive bodies of plagiogranites. A younger Aparan metamorphic series has been identified in the Mishkhan Massif. Both metamorphic complexes are cut by Late Jurassic– Early Cretaceous granitoids, transgressively covered by Upper Cretaceous limestone. To the south and south-west of the belt of metamorphic massifs, a Devonian-Triassic sequence of terrigenous (in the lower part) and carbonate (in the upper part) deposits are widely developed. At its top, coal-bearing beds are present. The Upper Carboniferous is missing, and Permian beds cover the Lower Carboniferous with a crust of weathering at their base. The Mesozoic sequence, lying unconformably on the Paleozoic-Triassic, includes Middle Jurassic marls, Upper Cretaceous limestone, Paleocene flysch and Eocene intermediate volcanics. This sequence of the Yerevan-Ordubad (Ararat-Djulfa) Zone is supplemented in the vicinity of Yerevan by a shallow marine sandy-clayey Oligocene.

Eocene magmatic activity was not confined solely to the Yerevan-Ordubad Zone, but rather was widespread throughout the central and southern Lesser Caucasus. Among the plutons, most belong to the granitoid family, e.g., the large Dalidag Pluton in the eastern part of the Sevan-Akera zone, or the even larger and polyphase Late Eocene-Miocene Megri-Ordubad Pluton in the northeastern part of the Yerevan-Ordubad Zone. A pluton of nepheline syenites has been identified in the Pambak Ridge to the west of Lake Sevan.

The formation of the modern structure of the Lesser Caucasus and its general uplift began toward the end of the Eocene. In the Early Miocene the formation of superimposed structures began. The largest of these basins is the Araks (Middle Araks) Basin. These basins contain lacustrine, shallow marine and evaporitic Miocene sediments with embryonic manifestations of halo kinesis.

In the latest Miocene, Pliocene and Pleistocene a vast area of the southwestern part of the Lesser Caucasus became an arena of strong volcanic activity, forming extensive lava plateaus continuing to the west in Eastern Anatolia. Several strato-volcanoes emerged amidst these plateaus; the highest is Mt. Aragats in Armenia (4,095 m).

The Adzharo-Trialet and Talysh Zones (Fig. 2.4) are set apart in northwest and southeast periphery of the Lesser Caucasus, respectively, but have much in common. The Adzharo-Trialet ridge, with summits up to 2,850 m was built up by Albian-Lower Senonian island-arc volcanics, Upper Senonian limestone, Paleocene-Lower Eocene tuffaceous flysh, and Middle-Upper Eocene subalkaline and alkaline intermediate volcanics. The latest Eocene folding was accompanied by small syenite-diorite intrusions. The Talysh Mountain fold Zone, which is elevated up to 2.4 km, does not belong to the Lesser Caucasus proper, because it is separated from it by the Lower Araks Oligocene-Quaternary northeastern strike depression, a branch of the Lower Kura Depression. The oldest rocks exposed in the Talysh are Upper Cretaceous limestone, followed by Paleocene-Lower Eocene tuffaceous flysch, in turn unconformably overlain by Middle-Upper Eocene subalkaline and alkaline andesite-basalt volcanics. The whole sequence is very similar to that of the Adzharo-Trialet Zone. Similarly, the Talysh experienced intense deformation, including the formation of nappes of northern vergence by the end of the Eocene. Thick Oligocene-Miocene molasses accumulated to the northwest and north of the uplifted part of the Talysh. They were folded and thrust to the north on more recent strata of the Lower Kura Depression in the latest Miocene era.

Many renowned specialists have studied the metallogeny of the Caucasus, including Abdullayev (Abdullayev et al. 1962), Magakyan (1961), Smirnov (1978), G. Tvalchrelidze (1976), A. Tvalchrelidze (2002), and their disciples. The Somkhet-Karabakh Zone (Fig. 2.4) with its numerous ore deposits and in particular the well-known Dashkesan-Kedabek (Gedabey) ore region have been the most extensively investigated. The copper-pyrite (sulphide) ores in the Alaverdy and Gedabey deposits, the magnetite ores in the Dashkesan iron deposit have been exploited for decades. The Dashkesan deposit was the sole source of cobalt for the Soviet industry during the Second World War. The pyrite stocks of the Gedabey deposits (in exocontact with the Gedabey intrusive) in the upper horizons converse to copper-pyrite and copper-zinc ores. Most geologists believe that the pyrite ores of the Lesser Caucasus are related to small Later Bajocian sub-volcanic rhyolite-dacite (quartz-plagioporphyry) bodies, whereas the copper-pyrite and copper-zinc ores are connected to the postmagmatic activity of the Gedabey intrusive.

Commercial iron, cobalt and alunite ores in Upper Jurassic carbonate-pyroclastic rocks are located in exocontact with the Dashkesan polyphase intrusive. A characteristic feature of the Dashkesan iron deposits is their virtually horizontal bedding of magnetite ores. Mustafayev (2001) suggested that this feature can be attributed to the formation of the Dashkesan intrusive in the synclinorium, unlike the similar intrusives in positive structures of the Lesser Caucasus; in the Dashkesan area, a strain in continental rifting generated sub-horizontal brecciated zones on contacts of rocks with different compositions. Sheet-like and lens-like iron bodies were formed from hydrothermal solutions within these weakening zones, and skarn-magnetite mineralization was overlapped on the earlier massive ores.

Modern studies of copper mineralization revealed the Karadagh-Kharkhar copper-porphyry field. This field is composed of the rocks of the Atabek-Slavyanka plagiogranite intrusive that is intruded by small dike-like and stock-like bodies of intermediate-basic composition. These small intrusives (quartz-diorite-porphyrites) correlate with sub-meridian fault zones and control the quartz-pyrite-molybdenite and quartz-pyrite-chalcopyrite vein-disseminated ores (Baba-Zadeh et al. 1990).

Gold mineralization in the Azerbaijan part of the Somkhet-Karabakh Zone kindled great interest when information became available about significant gold extraction (about 5–7 t) from the copper-sulphide ores of the Gedabey deposit by the German firm Siemens, which owned copper concessions until to the 1920s. However, no commercial gold-bearing mineralization was actually found. Only modern geological-geophysical efforts designed to study porphyry copper and other mineralization have located any significant sources, e.g., in the Chovdar site.

Gold in the form of admixtures is present in all the copper-porphyry deposits of the Lesser Caucasus. Gold mineralization is distributed broadly by auriferous gravels of the Upper Quaternary alluvial deposits of the rivers flowing from the northeastern slope of the Lesser Caucasus. Gold mineralization was also found in barite-polymetallic veins of the Chovdar deposit (northward Dashkesan) with barite resources of more than 1.5 million tons. The barite-bearing strip on the northeastern slope of the Lesser Caucasus has about 20 deposits and manifestations within fault and fracture zones in anticline arches; the host rocks are Middle-Upper Jurassic volcanics.

The Chovdar goldfield spatially coincides with barite deposits of the same name and series of copper-sulphide, hematite and polymetallic occurrences, as well as gold placers within the northern part of the Dashkesan Mining District. The Chovdar deposit is related to a volcanic-dome structure located at the intersection node of the northwestern and northeastern disjunctive dislocations. Sub-volcanic rhyolite and rhiolite-dacite bodies caused intensive hydrothermal alterations of the Middle and Upper Bajocian volcanics along fault zones with high angle dips. Ore concentrations are found within the porous-fractured secondary quartzite (primarily, quartz-porphyres) under tuffaceous or porphyritic compact rocks, or in barite/polymetallic veins. Intensive pyritization occurred over polymetallic mineralization (down to a depth of 50–80 m). The polymetallic ores contain sphalerite (mainly), galena (less) and chalcopyrite (infrequently). Recent prospecting work has revealed a new gold-barite-polymetallic mineralization near the villages of Laish and Chovdar within the Kheirachay (Chovdarchay) basins and Goshgarchay Rivers. In the first basin, linear-isometric ore-bodies of irregular shape were contoured within secondary quartzite with a gold content up to 14.4 g/t. Gold content up to 6.2 g/t was detected in barite veins localized in the ruptures of the northwestern strike. In the second basin, the same maximal gold content (6.2 g/t) was detected in vein-lens-like bodies.

The copper-gold mineralization of Kuroko (Lesser Caucasian, according to Tvalchrelidze (1976)) is represented by sheet-like and lens-like bodies, as a

Madneuli deposit in Georgia and in the Kyzyl-Bulakh ("Gold Spring") deposit in Azerbaijan. In similar deposits, the gold mineralization is concentrated in comparatively narrow and extended zones of hydrothermally altered rocks. Gold mineralization was also found in the polymetallic veins of the Mekhmana deposit near the intrusive of the same name. In general, gold-polymetallic mineralization is attracted to transverse depressions with relative young acid magmatism.

The Sevan-Akera ophiolite Zone represents the main Tethyan suture zone between Eurasia and the South Armenian Subplatform (Fig. 2.1) is characterized by chromite deposits and manifestations within ultrabasite bodies; mercury and gold are attracted to the intersections of these structures. Polymetals, molybdenum and rare metals are attracted to the Dalidag intrusive. In the Yerevan-Ordubad Zone with its well-known large Co-Mo deposits, there is a potential for copper, molybdenum and other ores connected at contacts of the Megri-Ordubad Pluton, and intersections of sub-meridian and sub-latitude fault zones.

2.2 A Brief History of Geophysical Studies in the Caucasus

2.2.1 Initial Stage

The first geophysical measurements in the Caucasus were conducted in the nineteenth century using pendulum instruments for absolute gravity acceleration at several locations including a point near Tbilisi (Gongadze 2006). Measurements of magnetic variations have been made in the Tiflis (Tbilisi) Geophysical Observatory since 1844. Applied geothermic studies in boreholes were initiated in the Absheron Peninsula by Batsevich (1881) and further work was done by Stopnevich (1913) and Golubyatnikov (1916).

The first findings to emerge from exploration geophysics were collected in Azerbaijan at the Dashkesan magnetite deposit (Somkhet-Karabakh Zone 1923–1924) by the Geological Committee of the USSR. Ortenberg (1930) carried out measurements of the vertical (Z) and horizontal (H) components of the magnetic field and assessed the iron reserves on the basis of interpretative data; the practically horizontal bedding of the magnetite ores required vector magnetic measurements (Fig. 2.5).

Self-potential (SP) measurements were carried out over the Chiragidzor sulfur deposit (central Azerbaijan) for several years (Fig. 2.6). This figure shows that the mining works in the underground shaft distort the observed SP field strongly at the earth's surface. This testifies to the tight correlation between mining processes and SP anomalies.

L. V. Sorokin, V. V. Fedynsky and A. I. Zaborousky carried out gravity (with an Eötvös variometer) and Z measurements over the Neogene structures of the Absheron Peninsula (Puta, Binagady) and the southeastern Shirvan in the Lower Kura Depression (Neftechala-Babazanan) from 1926 to 1930. The increased

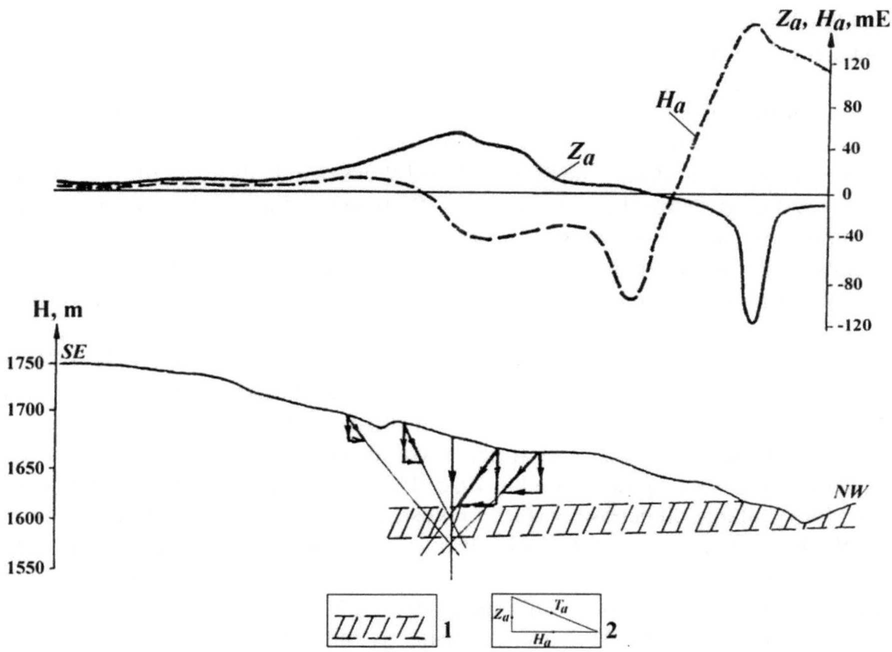

Fig. 2.5 Evaluation of depth of the magnetite bed by the vector technique in the Dashkesan deposit (After Ortenberg 1930). (*1*) magnetite bed; (*2*) components of the anomalous magnetic field (*T* is the total magnetic vector)

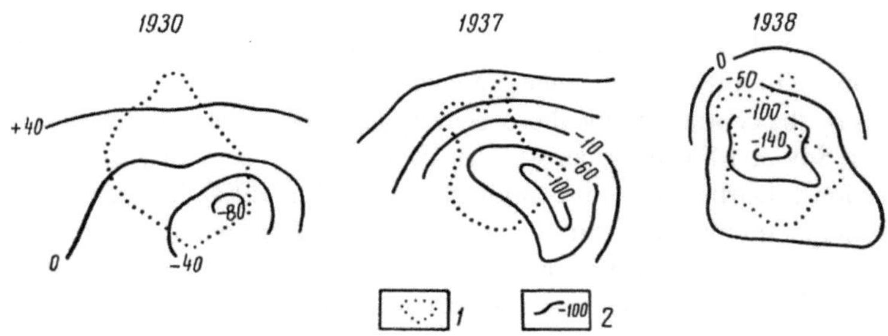

Fig. 2.6 Displacement of SP isolines during exploitation of a new shaft of the Chiragidzor sulfur deposit (Lesser Caucasus) (After Khesin 1969). (*1*) shaft contour; (*2*) isolines of SP field (in milliVolts)

magnetization of the Productive series, and the gravitational effect of fold arches enabled them to delineate anticline structures and trace disjunctive dislocations. The presence of a Hilly fold in the Lower Kura Depression, contoured by the magnetic prospecting, was then verified by drilling. Roughly at this time

(1929–1934), these pioneers conducted general pendulum gravity (Fig. 2.7) and magnetic *T* surveys.

The findings of these general surveys revealed the main regional peculiarities of the deep structure of the Caucasus (Arkhangelsky and Fedynsky 1932; Arkhangelsky 1933; Abakelia 1937; Fedynsky 1937; Nodia 1939).

Before as well as during the Second World War several other geophysical studies were conducted. Petrovsky and Skaryatin (1929) successfully applied an ondometric (radiowave) method to the translucent Chiragidzor pyrite stocks in Somkhet-Karabakh Zone (Azerbaijan). Slightly earlier, in the 1930s, a self-potential (SP) survey was also successful in studying these stocks (M. L. Ozerskaya, S. N. Kondrashev, A. G. Surikov).

A new (Toganaly) pyrite deposit was found in the area of SP anomaly in 1938 (Khesin 1962a). In the early 1930s, geophysicists from the French firm Schlumberger introduced the resistivity method of electric prospecting in the Absheron Peninsula. A Buzovny structure was detected here by a reflection survey of seismic prospecting that started in 1935–1936 in southeastern Shirvan (Pirsagat-Khydyrli); a large oil deposit was later revealed in the Busovny anticline (Kerimov 1996). Balavadze (1939) studied Akhaltsikhe coal deposits in Georgia using gravity prospecting. He identified cobalt-bearing veins in the Dashkesan deposit using electric resistivity profiling (Balavadze 1944). A correlation refraction survey method was used in the Absheron Peninsula as early as 1944 (G. A. Gamburtsev and Yu. V. Riznichenko).

2.2.2 Formative Stage

After the Second World War and practically until the final years of Gorbachev's "Perestroika" in the USSR, huge geophysical works were carried out in the Caucasus by the central and local prospecting services of the Oil-and-Gas, Geology and other Soviet Ministries, as well as by numerous scientific and educational organizations. During this time mid-scale (mainly, 1:200,000) gravity surveys and aeromagnetic/magnetic surveys were conducted elsewhere, depressions were studied by seismic and electric prospecting (primarily for petroleum exploration), and deep seismic sounding was used to study the Earth's crust. Hundreds of prospective oil-and-gas bearing structures were prepared for deep drilling by large-scale geophysical methods (1:50,000–1:25,000) as well as detailed geophysical studies (e.g., Kerimov 1996). Deep structures were primarily explored through seismic prospecting (Fig. 2.8).

This profile across the Taman mud volcano province shows diapirism attenuation with depth: in the near-arch part of the diapiric fold, dip angles of the reflectors decrease with increases in depth from 50–60° down to 10–5°.

At the same time that these indirect (structural) geophysical explorations of oil-and-gas were being pursued, geophysical methods for direct petroleum prospecting were initiated. The first successful attempts took place in Azerbaijan

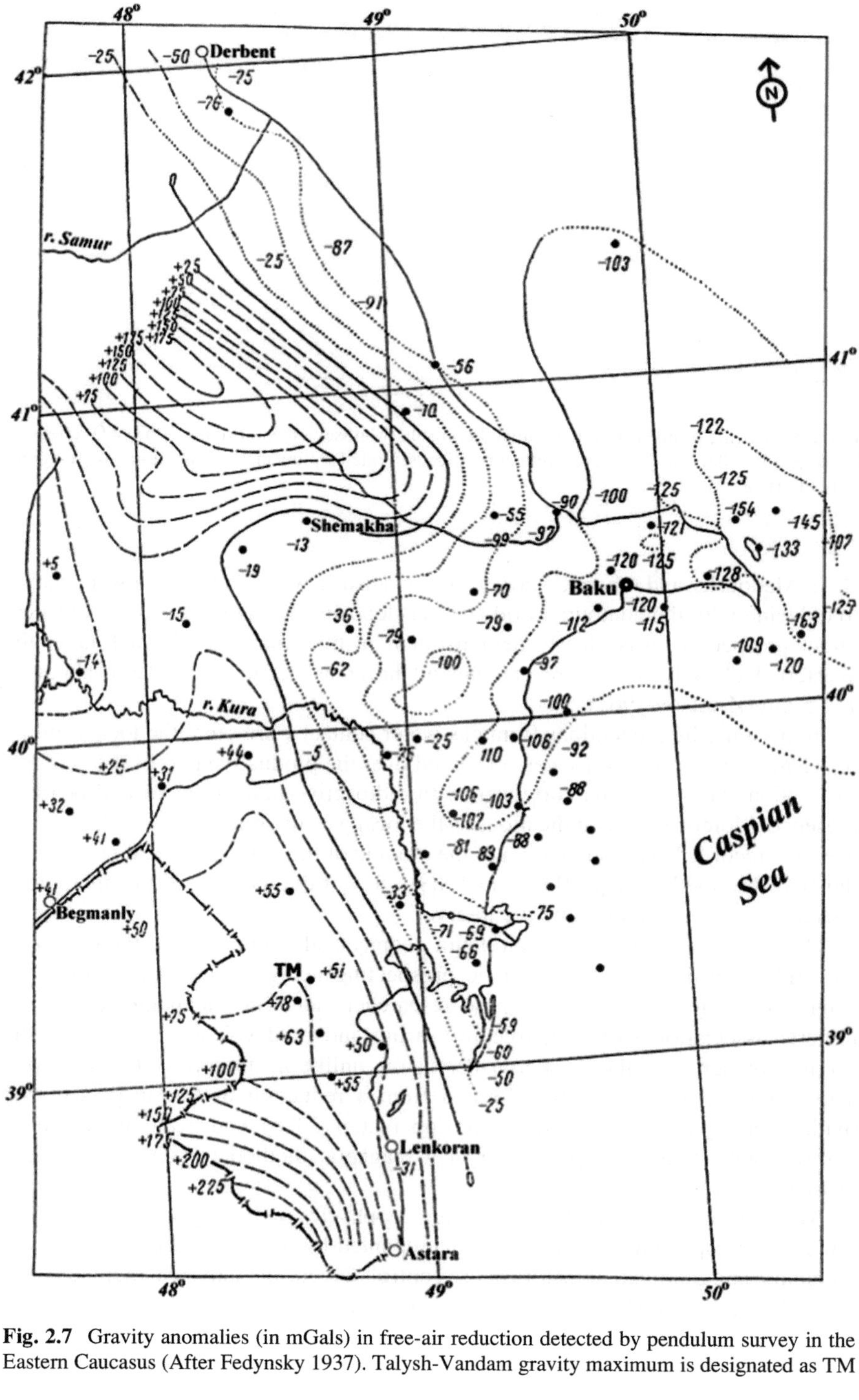

Fig. 2.7 Gravity anomalies (in mGals) in free-air reduction detected by pendulum survey in the Eastern Caucasus (After Fedynsky 1937). Talysh-Vandam gravity maximum is designated as TM

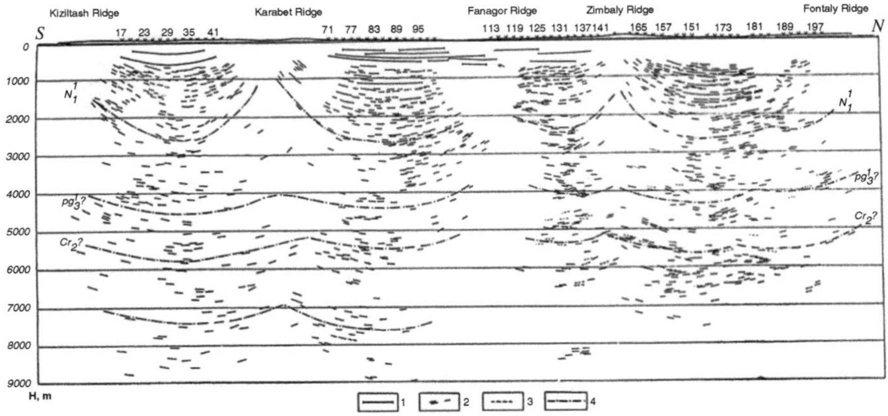

Fig. 2.8 Seismic profile across the Taman Peninsula (Matusevich and Priyma 1967). (*1*) reflection horizon; (*2*) reflection pieces; (*3*) reflection pieces plotted by uncertain reflections; (*4*) conventional horizon

(I. G. Medousky and others), where zones of seismic wave attenuation were shown to coincide with the contours of oil and gas deposits. The potential of geophysical direct prospecting was clearly demonstrated in the Anastasievskaya fold in the Western Kuban Basin, where gas and oil horizons occur at the depth about of 1–1.5 km (Zemtsov 1967).

In this anticline, reflections from gas-water contact were seen; the local minima of seismic amplitude curves were observed over the productive part of the geological section. These minima correlated with radioactive indications of hydrocarbon presence. Moreover, such indications as the local gravity minimum over the deposit was also recorded. The calculation of the module of the gravity vertical gradient by the Berezkin method (e.g., Berezkin 1988) showed the location of the anomalous source within the main productive interval.

The development of direct petroleum geophysical exploration in the Caucasus was aided considerably by magnetic prospecting using a method developed by Bagin and Malumyan (1976). These researchers identified a magnetite-siderite mineral association within petroleum-bearing and mud volcano deposits of the Productive Series of the Absheron Peninsula, unlike in non-productive deposits. This magnetic association was formed from a reduction of thin-dispersed iron oxides and hydroxides (Pilchin and Eppelbaum 2006) under the influence of hydrocarbons; thus, the application of a magnetic method for direct petroleum prospecting was feasible.

By contrast, the development of mining geophysics in the Caucasus started from direct ore prospecting and later was accompanied by the study of deep structural, magmatic and lithological factors of ore control. Diverse geophysical methods were applied to assess different mineral resources. For example, the SP survey was

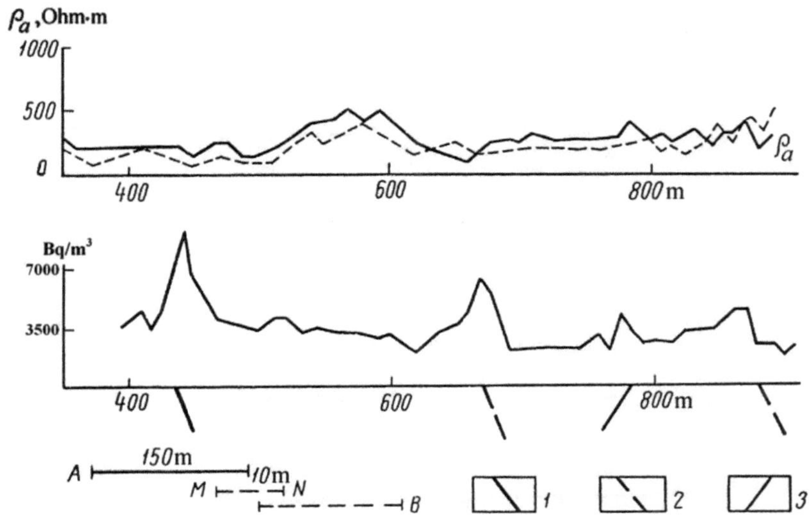

Fig. 2.9 Delineation of cobalt-bearing fragmentation zones by combining data on resistivity profiling and an emanation survey in the Dashkesan ore district (Lesser Caucasus) (After Khesin 1962b, 1969). (*1*) known cobalt-bearing fragmentation zone; (*2*) zones of cobalt-bearing fragmentation revealed by geophysical data; (*3*) contact of tuffaceous suites

continued, and SP anomalies were revealed over copper-sulphide and pyrite stocks of the Gedabey deposits. As of 1959, most of the geophysical studies by the Azerbaijan Geophysical Expedition (AGE) were concentrated in the northwestern part of the Lesser Caucasus in the Dashkesan-Gedabey ore region. In the Dashkesan area, new cobalt-bearing fractured zones were detected by combined electric profiling based on "ore" (direct, "conductive") intersections of apparent resistivity (ρ_a) graphs and the results of radioactive (radon) surveys (Fig. 2.9).

Combined electric profiling also helped resolve other prospecting problems, including the detection of caoline clays with a resistivity of 60 Ω m within the Bajocian effusives with a resistivity about of 1,000 Ω m (Fig. 2.10). Figure 2.10 also shows the dip of the kaolinized zones as a function of the displacement of "conductive" intersections for different separations of the current electrodes, i.e., for different probe depths.

During the study of bentonite clays, high-quality bentonites with a resistivity of 1–2 Ω m were separated from low-quality bentonites (~5 Ω m).

AGE geophysical surveys were usually integrated with geochemical (metallometric) studies that provided direct indications of different ores. One example is the detection of chromite ores in the Sevan-Akera Zone of the Lesser Caucasus (Fig. 2.11).

The Kuban and Armenian Geophysical Expeditions successfully carried out direct geophysical prospecting of ore deposits as well. For example, electric prospecting of copper-sulphide deposits within the volcanogenic strip of the Northern Caucasus was

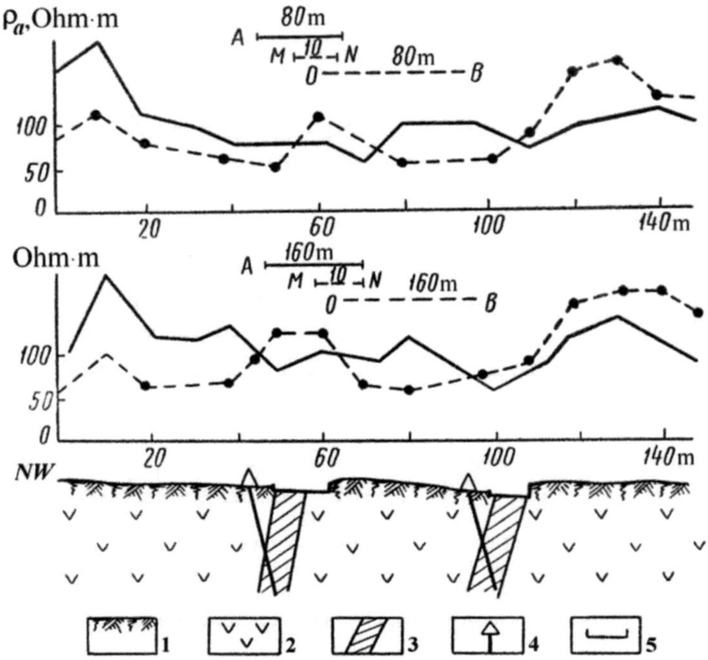

Fig. 2.10 1:5,000 combined profiling implemented for the detection of caolinite clays, in the Chardakhly deposit (Gedabey ore area, Lesser Caucasus) (After Khesin 1969). (*1*) loose deposits; (*2*) secondary quartzite; (*3*) caolinite and refractory clays; (*4*) borehole; (*5*) trench

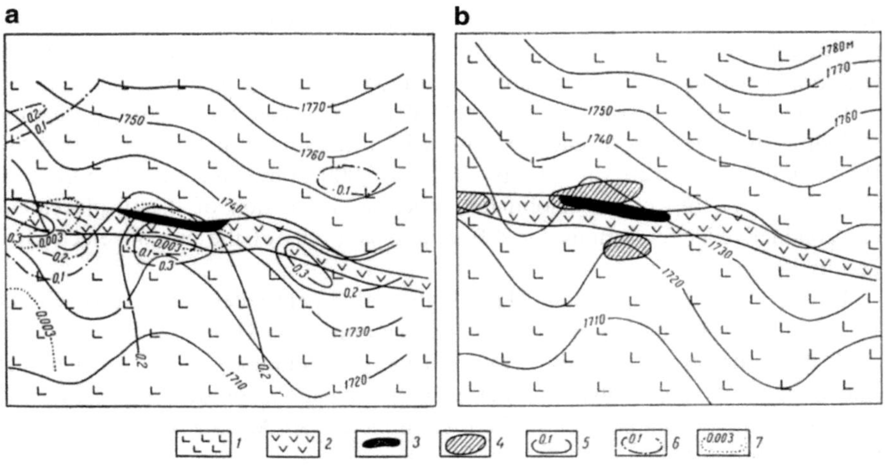

Fig. 2.11 Delineation of chromite-bearing dunites with metallometry (**a**) and gravity prospecting (**b**) data (Khesin and Muradkhanov 1962). (*1*) serpentinized peridotites; (*2*) serpentinized dunites; (*3*) chromite ore; (*4*) gravitational maximums, isoconcentrations (in%); (*5*) chrome; (*6*) nickel; (*7*) zinc

complicated by the presence of mineralized fault zones and graphitized schists of low resistivity. Nevertheless, the resistivity of massive ores ($\rho = 0.02$–$0.03\ \Omega\,\mathrm{m}$) is less in the often cited 1–2 ranges above false targets, and these ores were clearly detected at a depth of 70 m by the inductive non-grounding loop method (Fig. 2.12).

Furthermore, the integration of geophysical methods became the main tool for successful prospecting in the complex environments of mountainous ore regions with rugged topography, intense tectonics, a multitude of rock classes and other factors that make the interpretation of geophysical anomalies difficult. The history of discovery of the very large polymetallic-sulphide deposit in the Azerbaijan part of the southern slope of the Greater Caucasus, near the border of Georgia, followed by a new polymetallic province is one of the most impressive of these stories. The 1958 geological survey within the disrupted arch of a near-latitude anticline revealed two small ore outcrops in the Toarcian-Aalenian schists of the Tfan anticlinorium. These outcrops contained a series of ore minerals including pyrite, pyrrhotite, sphalerite, galena, chalcopyrite, and precious metals. An integrated detailed geophysical survey of the AGE in 1960 reported on the magnetic, SP and induced polarization (IP) measurements of apparent polarizability (η_a) (Mustafabeily et al. 1964). This survey showed that these outcrops represented a very large ore-body both in strike and dip. The SP negative anomaly (several hundreds of milliVolts) traced this Filizchay ("Ore river") ore-body to 1,200 m along the strike (Fig. 2.13).

The SP anomaly could have been caused in this area by a conductive schist, but the nature of the ore in the anomaly was shown to have magnetic anomalies of about 200 nT over the eastern part of the ore-body enriched by magnetic pyrrhotite. On the other hand, the magnetic anomaly in this area might have resulted from a magnetic diabasic (non-conductive) body, but the SP anomaly and the anomaly of η_a up to 30–40% reflected a conductive ore source. The quantitative interpretation of the magnetic anomaly indicated that the depth of the upper edge of the magnetic source was 40–45 m. Subsequent drilling confirmed this calculation to 10% accuracy (Khesin and Muradkhanov 1967).

On the basis of these results, integrated geophysical surveys of the AGE were made on a scale of 1:10,000 for the entire Filizchay ore field and on a large scale for the adjacent territory. These surveys revealed a series of prospective areas and sites, several of which were recognized later as polymetallic deposits. The location of these areas and sites was comparable to some of the features of the regional distribution of geophysical fields (Fig. 2.14).

In the central part of this ΔT map (Fig. 2.14) the near-latitude maximum (Guton maximum) can be seen. This maximum was interpreted as a hidden intrusive body that might affect the mineralization distribution. The largest deposits in the western margin of the anomaly correlated with the transverse dislocation reflected in the strike of the ΔT field. Here, a high dip was also detected in the southeastern flank of the gravity maximum. To evaluate the depth perspectives, area surveys were supplemented by reference regional profiles with IP soundings (Khesin 1969, 1976). This

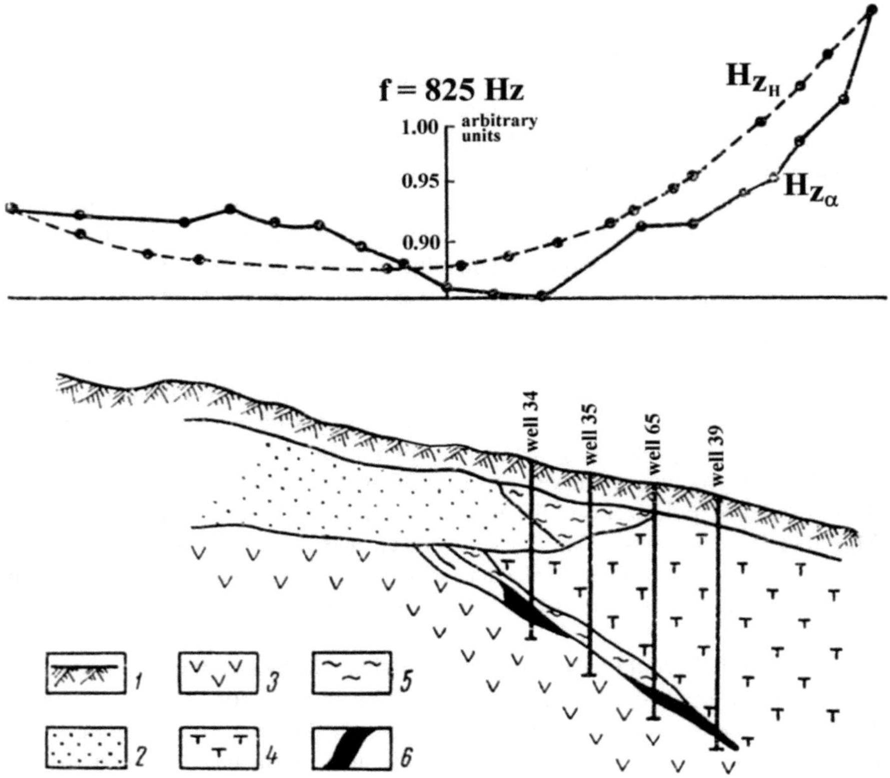

Fig. 2.12 Detection of copper-sulphide ores in the Northern Caucasus by the non-grounding loop method (After Vinogradov 1964). (*1*) modern deposits; (*2*) Jurassic deposits; (*3*) quartz-albitophyres; (*4*) tuffs of intermediate composition; (*5*) siliceous schists; (*6*) massive copper-sulphide ore

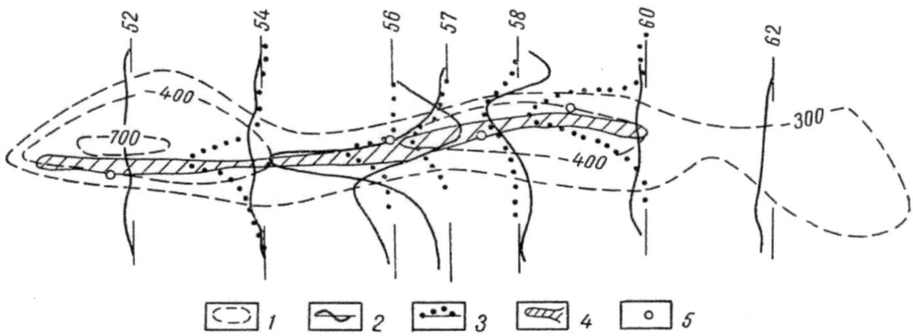

Fig. 2.13 Detection of the Filizchay ore deposit by the integration of geophysical methods (After Khesin and Muradkhanov 1967). (*1*) isopotentials of the negative SP field (in milliVolts); (*2*) ΔZ graphs; (*3*) η_a graphs; (*4*) projection of the upper edge of the ore-body on the plan; (*5*) boreholes, intersected polymetallic-sulphide ore

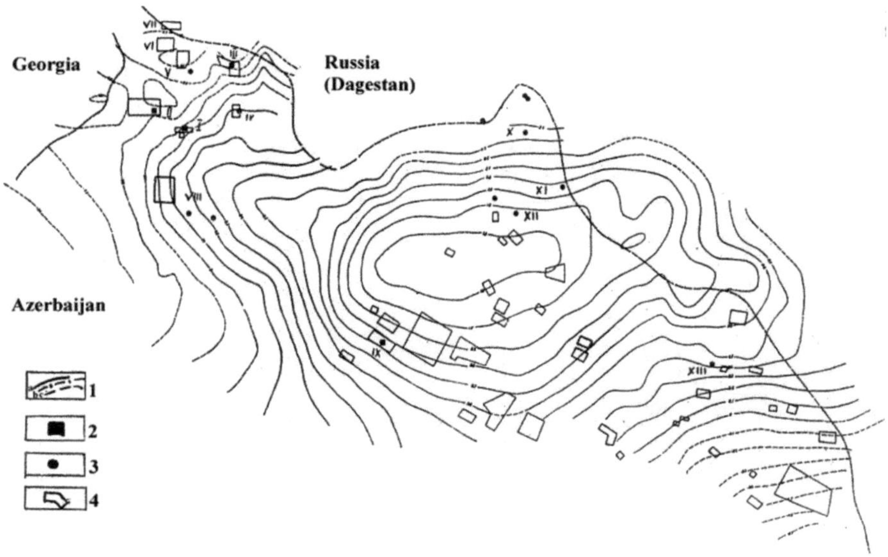

Fig. 2.14 Average results of helicopter magnetic survey and distribution of ore and prospective sites of the Belokan-Zakatala ore region (Modified from Khesin 1976). (*1*) isodynams (in nano-Tesla) of the modulus of the full magnetic vector (ΔT): (*a*) positive, (*b*) zero, (*c*) negative; (*2*) ore deposits; (*3*) ore manifestations; (*4*) anomalous sites revealed by large-scale ground and detailed geophysical surveys: (*I*) site of the Filizchay deposit, (*II*) site of the Katsdag deposit, (*III*) site of the Djikhikh deposit, (*IV*) site of the Katekh deposit

geophysical experience was implemented in adjacent territories and led to the discovery of a large copper-polymetallic deposit (Kyzyl-Dere) in Dagestan.

A similar combination of indirect and direct geophysical prospecting was undertaken in the Lesser Caucasus. In the Shamkhor (Shamkhir)-Gedabey-Dashkesan region of the Somkhet-Karabakh Zone (Fig. 2.4) many geologists have associated its ore potential with different phases of the Dashkesan and some other intrusives. Aeromagnetic data analysis made it possible to detect the hidden intrusive masses (Fig. 2.15).

Integration of geophysical methods including various modifications of seismic prospecting yielded abundant information about the hidden parts of intrusive bodies and other deep peculiarities (Fig. 2.16).

Figure 2.16 shows that seismic prospecting by the reflection and refraction methods clearly outlines the intrusive at a depth, whereas the local fault was detected by the diffraction method. Gravity and magnetic maxima additionally confirmed the presence of dense and magnetic basic intrusive rocks at a depth. Integrated interpretation of seismic and other geophysical data using the information approach (Khesin 1976) also identified a hidden intrusive body in the southern immersion of the Gedabey intrusive (reference profile III-III in Fig. 2.15).

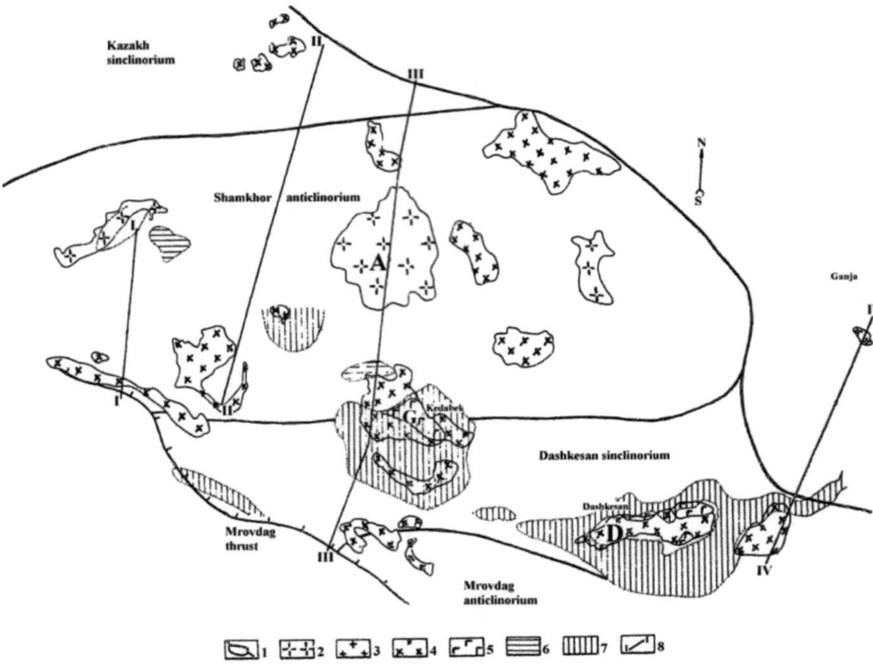

Fig. 2.15 Distribution of intrusive bodies in the northeastern Lesser Caucasus indicated by magnetic data (After Khesin 1976, with supplements). (*1*) borders of large geostructures; (*2*) plagiogranites; (*3*) granites; (*4*) granodiorites and diorites; (*5*) gabbroid rocks; (*6*) granodiorites and diorites; (*7*) gabbroid rocks detected down to a depth of 0.5 km by airmagnetic data interpretation; (*8*) location of interpreting profiles. Main intrusive bodies: *A* Atabek-Slavyanka, *D* Dashkesan, *G* Gedabey

2.2.3 Contemporary Period

In the last years of the USSR's existence and the first years after the formation of independent states which replaced the former republics of the USSR, financial and organizational problems curtailed geological-geophysical prospecting considerably. At the same time, growing geophysical data and methodological-technological innovations facilitated new trends. First, the analysis and generalization of the results of geophysical studies led to the publication of major works (e.g., Gugunava 1988; Ismail-Zadeh and Khesin 1989a,b; Kerimov et al. 1989). Second, underground geophysical investigations and the integration of traditional geophysical methods with direct techniques of mineralization detection (e.g., Ryss 1983) were extended. Third, the new kinds of mineral resources within known prospective areas began to be studied (e.g., gold prospecting in relation to porphyry copper studies in the Somkhet-Karabakh Zone), as well as cooperation with firms and organizations from the USA, Europe, Japan and Israel, especially in petroleum exploration.

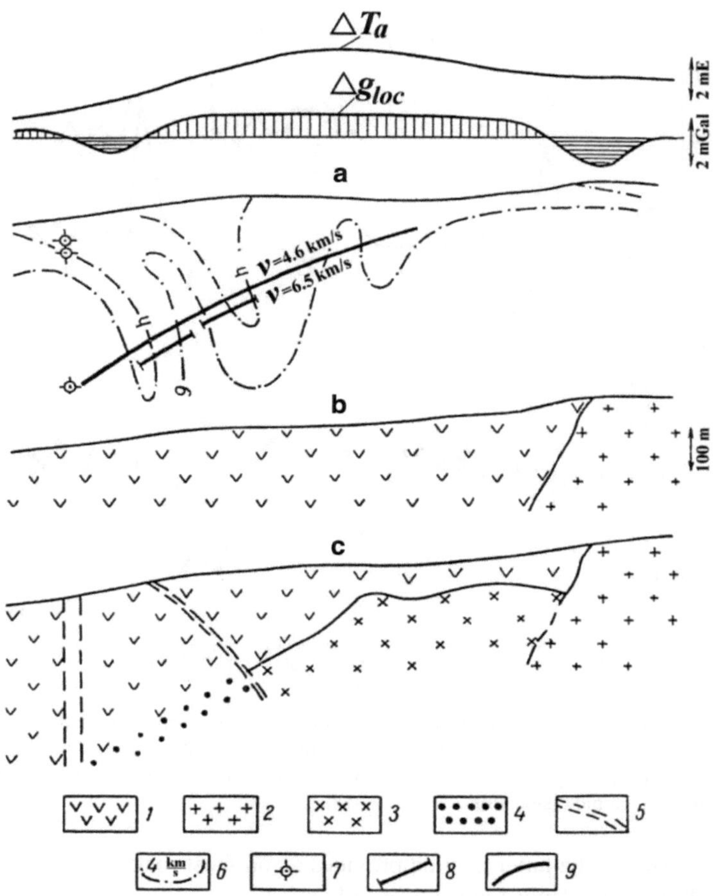

Fig. 2.16 Integration of seismic prospecting with other geophysical methods for the study of a hidden intrusive body to the NW of the Gedabey intrusive, near reference profile II-II in Fig. 2.15 (Khesin 1969, seismic data of A. Sh. Mamed-Zadeh, Yu. G. Shopin and S. A. Miri-Zadeh): (**a**) aeromagnetic and gravity anomalies over velocity section, (**b**) geological section, (**c**) result of interpretation of geophysical data. (*1*) porphyrites and tuffs of the Lower Bajoician; (*2*) intermediate intrusive rocks; (*3*) basic intrusive rocks; (*4*) zone of intense contact metamorphism; (*5*) disjunctive dislocation; (*6*) isolines of refraction wave velocities (km/s); (*7*) diffraction point; (*8*) reflection piece; (*9*) refractor

Underground geophysical studies revealed new ore-bodies in such old copper-sulphide deposits as Alaverdy in northern Armenia (Somkhet-Karabakh Zone). These copper-sulphide and pyrite ores have low resistivity (on average, 30 Ω m) and high polarizability (on average, 22%), whereas the host tuffites, porphyrites and volcanic breccias have high resistivity (on average, 1,300 Ω m) and low polarizability (Ovsepyan et al. 1986). The results of measurements of the IP, SP

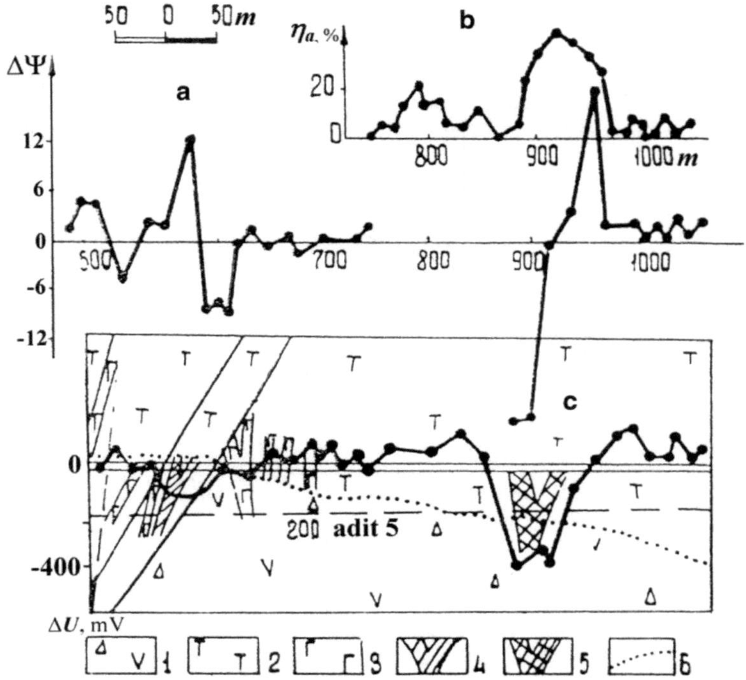

Fig. 2.17 Results of underground electric prospecting in the Alaverdy copper deposit: (**a**) stray currents, (**b**) IP, and (**c**) SP (Ovsepyan et al. 1986). (*1*) porphyritic breccias; (*2*) tuffites; (*3*) andesite-dacite porphyrites, veins; (*4*) known ore-body; (*5*) ore-body, revealed by electric prospecting; (*6*) tectonic dislocation

and stray currents (the latter method provides an ore/host rock conductivity ratio – $\Delta\Psi$) in adit 6 of the deposit are shown in Fig. 2.17.

Figure 2.17 shows that a known ore-body causes the geophysical anomalies in the 580–600 m interval. An intensive SP negative anomaly, $\Delta\Psi$ and η_a anomalies detected a hidden ore-body in the 900–980 m interval under adit 6. This conclusion was confirmed by mining. In this deposit, the polarization curve correlation method (Ryss 1983) enabled the authors to define the spatial distribution and content of copper mineralization with greater precision.

The Karadagh-Kharkhar copper-porphyry field is the most highly studied area of the Azerbaijan part of the Somkhet-Karabakh Zone (Kerimov 1996). Ore-bearing sites are located within hydrothermally altered rocks that were accurately detected by data showing a decrease in magnetic field and resistivity from an integrated geophysical survey on a 1:10,000 scale. The deposit area (400–900 m $\times$ 2,200 m) is outlined by a polarizability isoline of 7.5%. Within the contour, local η_a maxima (up to 10–18%) are related to the increase in the concentration of disseminated pyrite and certain other sulphides. These maxima coincide with the SP minima (down – 60 mV). Method of partial extraction of metals (Ryss 1983) applied to the

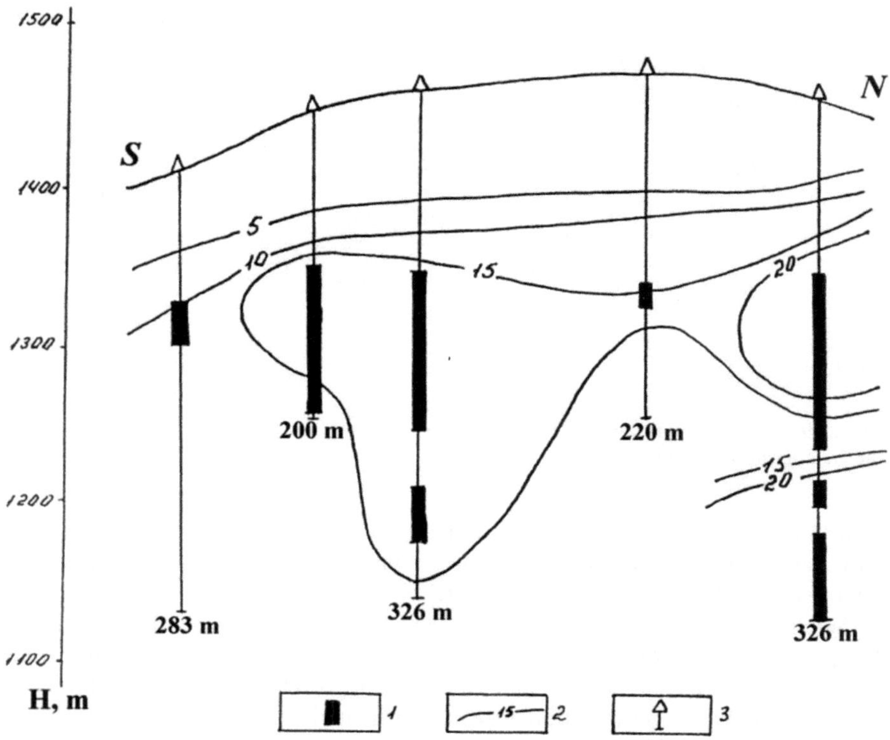

Fig. 2.18 Results of borehole η_a measurements in the Karadagh copper-porphyry deposit (After Bagirov et al. 1996). (*1*) intervals of commercial concentrations of copper in the drill core; (*2*) η_a isolines (in%); (*3*) borehole

sites of the local anomalies showed 30 μg of the copper, 5 μg of the lead and 500 μg of the iron on a background of 5, 2 and 200 μg, respectively. Nuclear geophysical methods were used here for drill core and well studies. They showed copper concentrations of 0.2–0.8%; concentrations of 1.1–1.4% were detected in several borehole intervals at thicknesses of up to 1.5 m. IP anomalies in the wells correlated closely with the ore cuttings (Fig. 2.18).

2.3 The Caucasus in the Light of Regional Geophysical Analysis

The key features of the deep structure of the Eastern Caucasus were first delineated by a pendulum survey that revealed the Talysh-Vandam gravity maximum (see Fig. 2.7). This maximum was shaped like a salient that extended from the

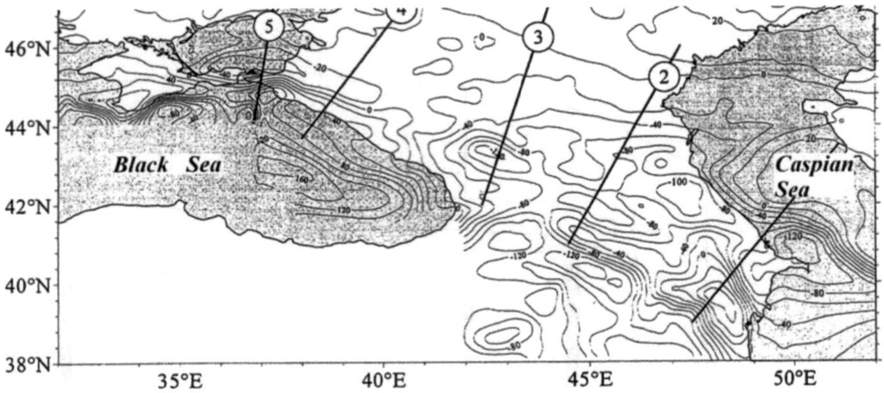

Fig. 2.19 Bouguer gravity anomalies of the Caucasus with isoline interval of 20 mGal (Data from the Geological Survey of Russia)

Near-Talysh Mugan steppe northward across the towns of Kyurdamir, Agsu and Geychay up to Lagich and Vandam. Latter, numerous gravimetric surveys provided a more complete picture of the gravity field (e.g., Fig. 2.19).

Satellite gravity data were obtained from the World Gravity DB as retracked from Geosat and ERS-1 altimetry (Sandwell and Smith 2009). Crucially, these observations were made with regular global 1-min grids (Sandwell and Smith 2009) and the gravity data computation error was estimated at 2–3 mGals. The compiled gravity map (Fig. 2.20) shows the intricate gravity pattern of this area (the isoline interval is 10 mGal; "zero" isoline is dashed and white bolded). This figure demonstrates that in specific cases it is useful to display the gravity field map without any reduction. The selected positive and negative gravity anomalies (Fig. 2.20) clearly reflect the main structural-geotectonical units of the region. Disparities between the free air, the Bouguer gravity map (Fig. 2.19) and the retracked gravity map (Fig. 2.20) can be used to estimate topographic corrections and solve various regional gravimetric problems.

The zones of gravity minima in Fig. 2.19 are related to mountainous edifices of the Greater and Lesser Caucasus, where the depth of the Moho discontinuity is increased. This is clear from the constructed map of Moho discontinuity (Fig. 2.21).

Models of deep structure of the Caucasus were coordinated with the results ofdeep seismic sounding (DSS); for example, the depths of the Moho discontinuity in the Greater Caucasus along the Grozny-Shamkhor (Shamkir) segment of Volgograd-Nakhichevan (Nakhchivan) DSS profile exceeds 50 km (e.g., Krasnopevtseva 1984).

The Talysh-Vandam gravity maximum was studied by different geophysical methods at different scales. Its characteristics elicited heated debate. Fedynsky (1937), Tzimelzon (1959), Gadjiev (1965) argued that this maximum resulted from the elevation of the crystalline basement and the development of magmatic rocks within the basement and sedimentary cover.

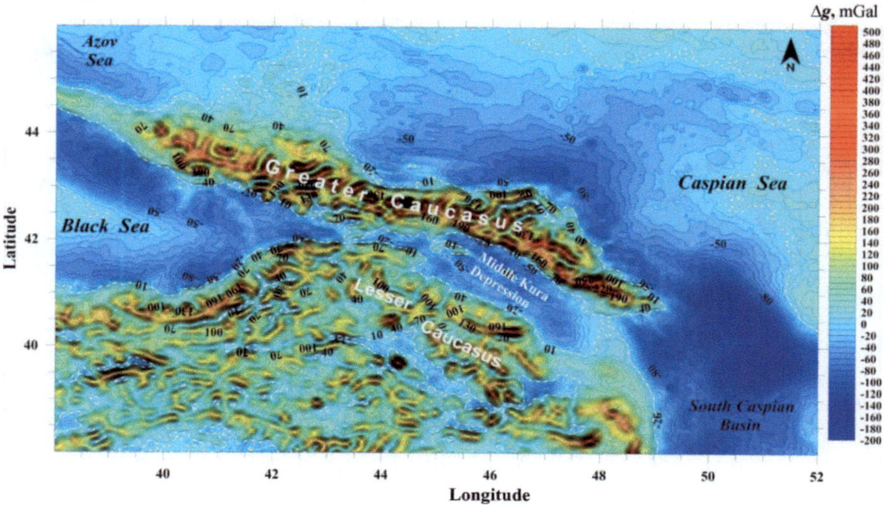

Fig. 2.20 Map of significant gravity anomalies of the Caucasus and adjacent areas (isolines are given in mGals)

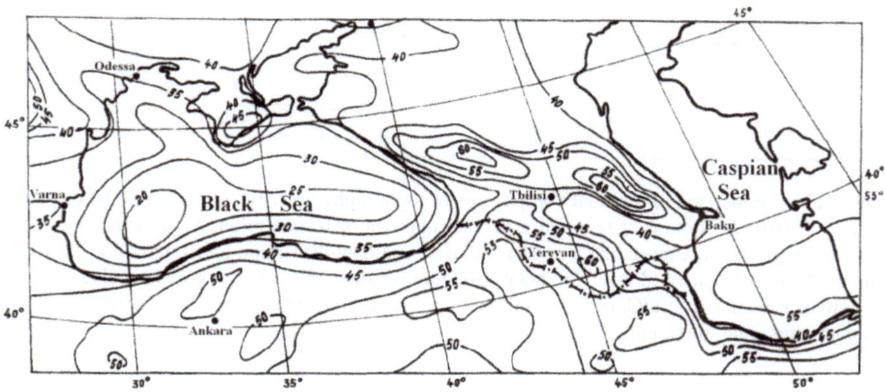

Fig. 2.21 Isodepths of the Moho discontinuity (in kilometers) in the Caucasus and adjacent regions (After Balavadze et al. 1979)

Based on the DSS data, this hidden basement arch explanation dominated. To test this hypothesis and reveal the Pre-Alpine basement, super-deep drilling was planned in the Saatly area, where the Kura and Araks Rivers merge. Examination of the gravity data (Tzimelzon 1970) and magnetic data analysis (Khesin et al. 1983) helped refine these predictions.

An upward continuation (recalculation onto the upper level) of the observed magnetic field to a height of 25 km contributed to showing that the Ganja regional

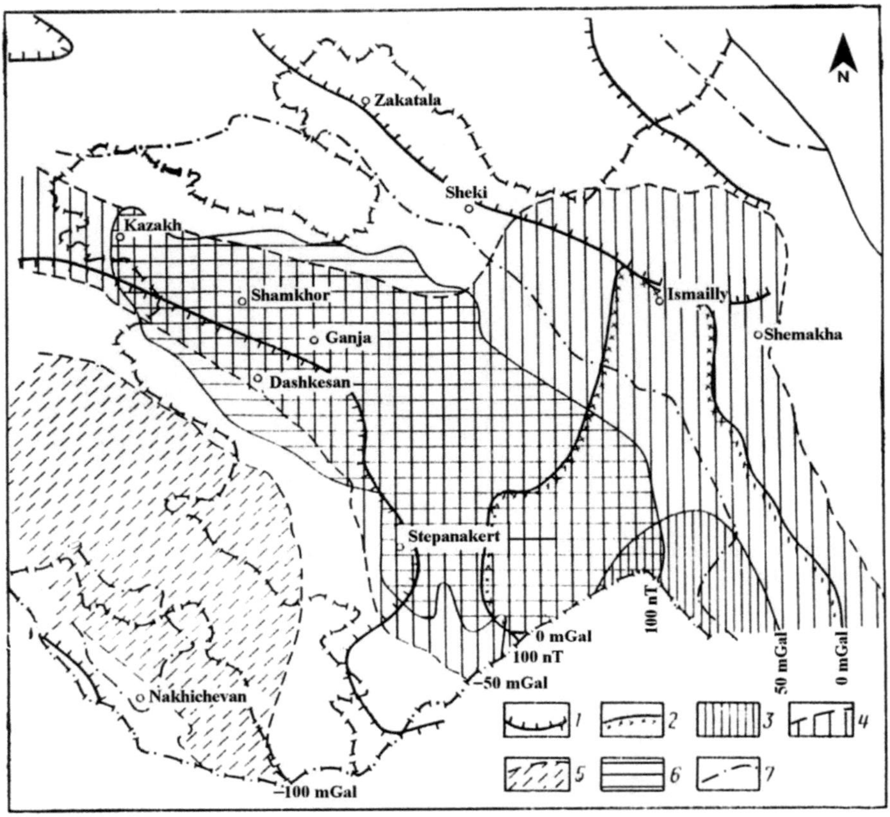

Fig. 2.22 Comparison of regional components of topography, gravity and magnetic fields (After Khesin 1976). (*1*) contour of regional mountain maxima of terrain relief; (*2*) zero isoline of the Bouguer regional maximum; (*3*) Bouguer intensive regional maximum (>50 mGal); (*4*) positive and moderately negative (down – 50 mGal) Bouguer regional field; (*5*) Bouguer intensive regional minimum (down – 100 mGal); (*6*) Ganja regional magnetic maximum; (*7*) zero isoline of the magnetic regional maximum

magnetic maximum primarily occupied the Middle Kura Depression and a significant part of the northeastern Lesser Caucasus (Khesin 1976). Regional components of the topography (elevation field) and gravity field were calculated as well (Fig. 2.22).

It is evident from Fig. 2.21 that the regional gravity maximum contains sub-meridian and sub-latitudinal components. A sub-meridian strike characterizes the most ancient crystalline complexes of the Talysh-Vandam (rather, the Talysh-Ismailly) maximum, whereas its western periphery is overlapped by the Ganja magnetic maximum of the Caucasian strike. Here, a common source of magnetic and gravity anomalies may be thick Mesozoic magmatic associations of increased magnetization and density at a depth of 3–4 km (Khesin 1976).

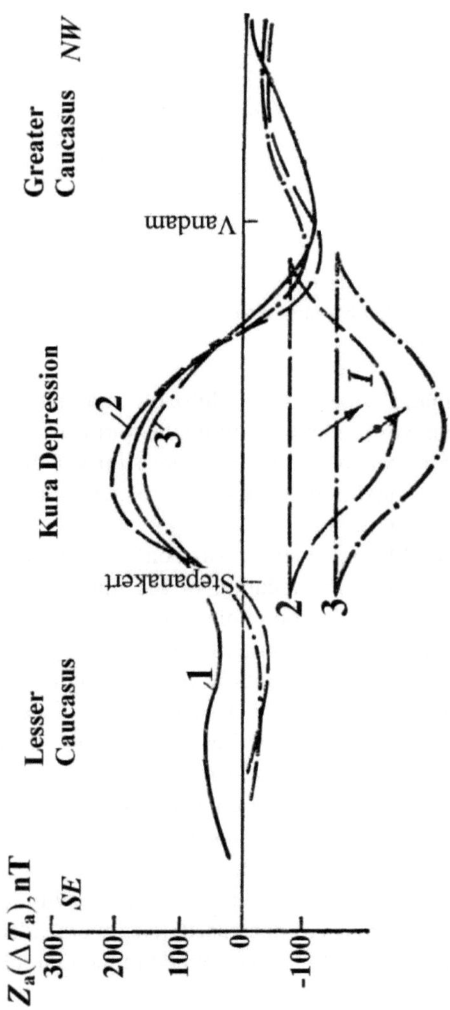

Fig. 2.23 Comparison of the results of upward continuation of the observed magnetic field to a height of 25 km (*1*) with modeled magnetic anomaly Z_a of the body occurring at a depth of 12.5 (2) and 25 km (3) (After Khesin 1976, with modification). *Arrows* show location of the magnetization vector ***J***

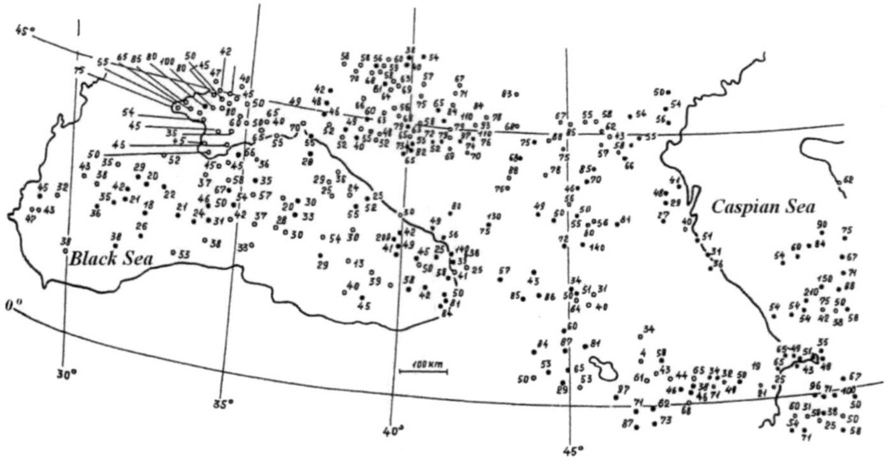

Fig. 2.24 Heat flow values at the Earth's surface for the Caucasus and adjoining areas (in mW/m^2) (After Alexidze et al. 1993,1995)

A Ganja tectonic-magmatic zone with a Mesozoic magmatic section of basic-intermediate composition was predicted to unite the northeastern part of the Lesser Caucasus and the adjacent part of the Kura Depression which corresponds to the regional Ganja magnetic maximum (Khesin 1976). A simple magnetic model verified this supposition (Fig. 2.23).

Later, a buried arch of Mesozoic strata was confirmed by drilling at the depth predicted by the calculation of the location of the upper edge of the magnetic source. A super-deep (8.2 km) SD-1 borehole, drilled on the Saatly arch, revealed that under the Cenozoic and Upper Senonian sedimentary deposits there was a thick sequence of island-arc Cretaceous and Jurassic volcanic, ending at the bottom of the borehole with Upper Bajocian quartz plagioporphyres. This sequence is very similar to that of the adjacent part of the Lesser Caucasus. It thus confirmed the supposition of the Ganja tectonic-magmatic zone.

Nevertheless, sole reliance on seismic data led to mistakes in the planning of deep drilling as well. In the northern Caucasus, in the area of the Tyrnyauz molybdenum-wolfram deposit, a deep well was planned to reveal basement rocks under granites. However, Khesin's interpretation of the gravity minimum over this acid intrusive showed that its rocks of decreased density extended down to a depth of more than 5 km. In fact, the drilling was stopped in the granite less than 5 km to the well bottom.

At the same time, the application of several geophysical methods provided important information on the deep structure of the Caucasus. For example, on the basis of combining geothermic studies (e.g., Alexidze et al. 1993, 1995) (Fig. 2.24) with magnetotelluric sounding (MTS) and electro-telluric surveys certain deep electric and thermal specificities were revealed. According to Gugunava (1981, 1988), the Caucasian crust contains several electrically conducting units: (1) the

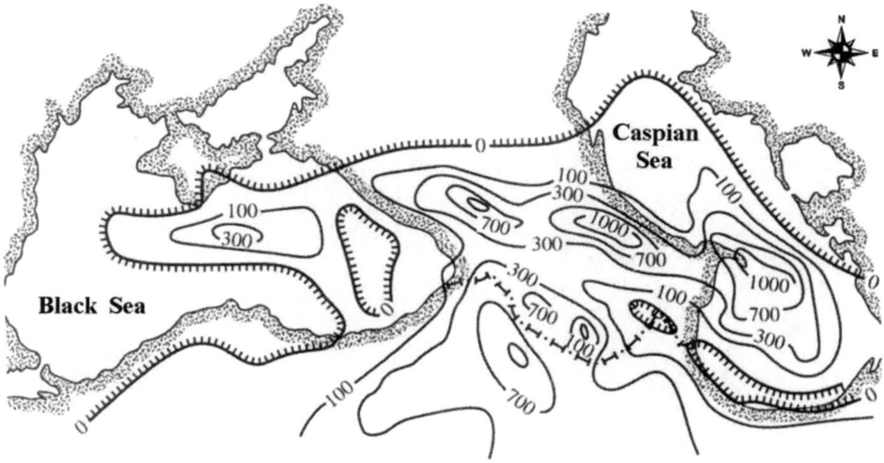

Fig. 2.25 Map showing the longitudinal conductance (in Siemens) of the crustal inverted layer in the Caucasus (After Gugunava 1988)

sedimentary cover (up to 15–20 km thick), (2) relicts of magma chambers within mountainous regions, and (3) the asthenosphere. This model was later confirmed by Spichak (1999). Chamber relicts were detected in the form of an oblate ellipsoid at a depth of about 20 km (Greater Caucasus) and as isolated lenses at a depth of 10–20 km (Lesser Caucasus). A crustal asthenosphere underlies the Transcaucasus (0–20 km thick) and attain maximal thickness beneath the Greater and Lesser Caucasus. High conductivity reflects the increased temperature at its depth; its highest values were detected in the southeastern prolongation of the Absheron Peninsula (Fig. 2.25), where increased geothermal gradients were measured (e.g., Khesin 1961; Kerimov et al. 1989).

More detailed deep structure according to the MTS data is shown for the chamber of the Elbrus volcano (Fig. 2.26).

Whereas the terrigeneous cover within the adjacent Scythian Plate is characterized by low resistivity (for example, 1–2 and even 0.5–0.6 Ω m for the Maykopian deposits), volcanic rocks on the slopes of the Elbrus and its immediate vicinity show a resistivity >1,000 Ω m, and basement Proterozoic rocks and Paleozoic granites measuring hundreds to thousands of Ohm·meters. The upper crust resistivity in the Elbrus area decreases to 40–25 Ω m at a depth of 5–10 km. Another low-resistivity zone (25–15 Ω m in its center) in the middle and low crust propagates to a sub-crustal depth (exceeding 55 km). These conductivity anomalies have been linked to the magma and parent chambers and are most probably related to partial melting of rocks. This conclusion is consistent with the low velocities that were located using the reflected earthquake wave method at these depths, and with a density defect at a depth of 5–15 interpreted from the gravity minimum under Elbrus (Gurbanov et al. 2004).

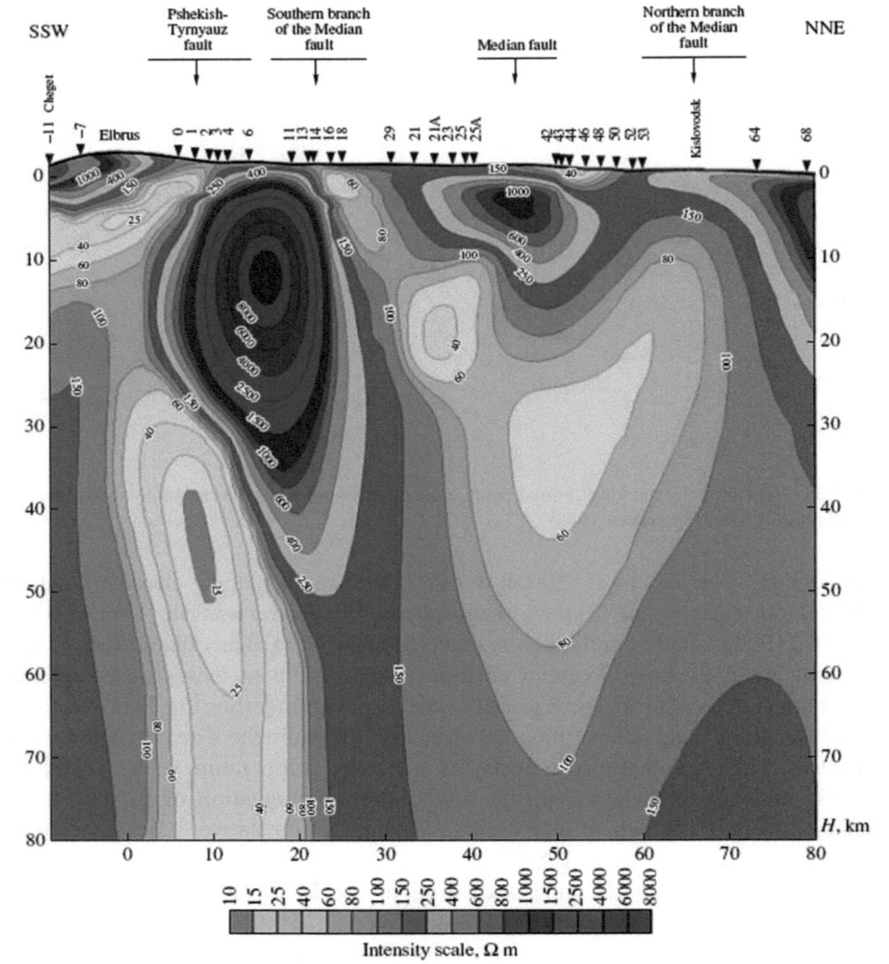

Fig. 2.26 The resistivity section according to MTS data along the North Elbrus profile (After Spichak et al. 2007)

At the same time, conductivity anomalies may be associated with deep faults. A high-gradient zone of conductivity in the interval of soundings No. 1 and No. 2 corresponds to the Pshekish-Tyrnyauz fault (Fig. 2.4), where highly mineralized fluids flow from the depths.

Paleomagnetic data are an important source for paleotectonic reconstructions (e.g., Gorodnitsky et al. 1978). The reduction of the location of the magnetic pole for different terranes at different times can serve to evaluate regional rotation and the relative displacement of these terranes. Issayeva and Khalafli (2006) carried out the last paleomagnetic study of this type. It was revealed that the regions of the

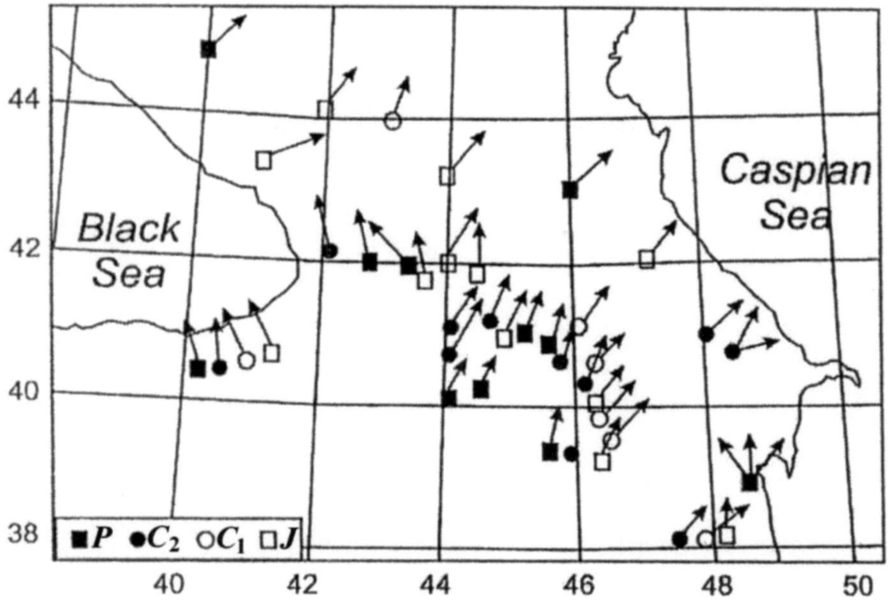

Fig. 2.27 Distribution of paleomagnetic vectors for the Mesozoic-Cenozoic rocks of the Lesser and Greater Caucasus, Iran and Turkey (After Issayeva and Khalafli 2006, with some modifications). *P* Paleogene, C_2 Late Cretaceous, C_1 Early Cretaceous, *J* Jurassic

Lesser and Greater Caucasus and Iran rotate clockwise through about 20–30° with respect to the magnetic meridian, and the North Anatolia rotates anticlockwise through about 30–40° (Fig. 2.27).

The average coordinates of the paleomagnetic poles for Azerbaijan, Georgia, Armenia, Turkey, and Iran were compared to the Russian platform and showed that these regions have been connected to the Eurasian continent since the Upper Eocene. The Lesser Caucasus has moved 5–7° to the south with respect to the Eurasian continent.

Chapter 3
Methodological Specificities of Geophysical Studies in the Complex Environments of the Caucasus

3.1 Specifics of Media and Geophysical Studies

3.1.1 Main Features: Advantages of Natural Field Studies

Geophysical studies of the Caucasus need to cope with the mountainous environments of many of its regions and their inclined (oblique) magnetization in temperate latitudes. The uneven topography not only impedes geophysical surveys in mountainous areas due to poor accessibility, but also distorts the measurement results.

Inclined (oblique) polarization distorts the pattern of magnetic and other geophysical fields. The procedure of "reduction to the pole" developed by Baranov (1957) is often used in conditions of oblique magnetization to convert magnetic anomalies into vertical magnetization. This transformation is actually a conversion from a magnetic field to a magnetic potential (Blakely 1995). Many authors have modified this approach, in particular Nabighian (1972). However, the reduction to the pole is constrained to situations where all the source bodies in the area are magnetized parallel to the geomagnetic field and have a subvertical dipping (Khesin et al. 1983, 1996). This is the only case where the magnetic field conversion is completely correct, graphs are symmetrical and can be used for further interpretation by conventional methods. Unfortunately these situations are fairly rare. Moreover, the polarization angle (which often coincides with body dip angle) is the one used to determine parameters in quantitative interpretation of geophysical anomalies. For these reasons, non-conventional inversion methods for complex environments have been developed and applied (Khesin et al. 1996).

The Caucasus is related to Alpine-type mountainous regions where solid heterogeneous associations approach the Earth's surface. These associations are multiply folded, with intensive rupture tectonics (including thrust tectonics). Rocks of various origin and composition with a broad range of physical properties rapidly

L.V. Eppelbaum and B.E. Khesin, *Geophysical Studies in the Caucasus*, Lecture Notes in Earth System Sciences, DOI 10.1007/978-3-540-76619-3_3,

change along both the vertical and lateral directions. This predetermines the complexity of the images of physical fields (Khesin 1978b).

At the same time, the mountainous conditions can hardly be treated solely as an obstacle. Deep erosion truncation and a lack or a thin layer loose deposits encourages the application of visual geological methods. These make it possible to obtain extensive geological evidence on the nature of anomaly sources, correlate them with geophysical data, as well as study physical properties of rocks and ores in natural and artificial outcrops. Rugged topography can serve to calculate effective physical parameters of exposed geological sections based on measurements of the corresponding fields on an uneven surface. It also facilitates geological applications of topography data. Outcroppings of mountainous regions promote an integrated application of conventional geophysical methods not only with visual geological, but also geochemical, petrophysical, physical-chemical and nuclear geophysical investigations (including areal studies), which considerably reduce the ambiguity of geophysical interpretation (Khesin 1976).

The use of cumbersome geophysical equipment is difficult in poorly accessible areas, and measurements of natural geophysical fields are preferable in mountain environments. Methods using natural phenomena are faster to do, environmentally friendly, and less expensive than methods using artificial energy sources (Khesin 2005). Natural types of geophysical fields are widely studied (Table 3.1).

The study of natural geophysical fields (electric, electromagnetic, magnetic, gravity and thermal fields, earthquake seismology) has important advantages. Relative geophysical methods are not expensive and non-invasive; they do not require cumbersome equipment, drilling, explosions or other signal stimulations. The investigated depth tends to be larger, because the path of useful signals from the natural field anomaly sources to the sensors is twice as short as the path of an artificial signal. In particular many natural modifications have been identified through electric prospecting (Table 3.2).

Magnetotelluric and telluric methods are used for the study of deep geological structures, whereas the SP method is mainly used for the solution of near-surface problems of mineral prospecting, engineering geology, hydrogeology, and in some cases for oil exploration.

The requirements for geophysical techniques adapted to mountainous conditions are less stringent for regional and detailed prospecting. In regional prospecting (airborne surveys), the measurement network may be tied in with comparatively stable elements of the topography. In ground surveys, it can be associated with roadways or riverbeds for geophysical measurements that involve cumbersome equipment. In this way, key intersections are obtained which are then used as reference sections for large-scale surveys of the area (Fig. 3.1).

In detailed prospecting, a small area of the districts and existing infrastructure facilitate transportation and organization and make it possible to use rather complex equipment.

The study of areas in the vicinity of wells and measurements in mines in mountainous regions is characterized by specific conditions that affect equipment transportation and the separation of wells and mines. The latter in, particular,

Table 3.1 Characteristic geophysical methods

Field studied	Natural	Artificial
Elastic waves	Seismology	Seismic prospecting
Electromagnetic waves	Magnetotelluric method	Frequency sounding
Electric	Self-potential	Electric profiling
Magnetic	Magnetic prospecting	Artificial additional magnetization
Gravity	Gravity prospecting	–
Thermal	Thermal prospecting	–
Radioactivity	Gamma-spectrometry survey (radon monitoring)	Nuclear magnetic resonance

Table 3.2 Classification of electric prospecting methods (After Khesin 2007, with modifications)

Investigated fields	Method	Source of field	Measured values	Frequency (period)
Natural	Self-potential (SP)	Oxidization of ore, filtration, resistivity contrast	Potentials	Quasi-stationary
	Telluric currents	Electromagnetic (EM) induction	Electrical components	0.5–600 s
	Magnetotelluric	"—"	EM components	"—"
	Audio-frequency magnetic fields (AFMAG)	Lighting strike discharge	Magnetic components	1–1,000 Hz
Quasi-natural[a]	Industrial currents	Electric motors, power lines, etc.	EM components	50–400 Hz
	Very low frequency[b] (VLF) fields	Remote transmitters	"—"	3–30 kHz
Artificial	Resistivity (direct current) methods	Battery/generator	Apparent resistivity	Direct current
	EM fields of low frequency	Generator	EM components	0.08–8 kHz
	EM fields of high frequency	Radio waves	Absorption	0.16–37.7 MHz
	Transient: INPUT	Transmitter loop of impulses	Potentials and EM components	0.3–1.9 ms
	TDEM		Resistivity	n (μs-ms)
	Induced polarization (IP)	Transmitter of impulses	Potential, resistivity, and time constant	0.3–12 s

[a]This term was introduced for methods employing investigator-independent sources (Khesin 1969)
[b]According to broadcasting bands

require a study of isolated wells. It is noteworthy that extended adits and a rugged topography make it possible to deploy rather complicated spatial systems for earth's surface/mine geophysical surveying. As a rule, these operations are carried out at the exploration stages (Borisovich and Eppelbaum 1988).

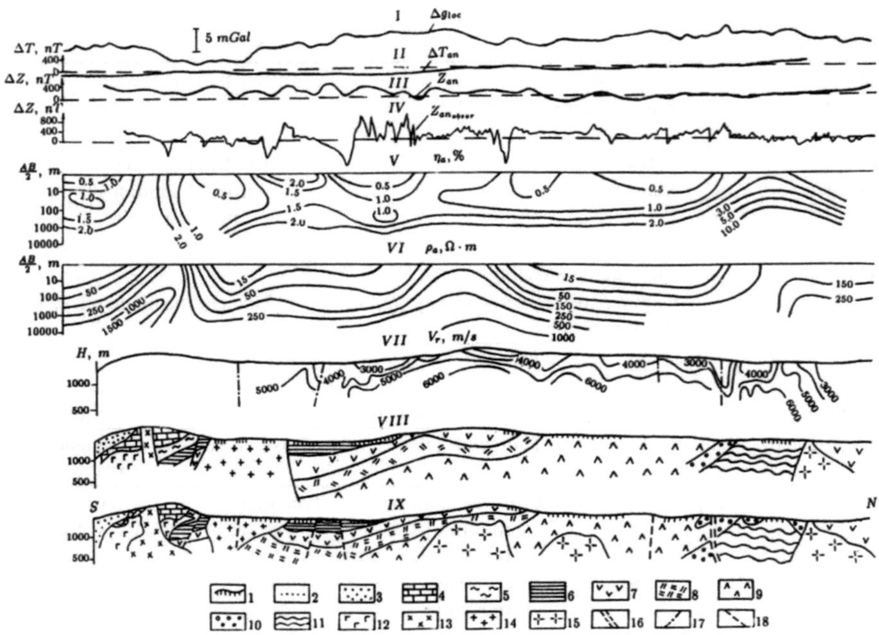

Fig. 3.1 Integrated geophysical studies of the Novoivanovka-Beyukkishlak reference profile (NE Lesser Caucasus); the northern margin of this profile crosses the Beyukkishlak outcrop of the Pre-Alpine basement. **(I)** Δg plot of local anomalies in incomplete Bouguer reduction (1 mGal = 10^{-5} m/s^2); **(II)** ΔT_{an} curve plotted by the map on a 1:50,000 scale; **(III)** Z_{an} curve plotted by the map on a 1:25,000 scale; **(IV)** plot of measured Z_{an} values; **(V)** η_a section; **(VI)** ρ_a section; **(VII)** velocity section obtained by refraction survey; **(VIII)** geological section; **(IX)** geological section compiled on the basis of geophysical data. (*1*) sandy-clay and gravel-pebble deposits (*Q*); (*2*) predominantly clastic formations (Ng_2 – *Q*); (*3*) mudstones, tuffaceous sandstones, tuff gritstones (Pg_2); (*4*) limestones (K_2cmp + maa); (*5*) mudstones and tuffaceous clastic rocks (K_2con); (*6*) tuffaceous clastic rocks and mudstones (J_3clv); (*7*) porphyrite and tuffaceous rocks (J_2bth); (*8*) quartzose plagioporphyres (J_2baj_1); (*9*) porphyrites and tuffs (J_2baj_1); (*10*) basal conglomerates, sandstones and shales (J_1); (*11*) metamorphic schists (*Pz*); (*12*) gabbro-diorites (γ Pg_3); (*13*) syenodiorites, monzodiorites ($\varepsilon\delta$ Pg_3); (*14*) subporphyritic quartz-syenodiorites; quartz-dioritic porphyrites ($\varepsilon\delta$ K_1); (*15*) plagiogranites ($\gamma\pi$ J_2baj_2(?)); (*16*) inferred zones of dikes; (*17*) fracture zones according to wave dynamic analysis; (*18*) disjunctive dislocations

3.1.2 General Characteristics of the Targets and Host Media

The physical properties of different targets and host media are very diverse (Table 3.3).

Geophysical studies of the mountainous areas are complicated by rapidly changing media and the great number of near-surface anomaly sources. Even for medium-scale areal surveys of the northeastern part of the Lesser Caucasus, up to five or six petrophysical varieties (according to the classification of the Institute of Geology, St. Petersburg) can be found in a 4 km^2 cell (Karkoshkin 1979). The petrophysical survey of the Gedabey mining district of the Lesser Caucasus showed that an area of 700 km^2 prospected on a scale of 1:50,000 had 17 petromagnetic and

Table 3.3 Physical properties of rocks, minerals and some media (On the basis of Parasnis (1986), Kobranova (1986), Erofeyev et al. (2009), with supplements)

Type of matter	Electric resistivity, Ohm·m	Polarizability, %	Magnetic susceptibility, SI unit	Thermal conductivity, W/(m·°C)	Density, g/cm^3	Compressional velocity, km/s
Granite	$5 \cdot 10^3$–10^6	0.5–2.5	(10–65) (without magnetite) 25–$5 \cdot 10^4$ (with magnetite)	1.9–3.2	2.5–2.7	4.6–7.0
Basalt	10^3–10^8	0.2–3.8	$1.5 \cdot 10^3$–$2.5 \cdot 10^4$	1.5–2.2	2.7–3.3	2.5–6.4
Dunite	10^6–10^7	0.3–1.9	10–300	3.7–5.2	2.5–3.30	7.45–8.1
Gabbro	10^3–10^8	0.5–1.8	$3.8 \cdot 10^3$–$9 \cdot 10^{4a}$	2.0–2.3	2.7–3.5	6.0–7.4
Diabase (dykes)	1,200–2,500	0.8–4.0	$(1.25$–$5) \cdot 10^3$	2.1–2.3	2.5–3.2	6.3–6.75
Gneiss	10^3–10^6	0.8–1.1	0–$3 \cdot 10^3$	1.9–3.7	2.6–2.9	5.2–6.3
Serpentinite	10^2–10^4	0.3–1	$(3$–$75)\ 10^3$	2.0–3.8	2.2–2.9	6.5–6.9
Shale	1–10^4	0.3–2.0	10–50	~7 (parallel) ~1 (perpendicular)	2.3–2.75	3.1–3.8
Sandstone	35–$4 \cdot 10^3$	0.25–2.0	25–$5 \cdot 10^3$	2.5–3.2	1.8–2.7	1.5–4.0
Limestone (compact)	120–$4 \cdot 10^3$	1.3	25–$3 \cdot 10^3$	2.0–3.0	2.6–2.7	3.5–6.5
Sand	1–$1 \cdot 10^6$	0.1–1.0	25–$5 \cdot 10^3$	0.3–2.95	1.3–2.0	0.3–1.5
Clay	1–120	0.1–1.7	25–$1 \cdot 10^3$	0.25–1.08	1.2–2.4	1.2–2.5
Gypsum	10^5–10^6	very low	−12.6	0.5–1.5	2.1–2.5	3.4–4.6
Rock salt	10^6–10^7	very low	−10	5.3–7.2	2.1–2.4	4.0–5.5
Iron	$1 \cdot 10^8$	18–20	$1 \cdot 10^6$–$1 \cdot 10^7$	18–20	7.8	4.6–5.9
Sulphide ore	$1 \cdot 10^{-4}$–$1 \cdot 10^{-3}$	12–60	10–40	5–6	3.3–3.7	6.0–6.4
Pyrrhotite	$1 \cdot 10^{-5}$–$1 \cdot 10^{-3}$	20–60	$1 \cdot 10^{-3}$–$1 \cdot 10^{-1}$	3.5	4.6	5.0
Chalcopyrite	$1 \cdot 10^{-4}$–$1 \cdot 10^{-1}$	10–25	10–30	10.7	4.2	4.85
Pyrite	$1 \cdot 10^{-4}$–$1 \cdot 10^{-1}$	50–90	35–60 (pure)	9.5–15	5.0	7. 9–8.0
Magnetite (ore)	$1 \cdot 10^{-2}$–$1 \cdot 10^1$	10–40	$7 \cdot 10^4$–$14 \cdot 10^6$	5.2	5.2	7.0–7.4
Galena	$1 \cdot 10^{-2}$–$3 \cdot 10^2$	10–30	−4.3	2.0	7.5	3.7–3.77
Sphalerite	$>1 \cdot 10^4$	12–50	−3.3	26.7	4.0	5.3–5.6

(continued)

Table 3.3 (continued)

Type of matter	Electric resistivity, Ohm·m	Polarizability, %	Magnetic susceptibility, SI unit	Thermal conductivity, W/(m·°C)	Density, g/cm^3	Compressional velocity, km/s
Gold	$2.4 \cdot 10^{-8}$	close to 0	−2.5	318	19.3	3.2
Graphite	$1 \cdot 10^{-4}$–$1 \cdot 10^{-2}$	30–90	−100 (diamagnetic)	1.15–17	2.2	4.1
Water	0.1–$1 \cdot 10^{5}$	close to 0	~0	0.50 (25°C)	1.0	1.43–1.59
Oil	$1 \cdot 10^{9}$–$1 \cdot 10^{14}$	close to 0	−10	0.11–0.16	0.9	1.15–1.3
Coal	$1 \cdot 10^{2}$–$1 \cdot 10^{4}$	3–15	10–20–$1 \cdot 10^{4}$ (fired)	0.13–2.2	1.2–1.5	2.4–2.7
Air	$1 \cdot 10^{14}$	0	~0	0.023 (0°C)	0.0012	0.33
Ice	$1 \cdot 10^{4}$–$1 \cdot 10^{8}$	0.5–20	−9 (0°C)	2.2 (0°C)	0.92	3.0–4.0
Pine wood	0.8–2.5	low	≈0	0.4–1.2	0.5–0.8	1.4–1.7

[a]Leucocratic gabbro of low magnetization (≤100 mA/m) occurs in some regions of the Caucasus.

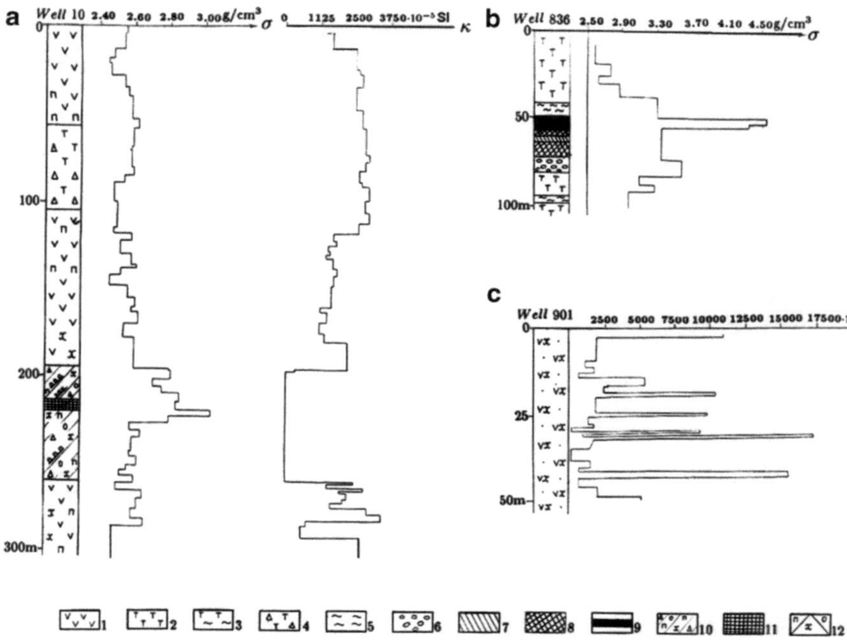

Fig. 3.2 Petrophysical variation in cores (NE Lesser Caucasus): (**a**) Kyzyl-Bulakh gold-pyrite deposit, (**b**) Dashkesan iron deposit, (**c**) Murut site. (*1*) porphyrites; (*2*) slightly altered tuffs; (*3*) hornfelsed tuffs; (*4*) tuff breccias; (*5*) hornfels; (*6*) garnet skarns; (*7*) magnetite-garnet skarns; (*8*) lean skarn-magnetite ore; (*9*) rich skarn-magnetite ore; (*10*) zone with phenocrysts, nests, and places of occurrence of pyrite and chalcopyrite massive ores; (*12*) pyritization, chloritization and silicification

11 petrodensity gradations. The changeability of mountainous regions along the lateral is accompanied by their rapid change along the vertical (Fig. 3.2).

Nevertheless, it is possible to distinguish a series of petrophysical discontinuities that have special value in regional studies (Fig. 3.3).

In mountainous regions (where the prospecting targets are ore-bodies and structures of small dimensions) most attention is focused on geophysical field regioning (to single out tectonic blocks) and the location of objects with relatively stable geophysical features. The small sizes of most of the objects dictate the choice of large-scale and detailed prospecting. For example, in half of the cases the thickness of steeply dipping blocks does not exceed 0.5 km in the Dashkesan mining district of the Lesser Caucasus (Fig. 3.4).

Regional and detailed prospecting is affected to a lesser extent by mountainous conditions than large-scale surveys. The regional ore controls are larger ore-bearing and ore-distributing structures and therefore are related to the corresponding regional boundaries of physical properties. We observed similar regularities in ordinary conditions during petroleum and underground water exploration. This makes it possible to study the behavior of the Pre-Alpine basement to detect large

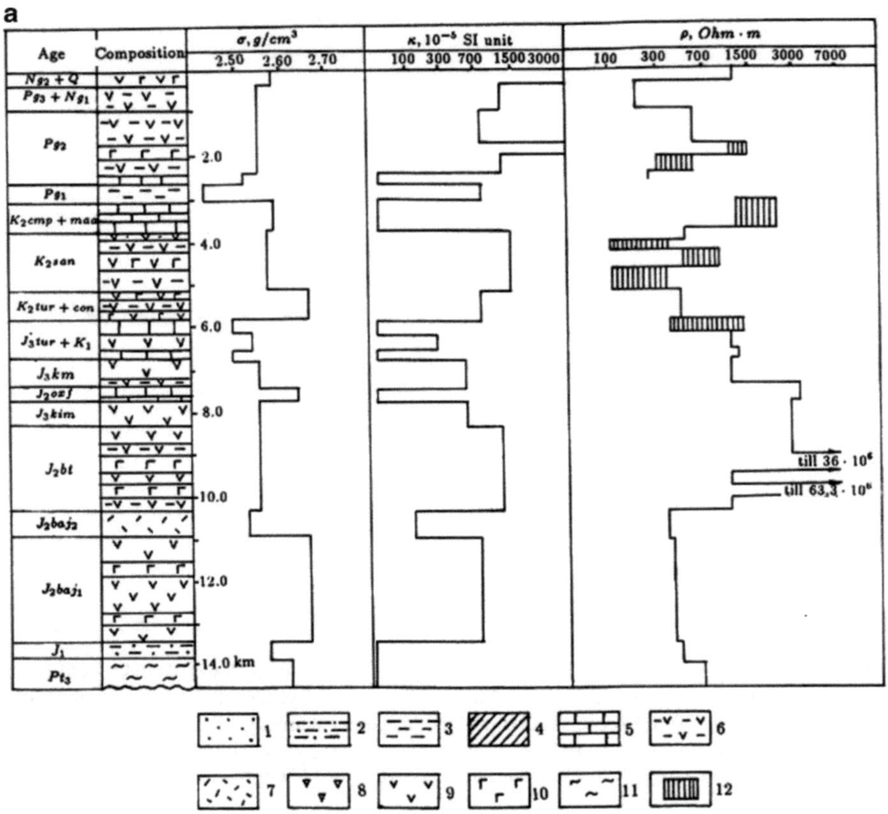

Fig. 3.3 (continued)

crust blocks with different mineral potentials. For regional prospecting, deep heterogeneities may be approximated by simple deterministic models developed on the basis of available data. This can serve to apply deterministic methods of quantitative interpretation developed for complex measurement surfaces and other complicating factors. Some lithologic and stratigraphic targets and tectonic lines, mineral deposits and zones of hydrothermal alterations may be also considered as reference points within areas under detailed study (Table 3.4).

3.1.3 *Typical Geophysical Noise Effects Under Mountainous Conditions*

The erosion of mountainous structures causes considerable changes in near-surface associations. The physical properties of the rocks occurring close to the Earth's

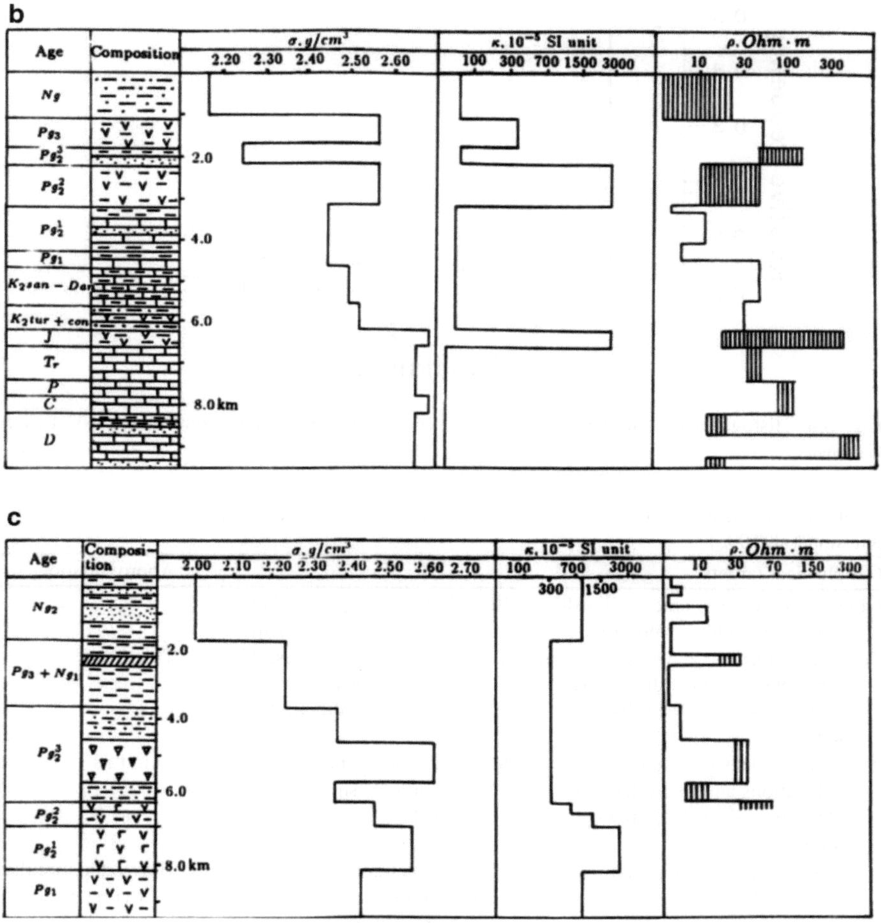

Fig. 3.3 Petrophysical columns: (**a**) Somkhet-Karabakh Zone, (**b**) Yerevan-Ordubad Zone, (**c**) Talysh Zone. (*1*) sands and sandstones; (*2*) sand and clay deposits; (*3*) clays; (*4*) marls; (*5*) limestones and dolomites; (*6*) volcanogenic-sedimentary rocks; (*7*) liparites and plagioliparites; (*8*) trachyandesites and trachybasalts; (*9*) andesites; (*10*) basalts, andesite-basalts, diabases; (*11*) metamorphic schists; (*12*) interval of resistivity mean values

surface may thus not be the same as those of deep-seated rocks with the same composition and age (Fig. 3.5).

Consequently, the measurements of physical properties of rocks located in the upper portion of a section should be treated with great care, especially when they are extrapolated to a depth. There is, however, a need for shallow surveys, since most deposits occur in the upper portion of a section. During a geophysical study of deeper horizons, the effect of this rapidly changing portion should be corrected by taking into account its true parameters. All these factors mean that special attention should be paid to the physical properties of rocks and ores measured on samples

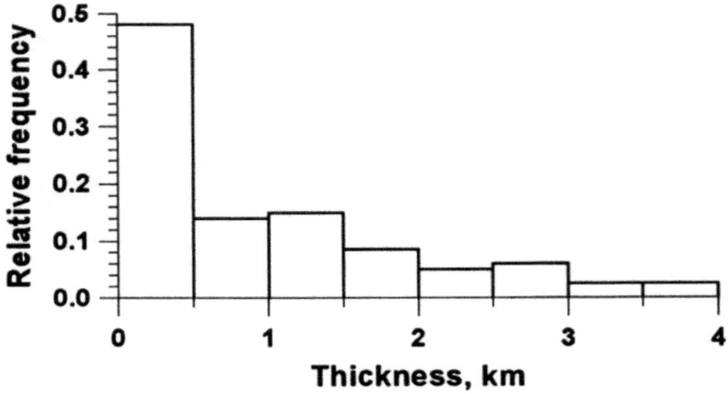

Fig. 3.4 Histogram of relative frequency of geological block dimensions in the NE Lesser Caucasus

Table 3.4 Typical approximation of geological objects by bodies of the simplest shape (After Khesin et al. 1996)

Geophysical targets		Approximation
Objects outcropping onto the Earth's surface and under overburden	Buried or cropping out when surveying by aerial method	
Tectonic-magmatic zones, sill-shaped intrusions and thick dikes, large fault zones, thick sheet-like ore deposits	Tectonic-magmatic zones, thick sheet intrusions and zones of hydrothermal alterations	Thick bed
Thin dikes, zones of disjunctive dislocations and hydrothermal alterations, sheet-like ore deposits, veins	Sheet intrusions, dikes, disjunctive dislocations, sheet-like ore deposits	Thin bed
Lens- and string-like deposits	Folded structures, elongated morphostructures, large mineral lenses	Horizontal circular cylinder
Pipes, vents of eruption, ore shoots	Intrusions (isometric in plane), pipes, vents of volcanoes, large ore shoots	Vertical and (inclined) circular cylinder or pivot
Karst cavities, hysterogenetic ore–bodies	Brachy-folds, isometric morphostructures, karst cavities, hysterogenetic ore–bodies	Sphere

taken from the surface, in their natural bedding, drill cores or mines. It is expedient to carry out their determination on the basis of measured geophysical fields. The variability of physical properties makes it possible to derive additional information, and single out the specific ore signs through petrophysical studies.

Geophysical field studies are also considerably complicated by the heterogeneity of geological sections, which can also be a target of investigation. The topography

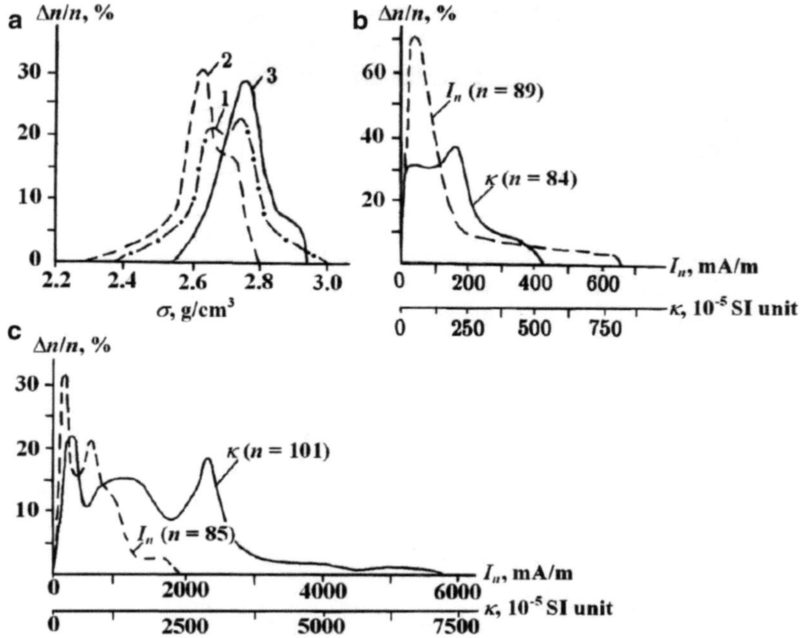

Fig. 3.5 Physical properties of crystalline schists from the Pre-Jurassic basement (**a**, **b**) and porphyritic tuffs from the Jurassic (**c**) in the north-eastern part of the Lesser Caucasus. (*1*) for all samples ($N = 159$); (*2*) for samples taken from the surface ($N = 87$); (*3*) for core samples ($N = 75$)

effect, the rapid variation in composition and thickness of eluvial-deluvial, proluvial, colluvial and alluvial formations, moistening and weathering, instrument sitting and signal generation conditions may also interfere with such studies. For instance, poor grounding when applying resistivity methods on stony mountain slopes can limit their application. Therefore, it is preferable to use inductive electric prospecting methods (e.g., Tarkhov 1980; Goldman 1990).

The results of an analysis of noise effects impeding geophysical prospecting under complicated environments are summarized in Fig. 3.6.

3.1.3.1 Artificial Noise

The *Industrial* component of noise mainly arises from power lines, cables, and various underground and transportation systems. The *Instrumental* component is associated with the technical properties of geophysical equipment (the most typical example is "shift zero" for gravimeters) and their spatial location. *Human* error, obviously, can accompany geophysical observations at any time. Finally, *Undocumented* (*poorly documented*) results of previous surveys can distort the development of the preliminary Physical-Geological Model (***PGM***) of the area under study.

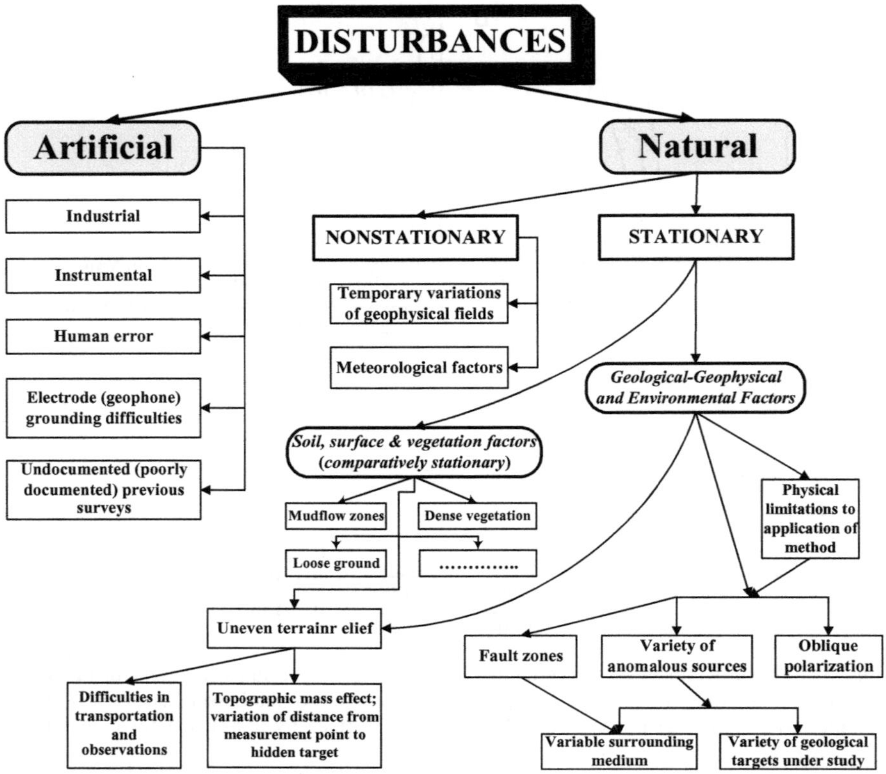

Fig. 3.6 Noise effects in geophysical studies in mountainous regions (After Eppelbaum 1989, with modifications)

3.1.3.2 Non-Stationary Noise

Non-stationary noise effects – the time variations of fields – are typical of natural (gravity, magnetic, SP, thermal) and quasi-natural (electromagnetic, VLF transmitter) fields. Several factors are responsible for these types of noise. In gravity prospecting these include lunar-solar and other gravity variations (Adler 1942), whereas in magnetic and VLF prospecting noise can result from changes in the ionosphere (Yanovsky 1978; Barr et al. 2000). Finally, in the SP method and near-surface thermal prospecting, meteorological factors such as diurnal and seasonal variations of temperature, pressure and humidity (Semenov 1974; Dobrynina et al. 1985; Ernstson and Scherer 1986) play a role. In the presence of strongly magnetized objects the time variations can produce *Secondary* transient effects in magnetic (and in some other) fields. Meteorological conditions (rain, lightning, snow, hurricanes, etc.) can also affect readings. *Surface, soil and vegetation factors* associated with certain surface conditions and soil types (e.g., mudflow zones or loose ground in deserts) and dense vegetation (e.g., forest areas are widely present in both the Southern and Northern Greater Caucasus, and Lesser Caucasus) may strongly hamper movement along the profile.

3.1.3.3 Geological-Geophysical and Environmental Factors

These constitute the most important physical-geological disturbances. The application of any geophysical method depends primarily upon the existence of physical properties relating the objects under study and the surrounding medium. The *Physical limitations* on method implementation include the constraints on assessing the physical contrast properties between the geological targets and the surrounding medium.

Uneven terrain relief is the main factor that complicates geophysical prospecting under mountainous conditions. It may negatively affect equipment transportation and geophysical data acquisition. Physically, this disturbance (for potential and quasi-potential fields) is generally twofold and involves the form and physical properties of the topographic features making up the relief as well as the effect of variations in the distance from the point of measurement to the hidden target (Khesin et al. 1996).

Fault zones are widely distributed in the Caucasus. The *Variety of anomalous sources* is composed of two factors: *the variable surrounding medium* and *the variety of geological targets under study*. Both these factors are crucial and greatly complicate the interpretation of geophysical data. The complex upper portion of the geological sections typical of mountainous regions determines a multitude of geophysical field anomaly sources. A considerable variety of anomaly sources and the superposition of fields of different origins require: (1) probabilistic-statistical or information-statistical methods for singling out the prospecting targets, and (2) methods for quantitative interpretation of anomalies under conditions of an unknown normal field level.

3.1.3.4 Oblique Polarization

Oblique polarization (*magnetization*) disturbs these geophysical fields in the following manner: the major extremum is shifted from the projection of the upper edge of the object onto the earth's surface, and an additional extremum may appear (Khesin et al. 1996). It should be noted that oblique magnetization is a characteristic feature of the Caucasus $\approx \left(58 - 61^{\circ}\right)$ due to its geographical location.

3.2 Terrain Correction and Utilization of Topography for Extraction of Geological Information

3.2.1 Problem of Terrain Correction: Two Aspects of This Problem

The elimination of the effects of terrain relief is crucially important for geophysical studies of mountainous regions. A rugged terrain relief has an effect not only on transportation, complicating geophysical prospecting under mountainous conditions (in particular, routes are not always located across the strike of the target, which limits the interpretation). The terrain relief effect also manifests itself in non-stationary and stationary topographic anomalies.

Non-stationary anomalies are customarily identified visually during fieldwork or by special measurement methods. Filtration SP anomalies are classified as non-stationary anomalies. Usually their gradient is considerably less than that of "ore" anomalies, and a mirror reflection of the topographical shape by the SP anomaly is used to reject false anomalies. When the SP survey profile crosses a mountainous river with a stony bed, a kinetic electric field is observed, which appears as a potential minimum against the background of its total increase due to filtration on the valley slopes. The potential of such a field grows along the stream. Magnetic anomalies on tops of hills, when associated with lightning (magnetization of outcropping rocks is severely altered in its vicinity) can also fall into this category, as well as thermal anomalies due to varying meteorological conditions (their effect can be excluded by repeated measurements at different time intervals).

The main features of terrain correction are similar to corrections for stationary noise effects incorporated into results of geophysical measurements. First, the form and physical properties of topographic masses (i.e. relief-forming morphostructures) are responsible for the effects of these masses in the anomalous field, which mask anomaly effects from hidden targets. For example, the attraction of topographical masses in the Lesser Caucasus reaches 10–15 mGal (Fig. 3.7), whereas large anomalous objects usually generate anomalies of several mGals.

Second, due to the uneven measurement line, the distance from the field recording point to the source varies, which manifests itself differently in anomalies from various objects. This requires: (1) calculation of that part of the field which is due to the known or "missing" topographic masses, and (2) conversion of the measured values to the horizontal reference plane (reduction to line).

Note that an error as regards a topographic effect results in the distortion not only of the details of the anomalous field but also the geological structure of the object and its presence as a whole. For example, in rugged terrain relief an ore-body may be located above the measurement point. It is only natural that magnetic, gravity and electric effects will appear to be opposite to that expected if the body were lying below the measurement plane.

As an example, consider the SP anomaly from a sulfide vein of a thin bed shape (Fig. 3.8a).

Proceeding from classical SP theory (Semenov 1974), the lower pole line (projected into point ***A***) of the polarized bed has a negative charge when the groundwater surface is parallel with the slope and intersects the ore deposit. Based on this theory a negative SP anomaly will be recorded on the earth's surface, but it will be considerably displaced from the ore-body projection (curve I). By contrast, if the groundwater table follows the horizontal plane and crosses the deposit, the upper pole line (projected into point ***B***) acquires a negative charge. In this case, the negative anomaly on the earth's surface becomes positive (curve II), and its actual characteristics are misinterpreted. The negative SP anomaly may completely disappear if the ore deposit occurs under low topography, which causes

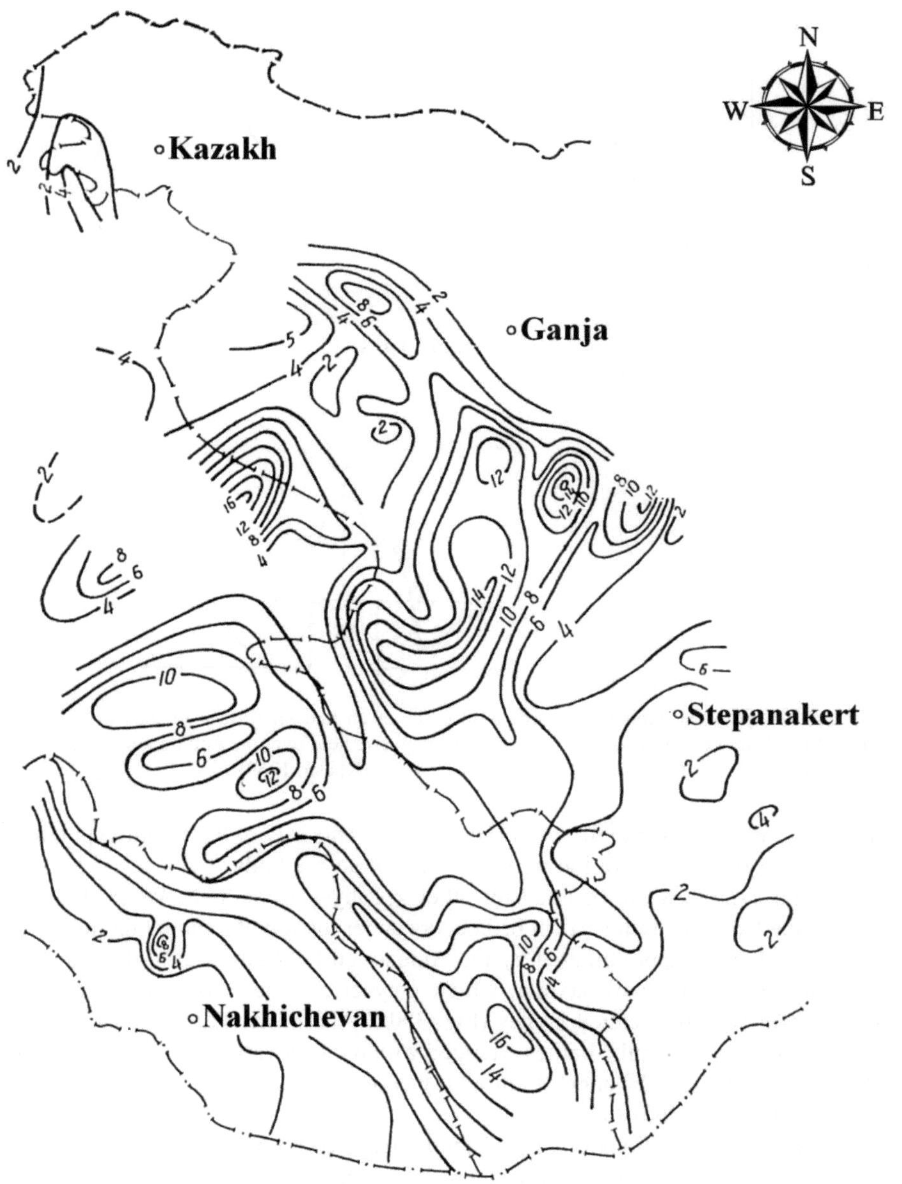

Fig. 3.7 Map of equal gravity terrain corrections (in mGals) for the Lesser Caucasus region in a radius of 200 km (After Khesin 1981)

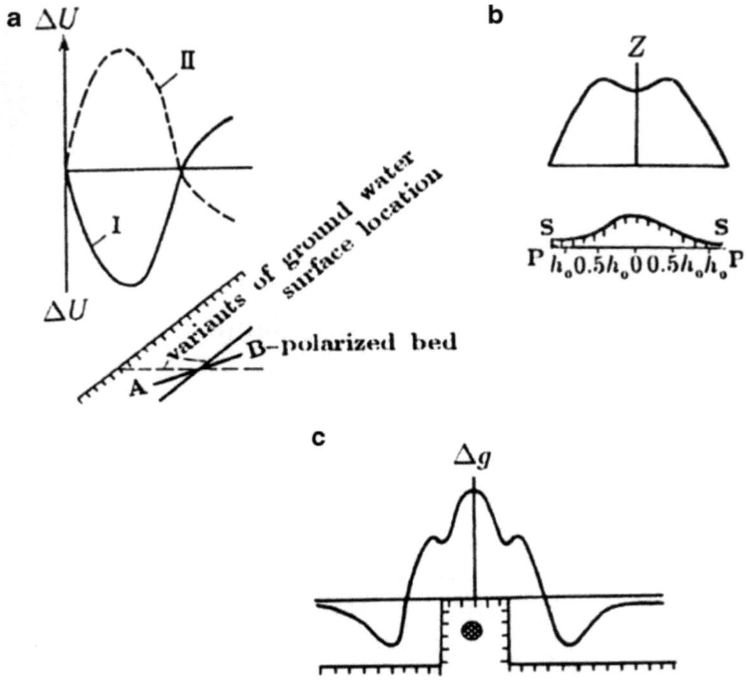

Fig. 3.8 Geophysical anomaly distortions in cases of an uneven surface of measurement for: (**a**) self-potential field over the *inclined bed*; (**b**) magnetic field over the infinite vertical *thin bed*; (**c**) gravity field over the deposit with a *circular section* (Khesin 1969)

a positive anomaly due to filtration. The negative anomaly of the filtration potential may appear over the elevation in the absence of an ore-body.

Similar examples can be cited for other geophysical methods. When the basic dike in the form of a thin vertical bed occurs under the top of a mountain range, the relief of which is approximated by the Aniezie curl (Mikov 1963), the magnetic anomaly splits into two maxima, which may create the erroneous idea of the presence of two magnetized bodies (Fig. 3.8b). A deposit with a circular section resembling some chromite bodies arranged within the elevation with vertical slopes (Fig. 3.7c), sets up a Δg anomaly of a very complicated form with four additional extrema (Lukavchenko 1961).

Thus, under conditions of rugged topography, commonly accepted notions of qualitative interpretation, according to which the number and location of geophysical anomalies are correlated with the number and location of geological objects may not apply (Khesin 1969).

It should be noted that under conditions of low-magnetic outcropping rocks, airborne magnetic surveys that assume the flowing of relief forms are much more effective than conventional air surveys along straight profiles. This can be seen from a comparison of the results of a survey using the two types of aeromagnetic

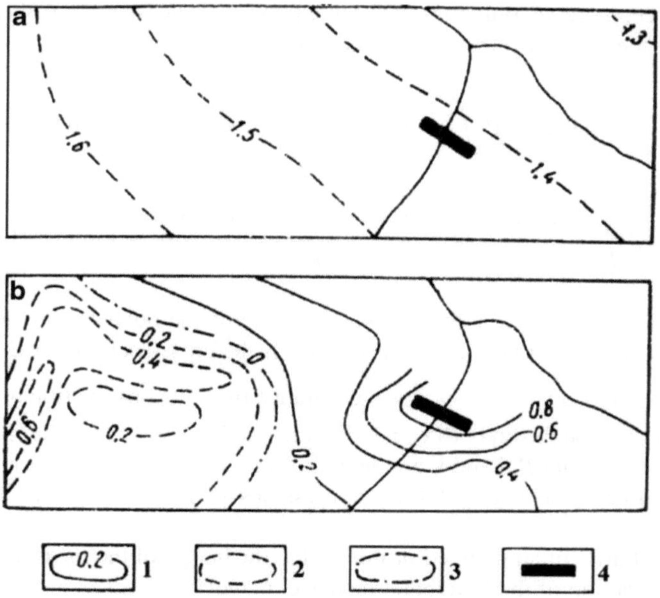

Fig. 3.9 Results of a helicopter magnetic survey in the area of the copper-pyrite deposit of Kyzyl-Dere in the Dagestan Mountains (Greater Caucasus): (**a**) along straight profiles, (**b**) topography flow. ΔT isolines in hundreds nanoTesla: (*1*) positive, (*2*) negative, (*3*) zero; (*4*) location of the ore deposit

measurements in the copper-pyrite deposit area of Kyzyl-Dere in the Dagestan Mountains (Khesin 1969).

The presence of pyrrhotite in the copper-pyrite ore provides redundant magnetization of this deposit occurring in low-magnetic sand-shale associations which an aeromagnetic survey along straight profiles over this deposit failed to detect (Fig. 3.9a). By contrast, a flowing airborne survey clearly showed the magnetic effect of the buried ore-body (Fig. 3.9b).

3.2.2 Common Correction Techniques for Different Measurement Heights

The effect of different heights of measurement points and the associated remedial techniques were first discussed by Alexeyev (1976) for magnetic prospecting. The problem was elaborated in Khesin et al. (1983, 1996). Similar techniques were discussed for the interpretation of gravity anomalies (Alexeyev and Khesin 1987). Theoretical analysis (Alexeyev 1976; Khesin et al. 1988, 1997; Eppelbaum 1989; Khesin and Eppelbaum 1994; Eppelbaum and Khesin 2002) showed that common correction techniques can be applied to all the geophysical methods under discussion for an inclined measurement line.

In essence, there are only two applicable types of general analytical expressions to describe these geophysical fields. They are:

$$U_1(x,z) = P\int_s \frac{(z_s - z)\cos\gamma_p + (x_s - x)\sin\gamma_p}{r^2}dx_s dz_s, \quad (3.1)$$

$$U_2(x,z) = P\int_s \frac{\left[(z_s - z)^2 - (x_s - x)^2\right]\cos\gamma_p + 2(x_s - x)(z_s - z)\sin\gamma_p}{r^4}dx_s dz_s, \quad (3.2)$$

where $\gamma_P = 90^\circ - \varphi_P, \varphi_P$ is the inclination angle of the polarization vector to the horizon, P is its value (a scalar in this case), S is the cross-section of the anomalous body, $r = \sqrt{(x_s - x)^2 + (z_s - z)^2}$ is the distance from the observation point $M(x,z)$ to a certain point on the body $P(x_s, z_s)$, $\mathbf{r} = (x_s - x)\mathbf{i} + (z_s - z)\mathbf{k}$ is the radius-vector from point M to point P, $\mathbf{i}$ and $\mathbf{k}$ are the unit vectors oriented along Ox and Oz axes.

To illustrate these manipulations expressions (3.1) and (3.2) can be used as examples.

The peculiarity of an inclined profile is that the height of the measurement point is a linear function of the horizontal distance, namely

$$z = x\tan\omega_0, \quad (3.3)$$

where ω_0 is the inclination angle of the measurement profile (Fig. 3.10).

The transformations are carried out in the following sequence. The inclined coordinate system $x'Oz'$ is introduced in such a way that

$$\left.\begin{aligned} x &= x'\cos\omega_0 - z'\sin\omega_0 \\ z &= x'\sin\omega_0 - z'\cos\omega_0 \end{aligned}\right\}. \quad (3.4)$$

The formulas for x_s and z_s are similar. The Ox'-axis in this system coincides with the inclined profile, which gives $z' = 0$ and results in formula (3.3). In the $x'Oz'$ system, formulas (3.1) and (3.2) are transformed to the following forms:

$$U_1(x,z) = P\int_s \frac{z'_s\cos(\gamma_p + \omega_0) + (x'_s - x')\sin(\gamma_p + \omega_0)}{r'^2}dx'_s dz'_s, \quad (3.5)$$

$$U_2(x,z) = P\int_s \frac{\left[z'^2_{s} - (x'_s - x')^2\right]\cos\widehat{\gamma}_p + 2(x'_s - x')z'_s\sin\widehat{\gamma}_p}{r'^4}dx'_s dz'_s, \quad (3.6)$$

where $r' = \left[(x'_s - x')^2 + (z'_s - z')^2\right]^{1/2}$ and $\widehat{\gamma}_p = \gamma_p + 2\omega_0$.

It is obvious that the right-hand sides of expressions (3.5) and (3.6) correspond to the functions $U'_1(x',0)$ and $U'_2(x',0)$ in the inclined system, but for a different inclination angle of the polarization vector. If a body in the initial system was

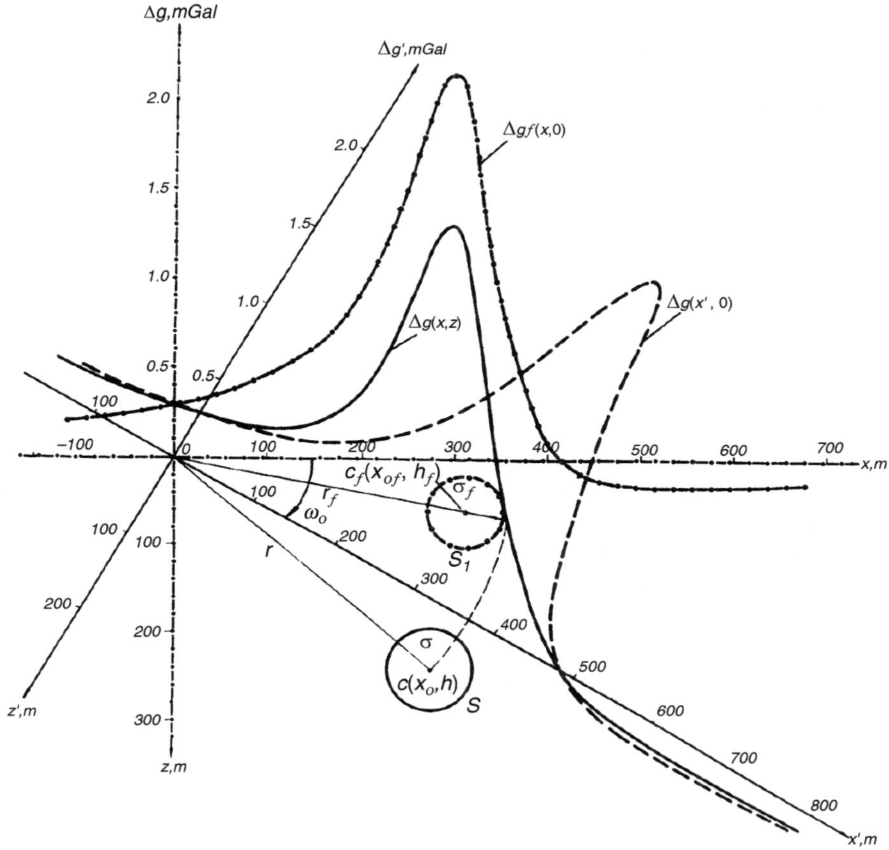

Fig. 3.10 Interpretation of gravity measurements for an inclined profile (Alexeyev et al. 1996)

vertically polarized, it turns out to be obliquely polarized (angles ω_0 or $2\omega_0$ respectively) in the inclined system, since the polarization vector does not intersect the Ox'-axis at a right angle.

Expressions (3.5) and (3.6) can essentially be interpreted in the inclined coordinate system by making use of the techniques developed for the horizontal profile, since the changing heights of the measurement points are not included there. To interpret in the initial system, the following sequence is required.

The entire space with the anomalous object and the polarization vector is rotated by an angle ω_0 and compressed at the compressibility coefficient of cos ω_0. In addition, when using formula (3.5), P is multiplied by secω_0. This done, the inclined profile Ox' coincides with the horizontal straight line, and the measurement points on the profile pass along the vertical into the corresponding points on the horizontal straight line. An anomaly plot constructed with these points (in horizontal projection) is a standard plot used routinely, although measurements are made on an inclined surface. However, after the space rotation and compression the anomalous object occupies a different position with respect to the initial one. Its cross-section is smaller than the initial one, but the outline is similar.

After this transformation, formulas (3.5) and (3.6) take on the following forms:

$$U_1(x,z) = U_{1f}(x,0) = P_f \int_{sf} \frac{z_{sf}\cos(\gamma_p+\omega_0) + (x_{sf}-x)\sin(\gamma_p+\omega_0)}{r_f^2} dx_{sf}dz_{sf}, \quad (3.7)$$

$$U_2(x,z) = U_{2f}(x,0) = P_f \int_{sf} \frac{\left[z_{sf}^2 - (x_{sf}-x)^2\right]\cos\widehat{\gamma}_p + 2(x_{sf}-x)z_{sf}\sin\widehat{\gamma}_p}{r_f^4} dx_{sf}dz_{sf}. \quad (3.8)$$

Here, the subscript "*f*" stands for the parameters of a fictitious body. When used with symbols U_1 and U_2, it denotes a fictitious body.

The interpretation of the curves U_{1f} and U_{2f} yields the parameters of a fictitious body, which are used to reconstruct those of a real body (denoted by an "*s*" subscript) by the following formulas of transition:

$$\left.\begin{aligned} z_r &= z_{sf} + x_{sf}\tan\omega_0, \\ x_r &= -z_{sf}\tan\omega_0 + x_{sf}, \\ S &= S_{sf}\sec^2\omega_0, \\ P &= P_f\cos\omega_0 (\text{by interpretation of } U_1), \\ P &= P_f (\text{by interpretation of } U_2), \\ \gamma_p &= \gamma_p - \omega_0 (\text{by interpretation of } U_1), \\ \gamma_p &= \gamma_p - 2\omega_0 (\text{by interpretation of } U_2). \end{aligned}\right\} \quad (3.9)$$

The analytical expression of the gravity anomaly caused by a certain anomalous body is:

$$\Delta g = 2G\sigma \int_s \frac{(z_s - z)\cos\gamma_g + (x_s - x)\sin\gamma_g}{r^2} ds, \quad (3.10)$$

where the value γ_g is an analog of the value γ_P in the expression (3.1), G is the universal gravity constant, and σ is the density.

This formula does not differ from expression (3.1) in terms of its structure. After the above substitutions it takes the following form:

$$\Delta g_f = 2G\sigma_f \int_s \frac{r_{gf}\cos\gamma_{gf} + (x_{sf}-x)\sin\gamma_{gf}}{r^2} ds_f, \quad (3.11)$$

such that

$$\gamma_{gf} = \gamma_g + \omega_0. \tag{3.12}$$

Taking into account that $\gamma_g = 0$, we obtain $\gamma_{gf} = \omega_0$. Hence, the anomaly of Δg observed on the inclined surface corresponds to that caused by a fictitious obliquely polarized body observed on the horizontal plane. The latter is affected both by vertical and horizontal gravity components. This is equivalent to the manifestation of "vector properties" of density on the inclined surface (see Fig. 3.10).

On the whole, the effect of the profile inclination is equivalent to an increased effect of oblique polarization, if vector **P** and the measurement surface have the same slope. We obtain a weaker effect if the measurement surface and vector **P** are inclined in opposite directions. This fact guides the unified techniques for interpreting obliquely polarized bodies observed on an inclined plane. It facilitates the analysis of the distortions due to the effect of sloping terrain relief, which can be completely attributed to the oblique polarization effect.

Thus, formulas (3.7) and (3.8) reduce the problem of interpretation of the potential fields observed on the inclined profile to the same problem for the horizontal profile. The formulas in expression (3.9) describe the transition from the parameters obtained for a fictitious body to those of a real body.

3.2.3 Reduction to Line

The above data substantiate the conversion in the inclined semi-space, since a field observed on a complex relief can be expressed using a specific inclined system of coordinates by reducing as in the normal system. The field in the measurement points of the relief is converted into an inclined straight line along the normal to the latter.

The advantage of this reduction is that the inclined plane of the reduction may be selected as close as possible not only to the highest points, but also to many points of the measurement surface (Fig. 3.11). As a result, the amplitude decreases appreciably less than when converting to the horizontal level of the highest topography points.

The use of the approach described in Sect. 3.2.2 for quantitative analysis of geophysical data along a separate profile is not always possible. However, the second feature of the topography effect (different distances between source body and an uneven measurement surface) is usually considerable. Measuring the topographic mass attraction may be useless, for example, for magnetic mapping aimed at detecting a magnetized body located in non-magnetic host rocks or in the case of electric exploration for conducting ores in a high-resistance medium. Nevertheless the distorting effect of a non-horizontal measurement line occurs when the target differs from the host medium in terms of its contrasting properties and causes the anomalous vertical gradient.

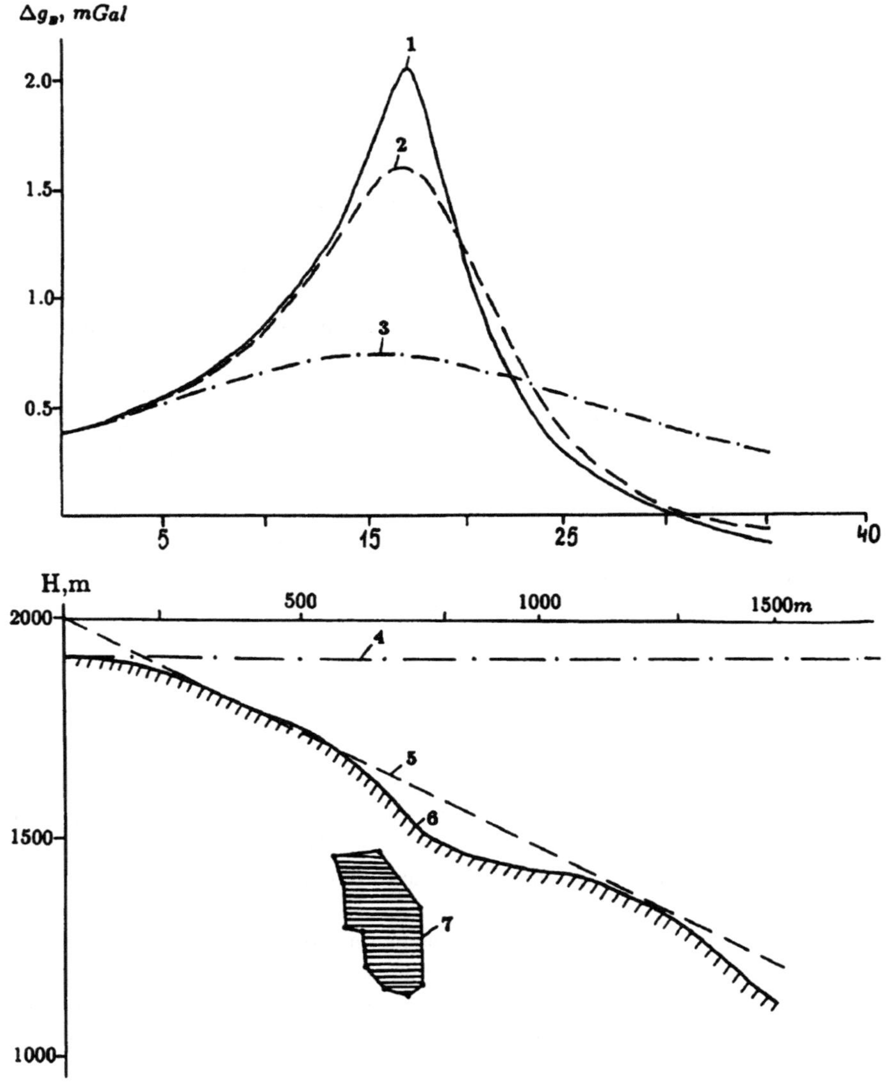

Fig. 3.11 Reduction of the geophysical field in inclined semi-space. (*1*) effect of anomalous body **7** on an uneven measurement surface **6**; (*2*) effect of anomalous body after reducing to the *inclined line* **5**; (*3*) effect of the anomalous body after reducing to the horizontal level **4** of the highest topography points

The comparison of anomalies Δg_{B} and ΔT from the local body observed on the inclined and horizontal terrain relief is illustrated in Fig. 3.12.

In general, to avoid the analytical continuation through the possible causative masses, these measurement results must be reduced to their common level, usually

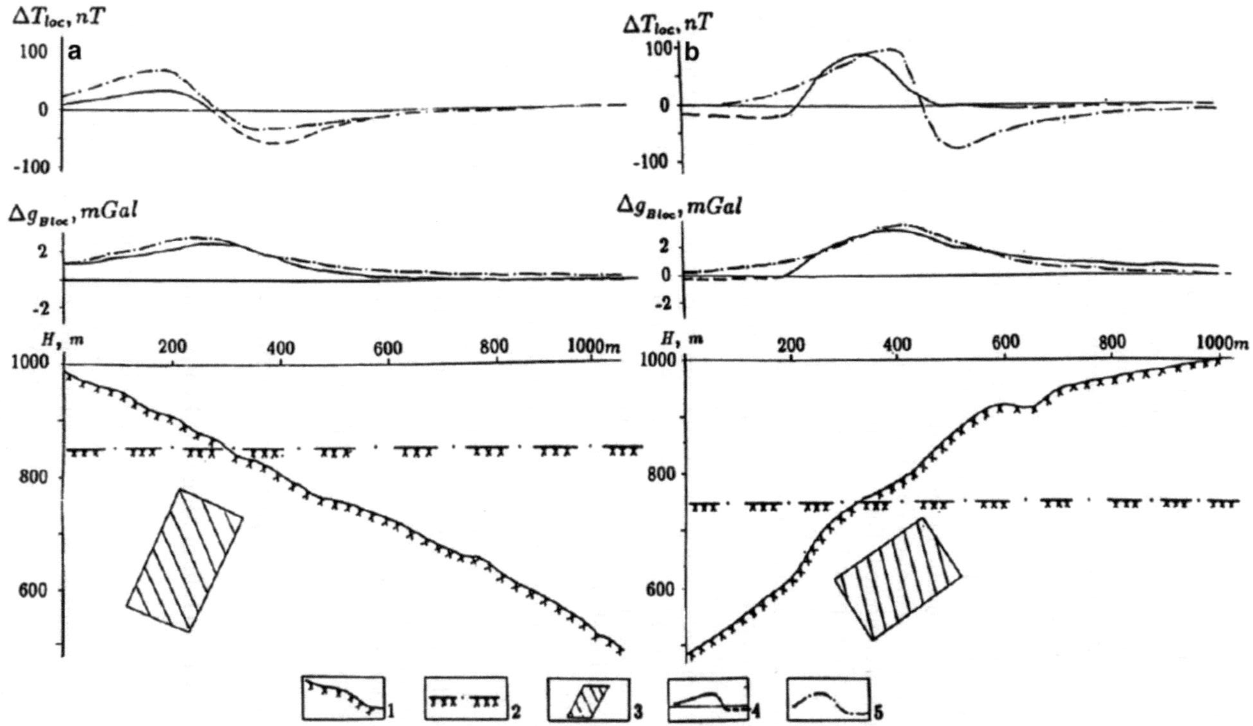

Fig. 3.12 Comparison of gravity and magnetic anomalies from a local anomalous body observed on *inclined* and *horizontal* profiles: **(a)** even slope, **(b)** complicated slope. (*1*) inclined profile; (*2*) horizontal profile; (*3*) anomalous body with redundant density $\sigma_{red} = 1.5\text{g/cm}^3$ and magnetization $J_{red} = 0.5\text{A/m}$; magnetic inclination $I = 60°$; anomalies of ΔT and Δg_B from the same body after topographic mass attraction correction: (*4*) on inclined profile; (*5*) on horizontal profile

the highest point level (reduction to line). This reduction means introducing a correction for an anomalous vertical field gradient, i.e. for the gradient component caused by anomalous bodies and the normal component of the vertical gradient.

Quite a number of methods for reducing to a horizontal level have been put forward (Andreyev and Klushin 1962; Berezhnaya and Telepin 1966; Roy 1970; Bhattacharyya and Chan 1977; Golizdra 1977; Khesin et al. 1983, 1996). The principle behind most of them consists of using the successive approximation method to select a field on the intermediate plane that coincides with the measured one when continued to the measurement surface. The observed field is assumed to be a zero approximation, i.e. it is thought of as being recorded on the intermediate plane. Then, this field is continued upward to the measurement line. The difference between the observed and calculated fields is the correction at each point, which is added to the zero approximation (observed field). This yields the first approximation. Next, the first approximation field is continued again to the measurement line, and the correction is computed. This provides the difference in the results between the first and the second upward continuations. This correction is added to the first approximation. After this manipulation, we obtain the second approximation, which is subjected to the same procedure until the correction approximates the survey error. The final approximation is used to continue to any horizontal plane. Of these techniques, Aronov's method (1976) is the best known.

These methods provide a calculation of the upward continuation. As a result, not only the anomalies caused by relief forms collapse or are attenuated, but also those from the objects under study (see, for example, Fig. 16.5 from Dobrin (1976)). Therefore, continuation should be estimated in terms of the type of work, problems and available data.

In aerial surveys of sinuous (in the vertical plane) routes and ground surveys, the sinuosity effect i.e. reducing measurements to horizontal or inclined planes is necessary in most cases. It solves problems related to quantitative determination of anomalous body parameters such as neglecting the effect of variation in the measurement point heights which can lead to interpretation errors. To correct for the anomalies under study, the researcher should choose a reduction level as close to the actual masses as possible and, if required, select several levels.

3.2.4 Correlation Technique for Terrain Correction

3.2.4.1 Terrain Correction in Magnetic Prospecting

There is a clear correlation between the elevations of the measurement point (H) and the magnetic field ΔZ for a relatively homogeneous magnetic medium. In the case of direct magnetization, the field maxima correspond to ridges of the "magnetic" relief, while the minima correspond to the valleys.

An analytical approach was suggested by Khesin (1965, 1978a) that applied the linear relation ΔZ (H) to a typical element of mountainous regions – a slope (inclined ledge, or step). All the main types of relief can be approximated by one or another combination of slopes. Thus although crude, this is a simple and effective method for eliminating the effect of magnetized rock relief. It only uses the data concerning the recorded field and terrain relief. To apply this technique, a correlation field is drawn up between ΔZ (ΔT) and H values, and then their average as a straight line is plotted. The terrain correction is determined by the regression equation

$$\Delta Z_r = c + bH, \tag{3.13}$$

where b and c are the factors of a linear equation computed using the least-square method (the b dimension is nT/m; the subscript "r" indicates that it is a relief correction).

The regression line (Eq. 3.13) is drawn on the basis of many measurement points from the data obtained under the conditions of a medium close to uniform and as distant as possible from sharp bends in the terrain relief. It was shown that

$$b = \frac{8J \cos \alpha}{R}, \tag{3.14}$$

where J is the topographic mass magnetization, α is an acute angle between the slope face and horizon, and R is the slope length across the strike.

Therefore, the J value can be determined by the angular coefficient of the regression line.

Elimination of the topographic effect by the correlation technique allows for practically complete smoothing out of the anomalies caused by morphostructures (Fig. 3.13a). When excluding the topographic effect from the observed field (Fig. 3.13b), the corrected ΔZ graph coincides closely with the magnetic field caused by a flat-dipping thin bed (magnetite deposit) under oblique magnetization.

Along with the linear approximation of the relationship between the field and the height, approximations in the form of the square trinomial (parabolic equation) can also be used. Judging by the type of the parabola, one can get useful interpretation evidence. If its vertex is at the top, the vertical gradient decreases with height, which is characteristic of a uniform medium. If, however, it is at the bottom, there are significant heterogeneities present in the section. The parabolic characteristics contain other information as well, including assessing the presence of a heterogeneity (i.e. additional source) and even its features. However, it has been statistically proven that the parabolic approximation $U(H)$ can be replaced by the linear approximation when the correlation between the magnetic field and the relief is known (Khesin 1981).

To reduce correlation distortions, in general it is recommended to plot approximation lines for each individual element of the topography. In regions with flat-dipping geological boundaries, rock outcropping onto the Earth's surface may differ markedly from those of the section on the whole. In such regions it is preferable to draw a unified approximating line by the correlation plotted for the whole surveying sheet. It is advisable to first determine the areas of correlation by visual analysis of geophysical and topographic maps (Fig. 3.14).

The results of the correlation terrain correction in the Mekhmana mining district are a good example (Fig. 3.15). Regioning of the terrain relief was carried out in this

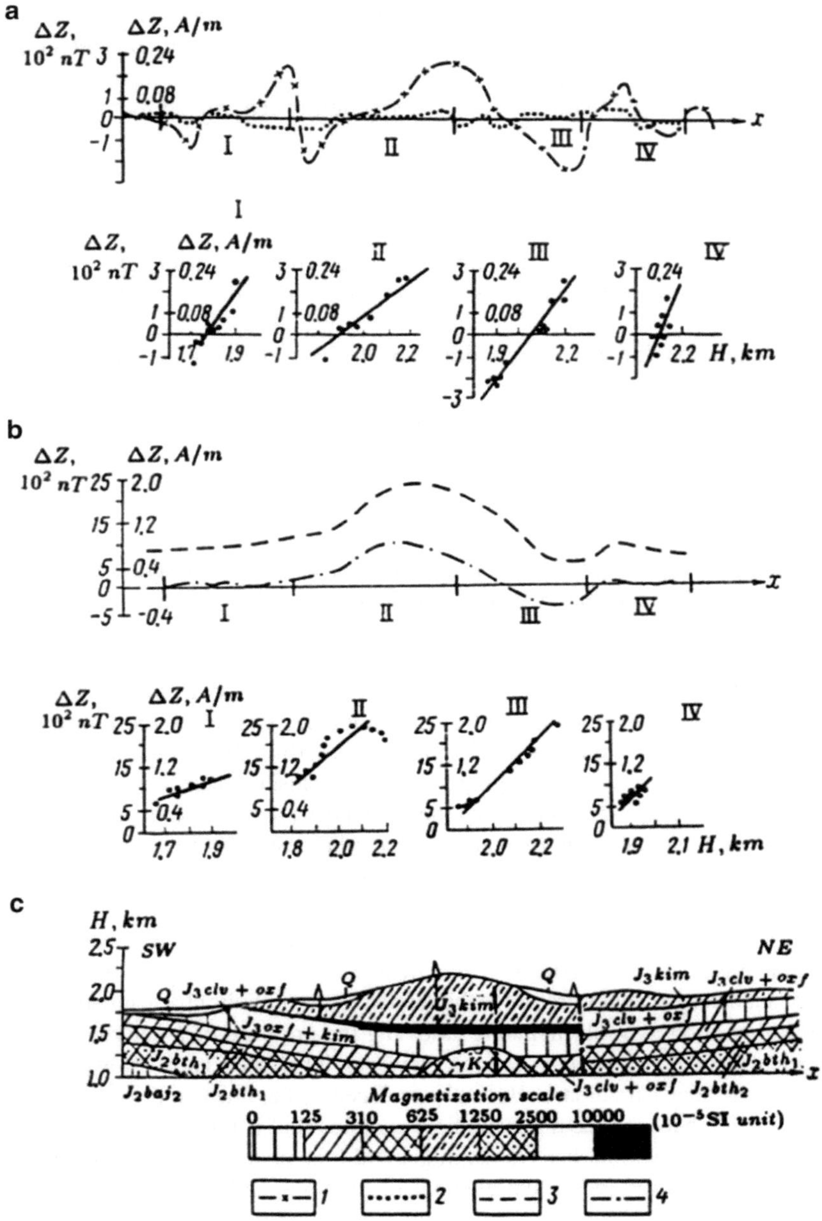

Fig. 3.13 Terrain correction by the correlation method for the ground profile (survey on a 1:25,000 scale) in the Dashkesan mining district, Lesser Caucasus: (**a**) calculated magnetic fields and approximating lines ΔZ (H) for the model **c**; (**b**) the same for the observed field; (**c**) petromagnetic section (γK is the hidden apophysis of the Dashkesan intrusive). I, II, III, IV are the profile portions where correlation ΔZ (H) was calculated. ΔZ field: (**1**) computed by the model **c**; (**2**) the same, corrected; (**3**) observed; (**4**) the same, corrected

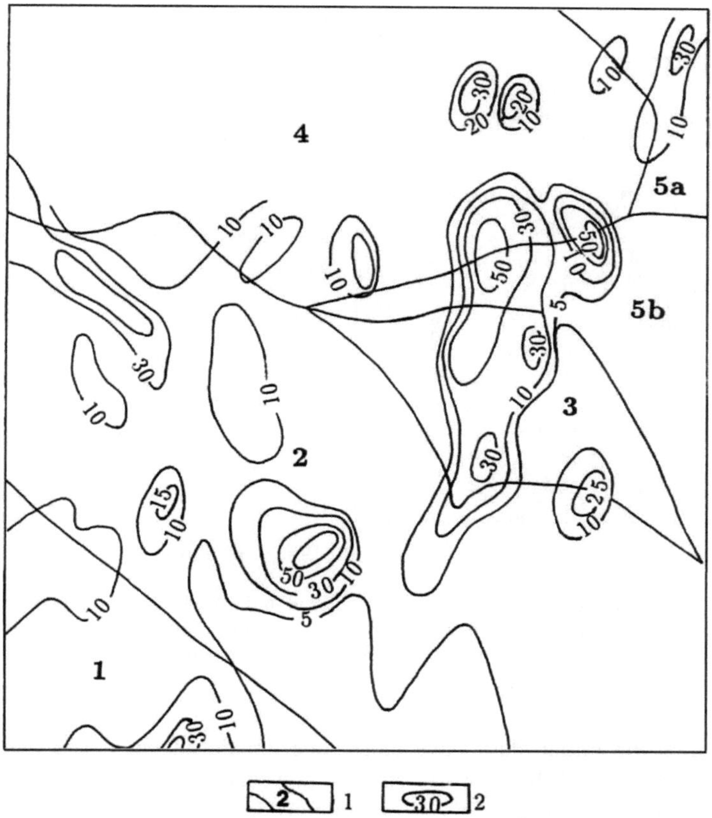

Fig. 3.14 Topography regioning in the Mekhmana mining district, NE Lesser Caucasus, by the inclination angles as compared to magnetic susceptibility. (*1*) area with similar inclination angles and number; (*2*) magnetic susceptibility isolines (in 10^{-3} SI unit)

district by comparing the dip angles with magnetic susceptibility data (see Fig. 3.14). No correlations were observed in areas 1, 4 and 5, which can be explained theoretically by the ratio of the inclination angles for the relief and magnetization. For area 2, the correlation takes the following form

$$\Delta Z = 0.765(H - 1150), \tag{3.15}$$

whereas for area 3 it takes the form

$$\Delta Z = 1.135(H - 1280), \tag{3.16}$$

where H is measured in meters and Z in nanoTesla.

In Fig. 3.15, the isogam map Z_c drawn up for areas 2 and 3 is compared with the initial field chart. The figure shows that the anomalies and other field features

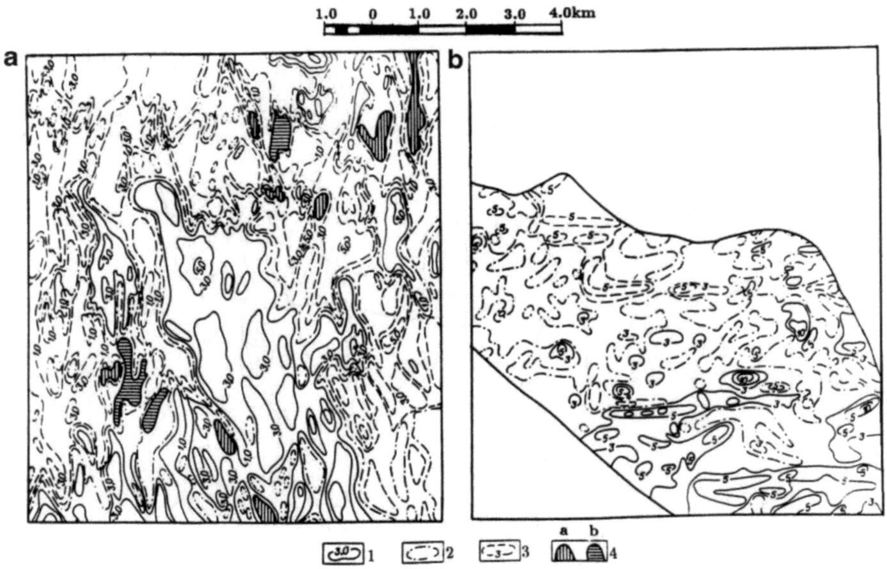

Fig. 3.15 Terrain correction by the correlation method in the Mekhmana mining district (Lesser Caucasus): (**a**) observed magnetic field Z_{an}, (**b**) magnetic field Z_c corrected for topographic effect. (*1–3*) isolines ($n{\cdot}100$ nT): (*1*) positive; (*2*) zero; (*3*) negative field; (*4*) sites of intense local anomalies: (*a*) positive, (*b*) negative

presented in the initial chart have undergone drastic changes. The field pattern becomes simpler and the anomaly strikes showed a marked deviation, such that some of them are smoothed. This allows for considerable reduction in interpretation errors, and primarily Type I errors (i.e. "false positives").

This method can be used to improve the choice of level for reducing the field to one plane. When plotting a correlation chart, the areas with the largest dispersion with respect to the averaging line correspond to the elevation of the points under which the targets are situated. The presence of dispersion in itself is indicative of the hidden source present in the section.

The advantage of the correlation technique is that it can rapidly reveal the influence of terrain relief as well as the presence and certain peculiarities of hidden sources, in addition to suppressing or reducing the noise caused by the terrain relief. It can also solve additional problems, such as determining the mean magnetization of the medium and the vertical gradients of magnetic field. Eppelbaum (2010b) suggested conducting a magnetic survey using Remote Operated Vehicles (ROV) to determine the upper geological section magnetization (including inclined ROV observations over flat surfaces).

Analysis of the spectral composition of the geophysical and height field makes the correlation method more precise (Erofeyev and Avteneyev 1971). The magnetic field spectrum is much more highly differentiated and wider than the height spectrum. For this reason, a higher correlation is obtained when using average values than the observed values of ΔZ or ΔT. If high-frequency magnetic components are involved in the correlation, the correlations turn out to be lower. These weaker correlations at high frequency testify to the presence of such geological sources as ore-bodies or dikes.

3.2.4.2 Terrain Correction in Other Geophysical Methods

The correlation between the measured fields and the terrain relief heights needs to be reflected in data obtained by different geophysical methods (Orlov 1984). Each elevation of the terrain relief with respect to the neighboring one can be considered as an additional pole or series of poles that change the field intensity at the point of measurement. The sum of the fields of elements making up the topography tends to increase with increasing height of the measurement point. Mathematical modeling confirms this relationship for an inclined ledge (except portions adjoining its horizontal faces) over which ΔT, gravity and thermal fields are computed (Fig. 3.16).

Theoretical support by direct problem solution presents a difficult mathematical problem in electric prospecting. Measurements and physical modeling have confirmed that the correlation approach proposed in (Khesin 1969) can be used for the IP method (Khesin 1981), VLF method (Eppelbaum 1984b), near-surface temperature and SP fields (Eppelbaum 1989). Figure 3.16 shows an application of correlation approach for SP (direct correlation) and thermal (inverse correlation) fields. Figure 3.17 illustrates the results of the correlation technique for an IP model field. Characteristically, an anomalous object appears as a cluster of points that is present in the correlation field.

Thus, the correlation technique makes it possible not only to easily eliminate the terrain relief effect, but also to reveal its influence and hidden geological sources; thus the nature and strength of the correlation can provide additional information. The approximate relationship plot contains information on the physical properties of the medium, which led to the development of the special technique described in paragraph 3.2.4.5.

3.2.4.3 Statistical Reduction of Gravity Anomalies

The correlation approach is particularly productive in gravity prospecting, and was first suggested by Nettleton (1940); today it takes the form of a statistical reduction.

A complete Bouguer correction (a *synonym* for topographical correction) should ideally account for the Earth's entire topography; however, the calculations are often complicated and, in practice, Bouguer corrections are rarely complete because the terrain correction is carried out only for a limited area, although a simplified method was suggested by Talwani (1998). In cases where the radius of the terrain correction area is too small, it introduces a substantial discrepancy between the Bouguer correction for an infinite slab and the gravitation effect of

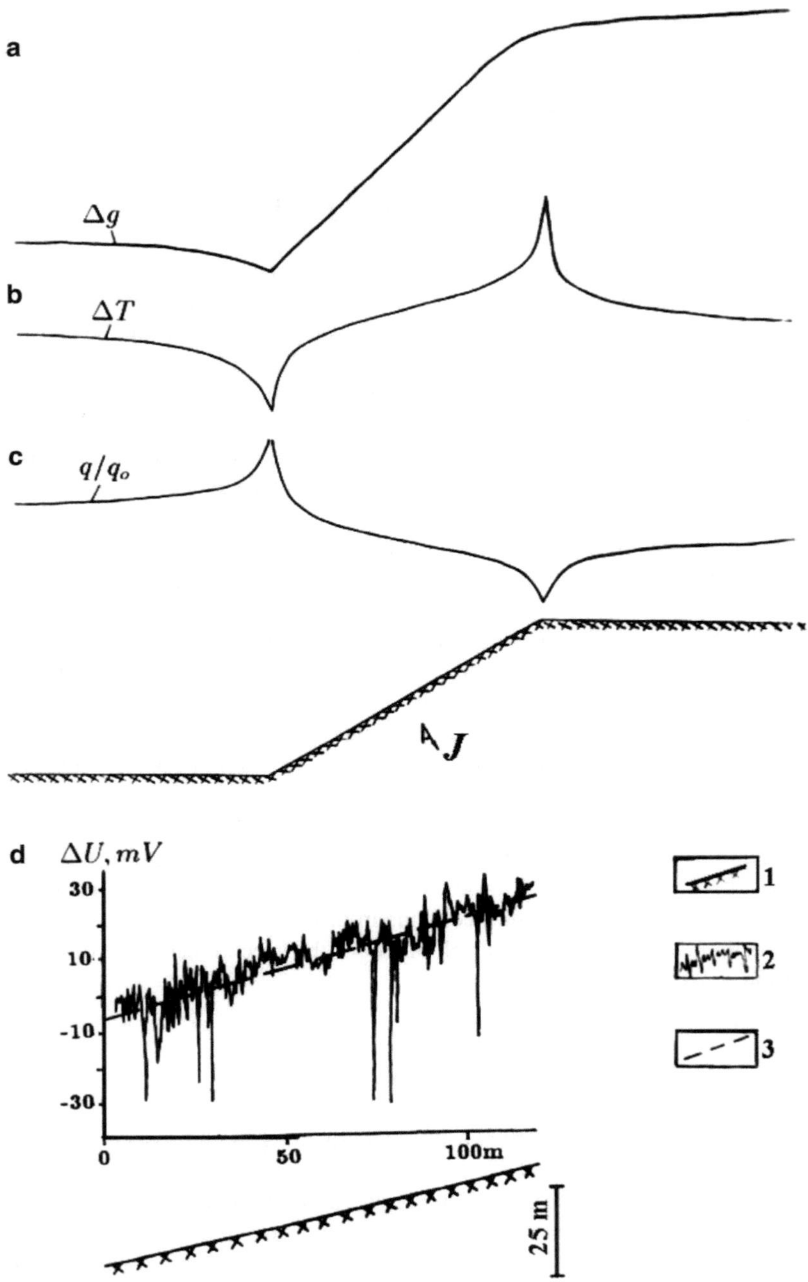

Fig. 3.16 Substantiation of the correlation technique for terrain correction in gravity (**a**), magnetic (**b**), thermal (**c**), and SP (**d**) prospecting. (**c**) and (**d**) after Lachenbruch (1968) and Ernstson and Scherer (1986), respectively; (**a**), (**b**) and (**c**) are model calculation results, (**d**) is real measurement data on a slope. (*1*) topography form; (*2*) SP measured; (*3*) straight line approximation ***J*** is the magnetization vector

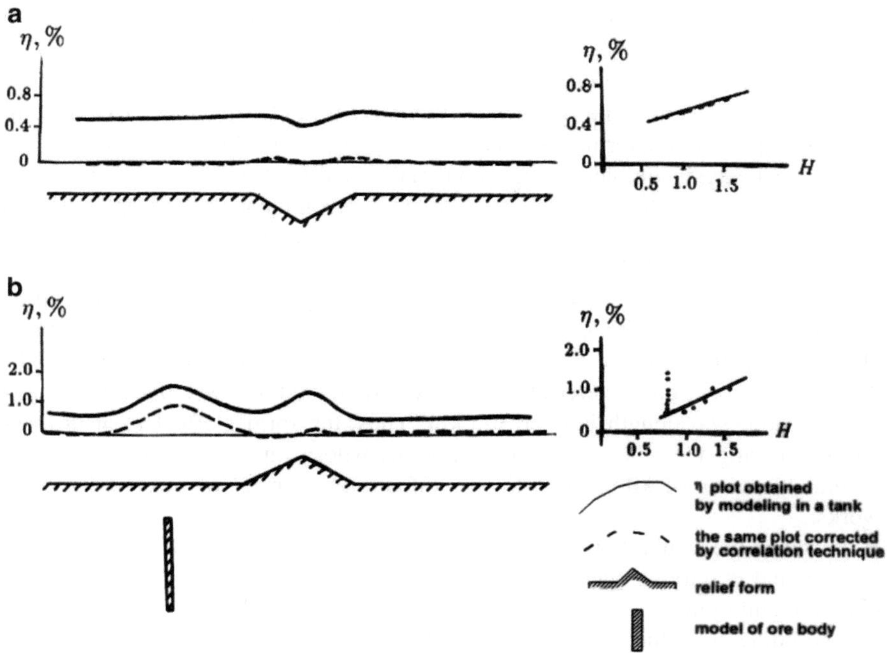

Fig. 3.17 Correlation technique testing for terrain correction in the IP method for a negative relief form (**a**), positive relief form with an ore-body within it (**b**); model after Polyakov (1969)

real topography, particularly of the Earth's curvature. LaFehr (1991) published a recent approach to a curvature correction. He suggests a way to calculate corrections up to Hayford zone O, i.e. 166.7 km. In fact, if the terrain corrections do not cover an appropriate area relative to the regional topography, the theoretical infinite slab may be left "hanging in the air" at the margins relative to the curved surface of the Earth. Obviously, in such cases, a density of 2.67 g/cm^3, which is usually assumed for the Bouguer correction, may be an overestimation and, consequently, Δg_B values may be underestimated (Khesin 1969). The correlation often observed between Δg_B and H values is to a large extent due to this artifact (Lustikh 1947; Nikolsky 1971). This effect is particularly strong in transition zones between the mountains and plains. Therefore, the terrain correction should be applied over a sufficiently large area in accordance with the dimensions of the rugged terrain, e.g., a 200 km radius for the mountainous regions of the Caucasus (Nikolsky 1971). If this approach is not used, complementary reductions to compensate for the size of the radius are required. Ideally, the Bouguer and terrain corrections should be integrated into one reduction method (Khesin 1975b; Parasnis 1986; Khesin et al. 1993a; Chapin 1996a; Eppelbaum and Khesin 2004). However, with several exceptions, this is very complicated.

The transition from highland to lowland is a common topographic feature on the Earth's surface. In light of the discussion above, it is clear that the gravity fields of such transitional zones are complicated, and the choice of an appropriate reduction method may be crucial for their analysis.

According to Nikolsky (1971), Bouguer anomalies in mountainous regions are practically equivalent to the Bouguer correction (e.g., 111.9 mGal per 1 km assuming the density of an intermediate slab of 2.67 g/cm^3). Lustikh (1947) showed that the ratio between incomplete Bouguer anomalies and H values in the Caucasus and Central Asia is 103.6 mGal/km, and therefore recommended the application of statistical reduction. Chapin (1996b) showed a similar relation in South America (~108 mGal per 1 km). This phenomenon is caused not only by the topography (the crust/atmosphere density contrast), but also by the depth of isostatic compensation (the mantle/crust density contrast). In a review of isostatic reduction, Chapin (1996b) indicates that the topography is dominant at higher frequencies of the $\Delta g_{\mathrm{f.a.}}$ power spectrum, whereas the isostasy is dominant at lower frequencies, and the threshold wavelength is over 300 km. Thus, the former effect is of primary concern during the study of separate regions whose dimensions do not exceed several hundreds of kilometers. Such a situation is typical of most prospecting works.

Based on the available Bouguer data, the Ruppel and McNutt (1990) analysis of the regional gravity compensation for the Greater Caucasus indicated some interpretation ambiguities.

According to a theoretical analysis (Khesin and Alexeyev 1986), statistical reduction has widespread applications in gravity prospecting for regional (Khesin 1975; Khesin et al. 2001) and especially for detailed studies (Khesin 1971; Khesin et al. 1988).

Using the statistical reduction approach for processing data, a detailed gravimetric survey in the Greater Caucasus revealed small Δg anomalies over polymetallic ore bodies (Khesin 1969). Nevertheless, statistical reduction is crucial primarily for processing the regional component of Δg. The Bouguer reduction artifact has relatively minor importance for residual Δg_{B} anomalies. This is because the regional components of the Bouguer anomaly $(\Delta g_{\mathrm{B}})_{\mathrm{reg}}$ and Δg_{B} both contain a similar effect that is practically eliminated in calculation of the residual anomaly $(\Delta g_{\mathrm{B}})_{\mathrm{res}} = \Delta g_{\mathrm{B}} - (\Delta g_{\mathrm{B}})_{\mathrm{reg}}$. However, this artifact is significant when the regional Δg_{B}component is used separately.

Statistical reduction, which automatically accounts for defects of the Bouguer reduction described in the literature and, particularly, in (Khesin 1976), employs a simple correlation technique, since except for the surrounding relief correction, all the corrections exhibit linear dependence on the observation point height. It can be shown that proceeding from the analytical expression of the inclined ledge (slope) gravitation, the attraction of typical relief forms is also linearly dependent on the height of the measurement point. According to Khesin and Alexeyev (1986)

$$\Delta g = c + bh, \tag{3.17}$$

where $c = 2G\sigma\alpha H_0$, $b = 2\pi G\sigma$, G is the gravity constant; σ is the density of rocks forming the ledge; α is an acute angle between the inclined face of the ledge and its base; H_0 is the vertical ledge thickness; h is the difference in levels between the observation point and the ledge base (if the ledge base is at the sea level, then $h = H$).

The relationship between Δg and the difference between heights h can be seen as a correlation because of the presence of density heterogeneities in the volume of the slope and beneath it, and also as a result of the deviation of the Earth's surface from the inclined step.

Further inspection of the gravity anomaly over a slope formed of homogeneous rocks shows that corrections for terrain relief g_r at all points of the slope are equal and proportional to the ledge thickness and to the inclination angle of the relief. Therefore, on a homogeneous slope Bouguer anomalies Δg_{B} are constant and equal to the attraction of the ledge to the point at its base

$$\Delta g_{\mathrm{B}} = 2G\sigma\alpha H_0. \tag{3.18}$$

Since $\alpha < 0$, then $\Delta g_{\mathrm{B}} < 0$. When $\sigma = 2.67$ g/cm^3 and

$$\alpha = -30^{\circ}, \Delta g_{\mathrm{B}} = -g_r = -18.6H_0,$$

where H_0 is expressed in kilometers.

The estimation of the $g_r$value thus obtained is supported by the data from terrain corrections. In practice, these corrections are added to Δg_{B}, which is usually employed as an alternative to the value of the anomaly in topographic reduction (Δg_{T}). Therefore, Bouguer anomalies (assumed to be local topographic anomalies) must be equal to zero, provided that the intermediate layer is accounted for correctly and geological sources are absent. Owing to the above defect of the Bouguer reduction, the correlation becomes $\Delta g = f(H)$ or $\Delta g = f(h)$.

The abovementioned article (Khesin and Alexeyev 1986) showed that the correlation technique is appropriate when analyzing the terrain relief effect on the free-air anomaly ($\Delta g_{\mathrm{f.a.}}$). This is confirmed by a rigorous formula derived by the authors

$$\Delta g_{\mathrm{f.a.}} = c_1 + bh, \tag{3.19}$$

where $c_1 = 2\pi G\sigma H_1 + c$, H_1 is the height of the ledge base.

If local geological and topographic heterogeneities are present, their gravitational effect (Δg_{an} and Δg_r, respectively) may be written as

$$\Delta g_{\mathrm{an}} + \Delta g_r = \Delta g_{\mathrm{obs}} - g_{\mathrm{corr}}, \tag{3.20}$$

where g_{corr} is the correction determined from the straight line plot of Eq. 3.17 type; Δg_{obs} is the relative observed value of the acceleration of free fall ($\Delta g_{\mathrm{obs}} = g_{\mathrm{obs}} - \gamma_0$). Here Δg_{obs} stands for the measured acceleration of free fall on the Earth's surface, and γ_0 stands for the normal acceleration of free fall on the Earth's surface.

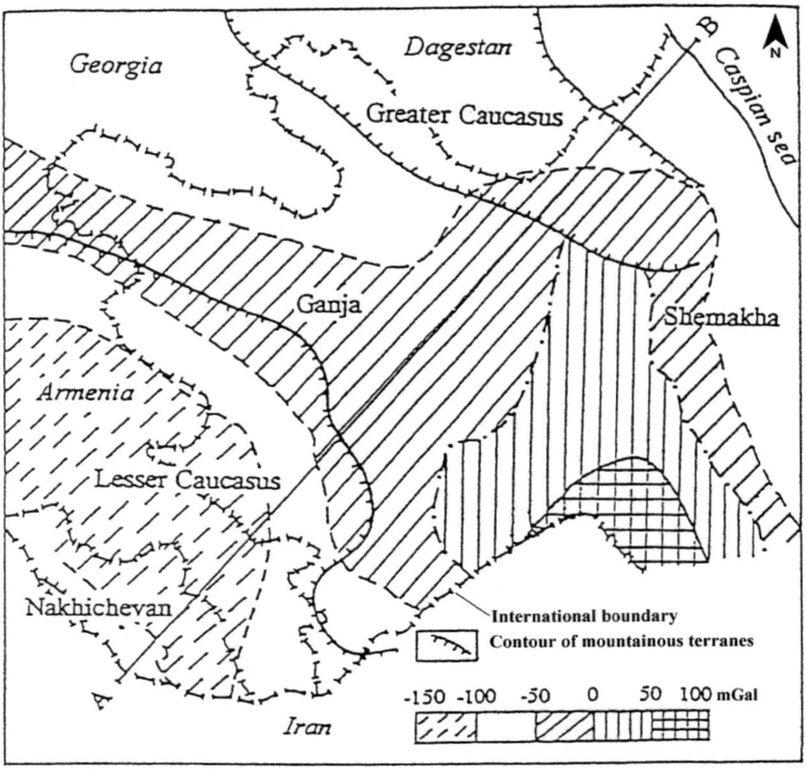

Fig. 3.18 Comparison of regional components of topography obtained by the averaging method and Δg_B obtained for the Eastern Caucasus by high-level conversion

Obviously, the density of the inclined slope forming rock can be determined using formulas (3.17)–(3.20). The absolute value of Δg_r varies slightly in detailed investigations. For instance, the comparison between the statistical anomalies

$$\Delta g_s = \Delta g_{\text{obs}} - g_{\text{corr}} \tag{3.21}$$

and the local topographic (Bouguer) anomalies provided a way to apply the statistical reduction (correlation technique) to rapidly single out anomalies in the area of the Katsdag ore deposit (southern slope of the Greater Caucasus).

The most illustrative results of statistical reduction were obtained in regional studies. Figures 3.18, 3.19 and 3.20 present the generalized topography, geological setting, regional Δg_Bfield, and gravity field after the statistical reduction for the Eastern Caucasus, and a NE-SW composite cross-section.

The Bouguer anomalies for the Eastern Caucasus (Azerbaijan and adjacent areas of Dagestan, Georgia and Armenia) were mainly calculated with a ~50 km radius terrain correction (Khesin 1975, 1976). Due to the limited radius used, the Bouguer field obtained was inadequate and did not reflect the main geological structures. The

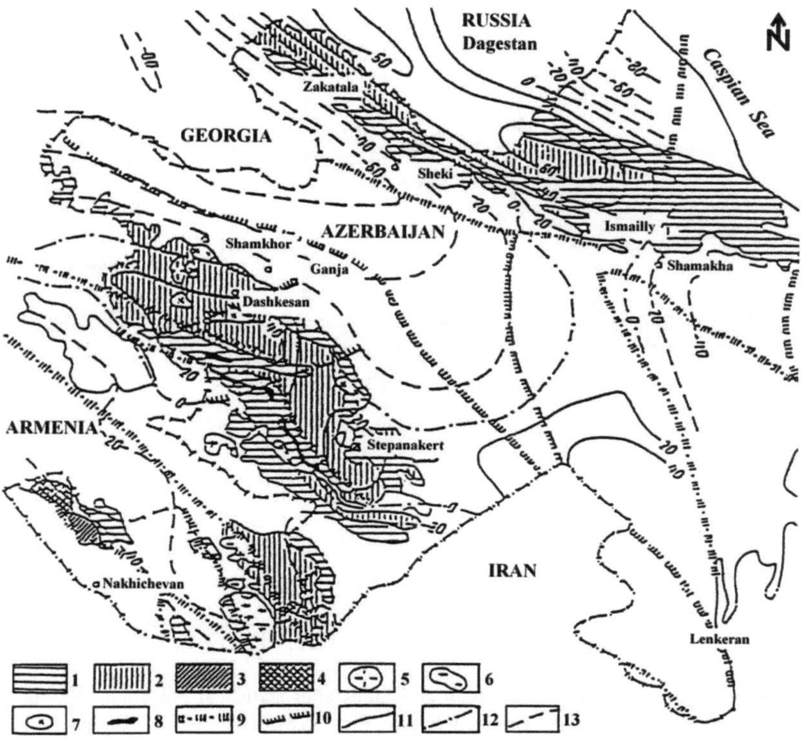

Fig. 3.19 Bouguer gravity field (regional component) after statistical reduction and generalized geology of the Eastern Caucasus. (*1*) Cretaceous; (*2*) Jurassic; (*3*) Triassic; (*4*) Paleozoic; (*5*) Mesozoic plagiogranites; (*6*) Cenozoic granitoids; (*7*) Mesozoic granitoids and gabbroids; (*8*) Mesozoic(?) ultrabasites; (*9*) deep faults; (*10*) flexures; (*11–13*) Δg_sisolines, mGal (*11* – positive, *12* – zero, *13* – negative)

drawback of Bouguer reduction in this region is evident: a strong negative regional Δg_Bcomponent is obtained (by both high level conversion and averaging methods) over the dense Mesozoic cores of the mega-anticlinoria of the Greater and Lesser Caucasus (Figs. 3.18 and 3.20).

In contrast, anomalies calculated by statistical reduction outlined the Jurassic cores of both Caucasus megastructures by positive isolines of several tens of milliGals (Figs. 3.19 and 3.20). In Fig. 3.20 it is worthwhile noting: (1) the good correlation between regional Bouguer anomalies Δg_{con} and Δg_{av} that were obtained by Δg_Bconverting and averaging, respectively; (2) the discrepancy between these anomalies and the geological structure; (3) the good correlation between Δg_s and geological structure. This suggests that in rugged topographic terrains, statistical anomalies reflect the geological section better.

In addition to better accounting for the topographic effect, the statistical reduction method also provides an approximation for the isostasy correction. Both isostatic and statistical reductions calculate gravity corrections are based on

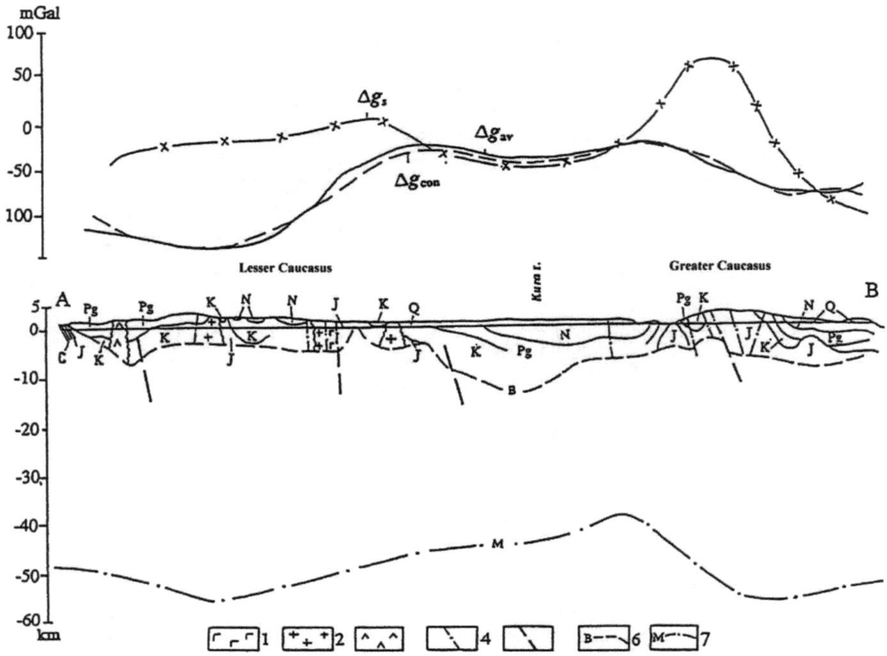

Fig. 3.20 Comparison of regional gravity anomalies after statistical reduction (Δg_s) and after Bouguer reduction along cross-section **a–b** (see Fig. 3.18). The regional Bouguer field was obtained by the Bouguer field converted to high level (Δg_{con}) and by averaging (Δg_{av}). *C*, *J*, *K*, *Pg*, *N*, *Q* are the Phanerozoic sedimentary, effusive and volcaniclastic rocks. (*1*) ultrabasites; (*2*) granitoids; (*3*) extrusive andesite-dacites; (*4*) faults; (*5*) deep faults; (*6*) roof of Pre-Alpine basement; (*7*) Moho discontinuity ((*5–7*) are based on the results of integrated interpretation of seismic and other geophysical data)

topography data, but there are some differences in the data utilization (e.g., Aiken 1982). With relatively moderate differences, anomalies obtained for the Caucasus by statistical reduction (Fig. 3.19) are similar to isostatic reduction anomalies calculated by Artem'yev and Balavadze (1973) and Artem'yev et al. (1985).

At the same, it should be noted that for qualitative analysis of the gravity field in complex regions it is sometimes useful to examine even observed gravity anomalies (e.g., Eppelbaum and Katz 2011).

3.2.4.4 The Correlation Procedure

By way of example, consider the procedure involved when using this technique in the VLF method (Eppelbaum 1991). The principle consists of obtaining a linear approximation of the relationship between the recorded field (U) and the relative height of the measurement point (h), and employing this approximation to calculate the topographic correction.

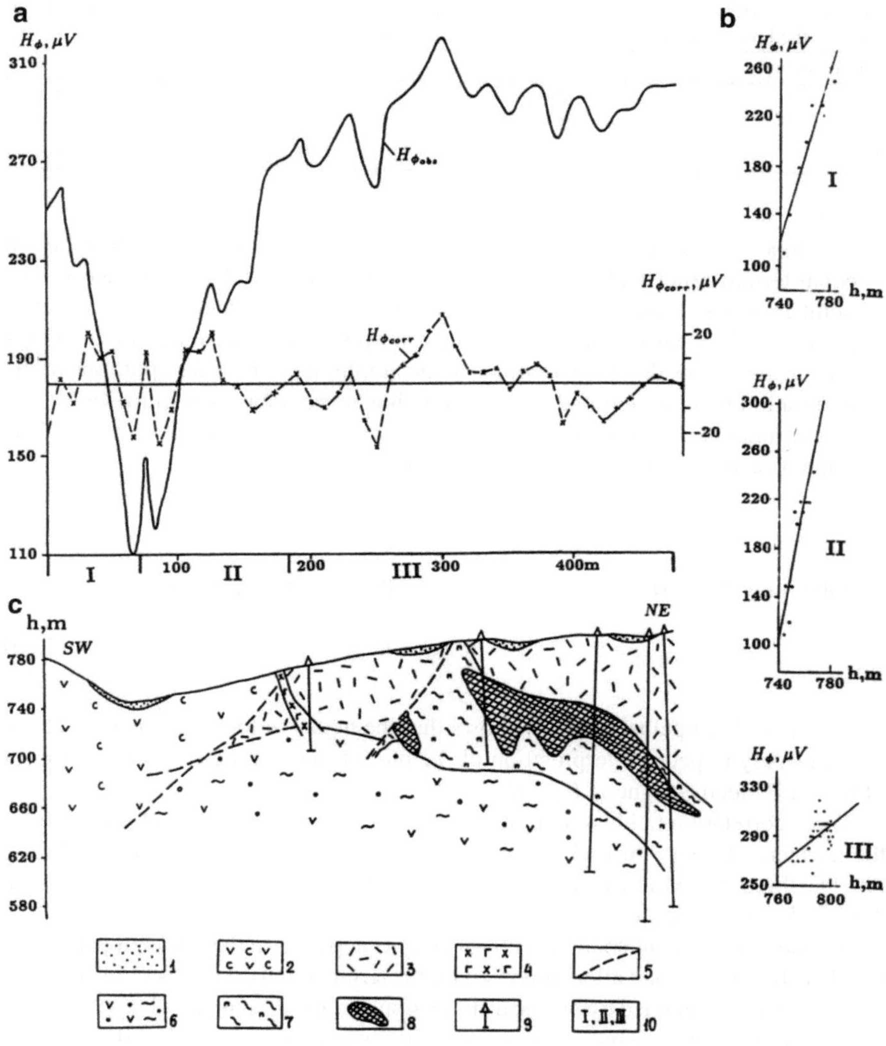

Fig. 3.21 Correlation technique for terrain correction in the VLF method in the Kyzyl-Bulakh gold-pyrite deposit (Lesser Caucasus): (**a**) plots of measured $H_{\varphi obs}$ and corrected $H_{\varphi corr}$ values, (**b**) correlation, (**c**) geological section. (*1*) deluvial deposits; (*2*) tuffs and lavas of andesitic porphyrites; (*3*) tuffs of liparite-dacitic porphyrites; (*4*) dike of andesite-basalts; (*5*) disjunctive dislocations; (*6*) pyritized and quartzited andesitic porphyrites; (*7*) zone of boudinage and crumpling; (*8*) ore-bodies; (*9*) prospecting boreholes; (*10*) profile intervals for plotting correlation field $H_\varphi = f(h)$

The elimination of the terrain relief effect involves three steps:

1. Plotting of the correlation field $U = f(h)$. For greater accuracy, the correlation fields for individual elements of the topography (for example, for two parts of the slope differing in inclination angle) are computed separately. The portions of

Table 3.5 Correlation coefficients (r) and regression equations for three portions of the profile crossing the Kyzyl-Bulakh gold-pyrite deposit (Lesser Caucasus)

Profile portion	r value	Regression equation
I	0.96	$H_{\varphi\,(\mathrm{appr})} = 118 + (3.75 \pm 1.16)h$
II	0.95	$H_{\varphi\,(\mathrm{appr})} = 103 + (5.60 \pm 1.80)h$
III	0.54	$H_{\varphi\,(\mathrm{appr})} = 263 + (0.88 \pm 1.37)h$

the slope with different geological and, hence, physical characteristics, but an equal terrain relief inclination angle should be also considered separately. The number of points used in the calculation must not be less than 15–20. The set of points within a narrow range of h values suggests a hidden source in this range indicated by considerable variations in the amplitude of U. These portions of the correlation field are ignored when computing approximate straight lines.

2. Determination of approximate line factors (regression equations) b and c. Terrain corrections are determined from the equation

$$U_{\mathrm{appr}} = c + bh. \tag{3.22}$$

3. Obtaining the corrected values of U_{corr} in the measurement points

$$U_{\mathrm{corr}} = U_{\mathrm{obs}} - U_{\mathrm{appr}}. \tag{3.23}$$

Figure 3.21 (Eppelbaum 1991) shows that the negative topographic anomaly of high intensity impedes interpretation of the plot of the total horizontal component of the VLF electromagnetic field $H_{\varphi(\mathrm{obs})}$.[1]

Three correlation fields were plotted here: one for the southern slope and two for the northern slope (with rock variations characterized by different conductivity). All the three regression equations were derived using the least-square method. Table 3.5 lists the results.

An anomaly from an anomalous (ore) body is apparent in the northern part of the profile in the $H_{\varphi(\mathrm{corr})}$ plot after correction for the terrain relief effect. Its form is similar to the anomalies obtained by physical simulation of a flat dipping thin bed (Gordeyev and Sedelnikov 1974), which is consistent with the available geological data.

3.2.4.5 Estimation of Magnetization Inclination and Value Using ΔT and ΔZ Measurements on an Inclined Surface

Two approaches are usually employed to determine inclination and the value of rock magnetization: (1) Sampling and measurements of oriented samples, (2)

[1] H_φ component values (as well as other components of the VLF magnetic field) are given in microVolts due to specific features of the Soviet (Russian) apparatus *SDVR*-4.

Effective magnetization ($\mathbf{J_e}$) vector selection according to the observed magnetic field (physical-geological modeling). The first method requires considerable additional work. In this case the negative effects of near-surface rock alterations and local $\mathbf{J_e}$ variations are not completely excluded. The efficiency of the second approach is conditioned on the reliability of the model of the medium. However, as a rule, reasonable data tend to be lacking at the initial stages of a new area study; that is, when an evaluation of the magnetization is needed to design the subsequent survey and interpret the magnetic anomalies. The method of $\mathbf{J_e}$ determination according to observed magnetic field using minimum assumptions concerning the geological structure and physical properties of rocks can resolve this problem well (Khesin 1998).

For an aeromagnetic survey of a small terrain clearance (elevation over the land surface) and for ground magnetic mapping, a linear dependence between the magnetic field and the heights of the ledge is defined in the central part of a slope (Khesin et al. 1983). In case of oblique magnetization, analytical expressions for the vertical Z_d component and ΔT field over a slope (inclined at an angle α) are the following:

$$Z_d = -2\pi J_e \sin\alpha\cos(\varphi_m - \alpha) + 8J_e \sin(\varphi_m - \alpha)\frac{h}{\mathrm{R}}, \tag{3.24}$$

$$\Delta T = -2\pi J_e \sin\alpha\cos(\varphi_m - \tau_0 - \alpha) + 8J_e C_0 \sin(\varphi_m \tau_0 - \alpha)\frac{h}{R}. \tag{3.25}$$

Here J_e is the projection of the magnetization vector on a vertical plane transverse to the strike of the ledge, φ_m is the angle of inclination of the magnetization projection J_{xz} to the horizon, h is the elevation of the observation point with respect to the middle of the ledge, R is the length of the ledge, $C_0 = \sin\frac{i_0}{\sin\varphi_0}$, i_0 is the geomagnetic inclination, $\tau_0 = 90^\circ - \varphi_0$ is the angle completing the geomagnetic inclination to the vertical in the plane of profile; φ_0 is the angle of inclination of the geomagnetic field vector projection on the plane of profile to the horizon.

We define:

$$c_d = -2\pi J_e \sin\alpha\cos(\varphi_m - \alpha),$$

$$b_d = \frac{8J_e \sin(\varphi_m - \alpha)}{R},$$

$$c = -2\pi J_e \sin\alpha\cos(\varphi_m \tau_0 - \alpha),$$

$$b = \frac{8J_e C_0 \sin(\varphi_m - \tau_0 - \alpha)}{R},$$

As a result we obtain:

$$Z_d = c_d + b_d h, \tag{3.26}$$

$$\Delta T = c + bh, \tag{3.27}$$

i.e. linear equations.

In natural conditions, due to fluctuation of heights and magnetic parameters, the linear equations (3.26) and (3.27) are determined by regression analysis. The coefficients in these equations can be calculated by the least squares method, or graphically on a correlation field. The coefficients provide an effective tool for determining the parameters of medium magnetization (the average values for the rock mass as a whole).

If the b value is calculated in units of nanoTesla/meter, the $J_e$value (in milliAmper/meter) must be multiplied by 0.1. Thus the expressions for b and c become:

$$c = -0.2\frac{\pi J_e \sin\alpha \cos(\varphi_m - \tau_0 - \alpha)}{R},$$

$$b = 0.8\frac{J_e C_0 \sin(\varphi_m - \tau_0 - \alpha)}{R}.$$

The coefficients c_d and b_d are calculated in a similar way.

If the Z_d and ΔT measurements were carried out, both the value and the direction of the magnetization vector can be found using these coefficients (Khesin and Alexeyev 1987) as follows:

$$\cot(\varphi_m - \alpha) = \frac{\left(\cos\tau_0 - \frac{b}{b_d C_0}\right)}{\sin\tau_0}. \tag{3.28}$$

This equation is simplified when the profile is oriented at an acute angle to the magnetic meridian ($C_0 \approx 1$).

It follows from Eq. 3.28 that this method (Khesin and Alexeyev 1987) does not require knowing the normal magnetic field. After defining the value of φ_m the J_e value can be calculated from the b and b_d coefficients. This technique was successfully applied in the Lesser Caucasus and Greater Caucasus as well as in other regions of the world.

Figure 3.22 shows such computations for the Mekhmana ore district of the Lesser Caucasus. The magnetic rock properties in this test site were known, including the Koenigsberger (Q) ratio, i.e. the remanent/induced magnetization ratio. The basic and intermediate volcanic rocks of the Bajocian were found to be magnetized parallel to the Earth's field, and the mean J_e value was also consistent with the results of measurements on samples.

In the Belokan-Zakatala ore field (Greater Caucasus) this method also showed that the shales and sandstones of the Aalenian are magnetized parallel to the Earth's field, and their magnetization (in the range of 30–70 mA/m) was closer to the upper limit of the results obtained from samples. This confirmed the underestimation of

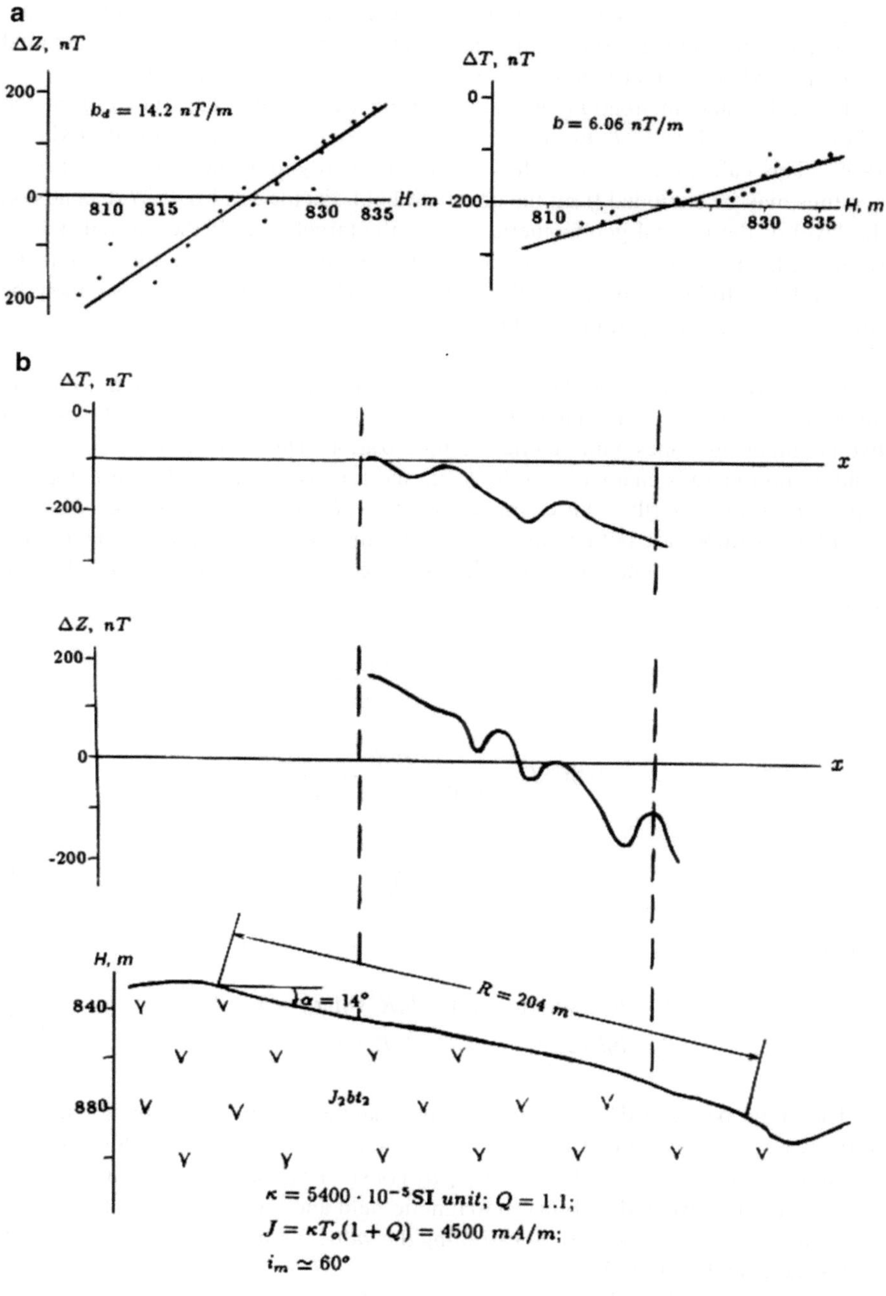

Fig. 3.22 Determination of the value and direction of rock magnetization in natural occurrence by the ΔT and ΔZ measurements on an inclined surface: (**a**) determination of coefficients b_d and b, (**b**) ΔT and ΔZ curves, and geological section along the measurement profile

magnetization that can take place during measurements of outcropped rocks. Actual magnetization can be determined more accurately with the use of the recommended technique, where a great volume of rock is studied in its natural setting.

The value and inclination of magnetization can be also calculated from the relationships in Eq. 3.27, even when there are only ΔT measurements (Khesin 1998). This calculation is possible when the normal geomagnetic field is known, and thus may be excluded from the observed field. Factor c in Eq. 3.27 depends on the level of the normal geomagnetic field in the target area. If the normal field is subtracted, the c factor becomes corrected and true. A formula for φ_m can be obtained by dividing the c by the b factors. The calculated φ_m can then be substituted into an equation for the b factor and subsequently the $\mathbf{J_e}$value can be obtained.

The above method cannot be used if the normal field is unknown. In this case an alternative technique is available (Khesin 1998). Consider the magnetic fields over two neighboring slopes with identical magnetization. The right slope (r) rises from south to north and is characterized by b_r,R_r and α_r($\alpha_r$is negative). The left slope (l) dips from south to north and is characterized by b_l,R_l and α_l (α_l is positive). The two neighboring slopes are both composed of the same rock with unknown magnetization $\mathbf{J_e}$; φ_m is also unknown. In this situation the following relations hold (Khesin 1998):

$$0.8C_0J_e = \frac{b_rR_r}{\sin(\varphi_m - \tau_0 - \alpha_r)}, \tag{3.29}$$

$$0.8C_0J_e = \frac{b_lR_l}{\sin(\varphi_m - \tau_0 - \alpha_l)}, \tag{3.30}$$

$$\tan\varphi_m = \frac{A}{B}, \tag{3.31}$$

where

$$\left.\begin{aligned} A &= b_lR_l\sin(\tau_0 + \alpha_r) - b_rR_r\sin(\tau_0 + \alpha_l), \\ B &= b_lR_l\cos(\tau_0 + \alpha_r) - b_rR_r\cos(\tau_0 + \alpha_l) \end{aligned}\right\}. \tag{3.32}$$

The angle between the inclination of the magnetization projection J_{xz} and the horizon (φ_m) is determined by Eq. 3.31, which has two possible solutions. It is possible to obtain a single solution for φ_m by considering the sign of φ_m and the type of correlation between the observed magnetic field and the measurement elevation (Khesin et al. 1996, Table 5.3). Substituting the value for φ_m in Eq. 3.29 and/or 3.30 yields the calculation of $\mathbf{J_e}$.

When the investigated profile is oriented to the north we obtain $\varphi_0 = i_0$, since

$$\cot\varphi_0 = \cot i_0 \cos\delta_0, \tag{3.33}$$

where δ_0 is the difference between the azimuth of $\mathbf{T}_0$ and the horizontal axis. In this case $C_0 = 1$. Hence, profiles oriented northwards are more appropriate for the primary investigation. To avoid mutual disturbances, the distance between the l and r ledges has to be sufficiently long. In addition the l slope has to be long enough so that the unambiguous linear correlation between ΔT and h in the central part of slope becomes evident.

3.3 Elimination of Field Variations with Time

Errors due to variations in the results of measurements of some geophysical fields can be represented as the sum of two components. The first (and the main) one is the contribution of the variation of the amplitude (and sometimes of the phase) to the measured field. The second effect is generated by the secondary variation effects from geological targets which differ from the host medium by their physical properties. As shown above, mountainous regions are characterized by a differentiation of physical properties along the vertical and the lateral, which may favor the generation of secondary fields.

Time variations of geophysical signals are observed in all natural geophysical fields. Methods to eliminate them have mostly been developed for magnetic and gravity prospecting.

Elimination of geomagnetic variations is vital for conducting high-accuracy magnetic surveys. The main principle involves continuous or discrete-continuous monitoring of magnetic field variations at the base (control) point (Logachev and Zakharov 1979; Telford et al. 1990). Continuous measurements are carried out in observatories by the use of magnetic variation stations (MVS), whereas discrete-continuous measurements are made at control points (*CP*) using MVS or magnetometers during field surveys. Only short-term and diurnal variations are essential for the field magnetic survey. They are taken into account by introducing corrections: the variations observed on the *CP* are subtracted from the magnetic field values measured on the profile. Secular variations of the magnetic field that change from one region to another, should be taken into account when associating surveys from different years (Nikitsky and Glebovsky 1990). Observations during magnetic storms are rejected.

The secondary effects of mean diurnal variations over strongly magnetized geological bodies were estimated by Dyadkov (1985). It was found, for example, that if the geological body magnetization is 5 A/m, variations cause a secondary effect of 2–3 nT. This makes it possible in most cases to ignore these secondary effects in conventional magnetic prospecting calculations.

Magnetic field variations should not be regarded solely as noise. For example, diurnal geomagnetic variations are used to examine the nature of geomagnetic anomalies (Terekhova 1977; Finkelstein and Eppelbaum 1997). In this way objects in certain classes can be singled out by their intensity and the shape of the variations. Deep structure studies using electromagnetic field variations are the basis for corresponding adaptations in electric prospecting (see Table 3.2).

Methods of elimination of gravity field variations are amply described in the literature (for instance, Adler 1942; Veselov 1986). The secondary effects of gravity field variations are negligible.

Self-potential variations depend on hydrogeological conditions, features of the soil and the vegetation cover and other difficult-to-consider factors (Semenov 1974; Ernstson and Scherer 1986). As a rule, these factors are much more difficult to eliminate.

Time variations can introduce a considerable amount of errors into the results of VLF techniques and near-surface thermal prospecting. In contrast to magnetic prospecting, where variations account for 10^{-3} to 10^{-4} of the constant magnetic field of the Earth, in the VLF method electromagnetic field variations are often comparable to the average intensity of the signal. Therefore, the effects caused not only by initial signal variations, but also by their secondary effects due to the heterogeneity of the section can arise. A ground survey method with synchronous variation measurements at the *CP*, which is similar to magnetic prospecting, was proposed for the VLF method (Eppelbaum and Mishne 1988). Application of a simple linear filtering can be used to obtain the VLF field value that eliminates variation for each profile point (Eppelbaum and Khesin 1992). This technique, described in Eppelbaum and Khesin (1992), makes it possible to eliminate the distortions caused by field variations in time and to reduce the measurement results to a common level.

Many methods for elimination of seasonal variations in near-surface thermal prospecting have been put forward (Chekalyuk et al. 1974; Khutorsky et al. 1983). A more accurate one is based on repeated temperature measurements along the measurement profile using the data on regional temperature field changes obtained from meteorological stations (Eppelbaum 1989). Using a combination of these data as a base, a system of linear algebraic equations can be written; solving these equations eliminates the effect of delayed temperature waves coming from the earth's surface (Eppelbaum 1999b).

3.4 Inverse Problem Solution in Complex Environments: The Example of a Magnetic Field

A special methodology for inversion of magnetic anomalies in complex environments (oblique magnetization, uneven topography and unknown level of the normal field) was developed by Khesin et al. (1983, 1996). A detailed description of this methodology (enhanced modifications of tangents, characteristic points and areas methods) is presented in Khesin et al. (1996) and Eppelbaum and Mishne (2011). Here we provide an introduction to these methods.

The model of the thin bed (Fig. 3.23) can be interpreted by the characteristic point method and the tangent method.

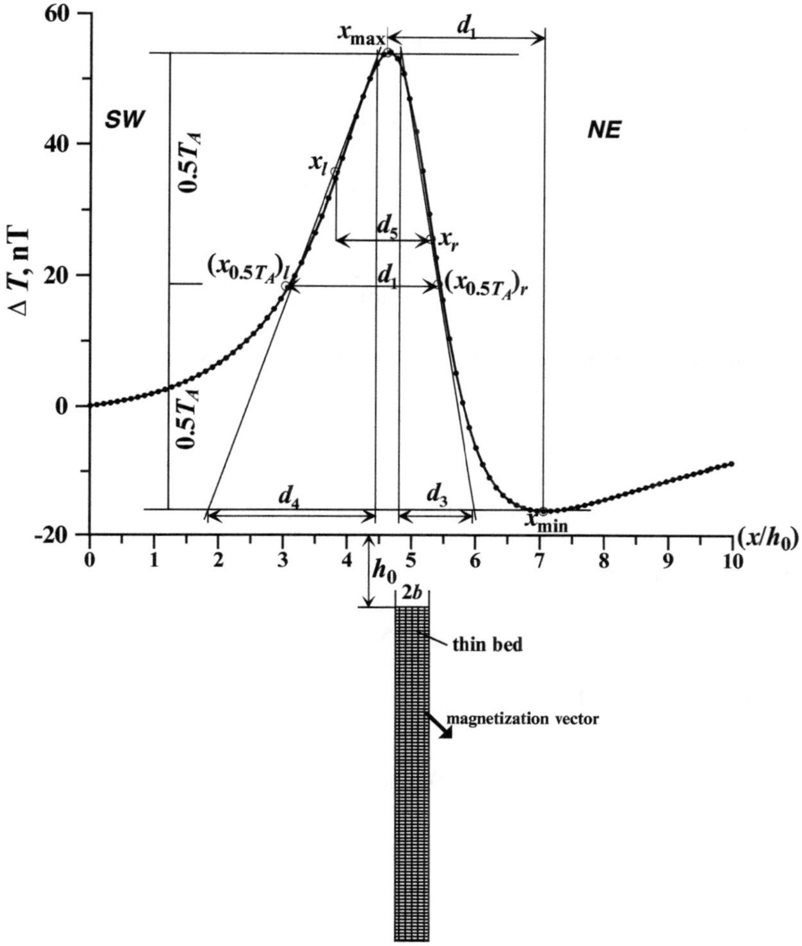

Fig. 3.23 Position of characteristic points and tangents on the ΔT anomaly due to an obliquely magnetized *thin bed*

3.4.1 Characteristic Point Method

The following parameters are taken from the anomaly plot (Fig. 3.23): d_1 is the difference of the semi-amplitude point abscissas, d_2 is the difference of extremum abscissas, d_5 is the difference of inflection point abscissas, $\Delta T_A = \Delta T_{max} - \Delta T_{min}$ is the difference of extremum ordinates (anomaly amplitude).

The following parameters are determined:

(a) Generalized angle θ :

$$\tan\theta = \frac{d_2}{d_1}, \tag{3.34}$$

$$\sin\left(\frac{\theta}{3}\right) = \frac{d_5}{\sqrt{3}d_1}, \tag{3.35}$$

(b) Depth h_0 to the upper edge of the thin bed:

$$h_0 = \frac{1}{2}\sqrt{d_1 d_2 \sin\theta \cos\theta}, \tag{3.36}$$

$$h_0 = \frac{d_5 \sin\theta}{2\sqrt{3}\sin(\theta/3)}, \tag{3.37}$$

(c) Location of the epicenter of the anomalous body relative to the greatest extremum:

$$x_0 = 0.5d_1 - h_0 \cot\theta, \tag{3.38}$$

$$x_0 = 0.5d_5 + h_0 \tan\theta, \tag{3.39}$$

(d) Effective magnetic moment M_e:

$$M_e = 0.5T_A h_0, \tag{3.40}$$

(e) Normal background level:

$$\Delta T_{\text{backrg}} = \Delta T_{\min} + T_{\text{A}} \frac{k_0}{1 + k_0}, \tag{3.41}$$

where $k_0 = \frac{1-\cos\theta}{1+\cos\theta}$.

3.4.2 *Tangent Method*

Here, four tangents are employed (see Fig. 3.23): two horizontal lines with respect to the anomaly extrema and two inclined lines passing through the points of the bend on the left- and right-hand branches of the anomaly plot. The following terms are taken from the plot: d_3 is the difference in abscissas of the points of intersection of an inclined tangent with horizontal tangents on one branch; d_4 is the same on the other branch (d_3 is selected from the plot branch with a conjugated extremum $d_3 \leqslant d_4$; the x-axis is oriented in this direction).

We obtain the following parameters.

(a) Generalized angle θ :

$$\tan\frac{\theta}{3} = \frac{\sqrt{3}\left(1 - \sqrt[3]{k_\theta}\right)}{\left(1 + \sqrt[3]{k_\theta}\right)}, \tag{3.42}$$

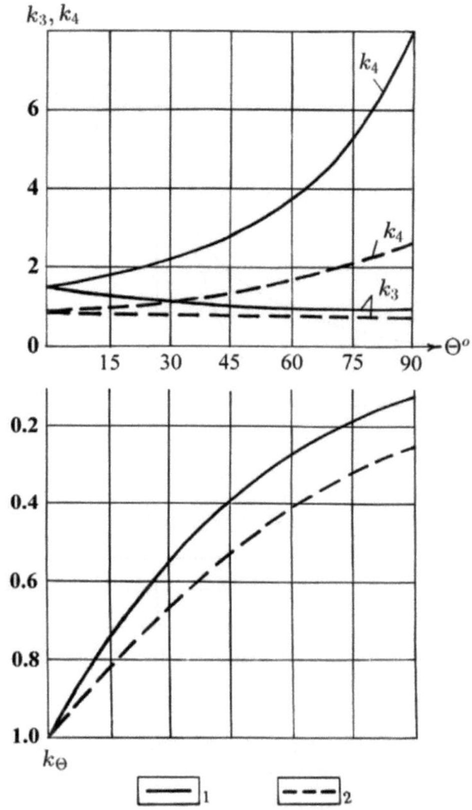

Fig. 3.24 Nomograph for the determination of the angle θ and coefficients k_3 and k_4 by the tangent method from the anomaly over a thin bed (*1*) and a horizontal circular cylinder (*2*)

where $k_\theta = \frac{d_3}{d_4}$.

The latter relation was utilized by formulating a simple nomograph for calculation of parameters θ and h_0 (Fig. 3.24).

(b) Depth h_0 to the upper edge of the thin bed or to the axis of the horizontal circular cylinder:

$$h_0 = \frac{d_i}{k_i}, \tag{3.43}$$

where i is equal to three or four.

The coefficients k_3 and k_4 stand for the values of d_3 and d_4 when $h_0 = 1$. They are functions of θ . After calculation of k_θ , the θ value is determined; then coefficients k_3 and k_4 can be calculated. Depths $(h_0)_3$ and $(h_0)_4$ are calculated as $\frac{d_3}{k_3}$ and $\frac{d_4}{k_4}$; parameter h_0 is the average value of $(h_0)_3$ and $(h_0)_4$.

(c) Depth h_0 to the upper edge of the thin bed is also found by using expressions (3.44a) and (3.44b):

$$(h_0)_3 = d_3 \sin^3\left(60^\circ + \frac{\theta}{3}\right), \tag{3.44a}$$

$$(h_0)_4 = d_4 \sin^3\left(60^\circ - \frac{\theta}{3}\right), \tag{3.44b}$$

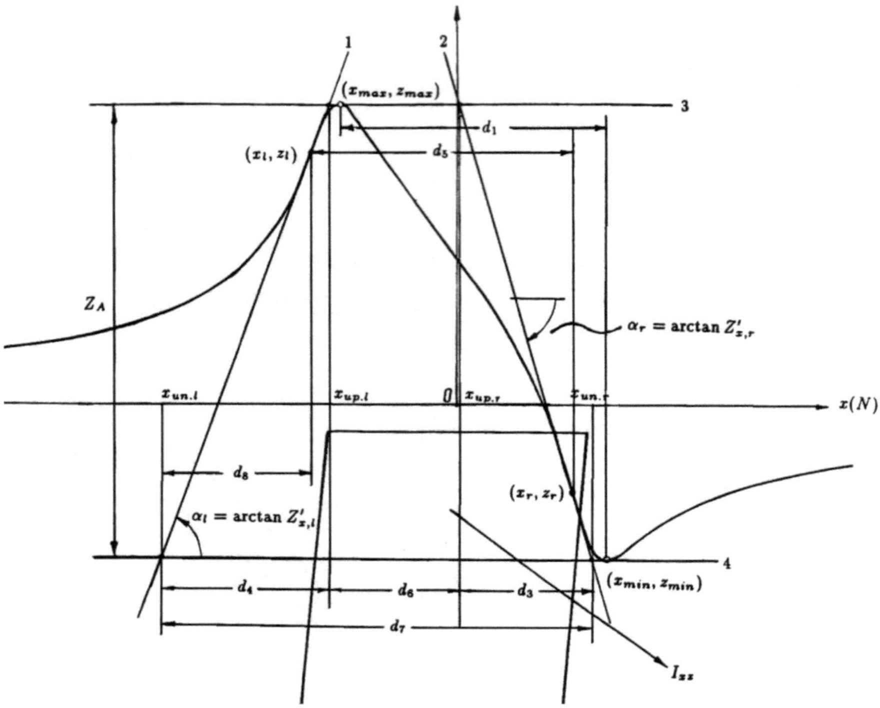

Fig. 3.25 Tangents and characteristic segments for the interpretation of a Z anomaly caused by the obliquely magnetized thick bed

The location of the epicenter of the anomalous body relative to the greatest extremum

$$x_0 = h_0 \tan\left(\frac{\theta}{2}\right). \tag{3.45}$$

Parameters M_e and ΔT_{backgr} are obtained from Eqs. 3.40 and 3.41.

The interpretation of anomalies generated by thick beds is a much more complex process because it takes into account a large number of parameters (Fig. 3.25). Note that interpretation process is the same for components Z, X, Y and total vector T. For details on the interpretation process see Khesin et al. (1996).

3.4.3 Interpretation of Magnetic Anomalies on an Inclined Surface

If anomalies are observed on an inclined profile, these parameters characterize a fictitious body. This problem was described in general in Sect. 3.2.2. The transition from fictitious body parameters to those of the real body is done through the formulas below which are analogical to the formulas in Eq. 3.9:

$$\left.\begin{aligned} h_r &= h_0 + x \tan \omega_0 \\ x_r &= -h_0 \tan \omega_0 + x_0 \end{aligned}\right\}, \tag{3.46}$$

where the subscript "r" stands for a parameter of the real body, h_0 is the depth of the upper edge of the source body, x_0 is the location of the source's projection to the plane relative to the extremum having the greatest magnitude, and ω_0 is the inclination angle of the measurement profile ($\omega_0 > 0$ when the inclination is in the positive direction of the x-axis).

Large-scale testing of the above methods on models and real situations indicates 8–20% accuracy, which is sufficient for practical applications.

3.5 Inversion of Other Natural (Gravity, Temperature, Self-Potential, and Seismicity) Fields

3.5.1 Gravity Field

The potential nature of a gravity field, which follows from the law of gravitation, is expressed by differential equations:

$$\mathrm{rot}\mathbf{F} = 0 \tag{3.47}$$

and

$$\mathbf{F} = \mathrm{grad}W, \tag{3.48}$$

where **F** is the gravity field intensity and W is the gravity potential.

The proportionality between the gravity field intensity and the substance density σ is determined by the expression:

$$\mathrm{div}F = -4\pi G\sigma, \tag{3.49}$$

where G is the gravity constant.

As follows from Eqs. 3.48 and 3.49, the gravity potential must satisfy Poisson's equation (Kaufman 1992):

$$\nabla^2 W = -4\pi G\sigma, \tag{3.50}$$

whereas points where the mass is zero ($\sigma = 0$)should satisfy Laplace's equation:

$$\Delta W = \frac{\partial^2 W}{\partial x^2} + \frac{\partial^2 W}{\partial y^2} + \frac{\partial^2 W}{\partial z^2}. \tag{3.51}$$

The magnetic field (for ΔT – when magnetic susceptibility is below 0.1 SI unit) is a potential field (e.g., Hughes and Pondrom 1947; Tafeyev and Sokolov 1981; Khesin et al. 1996) and is expressed as:

$$\mathbf{U}_{\mathrm{a}} = -\mathrm{grad}\, V, \tag{3.52}$$

where $\mathbf{U_a}$ is the anomalous magnetic field, and V represents the magnetic potential. This field satisfies Poisson's equation.

An overview of the literature (e.g., Dobrin 1976; Mironov 1980; Mudretsova 1981; Lyubimov 1983) indicates that there is no technique for solving the inverse problem of gravity anomalies observed under the above-mentioned complex conditions. For example, the tangent method described in (Mudretsova 1981) and the characteristic points method described in (Lyubimov 1983) call for a normal field level, which is often nonexistent in studies carried out in mountainous conditions. Interesting results reflecting a separate direction in potential field theory were obtained by Brodsky and Strakhov (1987). The applicability of this and similar procedures to complex geological conditions should be examined in concrete geological situations, and call for future studies.

Despite the lack of polarization effects as such in gravity prospecting, the dip of measurement lines is equivalent to the manifestation of "vector properties of the density", (see Sect. 3.2.2). When ignored this can produce fictitious sources of anomalies that are somewhat shifted from the real sources both on the plane and in depth.

For these reasons the rapid interpretation of gravity anomalies is elusive and there is an urgent need to develop selection methods that can be applied at the start of the interpretation process. However, the lack of an initial approximation model can lead to a dramatic increase in computation and is fraught with errors. It is highly important, therefore, for acceptable techniques of rapid interpretation to be developed (Khesin et al. 1988).

The most common models in gravity, resistivity, thermal and SP prospecting are the dipping thin bed (*DTB*), the horizontal circular cylinder (*HCC*) and the sphere (Zaborousky 1963; Lakhtionov and Tarkhov 1970; Semenov 1974; Avdevich and Fokin 1978; Parasnis 1986; Mironov 1980; Telford et al. 1990). A comparison of the analytical expressions for the fields of these models for the methods under consideration and magnetic prospecting (two-dimensional case) is presented in Table 3.6. The level of accuracy declines from the homogeneous to the heterogeneous host medium. However even in the latter case it is still about 15%. For mountainous conditions this degree of accuracy in rapid quantitative interpretation has been shown to be adequate.

It is easy to prove that the analytical expressions are proportional for a *DTB* model in magnetic prospecting and a *HCC* model in the gravity and SP methods and for a point source in magnetic prospecting and a sphere in gravity, thermal and SP methods.

The interpretation of gravity anomalies has the following special characteristics:

Table 3.6 Comparison of analytical expressions for natural geophysical fields

Field	Analytical expression	
Magnetic	*DTB*	Point source (rod)
	$Z_v = 2J2b\frac{z}{x^2+z^2}$	$Z_v = \frac{mz}{(x^2+z^2)^{3/2}}$
Gravity	*HCC*	Sphere
	$\Delta g = 2G\sigma\frac{z}{x^2+z^2}$	$\Delta g = GM\frac{z}{(x^2+z^2)^{3/2}}$
Self-potential	*HCC*	Sphere
	$\Delta U = 2\frac{\rho_1}{\rho_1+\rho_2}U_0 r_0\frac{z}{x^2+z^2}$	$\Delta U = \frac{2\rho_1}{2\rho_2+\rho_1}U_0 R^2\frac{z}{(x^2+z^2)^{3/2}}$
Temperature	–	Sphere
		$\Delta t = \frac{q}{\lambda_2}\frac{\mu-1}{\mu+2}\frac{R^3}{(x^2+z^2)^{3/2}}$

Here Z_v is the vertical magnetic field component at vertical magnetization; J is the magnetization; b is the horizontal semi-thickness of *DTB*; m is the magnetic mass; G is the gravity constant; σ is the density; M is the mass of the sphere; ρ_1 is the host medium resistivity; ρ_2 is the anomalous object (*HCC* or sphere) resistivity; U_0 is the potential jump at the source body/host medium interface; r_0 is the polarized cylinder radius; R is the sphere radius; t is temperature; q is the heat flow density; λ_2 is the thermal conductivity of the anomalous object; μ is the object-to-medium thermal conductivity ratio; x is the current coordinate; z is the depth of the upper *DTB* edge (*HCC* or sphere center) occurrence.

1. The angle θ determined from the nomograph (see Fig. 3.24) must be approximately equal to the angle ω_0 (ω_0 is the dip angle of the terrain relief; $\omega_0 > 0$, if the relief is inclined towards the positive direction of the axis x). This equality can be used to estimate the accuracy of the interpretation, since ω_0 is known.
2. Besides the geometric parameters of an object, the gravity moment can be also determined (Khesin et al. 1996):

$$M_{\Delta g} = \frac{1}{2}\Delta g_a h_0, \tag{3.53}$$

where Δg_a is the amplitude of the gravity anomaly (in mGal), h_0 is the occurrence depth of *HCC* axes (in meters).

It is possible to switch to the real source parameters as follows:

$$M_{\Delta g,r} = M_{\Delta g}\cos\omega_0, \tag{3.54}$$

where ω_0 is inclination angle of the measurement profile.

The model example below (Fig. 3.26) illustrates the rapid interpretation of a gravity anomaly revealed in rugged terrain relief by these improved methods. The Δg_B plot was computed by the improved version of the *GSFC* program (Eppelbaum et al. 1992). It is obvious that the application of a conventional approach to gravity field analysis (e.g., Mironov 1980; Parasnis 1986; Telford et al. 1990) would be unfeasible not only due to disturbances from the rugged terrain relief, but also to the superposition of gravity effects from different geological bodies. The results (an approximation model of the horizontal circular cylinder was employed) indicate that without calculating the terrain relief disturbance effect, the inversion results are

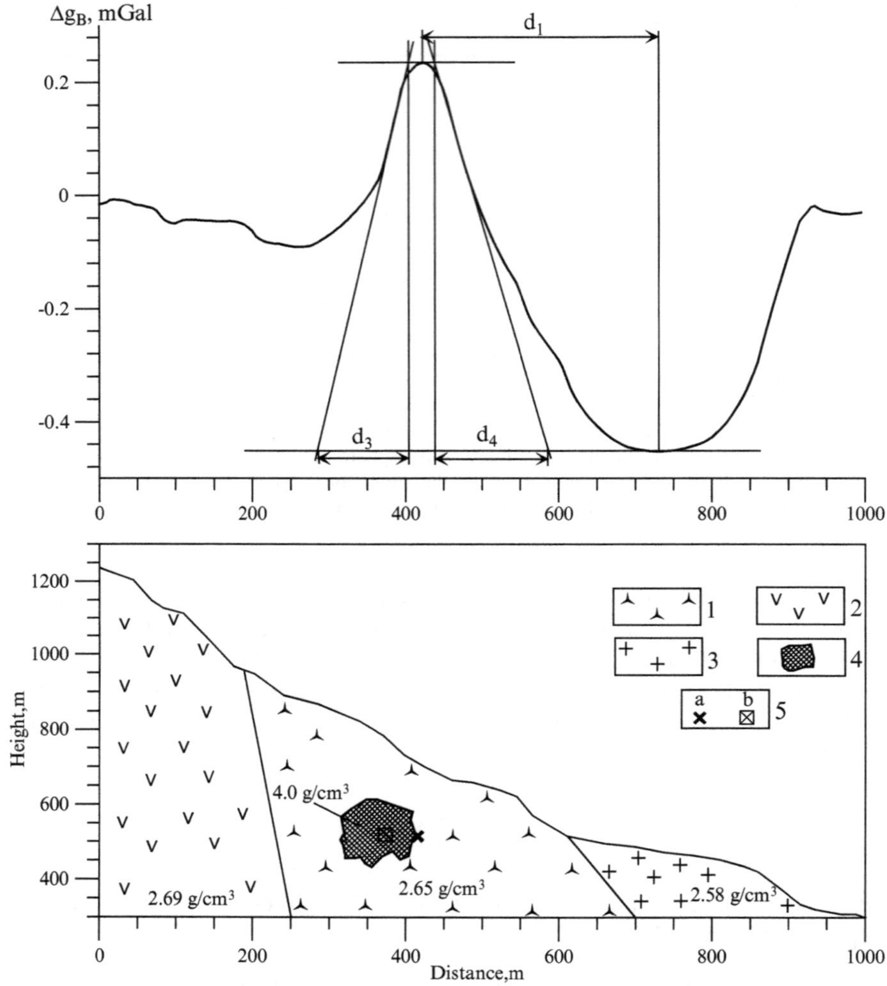

Fig. 3.26 Rapid interpretation of gravity anomaly using a model of the ore polymetallic body occurring in the complex host medium. (*1*) shale; (*2*) gneiss; (*3*) granites; (*4*) ore polymetallic body; (*5*) results of Δg analysis: (*a*) fictitious body, (*b*) real

not sufficient. However, application of the terrain relief correction can yield satisfactory results.

3.5.2 Temperature Field

Near-surface thermal prospecting was initially used by van den Bouwhuysen (1934) to study a fault structure near Vintersweek (in the Netherlands), and by Paul (1935) to study a salt stock near Hannover (Germany), and Rodionov and Sofronov (1935) in the Degtyarsk copper-pyrite deposit in the Middle Urals.

Near-surface thermal prospecting of ore deposits was further developed by Lakhtionov and Tarkhov (1967, 1970), Khutorsky (1982), Dmitriyeva et al. (1985), Dobrynina et al. (1985) and Eppelbaum (1989). The measurements were primarily subjected to a qualitative analysis. Basic temperature measurements were carried out in horizontal underground mines (adits) to lower the noise level. According to Leschak and Lewis (1983), many stages of near-surface thermal surveys based on the Shallow-Temp technique (including the reduction of types of noise) are automated. However, no details were given as to specific methods of noise reduction.

Analysis of thermal fields, based on approximate solutions to direct problems of the thermal conductivity of heterogeneous media by the finite-difference grid technique was covered in (Lyubimova 1968; Lyubimova et al. 1976; Ismail-Zadeh and Tackley 2010). However, solving these problems even with modern computers is prohibitive. A similar approach involved computer-aided selection of temperature fields by successively solving direct problems (Lyubimova et al. 1976; Eliseyeva 2007), as in other geophysical methods.

Suryaninov et al. (1983) derived certain functional dependencies between the thermal parameters of a geological body and the pattern of thermal anomalies arising from this body that occur in uniform host strata. A horizontal circular cylinder, of radius R, approximates the source body with the center located at a depth of h_0 from the Earth's surface. Suryaninov et al. (1983) showed that prospecting proves to be most effective for source bodies with parameters satisfying the following conditions:

$$0.1 \leqslant \frac{\lambda_2}{\lambda_1} \leqslant 10; \ \frac{h_0}{R} < 5, \tag{3.55}$$

where λ_2 and λ_1 are the thermal conductivity values for the host medium and the source body, respectively.

Simmons (1967) suggested a method for interpretation of heat flow anomalies analogous to those used in gravity prospecting. The similarity between gravity and temperature anomalies was also noted by Poley and Steveninck (1970) and Kappelmayer and Hänel (1974). However, in these studies, there are no examples of quantitative interpretation applications. Zorin and Lysak (1972) made an attempt to use the proportionality of analytical expressions for gravity and temperature fields of a point source. Their quantitative interpretation used Smith's inequalities developed for gravity prospecting (Parasnis 1966). However, the use of these inequalities under complex conditions was only partially successful.

In solids, heat flows propagate according to Fourier's law (Carslaw and Yaeger 1959):

$$\mathbf{q} = -\lambda \,\mathrm{grad}\, T = -\lambda \frac{\partial T}{\partial n}, \tag{3.56}$$

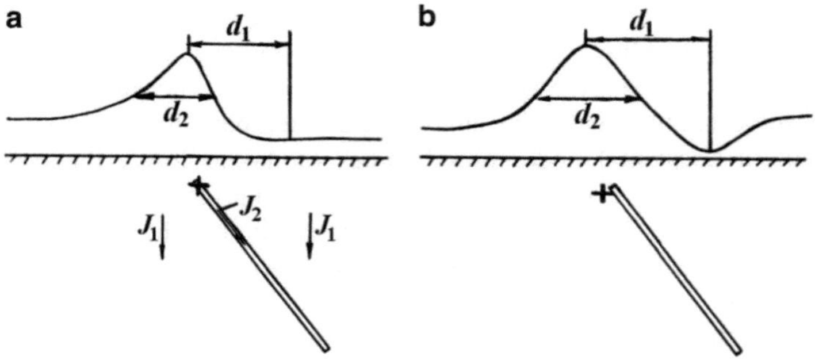

Fig. 3.27 Comparison of magnetic and temperature anomalies over a model of an inclined thin body. The *arrow* shows the direction of the magnetization vector ***J***. The "+" symbol marks the position of the *upper edge* of the *thin body*, as obtained from analysis of the anomaly profiles (Eppelbaum, 1989)

where T is the temperature, **q** is the heat flow density, λ is the thermal conductivity, and n is the outward normal to an isothermal surface.

If we assume that the heat transfer from the local sources is conductive and the temperature field is stationary, the temperature distribution in a space without any sources satisfies a Laplace's equation (Tikhonov and Samarsky 1963).

Taking into account the similarity of analytical expressions (Table 3.6), we can examine temperature anomalies by the interpretation methods developed for magnetic prospecting (Eppelbaum 1989; Khesin and Eppelbaum 1994). Figure 3.27 depicts the comparative quantitative analysis of magnetic (**a**) and temperature (**b**) graphs. The temperature graph is based on physical modeling and the magnetic graph was computed using 3D software (to create an exact analogy with the temperature field, the surrounding medium was computed with a vertical magnetization and the magnetization of the thin bed was selected with the same angle as the bed dipping).

One significant feature of the methodology should be noted: the angle of terrain relief is equivalent to that of the polarization vector on the magnetic field and the sign is inverted for the thermal field (Eppelbaum 1989). When interpreting plots of the temperature field recorded on an inclined profile, the measurement results and measurement point heights are inversely correlated (see Fig. 3.16c). This causes a reflection of the temperature anomaly plot about the vertical axis as compared to the magnetic anomaly presented in Fig. 3.23.

3.5.3 Self-Potential Field

SP polarization is brought about by the spontaneous manifestation of electric double layers on various geological formation contacts. The electric fields **E** of the electric double layer l caused by natural polarization are defined as the gradient of a scalar potential Π_i:

$$\mathbf{E}_{\mathrm{SP}} = -\mathrm{grad}\,\Pi_i. \tag{3.57}$$

The potential Π_i satisfies Laplace's equation everywhere outside the layer l (Zhdanov and Keller 1994).

The calculation of theoretical anomalies due to SP has long been based primarily on Petrovsky's well-known solution derived for a vertically polarized sphere (Zaborousky 1963). Later on, solutions for sheet-like bodies and inclined plates were obtained (Semenov 1974; Tarkhov 1980). The polarization vector was generally considered to be directed along the ore-body dip (along the longer axis of the conductive body).

To make quantitative interpretation of anomalies due to SP, a body with a simple geometrical shape approximates the anomaly source. Its parameters (i.e. the occurrence depth, the angle between the horizon and the direction of the polarization vector) are usually determined either graphically, using characteristic points of the anomaly plot (Semenov 1974), or by trial-and-error, by visually comparing the anomaly to the set of master curves (graticules) (Tarkhov 1980).

In the works of Zaborousky (1963), Semenov (1974), and Murty and Haricharan (1984) the SP anomaly generated by a plate and recorded along the profile across its strike is calculated by the following formula:

$$U(x) = \frac{j\rho}{2\pi}\ln\frac{r_1^2}{r_2^2}, \tag{3.58}$$

where j is the current per unit length, ρ is the host medium resistivity, r_1 and r_2 are the distances from the plate ends to the measurement points.

However, the techniques suggested in the above works require the normal field level to be known. They are also unacceptable for rugged terrain relief.

Fitterman (1984) presented a method to calculate SP anomalies for field sources of an arbitrary shape. The method is based on numerical integration using Green's function. This approach is highly computer intensive.

There are a number of recent interpretation techniques based on minimizing the difference between an observed anomaly and a theoretical one. The minimization is achieved by sequential optimization of the interpretation parameters through computer-aided iterations. These techniques are also complicated and time consuming.

Kilty (1984) published a paper which acknowledged the analogy between the current density of SP and magnetic induction. This author suggested interpreting SP anomalies on the basis of methods developed for magnetic prospecting. A similar approach was proposed earlier by Alexeyev (1971) and later expanded upon in Khesin et al. (1996).

Figure 3.28 presents the results of SP anomaly interpretation using these techniques (Eppelbaum et al. 2004b). The target parameters were determined by the characteristic point and tangent methods, using segments d_1, d_2, etc. For example, in the characteristic point method the inclination angle of the natural polarization vector ϕ_p is calculated from the expression for an inclined profile

$$\varphi_p = 90^{\circ} - \theta + \omega_o, \tag{3.59}$$

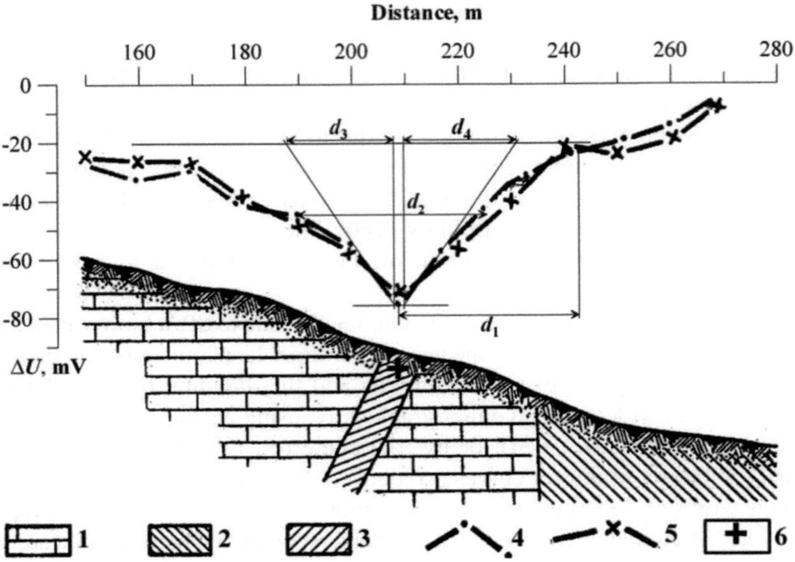

Fig. 3.28 Interpretation of the SP anomaly over the polymetallic vein by improved techniques (initial data from Yakubovsky and Lyakhov 1956). (*1*) limestone; (*2*) shales; (*3*) polymetallic vein; SP measurements; (*4*) July 1954; (*5*) end of August 1954; (*6*) middle of the upper edge of source body determined by the SP inversion

where $\theta = d_2/d_1$, ω_0 is the inclination angle of the profile.

In the tangent method the depth $h_0$is calculated by the expressions (3.43) and (3.44a,b).

3.5.4 Seismicity Field

The characteristics of elastic waves traveling in the Earth have been studied by seismologists specializing in earthquake signals (e.g., Aki and Richards 1980).

Recent developments in teleseismic tomography have characterized the main features of the Caucasian upper mantle (e.g., Al-Lazki et al. 2004; Tan and Taymaz 2006; Zor 2008).

A tomography method based on observed travel-time residuals is used to invert for P_n-wave velocity and anisotropy as well as event and station delays. The model assumes that P_n behaves like a diffracted head-wave that travels through the uppermost mantle (the crust layer over the half-space upper mantle). This tomography revealed a broad-scale (~500 km) zone of low (<8 km/s) P_n velocity anomaly underlying the Caucasus region and a smaller scale (~200 km) zone of very low (<7.8 km/s) P_n velocity underlying the Lesser Caucasus that may be associated with the latest stage of intense volcanism.

Very recently, different studies have attempted to utilize the natural field of microseisms to derive geological information. The microseismic sounding (MSS) method is the most highly developed (e.g., Gorbatikov and Stepanova 2008; Gorbatikov et al. 2008a, 2008b). This method is based on the inversion of the amplitude-frequency spatial distribution of the microseismic field. The authors (Institute of the Solid Earth Physics, Moscow) suggested that the vertical component of this field in the low-frequency band is determined mainly by fundamental modes of the Rayleigh wave. This low-frequency microseismic field reflects the structure of the Earth's crust. On the Earth's surface, amplitudes of certain frequency f are decreased over bodies of high velocity and increased over bodies of low velocity This frequency depends on body depth $h_0$and the velocity of the fundamental mode of the Rayleigh $V_R(f)$:

$$h_0 = \frac{0.5V_R(f)}{f}. \tag{3.60}$$

Thus, it is possible to distinguish different velocity features of geological sections and calculate their depths.

It is important to note that micro-tremors cause spectral anomalies over hydrocarbon reservoirs; Walker (2008) reviewed these studies.

3.6 Inversion of Artificial and Quasi-Natural (Resistivity, Induced Polarization, Very Low Frequency) Fields

3.6.1 Resistivity

The electric field **E** is the electrostatic field (div $\mathbf{j} = 0$, where **j** is the conduction current density). It is well known that the resistivity field is potential and satisfies the Laplace equation (Kaufman 1992):

$$E = -grad\ \Phi, \tag{3.61}$$

where Φ is the scalar electric potential.

There is a similarity between the analytical expressions applied in magnetic prospecting and those used in the resistivity method (Eppelbaum 1999a, 2007a). Thus, it is possible to apply the advanced methodology developed in magnetic prospecting (Khesin et al. 1996) to interpret resistivity anomalies.

Figure 3.29 illustrates an application of these methods for interpreting a theoretical anomaly over a model of an inclined thin bed.

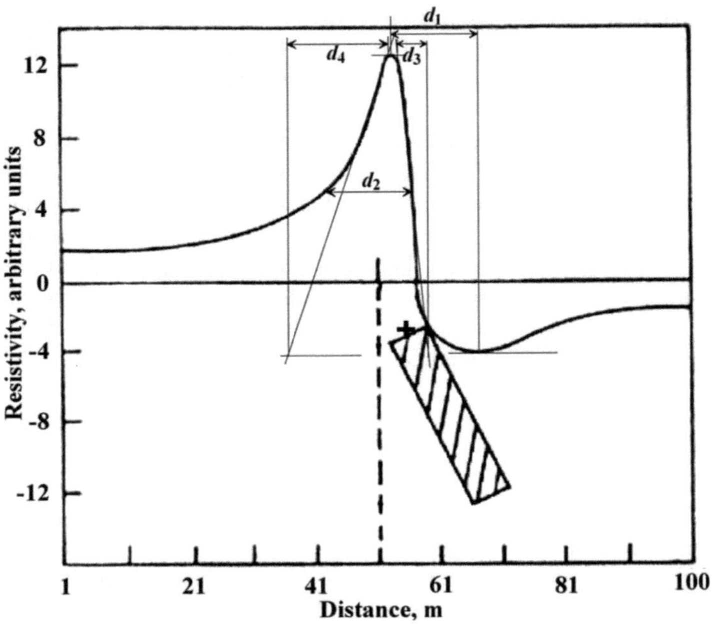

Fig. 3.29 Interpretation of a theoretical resistivity anomaly over a model of an inclined *thin bed* (initial section after Sundararajan et al. 2007). Symbol "+" designates the position of the middle of the upper edge of the anomalous body

3.6.2 Induced Polarization

The induced polarization method (Seigel 1959) is an effective geophysical tool in mineral prospecting, and can solve problems of hydrogeology and engineering geology. This method is widely used for ore prospecting (Komarov et al. 1965; Parasnis 1986; Telford et al. 1990). Its extensive application can be accounted for by its sensitivity to the presence of conductive minerals.

However, the quantitative interpretation of IP anomalies is significantly limited. We will consider IP studies in the time domain (data for the frequency domain is similar; e.g., Heinrichs and Ludwig 1966; Zonge et al. 1972; Gurevich et al. 1975; Komarov 1980; Johnson 1984; Telford et al. 1990; Duckworth and Calvert 1995). Among the works dealing with IP data interpretation, Ruderman and Golubkov (1985) discussed delineation of isometric anomalies, and Komarov (1980) put forward the tangent method. Of special interest is a paper by Quick (1974), which suggests interpreting IP anomalies from a gradient array by potential field theory. He recommends applying a conventional interpretation technique known in magnetic prospecting. All these techniques suffer from the same drawbacks which have been examined in detail in the sections above discussing the interpretation of other fields.

Certain conclusions concerning the parameters of the target can be made using three-layer curves of IP vertical sounding or by computer modeling (Komarov et al.

1965; Tarkhov 1980). However, specialized techniques are more suitable for rapid interpretation (Khesin et al. 1996).

Based on Alpin et al. (1985) and Smirnov (1975) and taking into account the work of Avdevich and Fokin (1978), IP field intensity can be written as follows:

$$\mathbf{E}_{\mathrm{IP}} = \frac{1}{2\pi\varepsilon_o} \int_S \frac{2(\mathbf{P}, \mathbf{r})\mathbf{r} - \mathbf{P}r^2}{r^4} ds, \tag{3.62}$$

where ε_o is the dielectric constant of free-space; S is the cross-sectional area of the source body; **P** is the polarization vector (dipole moment of a unit volume), $r = \sqrt{(x_s - x)^2 + (z_s - z)^2}$ is the distance from the measurement point $M(x, z)$ to a certain point $P(x_s, z_s)$, $\mathbf{r} = (x_s - x)\mathbf{i} + (z_s - z)\mathbf{k}$ is the radius-vector from point M to point P; **i** and **k** are unit vectors directed along Ox and Oz axes, respectively.

Then the components E_{IP} can be written as follows:

$$E_{\mathrm{IP},x} = -\frac{1}{2\pi\varepsilon_0} \int_S P_{\mathrm{IP}} \frac{(z_d^2 - x_d^2)\cos\varphi_{\mathrm{IP}} - 2z_d x_d \sin\varphi_{\mathrm{IP}}}{r^4} ds, \tag{3.63}$$

$$E_{\mathrm{IP},z} = \frac{1}{2\pi\varepsilon_0} \int_S P_{\mathrm{IP}} \frac{(z_d^2 - x_d^2)\sin\varphi_{\mathrm{IP}} + 2z_d x_d \cos\varphi_{\mathrm{IP}}}{r^4} ds, \tag{3.64}$$

where $z_d = z_s - z$ and $x_d = x_s - x$, φ_{IP} is the inclination angle of the IP vector.

At the same time, the horizontal and vertical components of anomalous magnetic field take the form (Khesin et al. 1996):

$$X_{\mathrm{an}} = \frac{\mu_0}{4\pi} \int_S J \frac{-(z_d^2 - x_d^2)\cos\varphi_m + 2z_d x_d \sin\varphi_m}{r^4} ds, \tag{3.65}$$

$$Z_{\mathrm{an}} = \frac{\mu_0}{4\pi} \int_S J \frac{(z_d^2 - x_d^2)\sin\varphi_m + 2z_d x_d \cos\varphi_m}{r^4} ds, \tag{3.66}$$

where μ_0 is the absolute permeability of free-space, and φ_m is the angle between the vector **J** and Ox-axis.

Obviously, Eqs. 3.63 and 3.65, 3.64 and 3.66, respectively, are proportional. This proportionality was used to prove the applicability of methods developed in magnetic prospecting for the interpretation of IP anomalies (Khesin et al. 1997; Eppelbaum and Khesin 2002).

Figure 3.30 shows the results of model curves η_a for a gradient and a potential array on an inclined profile. The figure indicates that boreholes drilled above the obtained fictitious center of *HCC* will not encounter the polarized body. As to the real sources, they reflect the anomalous object's position with sufficient accuracy.

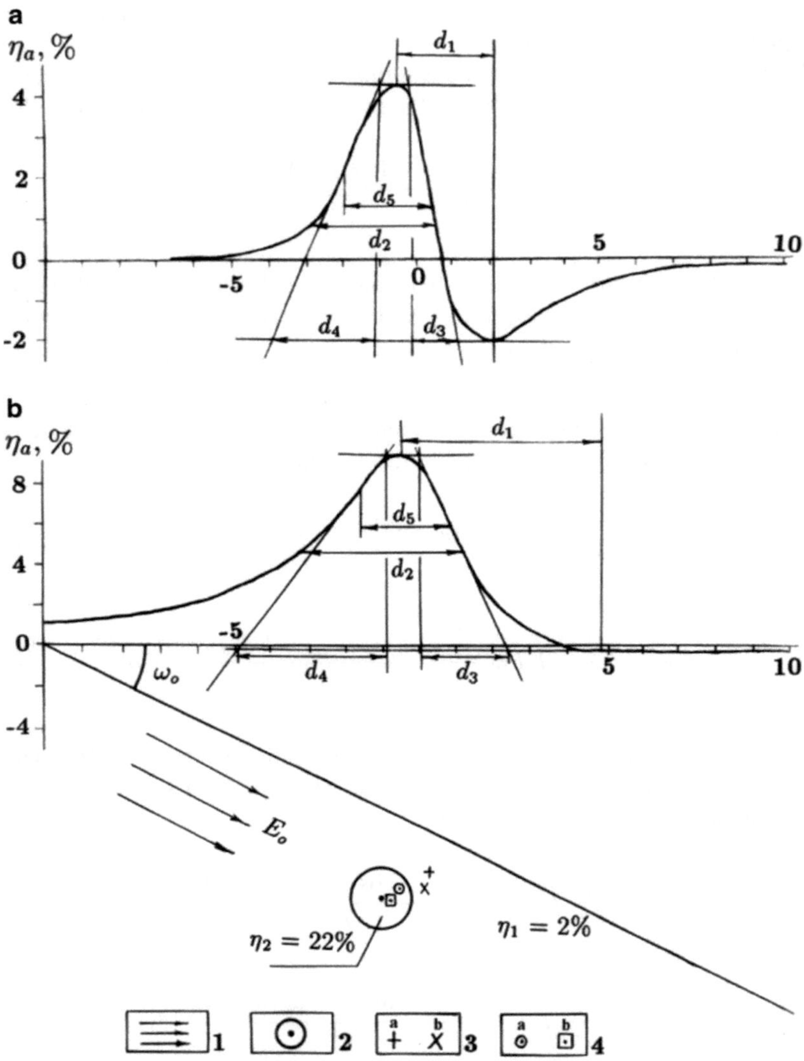

Fig. 3.30 Interpretation of model curves $\eta_a$on an inclined profile from a polarized cylindrical body: (**a**) for a gradient array, (**b**) for a potential array. (*1*) uniform polarizing field; (*2*) model polarized body; (*3*) center of the fictitious body: (*a*) by curve (**a**), (*b*) by curve (**b**); (*4*) center of the real body: (*a*) by curve (**a**), (*b*) by curve (**b**)

The inclination angle of the IP vector φ_{IP} may be calculated by the following expression:

$$\varphi_{\mathrm{IP}} = \phi_2 - \theta + 90^{\circ}, \tag{3.67}$$

where φ_2 is the dip angle of the bed (calculated from the vertical), $\theta = \arctan(d_2/d_1)$.

For the case of an inclined profile we have, respectively,

$$\varphi_{\mathrm{IP,s}} = \phi_2 - \theta + 90^{\circ} + \omega_0, \tag{3.68}$$

where ω_0 is the inclination angle of the profile.

3.6.3 VLF

To interpret VLF anomalies by the techniques described above, the VLF field's stationarity or quasi-stationarity need to be taken into account. An electromagnetic field can only be considered quasi-stationary, if it satisfies three conditions. These conditions were described by Landau and Lifshitz (1960), and applied to geophysics by Alpin et al. (1985). These conditions are as follows:

1. Slow field variation,
2. Closed currents,
3. To avoid an appreciable lag in the magnetic field variation, the region including the magnetic field generating currents and measurement points must not be too large.

These conditions are examined in more detail below.

1. The study of time variations of VLF fields has shown that abrupt changes in intensity are not characteristic of these fields. Nonetheless, these variations should be eliminated by a special procedure during the field survey (Eppelbaum and Khesin 1992; Khesin et al. 1996).
2. This condition is fulfilled (Eppelbaum and Khesin 1992), since div rot $\mathbf{H} = 0$ and, therefore, div $\mathbf{j} = 0$. Currents that are almost closed do not violate this condition.
3. This condition can be formulated in the following way: the length of the electromagnetic wave in the ground should basically exceed the length of the targets. The condition, as a rule, is sufficient for prospecting ore deposits, the study of faults and other engineering geophysics targets, and in other detailed studies.

Using the fundamental solution of Laplace's equation in the 2D case and the Helmholtz wave equation, Hänl et al. (1961) and Dmitriyev (1982) linked Green's solutions to the Laplace's and Helmholtz equations. This allowed them to extend the results obtained in potential theory to the Helmholtz equation (Eppelbaum 1989).

Dmitriyev et al. (1977) showed that associating the singular points from the VLF field anomaly plots with geometrical parameters of a bed is analogous to the well-known behavior of singular points in potential field anomalies.

Table 3.7 Comparison of analytical expressions for *DTB* in magnetic prospecting and the VLF method

Field	Analytical expression	
Magnetic	$Z_v = 2I2b\frac{z}{x^2+z^2}$	$X_v = 2I2b\frac{x}{x^2+z^2}$
VLF	$H_x = kH_0\frac{z}{x^2+z^2}$	$H_z = kH_0\frac{x}{x^2+z^2}$

In this table Z_v and X_v are the vertical and horizontal magnetic field components for vertical magnetization, respectively; H_x and H_z are the horizontal and vertical VLF magnetic field components, respectively; H_0 is the VLF primary field intensity; k is the coefficient involving the geometry and conductivity of the bed

Zhdanov (1984) carried out a theoretical study of the application of a set of Cauchy-type integral analogues to the problems of electromagnetism. He argued that the methods developed in the potential theory for analytical continuation, separation and quantitative interpretation could be applied to quasi-stationary electromagnetic anomaly analysis.

Thus, it is possible to interpret a VLF field as a quasi-stationary field (Eppelbaum 1989).

An overview of publications (Auzin 1977; Gordeyev et al. 1981; Tesmull and Crossley 1981; Olsson 1983; Ruderman and Golubkov 1985) indicates that a conductive inclined thin bed is the most common model employed in the VLF technique. The analytical expressions of the fields for the abovementioned model (two-dimensional case) are presented in Table 3.7.

Evidently, Z_v is proportional to H_x, while X_v is proportional to H_z. Similar results have been obtained for a HCC model (Eppelbaum 1989). Thus, the plots of magnetic fields of VLF transmitters can be interpreted by special methods elaborated in magnetic prospecting. The **E**-polarization vector, in the first approximation, is the analog of the magnetization vector (Eppelbaum and Khesin 1992).

It should be noted that the total horizontal component of the VLF magnetic field $H_\varphi = \sqrt{H_x^2 + H_y^2}$, the most frequently measured in practice, can be interpreted as the H_x component, since the contribution of H_y component is usually relatively small.

The essential distinctions reflecting the quasi-potential nature of the VLF electromagnetic field are as follows. In magnetic prospecting the induced magnetization vector of an inclined bed is approximately parallel to the geomagnetic field vector, irrespective of the bed dip direction, with a magnetic susceptibility that does not exceed 0.1 SI unit. In electric prospecting by the VLF method under **E**-polarization for highly conductive targets, the equivalent vector of polarization, causing the anomalies of H_x and H_z, approaches the body axis, with a slight deviation toward the vertical for gently sloping bodies. That is why the generalized angle θ, resulting from the interpretation by the proposed methods and representing the difference of the inclination angles for the bed and the polarization vector, can be used to estimate the bed dip angle by the following empirical formulas (Eppelbaum and Khesin 1992):

(a) For the H_x anomaly

$$\varphi_2 = 3\theta + 90^\circ; \tag{3.69}$$

(b) For the H_z anomaly

$$\varphi_2 = 3\theta - 180^\circ. \tag{3.70}$$

For the case of measurements on a profile inclined by angle ω_0, angle φ_2 is calculated as follows:

For the H_x anomaly

$$\varphi_2 = 3(\theta - \omega_0) + 90^\circ; \tag{3.71}$$

For the H_z anomaly

$$\varphi_2 = 3(\theta - \omega_0) - 180^\circ. \tag{3.72}$$

Figure 3.31 illustrates the application of the technique developed in magnetic prospecting for the interpretation of a VLF anomaly over a conductive ore-body (Eppelbaum and Mishne 2011).

3.7 Information and Probabilistic Interpretation Methods for the Detection of Hidden Targets

3.7.1 Entropy and Information

A quantitative estimation of the efficiency of geophysical prospecting, the exhaustivity of the geophysical data, is possible using reliable informational and statistical criteria. The first criterion is the quantity of information obtained by an individual method or by a set of various methods. The second criterion defines the risk of decision-making as a function of the geophysical results. The informational criterion is preferable, because geophysical prospecting is a permanent process of acquisition and processing of information; on information theory see the classic works by Shannon (1948) and Brillouin (1962). These criteria were further developed by certain researchers in the USSR (first by Khalfin (1958)). It has been shown (Khesin 1976, Khesin and Eppelbaum 1997) that informational and statistical approaches represent two aspects of a common approach. For instance, the solution to an identification problem using the criterion of minimal average risk or that of maximum information (minimum residual uncertainty), under certain conditions, yields the same expressions (Peresada 1970). A detailed review of the informational approach appeared in Svetov (1992).

Assume that there is only one deposit in the area under study. This area is divided into N equal cells. Assume further that each cell has the same probability of containing the ore-body. Thus, the probability of the ore-body appearance is equal

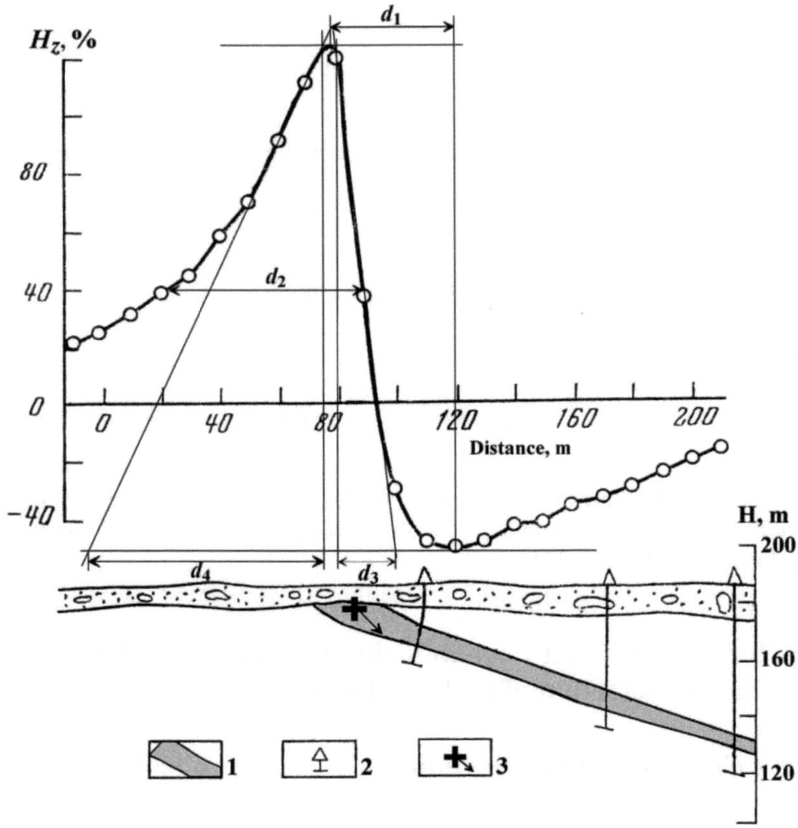

Fig. 3.31 Interpretation of the VLF curve over an inclined thin bed by the characteristic points method (initial VLF graph and geological section after Gordeyev et al. 1981). (*1*) ore-body; (*2*) borehole; (*3*) calculated upper edge of the source body and body dip

to $P = 1/N$ in each cell. Hence the entropy of the experiment β (discovery of an economic mineral) is log N. The entropy is determined using following expression:

$$H(\beta) = -\sum_{i=1}^{N} P(B_i) \cdot \log(B_i), \qquad (3.73)$$

where $P(B_i)$ is the probability of the outcome of B_i (B is the number of outcomes for experiment β).

In this situation $H(\beta)$ value (log N) is the maximum possible uncertainty. Experiment α (measurement of the geophysical field) provides additional information. The A value is the number of outcomes for experiment α.

The difference between uncertainties in the β results before and after the α experiment estimates the information of α as related to β:

$$I(\alpha, \beta) = H(\beta) - H(\beta|\alpha), \tag{3.74}$$

where $H(\beta|\alpha)$ is the conditional entropy of the experiment (under the condition that the experiment has been carried out).

Conditional entropy is the average value of the random variables taking the $H(B_i|A_i)$ value with probability of $P(A_i)$.

$$H(\beta|\alpha) = -\sum_{i=1}^{N} P(A_i)H(B_i|A_i). \tag{3.75}$$

What information do geophysical field measurements contain from an information theoretic point of view? We assume that a magnetic field is observed in range D, and the measurement precision is given by τ. Let us fit integers of small intervals ϕ into the intervals D and τ. After measurement, the value of the field in ϕ units (with an accuracy up to ϕ) fits the interval τ. Using expression (3.74) and considering that the entropy in this expression before and after measurement is expressed by logarithms D/ϕ and τ/ϕ, respectively, one easily obtains:

$$I(\alpha, \beta)_\phi = \log\left(\frac{D}{\tau}\right). \tag{3.76}$$

With decreasing ϕ, these entropies are unbounded, but the information remains unchanged. With an unrestricted increase in measurement precision information also increases to infinity, but slowly: an n-fold increase in accuracy only leads to log n information units more. An essential property of composite experiments is as follows. If tests Ψ, Λ and Θ are independent, experiments Ψ and Θ can have zero information about Λ. However, an integrated experiment Ψ and Θ can determine the outcome of the experiment Λ completely. Consequently, whereas separate geophysical methods provide no information on the presence (or absence) of ores or other targets, information can be obtained by integrating these methods. Given expression (3.76), it is obvious that the development of integrated studies is preferable to increasing the precision of separate methods.

3.7.2 *Information-Statistical Techniques for the Analysis of Single Geophysical Fields*

3.7.2.1 Revealing Hidden Intersections of Linear Structures

Singling out nodes of linear structures is a crucial problem, since such nodes control the location of economic ore, oil-and-gas and underground water deposits of various types. This problem is also essential to earthquake prediction (e.g., Keilis-Borok and Levshin 1984). This led to a method for direct localization of nodes (Khesin 1976, 2005).

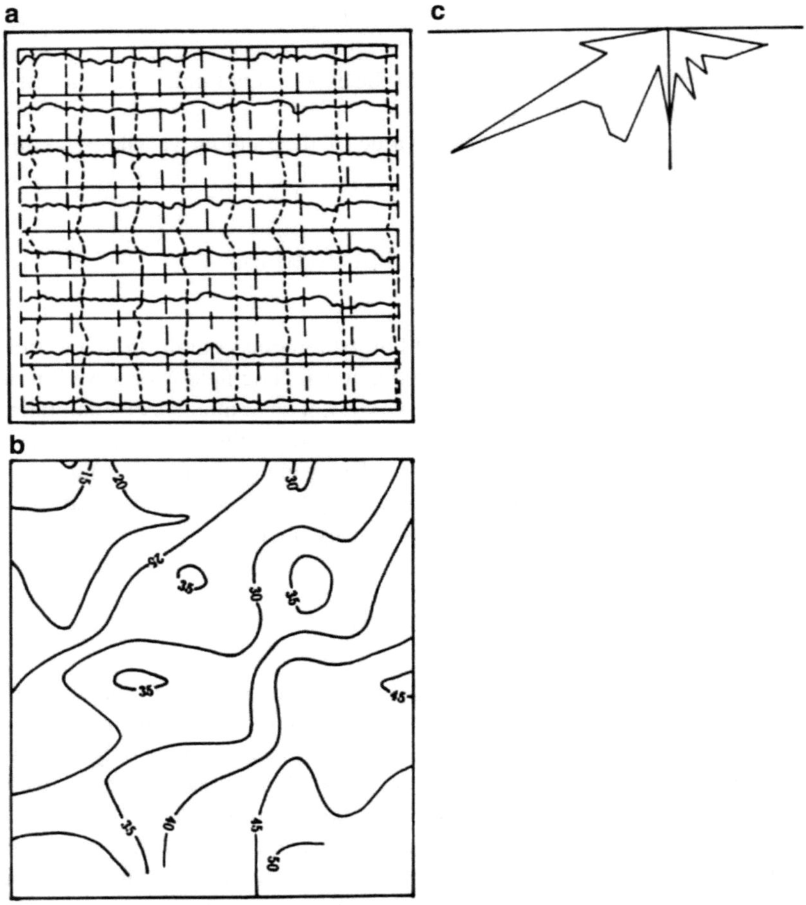

Fig. 3.32 Analysis of model magnetic field of mutually perpendicular, intersecting linear structures: (**a**) map of graphs, (**b**) map of isogams, (**c**) rose-diagram of isogam strikes

When structures with different strikes intersect, the field components caused by these structures superimpose. The observed pattern is complicated by random noise. Thus, the problem consists of suppressing random noise and revealing isoline strikes. Once the noise is removed from the intersection area, the strike of anomalous bodies can be approximated.

In the case of superposition of differently oriented anomalies, errors in determining strikes are frequent. The maps of graphs (Fig. 3.32a) and isolines (Fig. 3.33b) represent a model magnetic field of mutually perpendicular intersecting linear vertical beds. The rose-diagram (Fig. 3.33c) of the isoline strikes shows the false orientation clearly.

The proposed method distinguishes the true strikes.

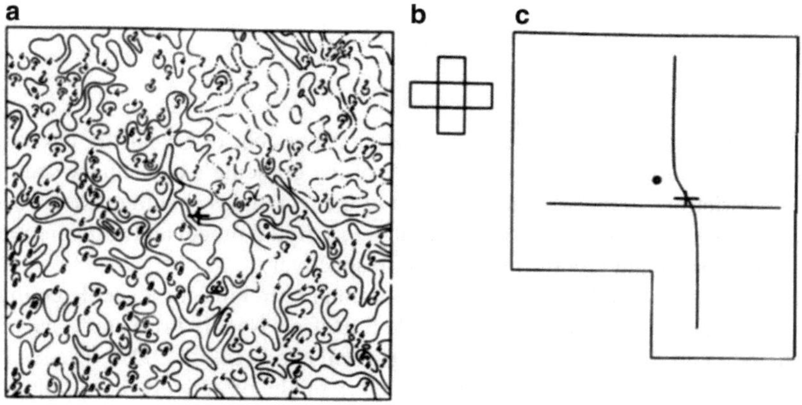

Fig. 3.33 Singling out intersection node: (**a**) synthetic magnetic field of mutually perpendicular intersecting linear structures, plus random noise, (**b**) cruciform graticule, (**c**) computation results

Figure 3.33a shows a synthetic magnetic field (dotted lines mark zero contours) of mutually perpendicular intersecting linear structures (ledges), plus random noise. Figure 3.33b shows a cruciform graticule used for computing the number of isolines per area. The results are presented in Fig. 3.33c. A small cross marks the location of the pre-set structure intersection. The solid lines are anomaly axes determined by an anisotropic graticule (Klushin and Tolstikhin 1961). The small disk is the point of ledge intersection defined according to the value of parameter I_{cr} (maximum information).

Each rectangle of the graticule (Fig. 3.33b) is made up of three elementary squares. The middle square, because it is shared by both rectangles, falls directly within the hypothetical area of intersection and is not involved in the computation. For each direction, isolines which cross the external line of the outer squares and the internal line parallel to it, are tabulated (μ and v). According to Brillouin (1962) and transforming the expression (3.76), the information from a physical measurement is calculated by

$$I_u = \log\left(\frac{L_u}{\Delta U}\right), \tag{3.77}$$

where L_u is the span of the U value, ΔU is the measurement error.

The isolines are computed with the intervals close to the ΔU error. Consequently, if μ or v isolines are present on a portion of the map, their numbers correspond to the $\frac{L_u}{\Delta U}$ values in Eq. 3.77. Thus the amount of information I_p on the presence of an intersection is expressed by the ratio of the amount of information over the presence of anomalies with different strikes $I_1 = \log_2\mu$ and $I_2 = \log_2 v$. In order to reduce the effect of high-gradient zones with a common strike, I_p is calculated by the formula

$$I_p = (I_1 + I_2)\frac{I_1}{I_2}, \tag{3.78}$$

where $I_1 \leqslant I_2$.

As shown Fig. 3.33c, this method makes it possible to reveal the intersection node to an accuracy of the size of the side of the graticule's elementary square.

The variations in the size and direction of the rectangles should make it possible to distinguish different node types. In practice, a digital model and software replace the isolines map and cruciform graticule, respectively.

It should be noted that the problem of optimal searching for a needle configuration in the 2D and 3D case – the so-called "Buffon Needle Problem" – is widely discussed in mathematics and physics (e.g., Dell and Franklin 2009). The solutions (observation grid, level of observation, etc.) might effectively be employed in applied geophysics.

To eliminate variable noise manifested as quasi-white noise, it is useful to transform the observed geophysical fields into informational features. Eppelbaum et al. (2003a) proposed the following expression:

$$J_i = U_i\left(\sum_i^n U_i\right)^{-1} \log\left[U_i\left(\sum_i^n U_i\right)^{-1}\right]K, \tag{3.79}$$

where U_i is the geophysical observation at the ith point ($U_i > 0$) in the area under study, n is the total number of observations, and K is the coefficient (K to enhance map visualization).

This algorithm has been successfully applied at large, medium and small scales to analyze gravity, magnetic, self-potential and VLF data.

3.7.2.2 Singling Out Concentric (Ring) Objects

Ring structures (RS) have been recognized as one of the major features on the Earth's surface (Khain 2000). These structures can be formed by a variety of crustal tectonic processes (terrestrial), or they can be impact structures (extraterrestrial); archaeological RS are a separate type (Fig. 3.34). According to the scheme (Fig. 3.34), terrestrial RS maybe subdivided into four general classes: magmatic, tectonic, metamorphic and erosional.

It is well known that numerous commercial petroleum and mineral deposits are associated with RS (e.g., Khain 2000; Pirajno 2000; Gudmundsson and Nilsen 2006). Concentric zonal structures such as ring and pipe-like shapes (such as diatremes, necks, collapsed caldera, ring and stock-like intrusive bodies and zones of their contacts) may contain deposits of diamonds, rare metals and gold, copper and other metals. These deposits differ in terms of their origin and type (magmatic and hydrothermal, pyrite, skarnous and carbonatite) but frequently have the same morphology. Oil-and-gas deposits are also associated with volcano-

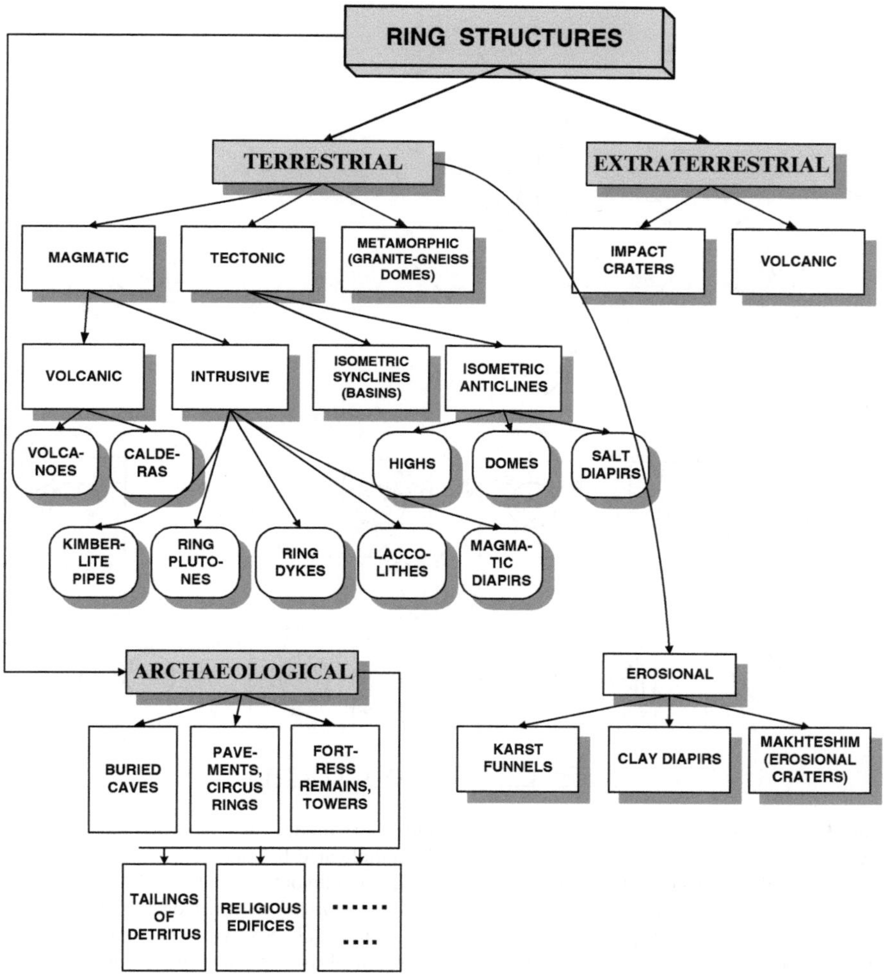

Fig. 3.34 Classification of Ring Structures (After Eppelbaum et al. 1998 and Eppelbaum 2007a).

tectonic structures and brachy-forms close to isometric sections such as brachy-folds, salt domes and diapires. However it is difficult to single out these objects in complex geophysical fields in mountainous regions, especially when the commonly employed rectangular network of measurements is used. Furthermore, anomalies caused by annular structures need to be differentiated from the features of the field on the periphery of stock-like bodies (central type volcanos, stock-like intrusives). For this purpose a special method for distinguishing concentric structures based on summing the horizontal gradients of the field using a circular graticule (zone chart) was developed (Khesin et al. 1996).

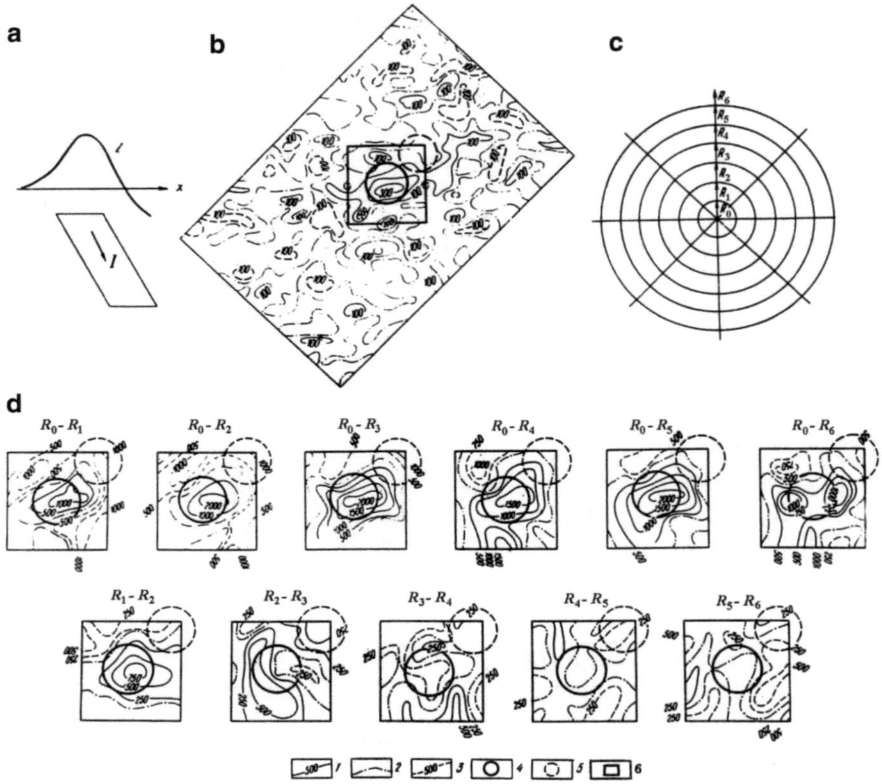

Fig. 3.35 Singling out a concentric (*ring*) structure: (**a**) model field of an *inclined circular cylinder* calculated along a profile, (**b**) model field complicated by random noise (in plane), (**c**) apparent graticule for concentric structure selection, (**d**) singling out a model body by summing the horizontal gradients of the field within apparent *circular graticule zones*. Isogams of the model field (**b**) and the sum of its gradients (**d**) in conventional units: (*1*) positive; (*2*) zero; (*3*) negative; cylinder edge projection: (*4*) upper; (*5*) lower; (*6*) contour of the portion treated on the (**b**)

Horizontal gradients are determined by the graticule radii drawn at intervals of 45°. When summing the gradients in various directions, the presence of a circular feature is highlighted and other signals are weaker. Here the correlation of the sum of gradients (or the average gradient) for a circle with radius R_n and a ring external to this circle limited by R_n and R_{n+1} radii makes it possible to determine whether the circular feature reflects a centric or a ring structure. The sum of gradients inside a circle tends to zero in the absence of a centric texture.

This method was successfully tested on a model of an inclined circular cylinder magnetized along its dip (Fig. 3.35a). The field of the cylinder was complicated by normally distributed random noise comparable in intensity to the useful signal. A random value with zero average and a standard deviation equal to 0.3 of the anomaly amplitude over the cylinder taken from the table of random numbers

(Krumbein and Graybill 1965) was added to the model field in all the points of its assignment.

Figure 3.35b presents the total field produced in such a way. A circular graticule was mounted on it; gradients were determined along the radii and subsequently summed over the circle and the ring zone external with respect to this circle. The target was singled out by the maximal horizontal gradient, whereas an irregular field was observed in the remote ring zones.

This example was described purely for purposes of illustration. This algorithm traces "apparent isolines" in response to the type of geophysical field. The relationships between the field gradients for various directions can be used to determine the criteria for the presence of a concentric structure and also to define it as a circle or ring.

A modification of this approach – computing vertical gradients of geophysical fields – was suggested in Eppelbaum et al. (2008).

3.7.2.3 Main Specificities of Transformation and Correlation Techniques

Singling out a hidden body is fairly often reduced to revealing a weak anomaly in the geophysical field. In other words, is said to have a mean square anomaly close to the mean square (dispersion) of random, normally distributed, and centered (zero mean value) noise. Frequently an anomaly is thought to be weak if its maximum amplitude is equal to three times the standard deviation. Any component of the field, its characteristics (ruggedness or dispersion), and the intersection of anomalies with different strikes can be considered an anomaly. In the latter case, the distribution of two components of the same field is studied; i.e. an anomaly in the 2D space of indicators. Calculating the parameter I_{cr} (I_p) to localize intersection nodes according to the data in one method (see Sect. 3.7.2.1) is a special procedure employed in the integrated interpretation described below.

Useful anomalies should be singled out against the background of regular field noise caused by the sources, which are of no interest to the researcher, and random noise due to technical errors, and fluctuations conditioned by minor changes in the medium. This is done by smoothing, averaging or upward continuation, or computation of local and circular (differential) anomalies, and horizontal and vertical gradients of the geophysical fields (Khesin et al. 1996; Eppelbaum et al. 2008). The notions of anomaly, regional background and random noise are relative. This is reflected in the initial model of the medium. In different problems, the same anomalous field components can be considered as noise, useful local anomalies, or alternatively a regional background. As a result of various transformations and correlation analysis, some components of the observed fields are singled out, and others, which are of no value for the given geological problem, are suppressed. These conversions, employing various statistical properties of useful signals and noise are aimed at solving problems of frequency or azimuth filtering.

Conversion techniques are based on integrating and calculating the moments of the observed field distribution. For instance, upward continuation is done by integration within certain limits. In this procedure the field values are multiplied

Table 3.8 The main types of transformations

Problem	Solution Method
Singling out regional isometric anomalies	Averaging (Tikhonov and Bulanzhe 1945)
	Sliding averaging
	Quantiles (including the median, see Borovko 1971)
	Upward continuation
	Smoothing with polynomials
Singling out local isometric anomalies	Higher order derivatives (Logachev and Zakharov 1979)
	Variations (Griffin 1949; Andreyev and Klushin 1962)
	Downward continuation (Strakhov 1960)
Revealing linear anomalies of a non-unidirectional type	Horizontal gradients
	Anisotropic graticules
	Self-setting filtering (Demura et al. 1974)
	Determination of predominant strikes
Singling out and tracing anomalies of approximately known shapes or strikes	Inverse probability
	Cross-correlation (Shaub 1963)
	Sign correlation (Alexeyev and Khesin 1971).
Revealing hidden nodes of anomalies of different strikes	Detection of structure intersections (Khesin 1976)

by weighting coefficients, which tend to diminish with increasing distance from the point where the transformation (convolution) takes place. Computing correlation functions is similar, when the values of the observed field (autocorrelation) or the field on the neighboring profile (cross-correlation) or the ordinates of the given signal (inverse probability method) serve as weighting coefficients.

The main types of conversion are listed in Table 3.8 with references to the first proposals which were then widely adopted with the advent of powerful computer technologies.

A set of results obtained by various transformations makes up the indicator space to be used in further analysis.

In most cases, any type of transformation may be used. A clear physical sense of the transformation by upward continuation was suggested by Andreyev (Andreyev and Klushin 1962), and its comparability with airborne surveys testifies to the advantages of this method. However, the discarded residual term in Poisson's integral may considerably affect the results.

In informal calculations (which are reasonable when the dataset is small) the type of conversion is governed by the fundamental advantages of each method. The median method is attractive from this standpoint. In fact, the assumption of compensation of random deviations from the mean within the limits of the transformation cell (averaging method) may not hold true in smooth fields, but for sharply varying fields polynomial smoothing cannot be used. In the sliding average method, as shown Borovko et al. (1969), the findings are affected to a greater extent by the anomaly amplitude than in the median method.

In order to single out general features of geophysical fields (and the height field, as well) it is often advisable to calculate tertiary indicators, such as the length of dislocations emerging from the results of the slide window (Khesin and Metaxas 1974).

To reveal local anomalies, certain methods compute higher order derivatives. This is the equivalent of the "difference" bodies developed by Logachev (Logachev and Zakharov 1979) as well as the variation method in 2D (Andreyev and Klushin 1962) and 3D (Griffin 1949) versions. The localization of anomalies in the analytical continuation downward can use the methods described in Strakhov (for instance, Strakhov 1976) and other authors. As the methods of analytical continuation are usually characterized by high sensitivity to surface heterogeneities, they are more suitable for plane conditions. The influence of oblique polarization is essential for their application.

Local anomalies are most often computed by the available data on regional anomalies. The most useful in this respect is the "ring" method involving the difference between the regional anomalies computed with different averaging radii. In this way, high-frequency and low-frequency components are filtered simultaneously, and the presence of anomalies of the desired classes can be determined (with the known frequency characteristic). Ring (difference) anomalies authorize important interpretative conclusions (Ismail-Zadeh and Khesin 1989b).

The localization of anomalies caused by extended bodies of various strikes calls for the application of anisotropic conversions. To do this, it is desirable to have information on the variation of the autocorrelation radius τ_e in different directions. The data on the τ_e value within localized areas of the stationary field are also employed in regioning. Given a favorable geological situation, they can be used to estimate the mean depth of anomalous bodies.

A method for sequential correlation and periodogram analysis of geophysical fields (revealing the latent periodicity) was developed by various authors for profile variants. The 2D autocorrelation analysis is employed for areas. In this case, using the self-setting filtering method (Demura et al. 1974), the sliding window dimensions are defined by calculating the 2D autocorrelation function. The method consists of orienting the window with respect to the direction of the profiles to reveal anomalies in different strikes in the area under survey.

The first methods developed were inverse probability and cross-correlation, which formed the basis for various commonly used variants of self-setting filtering. The inverse probability method (Demidovich 1969; Nikitin 1986) provides effective localization of weak anomalies of a specified form. In order to determine roughly the position of the expected anomaly $a(x_i)$by the observed field $U(x_i)$ corrected for the zero average, it suffices to plot the correlation sums of the type

$$S = \sum_{i=1}^{n} a(x_i)U(x_i). \tag{3.80}$$

The number of points n in the processing interval is determined by the width of the anomaly $a(x_i)$ and the processing step. Calculating the sums of the correlations is analogous to the method used to find the sliding average weighted by the expected anomaly ordinates.

The plausibility ratio (likelihood coefficient) is computed by the formula

$$\wedge_j = \exp\left[-\frac{1}{2\sigma^2}\sum_{i=1}^{n} a^2(x_i)\right]\exp\left(\frac{S}{\sigma^2}\right), \tag{3.81}$$

while the posterior probability of localizing the anomaly is found by

$$P_j(a>0|U_1, U_2, ..., U_n) = \frac{\wedge_j}{1+\wedge_j}, \tag{3.82}$$

where σ^2 is the noise dispersion, j is the number of the observation point (selected point).

It can easily be shown that the first exponent in expression (3.81) is a constant determined by the square of the target anomaly. If $\wedge_j > 1$ (or $P_j > 0.5$), then the anomaly is present here.

In many cases it is assumed that noise is normally distributed and not correlated to the geophysical signal. Analysis of geophysical data indicates that these assumptions are often erroneous (e.g., Eppelbaum and Mishne 1988; Bashirov et al. 1992; Eppelbaum and Kardashov 2001; Kardashov and Eppelbaum 2008).

Let us consider a trivial case. Sometimes a correlation between the useful signal and the dispersion of noise is observed (e.g., Eppelbaum 2011a). In this case a cross-correlation function B_{sn} between the signal and noise can be expressed as

$$B_{sn}(m) = E[(S_m - E(S_m))(n_m - E(n_m))], \tag{3.83a}$$

where E is the statistical expectation, S_m and n_m are the expected signal and noise, respectively.

Equation 3.83a can be rewritten as

$$B_{sn}(m) = E[S_m(n_m - E(n_m))]. \tag{3.83b}$$

By considering the properties of the dispersion as a random quantity, Eq. 3.83b can be transformed as

$$B_{sn}(m) = E\left[S_m\left(n_m - E(n_m) \pm \sqrt{D_n^2}\right)\right], \tag{3.84}$$

where D_n is the dispersion.

Taking into account that

$$E\left[S_m\left(n_m - E(n_m) - \sqrt{D_n^2}\right)\right] = 0,$$

Eq. 3.83b may be represented in the following form

$$B_{sn}(m) = E\left(S_m\sqrt{D_n^2}\right). \tag{3.85}$$

Thus, the cross-correlation function between the useful signal and noise dispersion may be described by Eq. 3.85. This relationship between the signal and noise makes it possible to utilize these features in a physical-geological model.

Studies have shown that the determination of a signal's position on a profile after processing is weakly dependent on the anomaly configuration details. More detailed results can be obtained using a bell-shaped (anomaly from many model bodies) or a rectangular impulse.

The cross-correlation method can reveal anomalies with the same direction. A modification of this method makes it possible to single out the systems of magnetic and gravity anomalies of different strikes (Kogan 1975).

The correlation function maximum cannot be assigned to a certain point on the profile. Therefore, Shaub (1963) suggested a modification in which the highest correlation is observed for a certain point in the survey sheet. It is expected that when averaging the product of the U_1 and U_2 field values for neighboring profiles 1 and 2, all random effects will tend to 0 in a fairly wide interval, and the useful signal effect can be singled out given their common source. The type of signal can also be determined by the kind of cross-correlation extremum $K(x)$: the maximum of $K(x)$ corresponds to an even signal (anomaly from a steeply dipping body, dike), and the minimum of $K(x)$ corresponds to an odd one (ledge anomaly).

The values of the $K(x)$ function referring to the middle of the j-th interval in the point between two neighboring profiles are computed by the formula

$$K_j(x) = \frac{1}{2n+1}\sum_{\frac{\tau}{\Delta x}=-n}^{n} U_1(x_i - \tau)U_2(x_i + \tau), \tag{3.86}$$

where Δx is the measurement step and τ is the displacement.

The cross-correlation method is approximate because some of its underlying assumptions (mainly that the cross-correlation function of the anomaly and noise is equal to 0 in the averaging interval) and the possibility of spurious extrema. Therefore, Alexeyev and Khesin (1971) proposed another approximate, but simpler technique of sign correlation, in which instead of the expression (3.86), the sign-correlation function is computed:

$$\omega_j(x) = \frac{1}{2n+1}\sum_{\frac{\tau}{\Delta x}=-n}^{n} \mathrm{sgn}U_1(x_i - \tau)\mathrm{sgn}U_2(x_i + \tau), \tag{3.87}$$

where sgn is the sign of the U_1 and U_2 fields for neighboring profiles reduced to the zero average value. All the other designations and the method of computation are identical to expression (3.86).

The calculations are highly efficient since they are reduced to the multiplication of signs followed by the summing of the results. Through the information measure, the cross-correlation method can be applied not only to the correlation of anomalies of the same type on parallel profiles, but also to reveal signals from a common source on one profile by different fields (Khesin et al. 1996).

Although the effectiveness of the methods described above is well known, in each case the conversion parameter (averaging radius, height of continuation, dimension of the sliding interval) needs to take geological and geophysical conditions and the aim of the investigation into account. In a series, the mean intensity of a geophysical field is fairly homogeneous, but is heterogeneous in terms of variability in local areas. Under such conditions amplitude filtering is ineffective and, therefore, methods for deriving information that exploit the differences in other characteristics of the targets, such as computation of the field ruggedness, predominant strike of isolines, entropy, and singling out the nodes of intersecting structures must be employed.

For qualitative interpretation of field isoline charts observed in mountainous regions, the field should be continued along the vertical without any displacement, since fictitious bodies are possible here, and information on the real source bodies cannot be determined before quantitative interpretation. In contrast, it is more expedient to analyze the spatial field structure on an inclined profile in inclined coordinates, thus making field continuation along the normal to the profile. This method yields the correct mutual arrangement of inclined continuation levels, and transforms different levels, which is not the case for the previous variant.

3.8 Integrated Interpretation

The majority of specialists discuss the conventional nature of the notions of qualitative and quantitative interpretation. In terms of analogy and modeling theory (Batoroyev 1974; Kondakov 1975), there are different levels of inference by analogy. Intuitive analogy is predominantly used in qualitative interpretation when a set of empirical facts is considered from the point of view of qualitative relations. The use of an empirical field/geological object relationship obtained by summarizing the previous experience makes it possible to relate the experimental field to the geological structure of the area under study and to come to new conclusions concerning its structural features. The new information thus obtained is immediately described in geological terms. When the amount of prior information is sufficient, qualitative relations can be expressed in a quantitative (statistical) form. In quantitative interpretation, a model analogy is usually employed; i.e., the information obtained from a model study is applied to the area under investigation.

The most complete description of the interpretative process structure was given by Strakhov (1976). An interpretation process may be roughly subdivided into the following stages: (1) summarizing prior information; (2) sequential analysis, and (3) geological synthesis.

3.8.1 Combined Information Formalization of Geophysical-Geological Processes

Theoretical errors in determining parameters on the basis of regression equations are characterized by argument variability and measurement errors (Akselrod and Putkaradze 1979). Theoretical errors can be estimated by analogy to integrated interpretation, if we test the hypothesis that a parameter can be classified in a certain class (or a group of similar values) of parameters. Integrated interpretation errors are the most damaging (Khesin et al. 1996) and can introduce highly incorrect data as regards the targets.

Integrated interpretation errors vary with the type of specific algorithm. For example, in the "Linear solving function" algorithm (Fotiadi 1970) that uses unequal covariance matrices, the classification error depends on the value of d. The latter is an analogue of the generalized distance between the compared images for identical covariance matrices. The probability of correct classification P is given by

$$P = \Phi(d),$$

where $\Phi(d)$ is a normal distribution function.

Zadeh (1983), who revolutionized by introducing the fuzzy logic approach to the analysis of complex systems divided all systems into two types: technical ("type 1") which applies equations, and expert ("type 2") where experience, knowledge and man-made solutions dominate.

The high complexity of expert systems led to the development of a special approach known as fuzzy logic for the analysis of similar systems. These systems do not only assume the usual quantitative variables, but also integrate qualitative, so-called linguistic variables, which may be words and expressions in natural or artificial languages. Geological-geophysical investigations are typical examples of an applied field where the greater part of most solutions depends on expert opinion.

All forms of geological-geophysical investigations take place in a definite sequence in time and space. In investigations, different geological methods are employed (geophysical investigations, geological mapping, geochemical analyses, mining works, drilling, etc.). The final goal of a geological-geophysical examination is the most accurate identification of targets in a given area within the limitations shown in Fig. 3.36.

The problem of rational integration of geological methods can be resolved through consideration of the following criteria:

1. Necessary expenditures for successful integration (cost criterion C),
2. Necessary time for integration (time criterion T),
3. Information value of the integration (informational criterion Π).

Criteria C and T can be easily determined by direct calculation whereas evaluation of criterion Π calls for complex analysis.

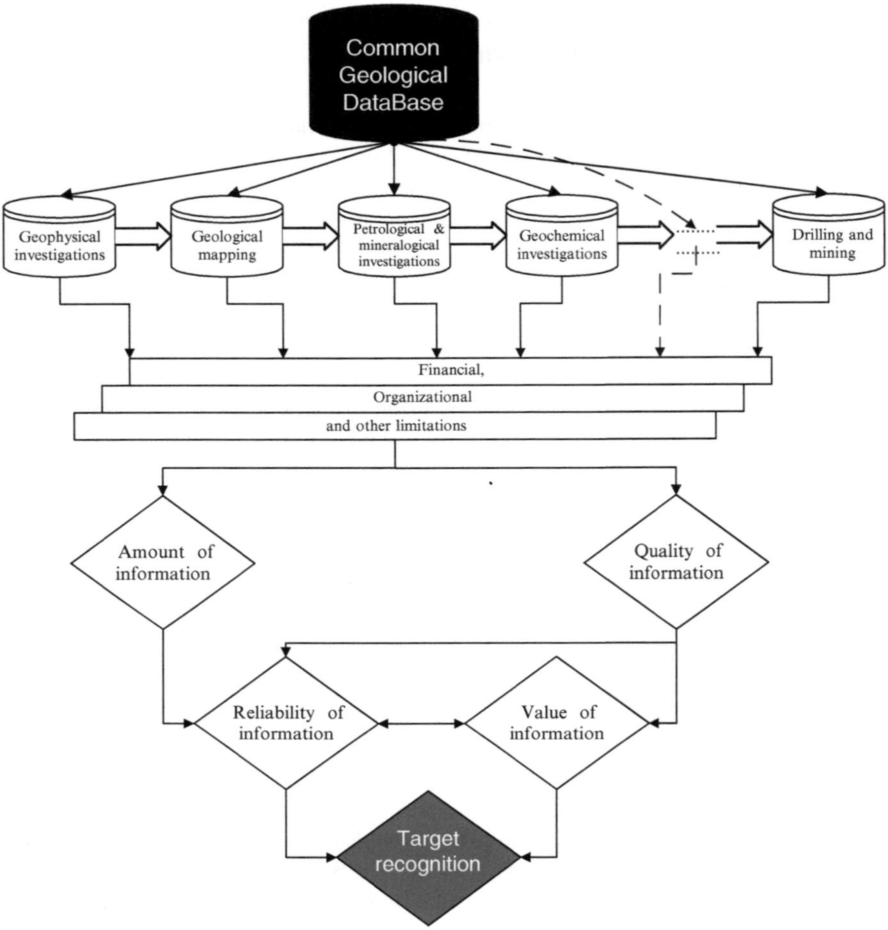

Fig. 3.36 Ways to evaluate geophysical-geological prospecting information (Eppelbaum et al. 2003b)

All available geological/environmental information can be represented in a classic three-level model composed of: (a) a syntactic component – the amount of information; (b) a semantic component – the nature of the information, and (c) a pragmatic component – the value of the information (Eppelbaum et al. 2003b).

This model is based on the fundamental terms of information theory combined with a structural (hierarchical) approach. This approach defines each geological indicator as a structure that reflects a set of typical situations. Next, depth of search is estimated and calculated using the informational approach. This strategy provides a quantitative calculation and effective control of geophysical-geological studies.

The main logical-heuristic model for geological information can be described by the following formula:

$$\Pi = Q \cup R \cup V, \tag{3.88}$$

where Q is the quantitative estimation of information, R is the estimation of informational reliability corresponding to the semantic criterion, V is the estimation of the informational value according to the pragmatic criterion, and $\cap$ is the symbol of unification.

In real conditions many random factors affect the results obtained by a set of geophysical tools (means). One of the essential problems is that it is impossible to obtain a satisfactory formalized description of factors influencing the results of local determinations. Similar situations are known to exist in decision making theory where a full mathematical formalization of a problem is complex. However using expert methods in many situations (a set of logical and math-statistical procedures) can contribute to greater effectiveness.

For purposes of illustration, take the reliability of geophysical prospecting at the level of local determination and the reliability of information obtained by separate means or a set of means. The main aim of such a study is a problem of determination of the set of means making up the notion of "geophysical prospecting" (relative to some fixed feature) by the assumed reliability of the means, (in contrast to Khesin and Eppelbaum (1997), where mainly statistical criteria were analyzed). Solving this problem will yield the most optimal combinations of geological means for different physical-geological conditions. Determination of the reliability of individual means can be obtained through control observations or on the basis of expert methods.

The notion of reliability of means S relative to feature R is a quantitative measure of the frequency of associating feature R (obtained by the use of means S) with its real value (Eppelbaum et al. 2003b). When feature R is determined by a set of means $S = \{S_i\}_{i=\overline{1,n}}$ has the following features. Any series of observations (investigations) found by using by a set S is defined, obviously, not by one alternative r_τ, but by a set of alternatives $r_{\tau 1}$, $r_{\tau 2}$, ..., $r_{\tau n}$ which, broadly speaking, are different. In this case one value of feature R is not generated by set S. This leads to the following question: having a set of obtained values $R - r_{\tau 1}$, $r_{\tau 2}$, ..., $r_{\tau n}$ which alternative r_τ needs to be recognized as value of feature R (Eppelbaum et al. 2003b).

Selection of the best hypothesis should use an algorithm (rule). Formally such a rule may be considered as mapping $\aleph$ by a set of possible indications of means $r_{\tau 1}$, $r_{\tau 2}$, ..., $r_{\tau n}$ to the set of values of features $\{r_t\}_{t=\overline{1,k}}$:

$$\aleph : \{(r_{\tau_1}, r_{\tau_2}, ..., r_{\tau_n})\}_{(\tau_1, \tau_2, ..., \tau_n)} \to \{r_t\}_{t=\overline{1,k}}. \tag{3.89}$$

If we associate some fixed rule $\aleph = \aleph(S, R)$with each feature R and set of means (Eq. 3.89), then S uniquely determines R. More specifically, the series of local observations of feature R defines the set of alternatives $r_{\tau 1}, r_{\tau 2}, ..., r_{\tau n}$ and rule $\aleph$

assigns the set one single value $\aleph(r_{\tau 1}, r_{\tau 2}, ..., r_{\tau n})$ of feature R (Eppelbaum et al. 2003b).

Formalizing problems associated with reliability is mathematically complex; if $\aleph_1$ and $\aleph_2$ are two different rules, there is no simple method for their comparison.

However a probabilistic approach to the definition of reliability can formulate a criterion for comparison of rules and simultaneously solve the problem of rule selection (in the sense of a criterion).

Assume that the real value of feature R is r_1. Determining R by means S may differ from r_1 because of inaccuracy in obtaining the results. Obviously, a set of possible indications of means S may be described using a probability distribution:

$$P_{11} = P\left(r_1^{\mathrm{o}} \middle| r_1^r\right), P_{12} = P\left(r_2^{\mathrm{o}} \middle| r_1^r\right), ..., P_{1k} = P\left(r_k^{\mathrm{o}} \middle| r_1^r\right), \tag{3.90}$$

where $P_{1\tau} = P\left(r_\tau^{\mathrm{o}} \middle| r_1^r\right)$ is the conditional probability of the results of determination r_τ if the real value of feature R is r_1 (indices “o” and “r” designate the “observed” and “real” values, respectively).

Generally speaking, value r_2 of feature R corresponds to another set of probabilities: $P_{21}, P_{22}, ..., P_{2k}$. In a typical case, probabilities $P_{t1}, P_{t2}, ..., P_{tk}$ depend on t (i.e., on the real value of feature R). We consider a matrix of conditional probabilities:

$$P_{tr} = P\left(r_\tau^0 \middle| r_t^r\right)_{t,\tau} = \overline{1}, \overline{k} \tag{3.91}$$

for each pair (S, R) by introducing expert methods since logical methods are practically impossible:

$$\Re(S, R) = \begin{pmatrix} P_{11} & \dots & P_{1k} \\ \dots & \dots & \dots \\ P_{k1} & \dots & P_{kk} \end{pmatrix}. \tag{3.92}$$

Note that the matrix depends not only on S and R but also on the concrete physical-geological conditions of geophysical prospecting. However, the relationship of $\aleph(S, R)$ to different geophysical (geological, mining, geochemical, etc.) factors is not discussed here since we assume the factors in real conditions are fixed.

It may seem that solving matrix $\aleph(S, R)$ containing k^2 numbers is practically insolvable. However, in real conditions the number of independent elements in this matrix is greatly reduced (Eppelbaum et al. 2003b). The application of the approach above leads to an optimal scheme for obtaining geophysical information on the medium that integrates all geological methods (surface mapping, drilling, geochemical analyses, etc.).

A flow-chart of this integrated comparative geophysical-geological analysis is presented in Fig. 3.37.

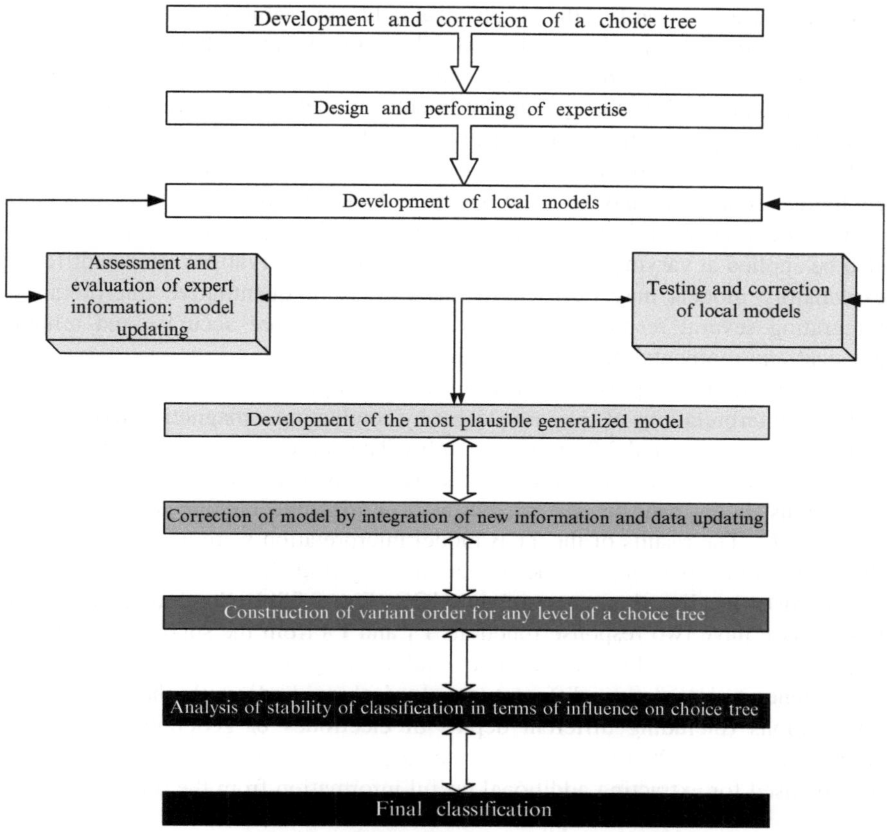

Fig. 3.37 Flow-chart of the geological-geophysical investigation process using information comparative analysis (Eppelbaum et al. 2003b)

3.8.2 Multimodel Approach to Geophysical Data Analysis (on Example of Magnetic Data Analysis)

The multimodel approach to geophysical data analysis may be clearly illustrated in the example below on detailed magnetic prospecting. The quantitative interpretation of magnetic anomalies (as is the case for many other geophysical methods) was traditionally oriented to a single model for buried object identification. In the case of several hypotheses relating to the parameters of the body causing the disturbance usually only one model was selected that roughly characterized the object in domain Ω_x of k-dimensional space of the physical-geological factors. However many geological features are disturbed by different types of geological, geodynamic activity as well as various environmental processes.

Additional sources of noise affecting interpretation include rugged terrain relief, oblique polarization of geological objects and a heterogeneous host medium. As a result response function Γ_i – the geophysical field – may ambiguously represent the target. Therefore, domain Ω_x should be divided into several subdomains $\Omega_1, \Omega_2, \ldots, \Omega_m$ such that a single model dominates in each (Eppelbaum 2010a). This leads to the development of *m* physical-environmental models of the same target, each corrected for separate subdomains $\Omega_1, \Omega_2, \ldots, \Omega_m$.

For instance, the remains of an ancient wall (in archaeogeophysics) in specific geological conditions may be identified as a thick bed (magnetic field), a *sphere* (gravity field) and/or a *horizontal plate* (resistivity). The multimodel approach can also be applied at varying levels of geophysical field observations. Hence different explanatory models may be used in the process of quantitative interpretation. Integrating several response functions Γ_I, yields a more accurate and reliable physical-environmental model of the buried target.

The simple model presented in Fig. 3.38 exemplifies the exploitation of two different interpretations of the same target by conducting a magnetic survey at two different levels (0.1 and 3.0 m, respectively). The survey at the 0.1 m level shows that it is a typical thick bed (TKB) model (Fig. 3.38a) but at the 3.0 m level, the anomalous body can be interpreted as a horizontal circular cylinder HCC (Fig. 3.38b). The results of the TKB model interpretation were used to determine the center of the upper edge of the anomalous body (Fig. 3.38a) and the HCC model was used to localize the center of the HCC (Fig. 3.38b). Combining these two models (we have two response functions Γ_1 and Γ_2 from the subdomains Ω_1 and Ω_2), yields a combined model of the anomalous body (Eppelbaum 2010a).

In general, integrating different geophysical methods with their multilevel observations (including different depths of electrodes or geophone grounding) (Eppelbaum et al. 2010) can generate a large number of combinations, which may be used for extracting additional useful information from the observed data.

3.8.3 Variants of Integrated Interpretation

There are different approaches to integrated interpretation, including those using traditional visual comparison of fields, and quantitative expression of the specificities of fields and media. The benefits of integrated interpretation become clearer when moving from qualitative integration to quantitative integration.

Quantitative integrated interpretation usually involves a quantitative estimation of parameters by a set of data. This concept appears to be more general; it primarily includes the determination of the type of target (quality) by a set of data employing quantitative criteria. There are different types of quantitative criteria used to evaluate the nature of the objects (Khesin 1976; Borovko 1979).

Integrated interpretation can be deterministic, probabilistic and mixed (probabilistic-deterministic). The latter actually includes elements of both types, as in the case of single-method interpretation. An interesting sophisticated description of the probabilistic-deterministic approach for a typical case was presented in Malyshev (1981).

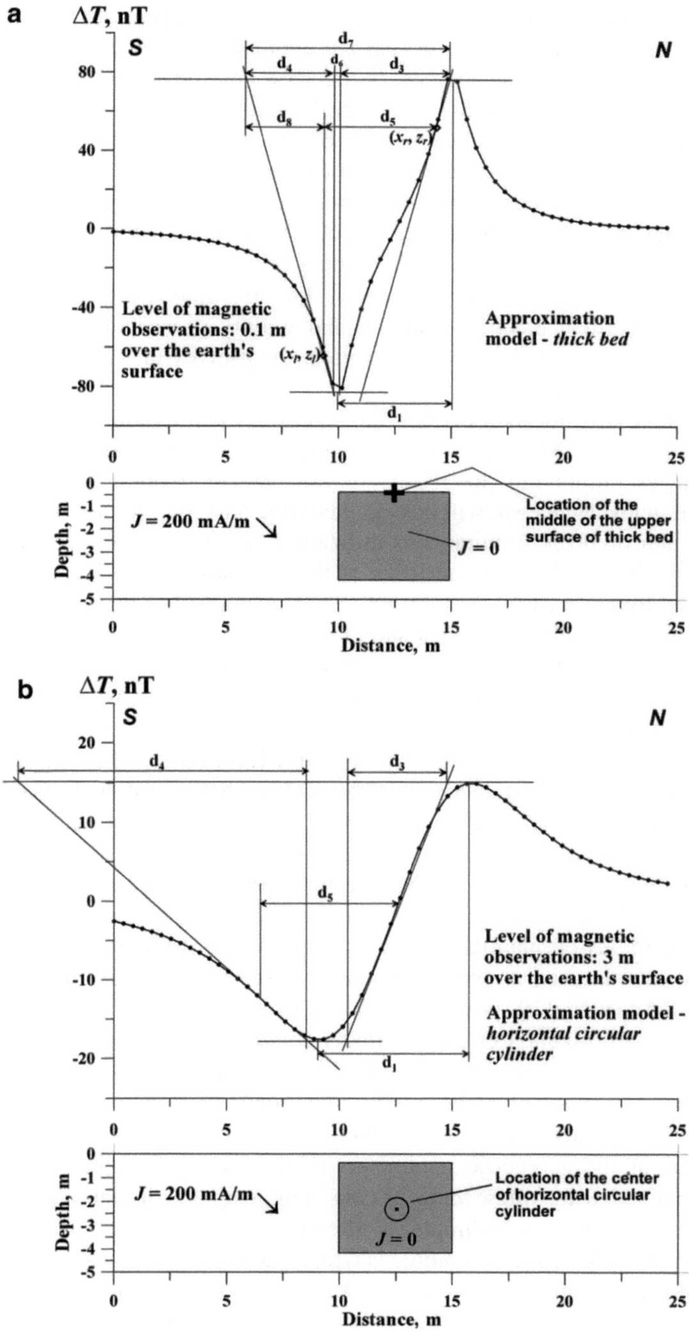

Fig. 3.38 Two-level observations with two different interpretation models: (**a**) model of a thick bed (TKB), (**b**) model of a horizontal circular cylinder (HCC). Magnetization is denoted by J

Integrated interpretation based on the deterministic approach can be successful provided the researcher is certain that the anomalies identified through different methods are from the same sources. In this case Poisson's law is applied between the magnetic and gravity fields (Napadensky and Sheremet 1976) and the linear programming technique (Shalayev 1972; Bulakh et al. 1976).

The most universal tool for deterministic interpretation is to select a geophysical field to model the medium obtained from data from another geophysical method, and compare the selected field with the observed one. This sequential integrated interpretation can yield results which cannot be derived from a separate study of data obtained through different methods. For example, by comparing the data obtained by vertical electric sounding and gravity prospecting, one can detect geoelectric horizons which do not manifest themselves in VES curves (Rempel and Parshukov 1976).

In a rough analogy with electric engineering, interpretation can take place in series or in parallel (Khesin 1981). However, a series of interpretation in with each of the stages employing the results of the preceding ones, actually interprets individual fields. Hence we will basically confine ourselves to the parallel interpretation, which involves simultaneous utilization of data obtained by different geophysical methods. The characteristics of this combined inversion of geophysical data can be found in Lines et al. (1988). This technique has potential as confirmed by investigations both in the West and in the former Soviet Union (Golizdra 1980; Starostenko et al. 1988). Other promising approach to integrated analysis of different geophysical fields involves a combination of advanced wavelet methodologies (Eppelbaum et al. 2011a).

Certain additional information can be derived from parallel integrated interpretation (these may be tentative values of vertical and horizontal thickness and the dip angle), if different approximations of a source body are admissible when interpreting anomalies caused by diverse fields. This is seen in the model of a pyrite-polymetallic deposit in the Middle Jurassic shale strata of the southern slope of the Greater Caucasus (Fig. 3.39).

Rapid methods of quantitative interpretation make it possible to determine the following parameters: position of the mass center of the anomaly-forming body by the plot of Δg (model **a**), position of the upper edge by the plot of ΔZ (model **b**), position of the horizontal circular cylinder center in the upper portion of the orebody at the ground water level by the plot of self-potential field (model **c**). The specific models thus obtained reflect the contrasting nature of the physical properties of the target and the host medium. They allow for a fairly exhaustive description of the geometric parameters of the target (Eppelbaum 1984a). The model obtained using these data is consistent with the initial (theoretical) one (Fig. 3.39d). A similar example is shown for a pyrite deposit occurring in volcanogenic rock mass (such model is typical for the Lesser Caucasus) (Fig. 3.40).

The results of rapid inversion provide a basis for a first approximation of the medium. This model is then used in an iterative process for direct problem solution. Such a process of deterministic integrated interpretation is characterized by many special features and primarily utilizes gravity and magnetic fields. For this reason the

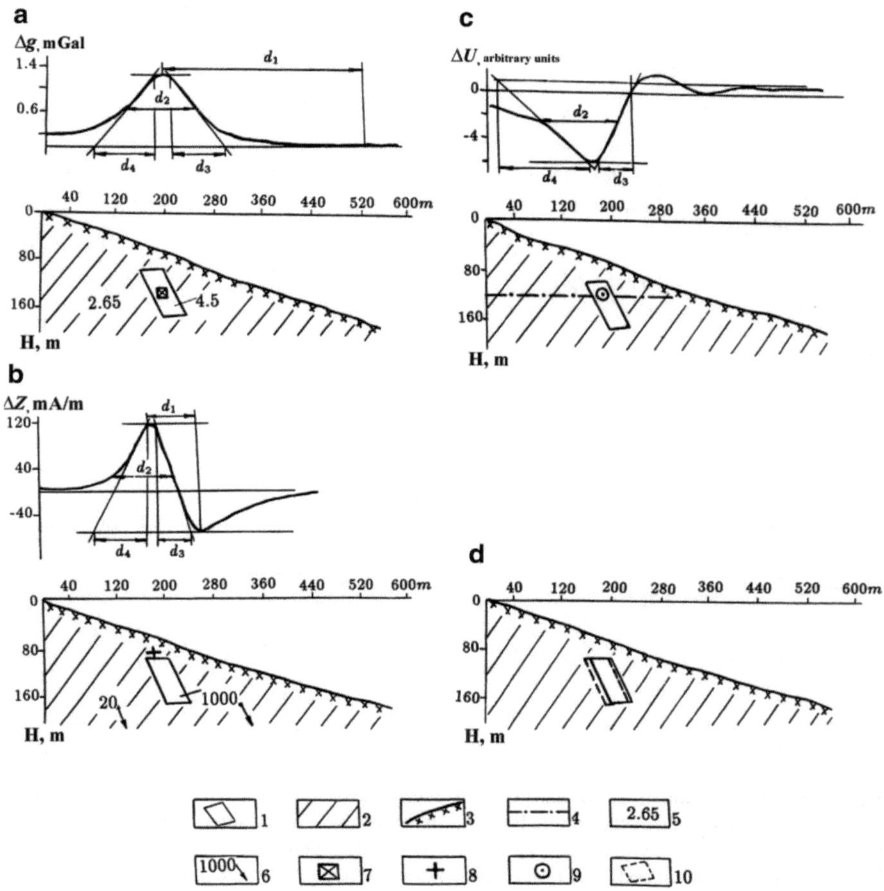

Fig. 3.39 Interpretation of Δg, ΔZ and ΔU_{SP} fields with deposits of the Filizchay type, under different approximations of the source body: (**a**, **b**, **c**) results of rapid interpretation of the model fields, (**d**) source body according to the integrated interpretation results (Eppelbaum and Khesin 1988). (*1*) source body; (*2*) host medium; (*3*) topography; (*4*) position of the groundwater level; physical properties: (*5*) density, g/cm^3; (*6*) magnetization, mA/m; (*7*) mass center of a circular horizontal cylinder by Δg plot; (*8*) mid-point of the upper edge of an inclined thin bed by ΔZ plot; (*9*) position of the center of HCC inscribed into the upper portion of the source body at the groundwater level by ΔU_{SP} plot; (*10*) contour of the source body obtained from the results of integrated quantitative interpretation

description of the final stage of interpretation sequence, i.e. integrated interpretation by physical-geological modeling of these fields is discussed separately in Chap. 4.

Regression-correlation integrated interpretation (e.g., drawing up structural maps from the gravity data on the basis of the correlation between these data and the position of seismic boundaries) is frequently used to solve problems in petroleum geology. This approach to interpretation became popular following a series of publications (e.g., Shraibman et al. 1977; Karatayev and Pashkevich 1986). This

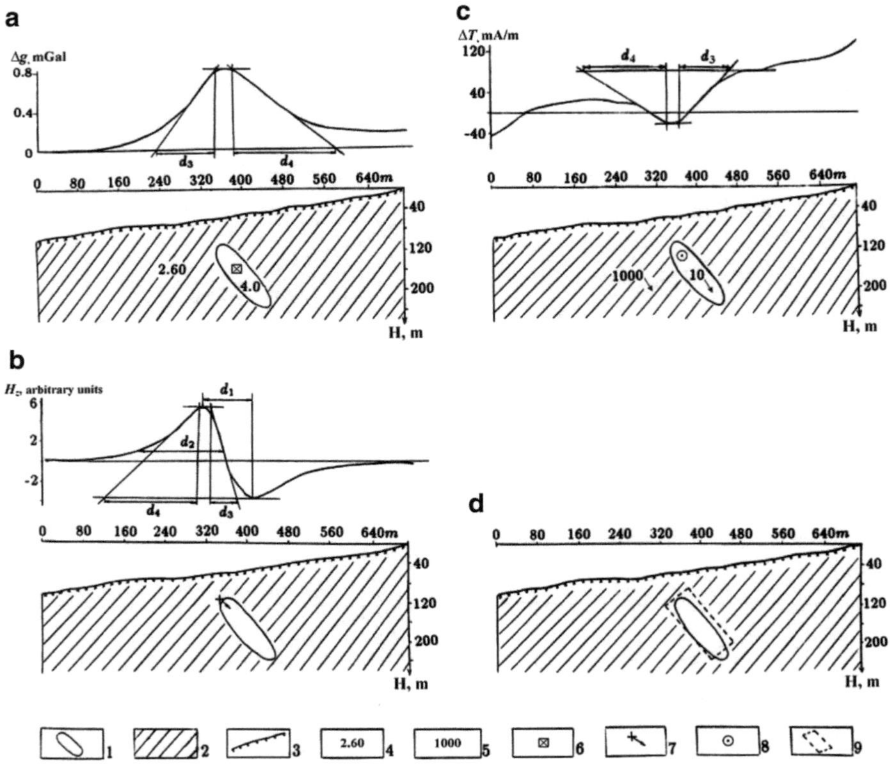

Fig. 3.40 Interpretation of Δg, H_z and ΔT fields with deposit of the Lesser-Caucasian type, under different approximations of the source body: (**a**, **b**, **c**) results of rapid interpretation of the model fields, (**d**) the source body according to the results of integrated interpretation (Eppelbaum and Khesin 1988). (*1*) source body; (*2*) host medium; (*3*) topography; physical properties: (*4*) density, g/cm^3; (*5*) magnetization, mA/m; (*6*) mass center of a circular horizontal cylinder by Δg plot; (*7*) mid-point of the upper edge and direction of an inclined thin bed by H_z plot; (*8*) position of the HCC center in the upper portion of the source body by ΔT plot; (*9*) the source body contour obtained by the results of integrated quantitative interpretation

kind of interpretation has features characteristic of probabilistic-deterministic interpretation. That is why the techniques based on computation of the correlation coefficient between the results obtained by different methods do not fall into this category if they have no data on the seismic boundary locations or other information obtained by the deterministic method. Rather, these techniques should be classified as information-statistical methods. For instance, Lyubavin and Shaub (1968) made use of the direct results of measurements to obtain the cross-correlation function of the electric and magnetic fields and its derivative during flight.

The most important problem in information-statistical (logical-statistical) integrated interpretation is determining the class of an object on the basis of

Table 3.9 Main types of quantitative integrated interpretation

Approach	Interpretation principle
Deterministic	Application of Poisson's relation between magnetic and gravity fields by joint analysis and other deterministic relations
Regression-correlation	Application of regression equations linking the results of geophysical studies with the known characteristics of targets
Information-statistical	Application of informational and statistical criteria to assign the objects under study to a certain class according to results obtained by different methods

Table 3.10 Varieties (groups of techniques) of information-statistical integrated interpretation

Group of techniques	Basic features	Techniques	Characteristic modifications (programs)
I	Learning on standard objects	Statistical	Discriminant functions (Fotiadi 1970)
		Empirical	"Crust-3" (Guberman 1987)
		Informational	Vysokoostrovskaya and Zelenetsky (1968)
II	Revealing anomalies and their groups	Filtration-correlational	Self-setting filtering (Detkova and Shopin 1969)
		Automatic classification	"Association" (Bugaets and Dudenko 1976)
III	Revealing targets with expected properties	Heuristic	Summation of complete gradients (Berezkin 1988)
		Logical-informational	Khesin (1969, 1976, 2005), Eppelbaum et al. (2003a)

probabilistic or informational criteria. The evaluation of the probable presence of geological bodies by formalized features facilitates decision-making based on the results of geophysical explorations. Under certain conditions it admits strict analysis of possible errors in the solution (Fotiadi 1970; Bugaets and Dudenko 1976).

The main types of quantitative interpretation are listed in Table 3.9.

3.8.4 Classification by Logical-Statistical (Information-Statistical) Techniques

There are three approaches to information-statistical interpretation: pattern recognition, classification of targets by association of indicators forming compact groups in the space of indicators, and identification of targets (classes of targets) with the expected properties. The first approach is limited by the presence of etalon targets. The second approach implies subsequent interpretation of the nature of each class, and, therefore, is applicable in the last stage of a thorough examination of all the specificities of an area. The third approach can provide a quick estimate of the presence of the target bodies on the reference profile or in the exploration area. Approaches **I**, **II** and **III** involve different groups of techniques (Table 3.10).

The techniques of group **I** listed in Table 3.10 are empirical or were termed heuristic (Fotiadi 1970). However, the more general notion of heuristic is more suitable for the techniques of group **III**. Empirical dependencies manifest themselves in statistical processing. Heuristic techniques use empirical dependencies along with theoretical abstractions (Raiffa 1968).

Clearly all the variants of integrated interpretation employ certain modeling concepts. This is well illustrated by the techniques of group **I**, where learning relies on natural standards. In the techniques of group **II,** field measurements of one or another standard are used. The techniques of group **III** are actually based on the utilization of predicted theoretical (model, ideal) standards.

3.8.5 Pattern Recognition by Standard and Control Sets of Targets

The best-known techniques of integrated interpretation are those based on the utilization of standard objects for supervised learning (known in Russian as "learning with a teacher") purposes. This, in fact, is the main limitation of the techniques of this group (group **I**), since there are very few standard large deposit targets, and a given area of investigations may completely lack such standards.

The techniques of group **I** include statistical, empirical and informational methods.

The statistical techniques (closer to game theory) are based on the statistical theory of decision making, which can lead to optimal solutions under conditions of uncertainty. These techniques include discriminant functions with linear and nonlinear rules, Shaw's algorithms and others (Duda and Hart 1973). Linear discriminant functions (Fotiadi 1970) are the best known.

The principal advantage of game theory approaches lies in the possibility to estimate recognition reliability theoretically. However, their application is restricted by a number of factors. In particular, application of discriminant functions calls for a normal distribution. This condition does not always hold in practice. Next, Shaw's algorithms can only be used when all the descriptions of standard samples have a non-zero probability of occurrence. These statistical techniques are highly efficient if independent indicators are employed. They techniques are totally useless when the data are represented solely by objects of a single class (e.g., ore deposits).

Empirical techniques make use of the decision rule (an integrated criterion), whose reliability is estimated by the number of empirical errors. Computer programs using such algorithms as "Crust-3", "Image-3", "Voting by deadlock testers" and others are based on this principle. They differ in the type of calculation, indicator selection and encoding, as well as the number of different classes and other features (Borovko 1972; Guberman 1987).

The "Crust-3" program implements Bongard's (1967) well known algorithm and classifies three indicator combinations that vote to attribute an object to a certain class after learning. The "Crust-3" program launched a number of similar programs. The advantage of this and similar programs is that they allow for associations between the data. However, these programs are difficult to apply in practice, as there are unsolvable theoretical problems, such as the choice of code and others. The "Crust-3" program, for example, is organized in such a way that the results obtained depend on the order in which the features were entered.

Approaches based on studying pattern recognition of distributed regional seismicity have been put forward by numerous authors (e.g., Keilis-Borok 1990; Molchan et al. 1990; Turcotte 1995; Chernobay and Gabsatarova 1999; Freund 2011).

If the data are independent, simple procedures to solve the recognition problem can be applied including selecting information-carrying indicators and generating an integrated criterion. A decision rule estimates the likelihood of prospects in the form of a function of the selected indicators. For example, the weights of different geophysical indicators of the target were defined by Golovin and Suprunenko (1971) as the ratio of the indicator distribution area to the total survey area. Borovko (1979) suggested more accurate definitions for indicator informativity in mineralization predictions. The integrated prospect criterion is defined as the sum of the informational weights in each cell.

Clearly geophysical work employing the techniques of group **I** has great practical benefits. However, the application of these techniques has certain shortcomings. The number of indicators and standards can at times reach a hundred or more. As pointed out by Duda and Hart (1973), the increasing number of indicators also increases the requirements for the amount of standard information. A great number of standards of the same type are only available in well-studied regions. However, in such regions quantitative prognosis seems less important than in regions that have not been studied properly. Small and non-commercial targets can prevail among standard ones, so similar deposits will be revealed as a result of predictions (Borovko 1979). As to large and unique deposits, which should be the main objectives of a search, their indicators can differ from those of standard deposits.

3.8.6 Classification of Targets into Compact Groups in an Indicator Space

Difficulties in the choice of reference objects have led to the development of Group II techniques based on the analysis of measured fields, anomaly correlations and differentiation between groups making similar contributions.

To single out anomalies, filtration-correlation techniques employ correlation analysis and different kinds of filters. In self-setting filtering (Demura et al. 1974) the anomaly parameters are not specified beforehand, but rather determined during processing. Areas where anomalies obtained by different methods form identical combinations are bounded by a common contour. Then a certain geological content is assigned to each contour. This method provides a vast amount of unbiased information. Nevertheless, because it is based on classification, it has certain limitations. Some anomalies singled out by this technique can be useless as far as predicting and prospecting problems are concerned.

Among the methods of automatic classification, Bugaets and Dudenko (1976) differentiated two large groups; namely hierarchical and cluster techniques. These serve to determine inter-group and intra-group characteristics of similarity, respectively. As a similarity measure, one can take the distance in a multi-dimensional space, the cosine of the angle between the vectors of indicators, or a potential, as in the potential function method. This method is known as a general method of constructing a differentiating surface by the data on standard images (supervised learning). However, it can be equally applied to dividing sets of objects into classes in unsupervised learning. Automatic differentiation of objects according to their typical indicators or their unique characteristics helps form an opinion as to their value. The outcome of application of automatic classification techniques depends, to a great extent, on the changes in the set of indicators, especially if they are numerous. Moreover, the analysis of the content of selected groups becomes computationally hard as in the case of filtration-correlation techniques.

3.8.7 Revealing Targets (Classes of Targets) with Expected Properties

Before moving to the techniques of group **III**, several issues should be discussed. Let us imagine a situation (which is practically nonexistent), where all the properties inherent to the objects to be classified are known exactly, and that the measurements and analysis of their results have been carried out without a single error. In this case an unambiguous division into classes is possible. Therefore, a deterministic solution of the recognition problem is a special case of the probabilistic one. By applying the above techniques, except the statistical ones, we can form an unambiguous opinion about the assignment of an object to one class or another. The theoretical probability of error in determining the nature of the object is taken to be zero. Some authors call such techniques logical (i.e., deterministic). The logical nature of the interpretation manifests itself when the techniques of group **III** are employed. However, under certain assumptions as to anomalies and noise, one can generate a decision rule with an estimate of the integrated interpretation reliability (Nikitin et al. 1971).

The techniques of group **III** are simpler than most of the methods listed above, but the estimates they yield are coarser. However when we want to single out, for example, the largest and richest ore-bodies against a background of noise, or identify typical ore controls such as deep faults, acid or basic intrusions, these techniques of geophysical data interpretation have proven to be highly promising, since they use the most common ("through") indicators for different kinds of targets.

One of the first techniques of integrated interpretation in group **III** was suggested in oil geophysics for direct prospecting for oil and gas by Medovsky (Berezkin et al. 1978). The technique (graphical integrated interpretation) and its digital modification were further developed by Berezkin (1988), who proposed a method of summation of complete normalized gradients.

In mining geophysics, following advances in integrated interpretation of geochemical data (Klichnikov 1970), the "reduction" of petrophysical or geophysical prospecting results was suggested (Vakhromeyev and Davydenko 1987), alongside the computation of a combined index ("index of contrast" and "generalized function").

The latter is the sum or product of centered and normalized anomalies of each indicator under examination. Good results have been obtained using this logical approach. The most precise quantitative content of this approach is obtained with summation or cross-correlation of information obtained by different geophysical methods (Khesin 1969, 1974, 2005).

At each observation point the amount of information I_i due to the application of the i-th method is

$$I_i = -\log P_j \tag{3.93}$$

or

$$I_i \approx \log\left|\frac{U_i}{\Delta U_i}\right|, \tag{3.94}$$

where P_j is the relative frequency of the j-th interval of the i-th indicator on the histogram of its distribution, U_i and ΔU_{i} are the amplitude and the error of this indicator's determination, respectively.

After summing the information elements which show a priori that the object of the desired class is present, random noise and components caused by different geological features are suppressed. To avoid singling out fictitious targets by the plot of $\frac{1}{n}\sum_{i=1}^{n} I_i$, which is possible in the case of a great amount of information contained in the data from only one or two methods, an additional integrated criterion should be computed. This depends on the number of significant indicators, if their relative influence cancels out:

$$I_{\text{integr}} = \sum_{k=1}^{\frac{n(n-1)}{2}} \frac{(I_p)_k}{(I_p)_{\max}}, \tag{3.95}$$

where I_p is determined using pair wise combinations of the results of n methods used by a formula of type (Eq. 3.78).

To avoid missing deep-seated objects, it is expedient in some cases to use relative frequencies of average values or average field estimates on a sliding average interval instead of P_j and U_i values, respectively.

Along with a simplified version based on summing information obtained from Eq. 3.95, the method has been realized as an "Integration" program. It computes the sums of information by Eqs. 3.93–3.95. I_{integr} is capable of revealing those objects characterized by the maximum number of indicators of different intensities and, at the same time, avoids missing an object which for some reason was not detected by any indicator. The combination of indices permits certain interpretative conclusions.

In practice I_i is usually replaced by the relative amount of information, also called the coefficient of informativity (Borovko 1979; Khesin 1974):

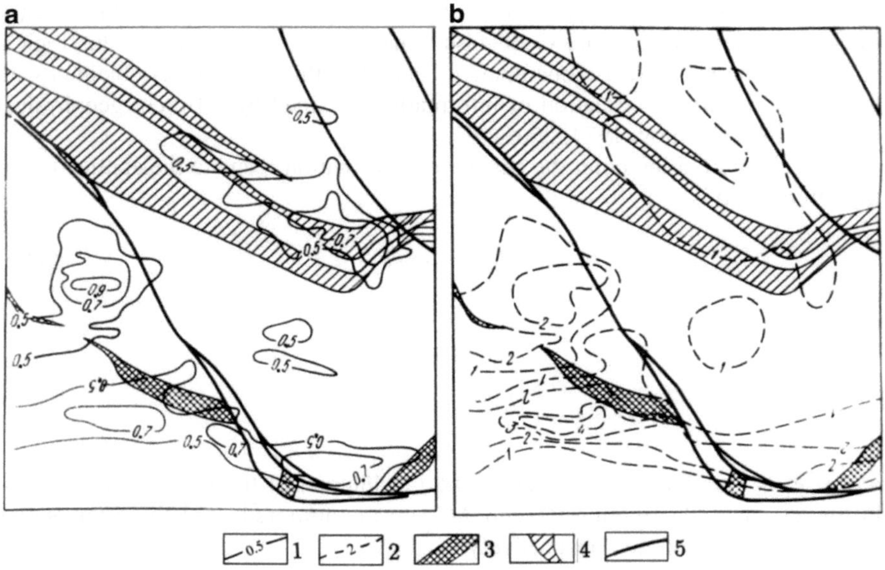

Fig. 3.41 Comparison of the indices $\frac{1}{4}\sum_i \frac{1}{4}K_i$(**a**) and I_{integr} (**b**) with ore zone location in the Katsdag pyrite-polymetallic deposit (southern slope of the Greater Caucasus). Isolines: (*1*) mean sums of informativity coefficients $\frac{1}{4}\sum_i \frac{1}{4}K_i$; (*2*) values of I_{integr}; (*3*) zone with known ore deposits; (*4*) a less studied ore zone; (*5*) disjunctive dislocations

$$K_i = \frac{I_i}{\bar{I}_i}. \tag{3.96}$$

The value of $\bar{I}_i$ determines the information obtained when the result of U_j falls into the x_j interval of histogram at an equal probability of falling into any of the R intervals. According to Ventsel (1969), it is equal to the average (complete) information, contained in the results of measuring by a single method.

$$\bar{I}_i = \log R. \tag{3.97}$$

Application of K_i makes it possible to take differences in the ranges of different fields into account. However, the use of formulas (3.93) and (3.95)–(3.97) may not be effective for scant sampling.

Figure 3.41 illustrates an example of the application of the abovementioned method at the Katsdag pyrite-polymetallic deposit (southern slope of the Greater Caucasus).

Interesting results were obtained for the Gedabey mining district (Lesser Caucasus) (Fig. 3.42a). Calculation of entropy (with application of Eqs. 3.96 and 3.97) by the use of density data showed that the maximal entropy corresponds to contact-changed rocks (Fig. 3.42b). It should be noted that fault zones and zones of

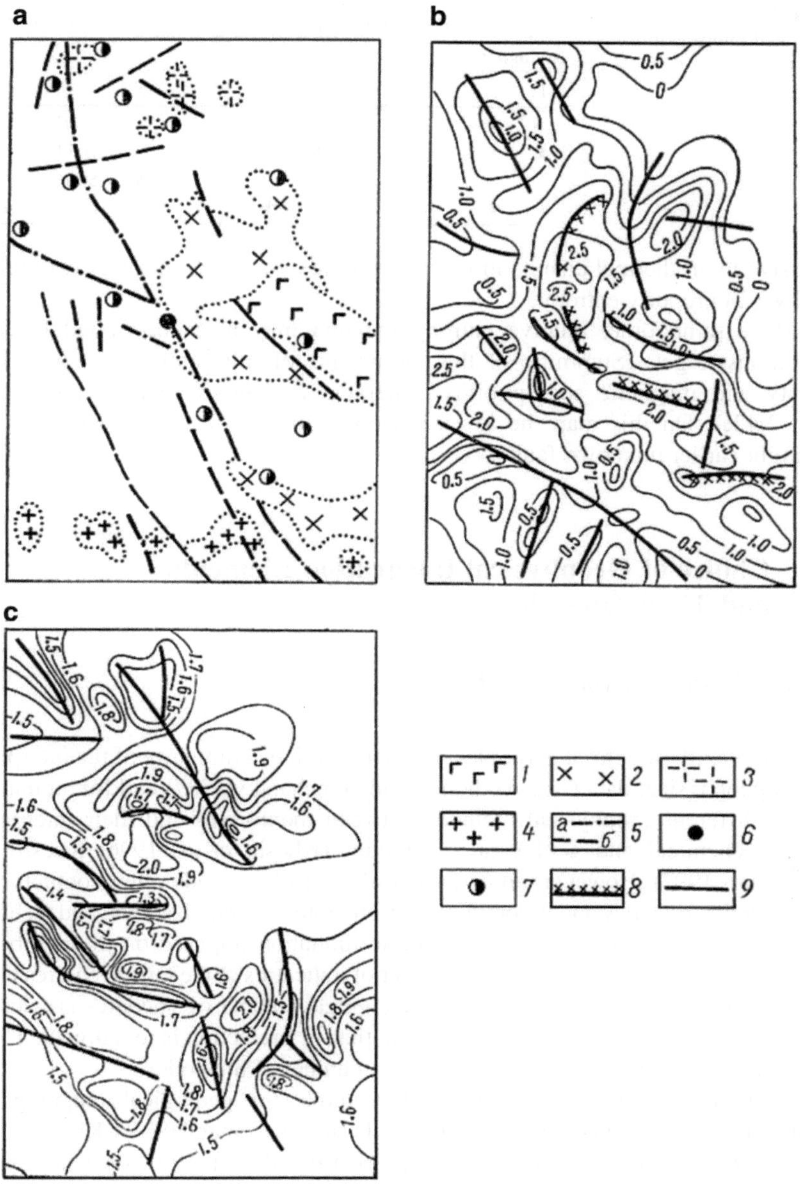

Fig. 3.42 Gedabey mining district: Elements of tectonics, magmatism and ore occurrence (**a**) compared with entropy of density (**b**) and integrated relative information (**c**) calculated by the use of rocks magnetic susceptibility and density (After Karkoshkin 1979). (*1*) gabbro, gabbro-norites and gabbro-diorites; (*2*) granodiorites and quartz diorites; (*3*) plagiogranites; (*4*) quartz diorite-porphirites; (*5*) faults: (*a*) delineated; (*b*) proposed; (*6*) Gedabey ore deposit; (*7*) ore shows; (*8*) active intrusive boundaries revealed by petrophysical data; (*9*) fault zones detected by petrophysical data

Table 3.11 An example of integrated interpretation

Typical field combination										Class of targets
Magnetic					Gravity					
1	2	3	4	5	1	2	3	4	5	
	♦					♦				Marl
♦					♦					Gabbroid
			♦						♦	Salt
	♦					♦				Fracture zone

intensive hydrothermal change in many cases are associated with relative minima of physical property entropy.

At the same time, relative entropy maxima may also correspond to some anomalous targets. Summing of the coefficients of informativity computed for density and magnetic susceptibility generated a map of relative information (Fig. 3.42c). On this map the lower entropy values indicate known faults and highlight earlier unknown faults.

3.9 Choice of Geophysical Integration Elements and Their Quantity

3.9.1 General Considerations

The complexity of current geological problems and the ambiguity of the interpretation of geophysical data (e.g., Tikhonov and Goncharsky 1987) call for an integration of different geophysical methods between themselves and their integration with geochemical and geological techniques (Khesin et al. 1996; Khesin and Eppelbaum 1997). However, an extension of a set of methods often conflicts with its economic efficiency and is complicated from both the organizational and technical points of view. In addition, there is a basic limitation imposed on the number of methods. As noted above, a growing number of indicators of the target requires larger amounts of standard information. However, a great number of standards is available only in well-explored provinces, where quantitative prediction is less urgent. Therefore, a survey set should involve a minimum number of methods.

The typical number of geophysical methods for geological mapping and prospecting various types of useful minerals in the former USSR (Khesin 1976; Nikitin 1986) and in the West (Parasnis 1986; Telford et al. 1990) does not exceed two or three. Here we omit exploration problems demanding repeated and successive application of various complex methods, for instance, 3D seismic prospecting.

Let us consider the results of the application of two methods (e.g., gravity and magnetic surveys). They are designated in Table 3.11 as follows: 1 = negative field, 2 = positive field, 3 = roughly zero field, 4 = alternating field, and 5 = high-gradient field.

Four combinations of two parameters can represent four classes of desired targets, each ranging from 1 to 5. The number of possible combinations of two parameters divided into five categories is 25. The geological features (each class) can be characterized by one of these 25 combinations. It is also necessary to take into account concrete geological situation. For example, in Table 3.11 value 11 can be obtained for aqueous mark (instead of gabbroid), and this may lead to an erroneous conclusion in the process of geological-geophysical interpretation. Therefore, a certain abundance of the set of geophysical methods sometimes is advisable.

The number of combinations can be increased at the expense of secondary parameters arising due to certain transformations of the fields (e.g., downward and upward continuation, various derivatives). Thus, measurements of two geophysical fields can provide sufficient data to resolve mapping problems.

3.9.2 Evaluation of Single Geophysical Method Efficiency

When selecting methods for integration, it is advisable to determine the amount of information provided by each method.

Suppose we have a well-studied area typical of the entire region, and 1/50 of it contains an ore mineral. In this section, the magnetic field is always positive, whereas in the part of the area without ores it may be either positive or negative with equal probability. In other words, it is known a priori that 2% and 98% of the area are ore-bearing and barren, respectively, and that in 49% and 51% of this area the magnetic field is, respectively, negative and positive. The result of the experiment β will be designated as follows: B – ore-bearing part of the area, $\bar{B}$ – barren part of the area. Thus, $P(B) = 0.02$, $P(B) = 0.98$. According to formula (3.73) $H(\beta) \cong 0.14$.

The result of experiment α will be designated as follows: A is the positive field, $\bar{A}$ is the negative field. The relative partial entropy (after recording the positive magnetic field at the measurement point) can be calculated in the following way: $P(A) = 0.51$, $P(\bar{A}) = 0.49$, $P(B|A) = \frac{2}{51}$, $P(\bar{B}|A) = 1 - P(B|A) = \frac{49}{51}$. $H(\beta|\alpha) \approx 0.24$. According to the recorded negative magnetic field, the district is certainly barren: $H(B|\bar{A}) = 0$. Consequently, the partial entropy for experiment β under the conditions of α, in Eq. 3.75, is $H(\beta|\alpha) \approx 0.12$. Thus, the uncertainty of the determination of the ore-bearing area decreases from 0.14 to 0.12 *bit*s after the magnetic field measurement

The effectiveness of the respective methods can be estimated as well in terms of their cost by comparing the values (costs) of the information units supplied by each method. On this basis, a valid solution for the integration of methods can be made.

In some cases a simple model can be used (Khesin 1976). If a set of methods is focused on some independent indicators of equal value, the anomaly detection reliability γ can be described by an error function (probability integral) (Nikitin 1993) as:

$$\gamma = erf\left(\frac{\sqrt{\sum_i v_i}}{2}\right) \tag{3.98}$$

where v_i is the ratio of the anomaly squared to the noise dispersion for each i-th geophysical field.

Now let us assume that the anomaly is indicated by three points and that the mean square of the anomaly for each field is equal to the noise dispersion. For a single method, the reliability of the detection of an anomaly of a known form and intensity by Kotelnikov's criterion (Ventsel 1969) is expressed by $erf\left(\frac{\sqrt{v_i}}{2}\right)$. Thus, the reliability for the individual methods is 0.61and the values for a set of two or three methods is 0.77 and 0.87, respectively (according to Eq. 3.98). This means that the q value (the risk of a false positive) by the integration of two or three methods decreases by factors of 1.7 and 3, respectively. A comparison of the risk with the expenditures yields the optimum set of methods.

3.9.3 *Estimating Information by Indicator (Field) Gradations*

Gradations of indicators can be also used to obtain information about the existence of mineralized zones. A physical field, the quantity of metal in a secondary dissemination halo, and similar data may serve as indicators. Let $P(A_i|B)$ denote the posterior probability of finding the i-th gradation of the indicator A over the targets B, and $P(A_i)$ be the prior probability of finding the same gradation in a survey sheet. Thus the partial information on the presence of target $I_{A_i \to B}$ contained in the A_i gradation takes the form:

$$I_{A_i \to B} = \log\left[\frac{P(A_i|B)}{P(A_i)}\right], \tag{3.99}$$

as the uncertainty in A_i recordings before the survey was $\log P(A_i)$ and after the survey was $\log P(A_i|B)$.

Similarly,

$$I_{A_i \to \bar{B}} = \log\left[\frac{P(A_i|\bar{B})}{P(A_i)}\right]. \tag{3.100}$$

Statistically, the probabilities (or, to be more precise, the relative frequencies) are expressed by the ratios of the areas: S is the sheet area, S_i is the area occupied by the i-th interval of isolines A_i, S_p is the total area of the prospecting target roofs (upper edges), S_{pi} is the area common to interval A_i and the target roofs.

Hence

$$P(A_i) = \frac{S_i}{S}, P(A_i|B) = \frac{S_{pi}}{S_p}, P(A_i|\bar{B}) = \frac{S_i - S_{pi}}{S - S_p}.$$

After calculating the areas (probability estimates), the increment of information contained in the A_i gradation can be calculated as follows:

Table 3.12 Example of information estimates (Filizchay ore field)

Indicator		ΔI_i, bit
Contents of metals in the secondary haloes	Cu	5.33
	Pb	3.46
Intervals of SP isopotential, mV	>−20	
	(−20)÷(−70)	
	<−70	
Intervals of magnetic field isogams, nT	40–60	0
	60–80	−2.634
	80–100	−1.334
	100–120	−0.739
	12–140	−2.386
	140–160	−0.018
	160–180	2.297
	180–200	0.591
	200–220	−0.805
	220–240	−0.488
	240–260	−0.378
	260–280	0.596
	280–300	1.699
	300–320	−0.263
	>320	0

Table 3.13 Informational significance of integration elements in the Filizchay ore field

Indicator	$I'_{A_i \to B}$	Q, expenditure unit	$\frac{I'_{A_i \to B}}{Q}$
SP	2.41	26	0.093
Concentration of 0.03% in secondary haloes:			
Lead	1.96		
		23	0.125
Copper	0.91		
		22	0.011
Magnetic field	0.25		

Notes: (1) Q is the basic expenditure for the survey for an area of 1 km^2 with a grid of 250 m × 50 m. (2) Since the sampling of metallometric samples and half-quantitative analysis are executed simultaneously, the calculation of commercial effectiveness of spectrometallometry is given as for a single method

$$\Delta I_i = I_{A_i \to B} - I_{A_i \to \bar{B}} = \log \left[\frac{P(A_i|B)}{P(A_i|\bar{B})} \right]. \tag{3.101}$$

This example is based on the results of an integrated geochemical (spectrometallometry) and geophysical (self-potential and magnetic methods) survey of the Filizchay ore field, on the southern slope of the Greater Caucasus (Table 3.12). The Filizchay ore field includes the Filizchay, Katsdag, Katekh and Djikhikh deposits and the ore manifestation of Kegnya-Medan (Khain and Alizadeh 2005).

It turns out that the largest increments of information on the presence of ore are contained in the metallometric data (anomalies of copper and lead), in intense negative SP anomalies, and isogam gradations of 160–180 and 280–300 nT.

For the respective estimates of informational significance (Vysokoostrovskaya and Zelenetsky 1968), the following formula was used:

$$I'_{A_i \to B} = \sum_i \left[P(A_i) \cdot I_{A_i \to B}\right]. \tag{3.102}$$

The results using Eq. 3.102 are given in Table 3.13.

These estimates substantiated the above-mentioned rational integration. Using this integration, large-scale surveys in the entire promising region of the Eastern Caucasus were performed. As a result, several copper-polymetallic deposits were discovered (Khesin et al. 1993b). It should be noted that the expression of geophysical results in informational units provides a clear picture of a complex prospect indicator for the area.

3.9.4 Estimating Geophysical Integration Efficiency Using Type I and Type II Error Probabilities

Classification efficiency can be estimated quantitatively not only for separate methods, but also for geophysical integration by calculating the reliability of revealing geological targets. On the southern slope of the Greater Caucasus, more than 50 anomalous areas were selected by integration of SP, magnetic, and metallometric prospecting. The areas were divided into three groups with various prospects. To test them, more complex integration works were carried out including induced polarization, transient electromagnetic and gravity prospecting. Assuming that the results of the second set of applications were absolutely reliable, the reliability of classification using the first set could be estimated. For this purpose, the likelihood of type I and type II errors were calculated (Khesin 1981).

The probability of a type II error (M_2) is expressed as the relative frequency of an erroneous diagnosis for objects from sampling B (geological targets). The probability a type I error (M_1) is expressed as the relative frequency of an erroneous diagnosis of objects from sampling $\bar{B}$ (the rest of the objects). These errors are used to determine the total unconditional error of the separation q between B and $\bar{B}$ classes (i.e. the risk of an erroneous solution):

$$q = M_2 P(B) + M_1 P(\bar{B}), \tag{3.103}$$

where $P(B)$ and $P(\bar{B})$ are prior probabilities of the appearance of the objects of the first and second types, respectively.

If $P(B) = P(\bar{B}) = 0.5,$ the q value corresponds to the intersection area of the distribution densities $P(X|B)$and $P(X|\bar{B})$. Here X is the separation index. It can represent field amplitude, value of the complex indicator, etc. The separation reliability (γ) is:

$$\gamma = 1 - q. \tag{3.104}$$

The total empirical error should be compared to a theoretical one. The approximation of the error testifies to a correct assumption and high reliability of identification. Using logical-informational methods, the classification reliability is estimated solely by the empirical errors.

Errors due to the assignment of the measurement results to either class (productive or barren) can be determined as follows. An absence of anomalies of the complex indicator in a known target-bearing area is a type II error or "omission of target". The presence of these anomalies in the barren part of this area is a type I error, a false positive.

By comparing the conclusions obtained by integration of rapid methods with the results of a more complex integration, it is also possible to estimate the respective errors and the reliability of classification. For 12 prospective areas in the southern slopes of the Greater Caucasus, the ore nature of 11 was confirmed by more complex geophysical integration. New anomaly areas that are barren were not revealed by complex integration. Thus, $M_1 = 0,\ M_2 = 1/12$. Assuming that $P(B) = P(\bar{B}) = 0.5$, and taking into account expressions (3.103) and (3.104), then $q = 0.04$ and $\gamma = 0.96$. Thus, the high reliability of integration of rapid methods is also confirmed by this type of estimate.

3.9.5 Minimization of the Number of Combined Methods by Solving the "Four Colors Problem"

The practical use of a small number of integration elements can be theoretically substantiated using the solution to the well-known mathematical and cartographic problem of "four colors" (Appel and Haken 1989) for integrating geophysical methods during regioning and prospecting useful minerals (Khesin and Eppelbaum 1986; Eppelbaum 1989).

A geophysical study is usually a multistage procedure beginning with areal mapping and ending with exploration. For the sake of simplicity, it will be assumed that the goal of each prospecting stage is the selection of an area for more detailed operations at the next stage. The result of prospecting is primarily a substantiated evaluation of the areas under study and their classification into two groups: those worthy and unworthy of further study. The objective of prospecting is to obtain maximum information at a set cost.

Let us now examine the "four colors theorem" from this standpoint. Using elementary notions from graph theory, the problem can be formulated as follows: prove that all the vertices of an arbitrary planar graph can be colored with four

Table 3.14 Subdivision of an area according to geophysical survey results

Level of knowledge of the area	Geophysical method		Combination (color) number
	First	Second	
High	+	+	1
Medium	+	–	2
Low	–	–	3
Unknown	No necessary data		4

colors in such a way that no two vertices joined by a common edge are of the same color. It was proved as early as the middle of the nineteenth century that four colors suffice to color different counties on the map of England. However, a general solution to this theorem was found only relatively recently (Appel and Haken 1989) and then updated by Thomas (1998). The authors subdivided all possible maps into almost 2,000 types and developed a computer program for their investigation. For each type the same problem was solved; namely whether a map which cannot be colored with four colors can be found among the set of maps. After lengthy investigations, an answer of "no" was obtained for all the types, and this fact confirms the above solution.

Any region under study can be divided into separate geological areas according to certain indicators. The following system of classification according to their prospects has been adopted in the USA (Suggestions 1978): high (*H*), medium (*M*), low (*L*) and unknown (*U*). The objective of the prediction is to single out promising areas (if any) from the whole set by means of an integrated geophysical survey. Colors refer to various combinations of geophysical methods. A positive conclusion from the data of a certain prospecting method is labeled (+), and a negative conclusion as (−). It is evident that a combination of at least two independent geophysical methods is necessary for the first three gradations (*H*, *M* and *L*); the gradation *U* implies no application of the method set (on the scale or not at all) in the area under study (Table 3.14).

The geophysical methods employed are a priori assumed to be of equal importance. The threshold field values (the borderline between plus and minus) and particular types of geophysical studies are determined according to prospecting results for similar objects investigated previously and other geological and geophysical considerations. The term borderline refers to a certain line (surface), where specified physical characteristics exceed threshold field values. These physical characteristics may, for example, include values of observed fields, field gradients or indicators of field variability.

Hence an optimal geophysical set consists of two independent geophysical methods. A map of geophysical results colored with four colors by the above technique can serve as a basis for more detailed study. It should, however, be kept in mind that this geophysical set is usually oriented toward a particular problem and substantiated by a corresponding physical-geological model of the medium. Any change in the problem (e.g., an increase in the required depth of investigation) or in geological and geophysical patterns of the area may bring about a change in the set of methods. Thus a certain redundancy in the set is needed.

Chapter 4
Regional Physical-Geological Models and Regioning

4.1 Utilization of Available Geological, Petrophysical and Geophysical Data

4.1.1 Use of Geological Data

A preliminary model is devised to represent the geological objective when preparing a geophysical project for the given area. Otherwise, it is impossible to select the set of methods and interpretation procedure. The latter is revised before interpreting, if necessary. The initial development of a model of the medium is the most important stage, since the results of interpretation and the investigation in general depend to a considerable degree upon its quality.

Any model must comply with two conflicting requirements, i.e. (a) reflect the essential features (indicators) of the object for modeling and (b) be sufficiently simple to use.

A model of the medium is an integrated and if possible, formalized representation of the geological and geophysical data concerning the target and host rocks (Kozlovsky and Krivtsov 1988; Movshovich et al. 1987). Hence, a model of the medium can be seen as a symbolic informational model. The background is no less important than the target object itself and should also be considered in the model in detail. The details should correspond to the scale of investigation. In addition, various noise effects should be taken into account.

In many cases the characteristics of geological media for open (one-stage) regions testify to the expediency of employing stochastic models. However, geophysical fields as random functions may contain regular components of an impulse or periodic nature. If we ignore tasks involving area mapping by statistical characteristics of the field, which as a whole are considered random, the most important geological problems remain. These involve the extraction of a regular component from its associated noise. This component may correspond to magnetized gabbroid or low-magnetic acid intrusions, a zone of hydrothermal alterations, skarn-magnetite or pyrrhotite containing lens, etc.

L.V. Eppelbaum and B.E. Khesin, *Geophysical Studies in the Caucasus*,
Lecture Notes in Earth System Sciences, DOI 10.1007/978-3-540-76619-3_4,

The regular component cannot be revealed without applying certain modeling notions. Such targets as large ore bodies are usually rare, and it is always a problem to determine their statistical characteristics. This has generated the need for developing a special system of processing and interpretation of the results of geophysical investigations. Such a system makes it possible to extract anomalies from a combination anomalies and noise on the basis of a deterministic approach to the object under study (the anomaly) and a probabilistic approach to the remaining salient features of the medium and geophysical observations regarded as noise.

The deterministic study of geophysical targets, as presented in Sect. 3.1.2, is facilitated by restricting the geometric approximations of the objects belonging to the main classes under investigation (see Table 3.4).

4.1.2 Use of Petrophysical Data

The sources used for devising a model of this medium include geological data, information from studies of physical properties, geophysical observations obtained in adjacent areas and the results of solving problems of the same type under similar conditions.

When studying geological descriptions and documentation obtained from the region and in other areas with the target object, the key geological factors which define the geological pattern need to be identified. The model must take all these factors into account. Any well-explored geological object related to the target (structural-facial zone, tectonic block, structure, ore field, orebody, etc.) may serve as a model for the object under investigation. Thus an examination of the geological data can provide a concrete idea of the possible classes of objects occurring in the area under study, their geometry, composition and relationships to the host rocks.

The regional ore controls are larger ore-bearing and ore-distributing structures and therefore are related to their corresponding regional boundaries of physical properties (see Fig. 3.3).

This makes it possible to study the behavior of the Pre-Alpine basement to detect large metallogenic crust blocks. For regional prospecting, deep inhomogeneities may be approximated by simple deterministic models as a function of available knowledge. It serves to apply deterministic methods of quantitative interpretation provided they are adapted to the complex observation surface and other complicating factors. Some lithologic and stratigraphic boundaries and tectonic lines, mineral deposit borders and intervals of hydrothermal alterations may also be considered as references within the areas under detailed investigation (see Fig. 3.2).

It should be emphasized that many peculiarities of geological structure (for example, plicative and disjunctive dislocations) affect different mineral resources, including oil and gas (Khesin 1991) and underground water (Shakhnazaryan et al. 1986).

Analysis of the data shows that in most cases the types of geometric approximations of objects are restricted to a comparatively small number of classes (see Table 3.4).

Pronounced gravity and magnetic anomalies are typical bodies characterizing the majority of the objects in Table 3.4, whereas ore and the zones of hydrothermal alterations are also distinguished by SP and IP anomalies. The objects in these two classes and disjunctives are often detected by VLF and thermal methods. In most cases, the objects are reflected by the anomalies recorded by several methods. This determines the shared features of their models and the possibility of an integrated approach to their study.

Geophysical investigation of the orogens is complicated by rapidly changing media and a great number of near-surface anomaly sources. The changeability of mountainous regions along the lateral is accompanied by their rapid change along the vertical (see Fig. 3.2).

Characteristically, petrophysical variability is typical of geologically identical rocks (see Fig. 3.5). The erosion of young mountainous structures causes a considerable change in near-surface associations. Physical properties of rocks occurring close to the Earth's surface may not, therefore, be the same as those of deep-seated rocks of the same composition and age (see Fig. 3.5a). Consequently, the measurements of physical properties of the rocks located in the upper portion of a section should be treated with great care, especially when they are extrapolated to the depth. There is, however, a need for a shallow survey, since the majority of the deposits occur in the upper portion of a section. When investigating deeper horizons, the effect of this rapidly changing portion should be corrected by taking into account its true parameters. All these factors call for special attention to the physical properties of rocks and ores measured on samples taken from the surface, in their natural bedding, drill cores or mines. Their determination should be carried out on the basis of the observed geophysical fields. The variability of physical properties makes it possible to derive additional information, and single out the specific ore characteristics by petrophysical investigations.

Through an integrated approach to investigation and interpretation, the physical properties of the material can be used to develop an integrated petrophysical model of the medium. For example, the process of compiling a petromagnetic model is analogous to the use of data on other physical properties for a similar purpose.

In open areas, if the petromagnetic study is conducted at the same time as the magnetic survey on the same scale, and a petromagnetic map is drawn up, the latter represents a petromagnetic model of the surface area under investigation (Ismail-Zadeh et al. 1983a,b). The need to construct several petromagnetic sections presents no problems in this case. The petromagnetic data obtained from core samples and deep mines are especially crucial, since they are not distorted by near-surface alterations. The determination of magnetic properties by correlating the observed field with the known geological section deserves special attention.

When no petromagnetic data are available for the area under investigation (which will be an exception to the rule in the future), the data on magnetic properties of similar rocks in adjacent areas, and on objects analogous with the target can be used, including those obtained from interpretation results. Naturally, the quality of a petromagnetic model in this case will be markedly lower. This may affect the quality of interpretation and progress in solving the problem overall. A range of variability of physical properties for different objects is shown in Table 3.3.

A petromagnetic model of a medium must yield the magnetic characteristics of all the rocks and their varieties occurring in the region. All stratified rock masses and compact geological bodies (including mineral deposits) should have corresponding average values of magnetic parameters with confidence limits. The same applies to the zones of altered rocks. The magnetic characteristics of the geological objects which do not outcrop onto the area under exploration, but are possible at a certain depth and are known to occur in adjacent areas should not be ignored either. When geophysical data obtained in neighboring or similar areas are employed, this can be used to develop a model of the medium to define deeper portions of the section in the area under investigation (even on a smaller scale), when the intensity and shape of magnetic anomalies over the objects are similar to those sought for.

To develop a petromagnetic model, the total magnetization should be computed of the individual structural-material associations using the available geological classification. However, if the classification does not conform to the results of the analysis of petro- and paleomagnetic data on the section, a model must be devised that incorporates these petromagnetic features. In this case, the weighted average magnetization is calculated for certain rock complexes that have comparable magnetic property values or magnetization direction (within the same paleomagnetic zone). The explosive index of magmatic rocks, i.e. the pyroclast percentage may be of use when computing the weighted average values of their total magnetization, since pyroclastic rocks are characterized by a lower magnetic susceptibility κ and natural remanent magnetization J_n, as compared to the lavas.

When dealing with oblique magnetization, it is crucial to determine the direction of magnetization: the intensity and especially the type of anomaly depend on the inclination angle of the object's redundant magnetization.

Alexeyev (1976) examined the magnetization of an arbitrary triaxial ellipsoid in the Earth's magnetic field $\mathbf{T}_0$. Under certain simplifications, the limiting cases for this body are a vertical bed, a thick horizontal bed and a horizontal circular cylinder. The above bodies had sufficiently high magnetic susceptibility to be magnetized almost in parallel to the active field. The angular error was less than 1.5° at $\kappa = 0.05$ SI unit and less than 3° at $\kappa = 0.1$ SI unit.

Therefore, we can assume that for all examined values and any shape of a body, a vector equality for the induced magnetization ($\mathbf{J}_\mathrm{i}$) is valid:

$$\mathbf{J}_\mathrm{i} = \kappa \mathbf{T}_0. \tag{4.1}$$

Note that for higher values of magnetic susceptibility ($\kappa > 0.08 \div 0.10$ SI unit) vector $\mathbf{J}_\mathrm{i}$ is not parallel to vector $\mathbf{T}_0$. As is generally known, a sphere is magnetized parallel to the active field independently of the magnetic susceptibility value.

4.1.2.1 Analysis of Some Paleomagnetic Parameters

Paleomagnetic data investigations can contribute considerably to the analysis of magnetic survey results since they characterize certain general features of the magnetization distribution. For example, data from a detailed paleomagnetic examination of the Lesser Caucasus indicate that the remanent magnetization of the Mesozoic volcanites is parallel or antiparallel to the induced magnetization, and the zones of the inverse polarity ($\mathbf{J_n}$) usually correspond to relatively weakly magnetized intervals of the geological section.

The Cenozoic volcanogenic series of the Talysh and Pre-Talysh have other characteristics. Here, direct polarity is dominant in the Eocene section. However, the most magnetized basaltoids of the Kosmalyan series are divided into two parts: (1) trachyandesites, andesite-basalts, basalts and tufogens with a total average thickness of 0.7 km of direct magnetization (Middle Eocene), and (2) andesite-basalts, gabbro-dolerites with a total average thickness of 0.8 km of inverse magnetization (Lower Eocene). The vector sum of these values determines the presence of a very complex magnetic field ΔT observed on the Earth's surface and at various levels above it.

At the same time, the Neogene-Quaternary lava sheets in the central part of the Lesser Caucasus are characterized by frequent alteration of paleomagnetic zones of different polarities. Magnetic field modeling indicated that by upward continuation of the magnetic field (computed on the basis of Poisson's integral) and an airborne magnetic survey carried out at certain altitudes (also computed on the basis of 3D modeling), the magnetic anomalies caused by the flat-laying Cenozoic bodies with different directions of natural remanent magnetization (NRM) were significantly attenuated (Khesin et al. 1996). This conclusion is congruent with the observed magnetic data analysis and confirms that paleomagnetic data can be employed for an overall estimate of magnetic prospecting possibilities and for qualitative and quantitative examination of magnetic data.

When analyzing magnetic properties, it is common practice to use the induced magnetization $\mathbf{J_i}$ alone, since the remanent $\mathbf{J_n}$ is assumed to be comparatively small. In this case it is sufficient to have information on the magnetic susceptibility of the host rocks and the target objects. However, in contrast to current opinion, the presence of remanent magnetization is often the general rule, rather than the exception to the rule (Affleck 1964; Khesin et al. 1983; Eppelbaum et al. 2004b), especially for volcanogenic and intrusive rocks. Here it may take on a higher value and a direction different from that of the modern geomagnetic field. The remanent magnetization is vector added to the induced magnetization.

With the advent of induction susceptibility meters (kappameters), magnetic susceptibility became easier to measure. These measurements have now been generalized to common practice. Remanent magnetization has become the subject of paleomagnetic investigations. The aims, objects and nature of the results of these investigations do not adhere to the requirements of magnetic prospecting involving the value and the direction of the total vector

$$\mathbf{J} = \mathbf{J_i} + \mathbf{J_n} \tag{4.2}$$

Here $\mathbf{J_n}$ is the NRM. Interestingly, many papers describing the results of paleomagnetic investigations do not present any data on NRM.

As paleomagnetic experience has shown, scattering is typical of the initial directions of $\mathbf{J_n}$, i.e., directions of NRM on a stereoprojection. They are often scattered over the whole field of stereoprojection, and sometimes reverse orientations occur. After eliminating the unstable components of remanent magnetization by various methods (such as demagnetization, thermomagnetic cleaning) the directions come together to form a rather narrow area of the stereoprojection. The average direction defines the orientation of the ancient magnetic field in the rock formation period, which is different from the modern field $\mathbf{T}_0$.

However, the chaotic behavior of the initial directions of $\mathbf{J_n}$ suggests that the average total magnetization defined from formula (4.2) will not appreciably deviate from the direction of $\mathbf{T}_0$ in terms of direction averaging. This supposition was confirmed by a paleomagnetic investigation in the Gedabey district of the Lesser Caucasus (Khesin et al. 1983). Over the course of this study, the initial materials were processed using formula (4.2) for 600 samples of volcanogenic and intrusive rocks of acid to basic composition from the Middle and Upper Jurassic, which are the primary sources of anomalies in the region (Fig. 4.1).

The total vector of magnetization $\mathbf{J}$, in spite of its chaotic nature and the presence of reverse directions of $\mathbf{J_n}$, lies nearly parallel to the modern geomagnetic field $\mathbf{T}_0$ (the average deviation is 5° for declination, and 1° for inclination). Similar calculations employing paleomagnetic data for magmatic rocks of the Mesozoic and Paleogene within the area of the whole Somkhet-Agdam Zone (Lesser Caucasus), and the Gedabey district within it yielded similar results. The magnetization of acid and basic intrusives along the field $\mathbf{T}_0$ was reinforced in the literature (Dukhovsky et al. 1970). This is probably a general postulate. It is confirmed by many years' experience studying magnetite deposits in the Krasnoyarsk area of the Altai-Sayan orogen (B.M. Afanasiev, private communication). Thus the inclination angle of magnetization may be regarded as a fixed parameter for interpretation under conditions of oblique magnetization.

Investigators should make sure that the condition

$$\mathbf{J} \| \mathbf{T}_0 \tag{4.3}$$

is fulfilled for typical rocks in each region subjected to magnetic survey. If not both the magnetic susceptibility and the NRM vector must be studied.

The total magnetization direction can be estimated if we know the factor $Q = J_n/J_i$ and the rough direction of $\mathbf{J_i}$. Formula (4.2) makes it easy to obtain the expression:

$$\tau = \arctan\frac{Q \sin \upsilon}{1 + \cos \upsilon}, \tag{4.4}$$

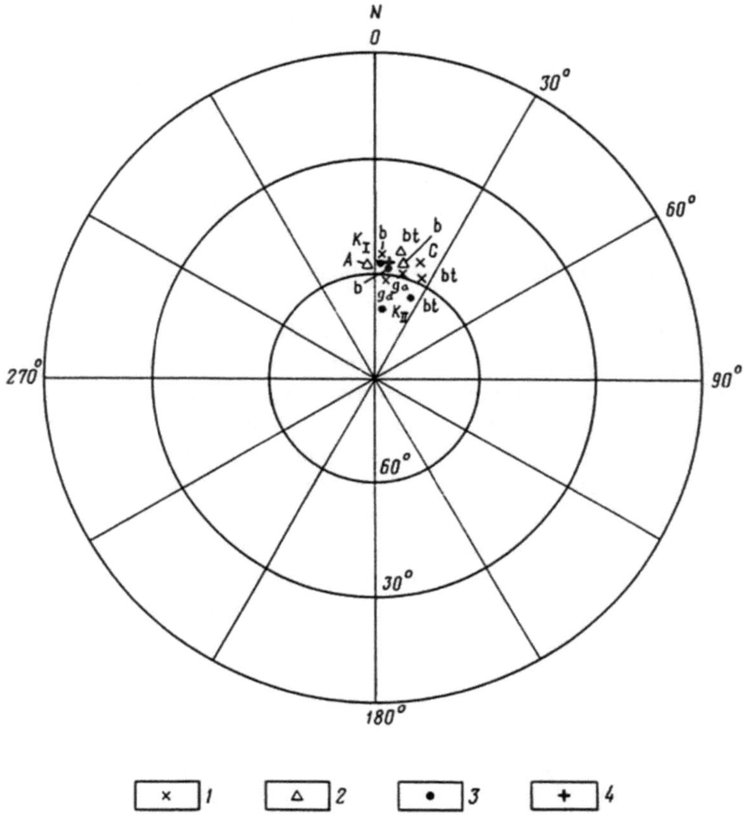

Fig. 4.1 Stereoprojections of average directions of magnetization $\mathbf{J} = \mathbf{J_i} + \mathbf{J_n}$ for volcaniclastic and intrusive rocks in the Gedabey district of the Lesser Caucasus. Directions of rock magnetization at the following values of J, A/m: (*1*) < 0.2 (< 200·10^{-6} CGS); (*2*) 0.2 ÷ 1.0 (200 ÷ 1,000·10^{-6} CGS); (*3*) >1.0 (>1,000·10^{-6} CGS); (*4*) direction of the modern geomagnetic field. *b* and *bt* are effusive and pyroclastic Bajocian and Bathonian rocks, respectively; *C* are the plagiogranites of Slavyansky intrusion; K_I are the gabbro and gabbro-diorites of phase I of the Gedabey intrusion; K_{II} are the diorites and quartz-diorites of phase II; *A* is the quartz diorites and diorites of Aitaly intrusion; g_d and g_a are granodiorites and granoaplites of sheet intrusions, respectively

where υ is the angle between vector $\mathbf{J_n}$ and vector$\mathbf{T}_0$, and τ is the angle between vector **J** and vector $\mathbf{T}_0$.

To determine angle τ, we plotted a nomograph (Fig. 4.2) for values $\upsilon = 0 \div 30^\circ$ ($\mathbf{J_n}$ and $\mathbf{T}_0$ are roughly parallel) and for $\upsilon = 150 \div 180^\circ$ ($\mathbf{J_n}$ and $\mathbf{T}_0$ are roughly antiparallel). In this situation, the direction of $\mathbf{J_n}$ with an angle of 30° does not fall outside the cone around $\mathbf{T}_0$. From the nomograph it follows that in the first case the direction of the total vector (**J**) is within the cone since it has a more acute angle ($\tau<\upsilon$) for all values of Q; the lower the Q factor, the smaller the angle is. In the second case antiparallelism holds for higher values of Q, whereas for smaller ones

J is roughly parallel to the vector $\mathbf{T}_0$. The most unfavorable situation is at $Q = 0.7 \div 1.5$, when the direction of vector **J** is markedly different from the $\mathbf{T}_0$ direction. For rocks characterized by such Q values and a reversed polarity of vector $\mathbf{J_n}$, the total vector is calculated by formula (4.2). For the first and second cases, it can be computed using a simple approximate formula

$$\mathbf{J} = \mathbf{J}_{\mathrm{i}}\,(1 \pm Q) = \kappa \mathbf{T}_0(1 \pm Q). \tag{4.5}$$

The parallel vectors $\mathbf{J_n}$ and $\mathbf{T}_0$ have a positive sign, and a negative sign when they are antiparallel.

Analysis of magnetic survey results and calculation of the direct magnetic effects require treatment of the local anomalous object's redundant magnetization with respect to the host medium. In contrast to excess density, redundant magnetization is a vector quantity. For purposes of calculation, all the components of magnetization vectors for both the local bodies and the host medium must be known. If the magnetization vectors of the body and the medium (of two adjacent objects) are parallel, and more specifically if condition (4.3) is fulfilled, the redundant magnetization modulus is the simple difference in absolute values of these vectors, and its direction is parallel to the direction of the initial vectors. When the magnetization vectors for the body and the medium are not parallel, the redundant magnetization components must be computed.

The NRM direction in rocks depends on the influence of quite a number of components which were formed during different geological time intervals. These can be altered as a result of certain physical and chemical processes. Therefore, care is required when selecting the magnetization direction as a fixed parameter for interpretation, and needs to be confirmed on a statistically significant sample. The constancy of the magnetization direction (accurate up to 20°) must be verified not only for the search objects, but also for the host rocks. The above is easy due to the fact that induced magnetization has a constant direction, and NRM, when measured on rock samples, varies randomly with respect to a certain average value. Therefore, the direction of their average sum, especially if $Q < 1$, will be close to that of the present geomagnetic field in the area under investigation.

For a better understanding of the appearance and further evolution of the NRM over the evolution of the Earth, it is important to analyze the conditions during the early Earth's formation and evolution. This analysis can provide a clear picture of the causes of primary magnetization formation and its possible changes (Pilchin and Eppelbaum 2007).

4.1.2.2 Results of Some Regional Paleomagnetic Examinations

The results of paleomagnetic investigations were compiled for the general paleomagnetic scale of the Mesozoic-Cenozoic of Azerbaijan (Ismail-Zadeh 1983b). This scale was updated by Guseinov (1988) on the basis of analysis of the Lower

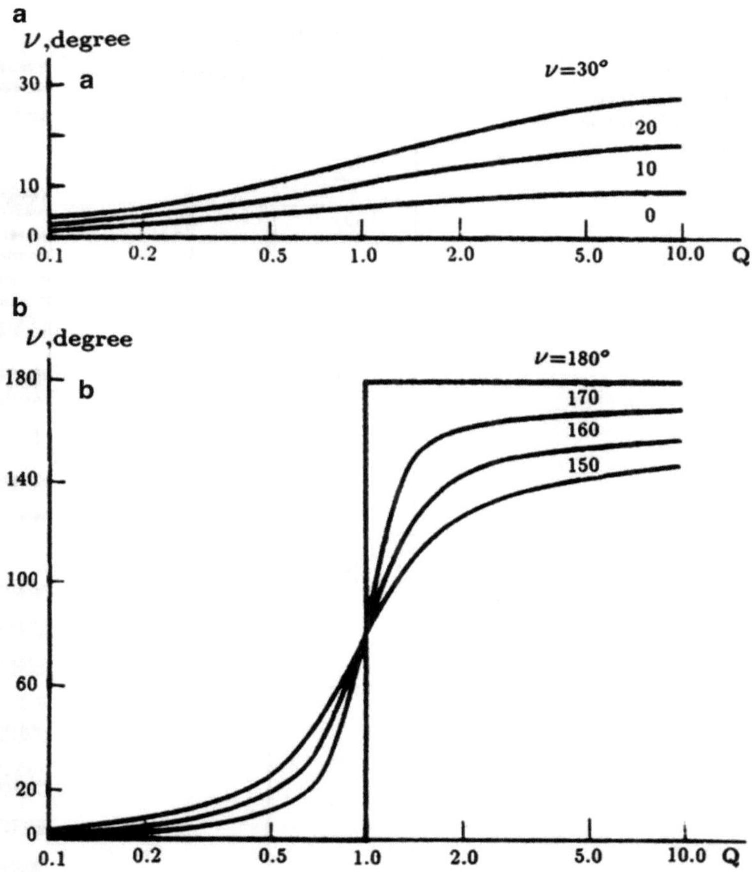

Fig. 4.2 Nomograph to determine the deflection angle τ of the magnetization vector **J** from the geomagnetic field vector $\mathbf{T}_0$ by a known angle τ between $\mathbf{J_n}$ and $\mathbf{T_0}$ vectors, and the Q factor $\left(Q = \frac{J_n}{\kappa T_0}\right)$: (**a**) $0^{\circ} \leq \upsilon \leq 30^{\circ}$; (**b**) $150^{\circ} \leq \upsilon \leq 180^{\circ}$

Jurassic and Aalenian deposits of the Lesser Caucasus (the generalized data are shown in Fig. 4.3).

The Jurassic Cretaceous, Paleogene and Neogene deposits are divided into 19, 22, 6 and 5 paleomagnetic zones of direct and inverse magnetization, respectively. This can be used to define the geological sections and stratigraphic position of some floors more accurately. The most significant was the correlations for the barren productive series of the Middle Pliocene (Balakhany floor) within western Absheron, the Lower Kura Depression and its northern frame (Jeirankechmez Depression). It was shown that the thicknesses of the same paleomagnetic zones in the above depressions is much greater than on the Western Absheron with its

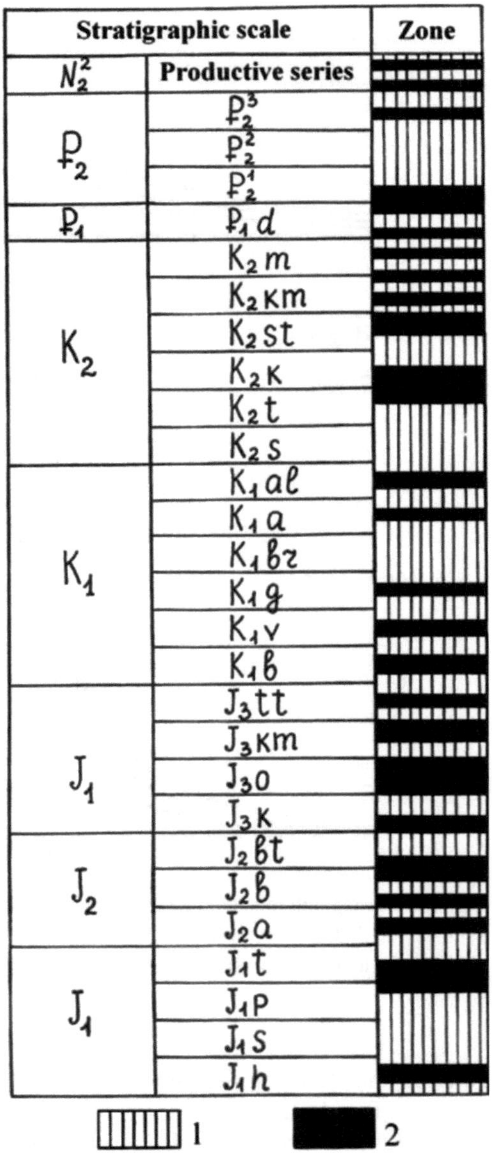

Fig. 4.3 Generalized paleomagnetic scale of the Mesozoic-Cenozoic of Azerbaijan. (*1*) direct magnetization; (*2*) inverse magnetization

commercial multihorizon oil, gas, and gas-condensate deposits. Thus, the oil and gas potential of these depressions noted by geologists was supported by the paleomagnetic examination. Regional paleomagnetic analysis testifies to the relationship between the productive series of Azerbaijan, the Shirak series of Georgia and the red sequence of Turkmenistan. This provides additional data for investigation of oil and gas in these regions given the relationships and spatial changes of the series identified by paleomagnetic techniques. Paleomagnetic identification of the Upper Bajocian deposits of the Lesser Caucasus (where an inverse magnetization prevailed) and the same zones of inverse polarity identified in the Northern Caucasus are presented in Fig. 4.4.

An interesting example of paleomagnetic correlations can be found between the Lower Eocene deposits and the most distant regions of Azerbaijan, i.e., Nakhichevan in south-west and Talysh in south-east. It was shown through the strong Paleogene correlations for these regions, that the lower part of the Lower Eocene is absent in Talysh.

The use of paleomagnetic data for paleotectonic reconstructions in the Caucasus is of great importance (e.g., Asanidze and Pecherskiy 1979; Asanidze et al. 1980; Camps et al. 1996). Table 4.1 lists results obtained by paleomagnetic groups under the supervision of T. A. Ismail-Zadeh for the Lesser Caucasus, the southern slopes of the Greater Caucasus and the Northern Caucasus (here D is declination and I is the inclination).

Table 4.1 provides the coordinates of the ancient magnetic poles and paleolatitudes for areas of the Lesser Caucasus: 22–26°. Thus, in the Middle-Late Jurassic the Lesser Caucasus was 14–18° more to the south that its modern location. In terms of longitudes the Lesser Caucasus can be seen as separate blocks that rotated in relation to each other. Data presented by Asanidze and Pecherskiy (1979) confirmed that in the Jurassic and Early Cretaceous the Greater and Lesser Caucasus wee located about 15–20° to the south of their modern location.

The paleomagnetic directions of the Lower Jurassic rocks of the Lesser Caucasian region and Jurassic rocks of the Greater Caucasian region are close to those of the Eurasian continent. Analysis of the available paleomagnetic data suggests that the deviation between the Jurassic and Lower Cretaceous declinations may have been caused by the turning of the Lesser Caucasian region anticlockwise towards the Greater Caucasian region.

The values of the paleolatitudes of the Jurassic and Cretaceous rocks of the Lesser Caucasian region are similar and differ from the paleolatitudes of the Greater Caucasian region by 8–10°. Consequently, during the Middle and Later Mesozoic the Lesser Caucasian region practically did not change its latitude position.

Paleomagnetic examination of the Paleogene and Eocene deposits in Talysh and Nakhichevan shed light on the rotation of the Lesser Caucasian plate anticlockwise 30–40° towards the Greater Caucasus. In addition, the plate was divided into separate blocks which turned relative to each other (confirmed later by Khalafli (2006)); obviously the horizontal transitional displacements of the Lesser Caucasian region which finished in the Jurassic were not renewed. Paleomagnetic research in the Lesser Caucasus within Turkey (Van der Voo and Channel 1980), indicates that the Turkish territory rotated toward the north of Eurasia. Thus, all the territory of the Lesser Caucasus, beginning from the Jurassic, turned at the angle of 30–40° anticlockwise to the Greater Caucasus. This rotation ended for the most part by the beginning of the Late Eocene.

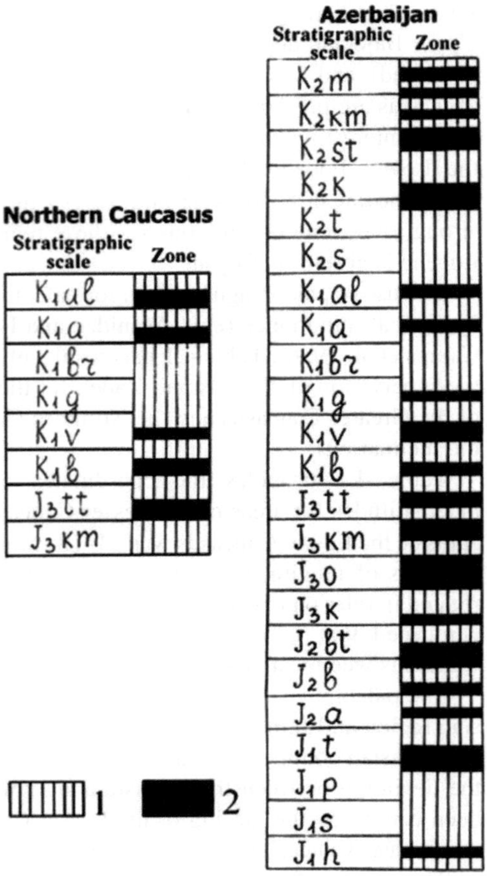

Fig. 4.4 Regional correlation of the Mesozoic paleomagnetic sections. (*1*) direct magnetization; (*2*) inverse magnetization

4.1.2.3 Analysis of Other Physical Parameters

The procedures for assessing the density σ, resistivity ρ, polarizability η and other physical non-vector properties are generally analogous to the procedures described above for petromagnetic data, but are simpler. Note once again that the complexity of the terrain relief may be of use in determining such an important feature as the effective density of the medium under study. This forms the basis for the well-known method suggested by Nettleton (1940), in which the Bouguer anomalies are calculated for different intermediate layer densities. The curve which is the least correlated with the relief profile, corresponds to the mean density of rocks in the upper portion of the section. Various methods for determining the density with respect to mountainous conditions were described by Parasnis (1966), Berezkin (1967), Varlamov and Filatov (1983), and as for other physical properties, they are described in various studies including reference books (Tarkhov 1980; Kobranova 1986;

Table 4.1 Average directions of ancient magnetizationfor different regions of the Caucasus

Sampling area	Age	Average direction of J_n, degrees	
		D	I
Lesser caucasus			
Dashkesan	J_2–J_3	0	40
Karabakh	J_2–J_3	25	40
Martuni	J_2–J_3	20	48
Lachin	J_3–K_1	30	50
Southern slope of the greater caucasus			
Xizi	K_1–K_2	350	45
Northern caucasus			
Krasnogorskii	J_1–K_1	343	55
Kislovodsk	K_1	340	50

Nikitsky and Glebovsky 1990). Density can be also determined by making use of the gravimetric measurement grouping method. This involves a group of additional measurement points located around a point of the network (Alexeyev and Khesin 1989).

The possibility of determining the density by the results of measurements at different levels of horizontal mines (adits, drifts) and by borehole gravimetric measurements is also worth mentioning (Veselov 1986).

Electric parameters can best be estimated by parametric measurements at outcrops (in situ) and electric sounding in the investigated sections.

In many cases, it is advisable not to estimate physical properties by their correlation with others. For instance, when dealing with rapidly changing rocks of various composition in mountainous regions, the correlation coefficient between the velocity and density for intrusive and volcanogenic rocks in the Lesser Caucasus varies from $0.1 \div 0.2$ to 0.5; i.e., the correlation is negligible (Detkova and Shopin 1969). This is related to the pronounced porosity of igneous associations.

The correlation between the physical properties of metamorphic rocks also breaks down. For instance, ultrabasic rock serpentinization leads to a sharp increase in their magnetization and to a simultaneous decrease in their density. Generally, it is not worthwhile to compute the density according to the density/velocity correlation for regions abundant in igneous and metamorphic rocks. The computation of the gravitational field by this type of density model will however be effective in the case of sediment section, where this correlation is reliable enough.

To determine physical properties, the position of the sampling interval (see Fig. 3.5), the representivity of the sample and the number of measurements for statistical processing and stable results all need to be taken into account. For instance, according to Kotlyarevsky, who investigated 20 intrusive massifs in Uzbekistan, the effective magnetization obtained by aeromagnetic prospecting is consistent with the magnetization calculated from at least 200 samples (Khesin 1972).

Statistical processing makes it possible to determine the uniformity of the data sample; it indicates whether to make additional petrophysical differentiations or, on the contrary, to combine certain sub-associations and associations according to their physical properties (Gadjiev et al. 1984, 1985; Ismail-Zadeh et al. 1983a, 1983b). The average characteristics (arithmetic average for a normal distribution and geometric average for a lognormal one) and their confidence intervals are selected on the basis of testing the hypotheses concerning the laws of distribution. The weighted values of the physical properties thus obtained can be attributed to their corresponding geological associations or structural-material complexes, and their distribution can be refined later by physical-geological modeling (Vakhromeyev and Davydenko 1987).

4.1.3 The Formation of an Indicator Space

When developing a model of a medium, the most important issue is to compute the anomalous effects from objects of various classes against the background of a host medium field (or to apply suitable data available in the literature) and to compare them with observed anomalies from similar objects. Comparison of the results serves to correct the model and determine the criteria for singling out the targets. This criterion is a quantitative indicator (such as field value, dissection, gradient, etc.) exceeding the critical level only in the presence of the target.

Effects caused by the targets are often weak under complex conditions and, therefore, one has to use poor criteria. Hence, the probability of misinterpretation (i.e., missing the target or erroneous conclusion as to its presence) increases. By implementing an integrated approach to geophysical study, the risk of making the wrong decision can be considerably lessened. For example, if a weak anomaly[1] is detected using three points, and the mean square of the anomaly in each method is equal to the noise dispersion, the interpretation reliability for separate methods amounts to 0.61 and for the set of three methods it reaches 0.87 (see also Sect. 3.9.2).

The use of indirect indicators; i.e., the effects caused not by the objects proper (for example, ore bodies), but by larger geological bodies (ore-controlling structures, etc.) which are spatially or genetically related to them (see Table 3.4), also reduces the interpretation error. Finally, secondary indicators, in other words, derived quantities (Borovko 1979) such as various initial field transforms and their combinations (smoothed values of a field, its variability, predominant strike of isolines, etc.) are also of major importance. In many cases the secondary indicators are far more important than the initial observed field. It is common knowledge that a

---

[1] A weak anomaly is a relative notion. Under very strong noise typical of open areas the anomaly may have a large amplitude, but is considered to be weak, since it has to be separated from the noise field background of the same intensity.

rare combination of many features is a salient characteristic of large deposits and their significant secondary indicator.

However, too many indicators is a negative factor which affects both labor output and interpretation (see Sect. 3.8.5). At the same time, there are cases where the number of indicators or standards reaches 100 or more. This is the reason why the selection of the most essential features is crucial in developing a model of the medium. In this situation, after critical analysis, geophysical experience in solving similar problems under similar conditions is called for. It is worth mentioning that as a rule, only a few methods are integrated in practice; nevertheless, positive results of integration are well-known (Berezkin et al. 1978; Brodovoi et al. 1989; Khesin 1972; Parasnis 1986; Nikitin (1983); Eppelbaum et al. 2004, 2011a). As shown in Sect. 1.3, a simple analysis testifies to the great interpretative capabilities of a set consisting of three relatively simple methods.

In regional investigations, one and the same set of methods can in principle be employed to study objects of various classes which control different types of mineral resources. Of special interest is a shared approach to predicting the presence of ore and oil-and-gas on the basis of geophysical data, by relating discoveries of non-conventional oil-and-gas deposits in mountainous regions to igneous and metamorphic findings (Khesin 1991). Table 4.2 illustrates these fundamental potentialities.

Secondary indicators are usually obtained from initial field transformations. Therefore, when developing a model of a medium, not only secondary indicators should be selected. Methods of obtaining them, their types and especially their transformation parameters should be specified as well. Experience and theoretical calculations show that the results of transformation depend on parameters, rather than on the type of transformation. Generally, the parameter selection must comply with requirements of maximum efficiency of the ultimate solution to the interpretation problem. But since concrete requirements entail certain difficulties, parameter selection is usually confined to the analysis of changes in the model signal during transformation, and to the evaluation of the related cost.

The choice is determined by the increase in the signal/noise ratio and preservation of as much accumulated information as possible. Of prime importance are indicators that are strongly correlated in the presence of the target and are weakly so (up to negligible) in its absence. The difference in the correlation of indicators is the major source of information about their presence or absence.

Thus, when selecting standards, we are interested in the objects which are well explored not only by geological, but also by geophysical methods involving various transformations, and the study of the relations between indicators, which enables the development of a standard indicator space. As a rule, there are not a large number of objects, but they should be found and used to devise a model of the medium. This highlights the extreme importance of the accumulation and analysis of prior geological and geophysical information for developing a model of a medium. Another approach is computer modeling of geological situations and their respective indicator spaces; i.e. the creation of simulation models.

Table 4.2 Revealing common controls of ore and oil-and-gas deposits by geophysical methods

Geological conrols (fluid conductors and hosts of mineral deposits)	*Predominant location*		*Typical geophysical methods and their results*	
	Mineralization	Hydrocarbons	Methods	Field patterns and geometry of sources
Deep faults	Pinnate joining	Faults zones to neighborhood	Gravity and magnetic prospecting	Gradient zones, local anomaly chains, linearly elongated anomalies
Overlap-overthrust structures	Volcanogenic and sedimentary rock masses	Sedimentary rock masses	Seismic exploration	Gently sloping reflecting boundaries
			Gravity prospecting	Extrema on cover edges
Intersection of longitudinal and transcurrent fractures	Local structures	Regional structures	Gravity and magnetic prospecting	Interference pattern
Igneous rocks concealed highs, their contacts with sedimentary rocks	Exomorphic and endomorphic zones of intrusives	Eroded volcanic structures, sedimentary rock pinching-out zones	Magnetic and gravity prospecting	Anomalies of circular and ring shapes
			Seismic exploration	Zones of zero reflection
Porous and fractured metamorphites (secondary quartzite, serpentinites, shales)	Scattered bodies	Continuous filling	Magnetic and gravity prospecting	Field decreasing (excluding magnetic maxima over serpentinites)
			Electric prospecting	Conductivity anomalies
Hydrothermal alteration and pyritizaton zones	Pyritizaton zones	Below pyritizaton zone	Magnetic prospecting	Linear field decrease
			Electric prospecting	Increased polarizations
Brachy-anticlines flexures in terrigeneous-carbonate deposits	North-eastern flanks of structures	Eastern and north-eastern flanks of structures	Electric prospecting	Contact flexures of media with various physical properties
			Seismic exploration	

Table 4.3 Magnetic and gravity fields over faults

Types of fractures	Geological characteristics	Reflections in fields
Faults – channels for magma flow	Fractured zones filled with basic rocks	Linearly elongated positive anomalies
	Intrusions or volcanic centers localized along fractured zone	Chain of near-isometric maxima, or sometimes minima
Contemporaneous (growth) faults	Abrupt changes in lithological composition, facies and deposit thickness on both sides of a fault	Change of sign or behavior of the same sign field at the fault
Faults fixing vertical displacement of blocks	Abrupt changes of boundary positions separating a section into individual structural facial complexes	Zone of high field gradients (an additional criterion is an abrupt change in occurrence depth for upper or lower anomalous mass edges)
	Crush zone due to differentiated movement of conjugated blocks along the fault	Chain of linearly elongated magnetic and gravity minima coincident in plan
Faults fixing horizontal displacement of blocks	Horizontal rock displacement determined by comparing age, and structural and facies features of rocks on both sides of a fault	Rupture and echelon displacement of zones with linearly elongated anomalies, abruptly inflected isolines

We will not dwell on the quantitative criteria for integrated interpretation. It is simply worth noting that routine qualitative criteria are also integrated, since geological problems that can be solved using one geophysical method are very rare, whereas a combination of only two methods considerably expands the possibilities of geophysical data interpretation. A number of papers dealing, for the most part, with the principles of joint interpretation of gravity and magnetic fields have made this point clear cut. For instance, it is expedient to consider qualitative criteria to single out and trace faults, which are among the main targets of geophysical investigation. It is possible not only to isolate and trace the zones of fracture by some typical indicators in the magnetic and gravity fields, but also to assess their types. The main indicators of various faults reflected in magnetic and gravity fields are summarized in Table 4.3.

Thus, the results generated from a model of the medium should be as follows:

1. Specification of classes of objects under investigation the definition of the geological, petrophysical and geometrical characteristics of typical objects in these classes and corresponding characteristics of host media,
2. Identification of the features from the observed field (and from other geophysical fields) and a set of indicators associated with the targets; determination of the methods and estimates of their anomalous values,
3. Interpretation criteria for singling out the targets according to the given field (and other geophysical fields) and/or to a set of indicators.

It is generally good practice to summarize the results of a model of medium in a table containing geological, petrophysical, geometric and geophysical data on the targets and host media, the key indicators of these objects and their identification criteria. These should be presented as generalized sections with the computed field charts placed above them.

It is also worthwhile presenting parts of the model in the form of petrophysical and/or schematic geological maps that highlight the geological objects that are the probable sources of anomalies and exclude all other objects. These complete data graphic models can be presented as block diagrams.

Naturally, a model of the medium can only partially reflect reality. The greater the accuracy and completeness of the analysis, and the better the use of existing data, the more reliable it is and the more robust the analogies drawn from it. Clearly, a model of a medium is a complex hypothesis of unknown features in the geological structure in the investigation area and its likelihood of containing a certain mineral. Uncertainty is always present (and must be present) in a model. The aim of subsequent field data interpretation is to reduce this uncertainty and to test the hypotheses by applying the criteria (Goltzman 1971).

In the petrophysical sequence of the Greater Caucasus mega-anticlinorium (together with the adjacent area of the Pre-Caucasus) (Fig. 4.5) five density steps were defined (Khesin and Eppelbaum 2007): (1) Quaternary ($\sigma_{aver} = 1.95$ g/cm^3), (2) Paleogene-Neogene ($\sigma_{aver} = 2.23$ g/cm^3), (3) Cretaceous ($\sigma_{aver} = 2.48$ g/cm^3), Bajocian – Upper Jurassic ($\sigma_{aver} = 2.62$ g/cm^3), and Lower Jurassic – Aalenian ($\sigma_{aver} = 2.72$ g/cm^3). Thus, in the section four petrodensity boundaries were found with the following density contrasts at the interfaces (top-down): +0.28, +0.25, +0.14 and +0.10 g/cm^3.

As a whole, the geomagnetic sequence of the Greater Caucasus mega-anticlinorium is characterized by low values of magnetic susceptibility ($\kappa_{aver} = 80{\cdot}10^{-6}$ CGS unit), which increase sharply only for volcanogenic and volcanogenic-sedimentary associations. An analysis of magnetic susceptibility behavior revealed three petromagnetic floors: (1) Pliocene-Quaternary (coarse molasse) with $\kappa_{aver} = 100{\cdot}10^{-6}$ CGS unit, Eocene-Miocene (fine molasse) – $\kappa_{aver} = 60{\cdot}10^{-6}$ CGS unit and Jurassic-Cretaceous – $\kappa_{aver} = 30{\cdot}10^{-6}$ CGS unit.

The upper (uppermost) part of the Kura Depression sequence is represented (Fig. 4.6) by rocks of low density and Jurassic-Cretaceous associations relate to the middle density class only at the bottom of the sequence. Here three petrodensity floors could be identified: (1) Quaternary ($\sigma_{aver} = 1.98$ g/cm^3), (2) Palaeogene-Neogene ($\sigma_{aver} = 2.18$ g/cm^3), and Jurassic-Cretaceous ($\sigma_{aver} = 2.62$ g/cm^3). These floors are interrupted by two petrodensity boundaries (top-down): +(0.20 and 0.44) g/cm^3.

The current stage of the Kura Depression investigation reveals three petromagnetic floors: (1) Pliocene-Quaternary ($\kappa_{aver} = 300{\cdot}10^{-6}$ CGS unit), (2) Eocene-Miocene ($\kappa_{aver} = 100{\cdot}10^{-6}$ CGS unit), and (3) Cretaceous (volcanogenic and volcanogenic-sedimentary deposits with $\kappa_{aver} = 1{,}500{\cdot}10^{-6}$ CGS unit). It should be noted that the first two floors are similar to the floors found for the Greater Caucasus mega-anticlinorium, but the magnetic susceptibility of the corresponding geological associations in the Kura Depression is significantly higher.

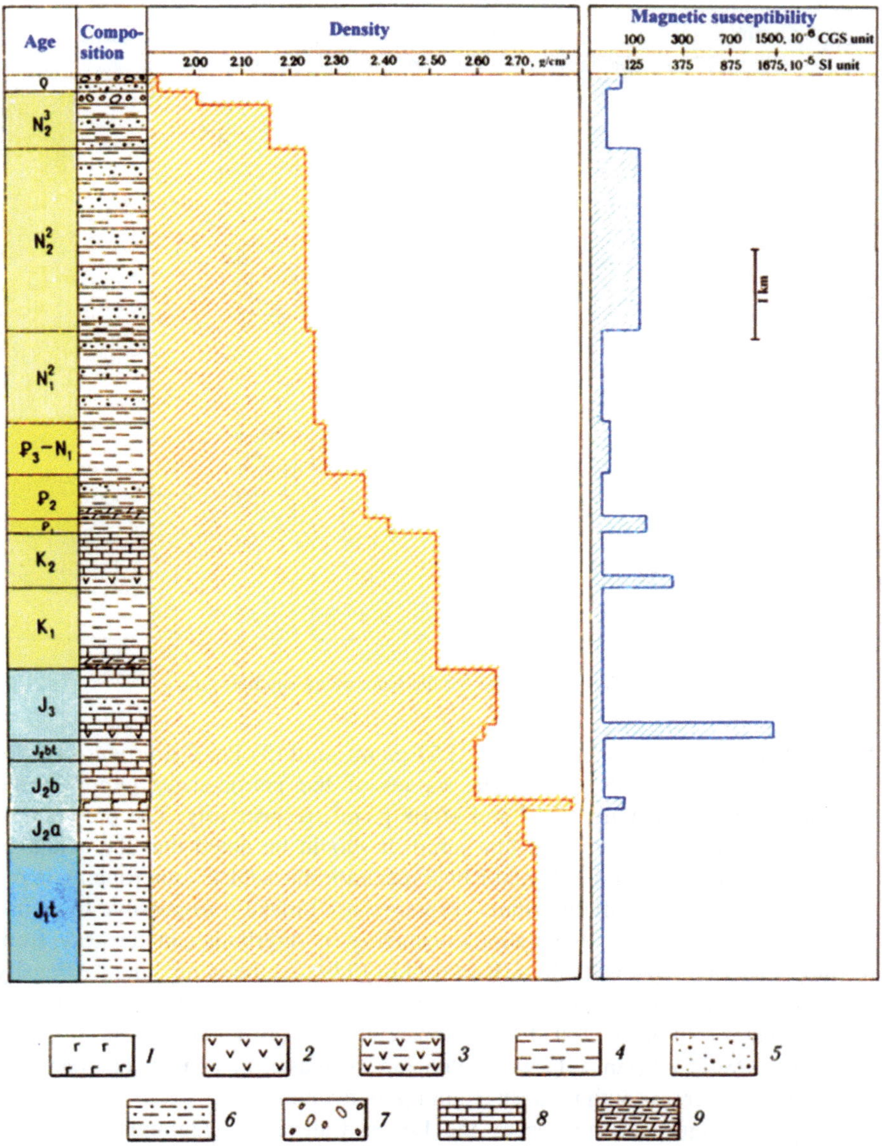

Fig. 4.5 Generalized petrophysical column for the Greater Caucasus mega-anticlinorium and the adjacent area of the Pre-Caucasus. (*1*) basalts, andesito-basalts and diabases; (*2*) andesites; (*3*) volcanogenic-sedimentary rocks; (*4*) clays; (*5*) sands; (*6*) sandy-clay deposits; (*7*) coarse gravels and conglomerates; (*8*) limestone and dolomites; (*9*) marl

The sequence of the Lesser Caucasus mega-anticlinorium is characterized by low differentiation of density at the vertical (Fig. 4.7) that complicates clear identification of the petrodensity boundaries. Initially, three petrodensity floors were described: (1) Quaternary (molasses and river deposits) – $\sigma_{aver} = 1.98$ g/cm^3, (2)

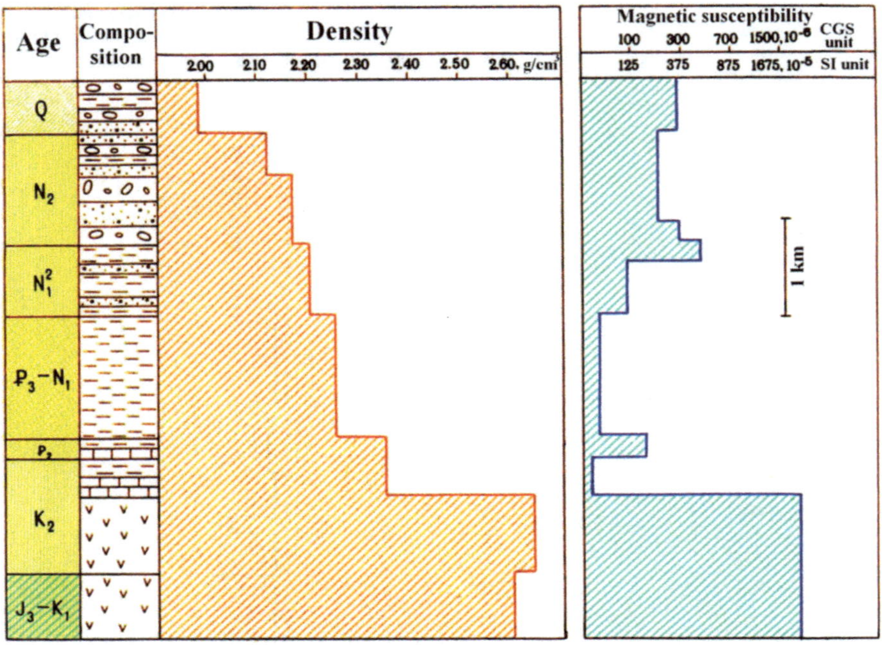

Fig. 4.6 Generalized petrophysical column for the Kura mega-sinclinorium

Upper Bajocian – Negeone – $\sigma_{aver} = 2.57$ g/cm^3, and (3) Pre-Cambrian – Lower Bajocian – $\sigma_{aver} = 2.65$ g/cm^3. The values of the petrodensity jumps at the boundaries (top-down) are +0.57 and +0.08 g/cm^3, respectively. The presence of Quaternary lavas eliminates the first jump. It should be noted that the available density characteristics of the Beyukkishlak metamorphic schists (outcropping in the core of the Shamkhor anticlinorium – Proterozoic ?) make it impossible to distinguish these very ancient associations occurring in Azerbaijanian territory from the independent petrodensity floor (Karkoshkin 1979). At the same time, the density of these associations in Armenian territory testifies to the heightened values of this parameter (Nikolsky et al. 1975). Apparently, as was noted in (Khesin 1976), the Upper Pre-Cambrian – Lower Paleozoic decreases in density compared to overlying rocks from south to north (right up to inversion).

The petromagnetic sequence of the Lesser Caucasus mega-anticlinorium is highly differentiated and is abundant in magnetic and high-magnetic associations (Fig. 4.7). In the sequence there are six petromagnetic floors: (1) Eocene-Quaternary ($\kappa_{aver} = 700 \cdot 10^{-6}$ CGS unit), (2) Upper Cretaceous – Paleocene ($\kappa_{aver} = 100 \cdot 10^{-6}$ CGS unit), (3) Upper Cretaceous ($\kappa_{aver} = 700 \cdot 10^{-6}$ CGS unit), (4) Upper Jurassic – Lower Cretaceous ($\kappa_{aver} = 250 \cdot 10^{-6}$ CGS unit), (5) Middle Jurassic ($\kappa_{aver} = 600 \cdot 10^{-6}$ CGS unit), and (6) Paleozoic – Lower Jurassic ($\kappa_{aver} = 10 \cdot 10^{-6}$ CGS unit).

In the Nakhichevan Folded region three petrodensity floors were identified (Fig. 4.8): (1) Neogene ($\sigma_{aver} = 2.17$ g/cm^3), (2) Upper Cretaceous – Oligocene ($\sigma_{aver} = 2.51$ g/cm^3), and (3) Paleozoic – Jurassic ($\sigma_{aver} = 2.65$ g/cm^3). The jumps

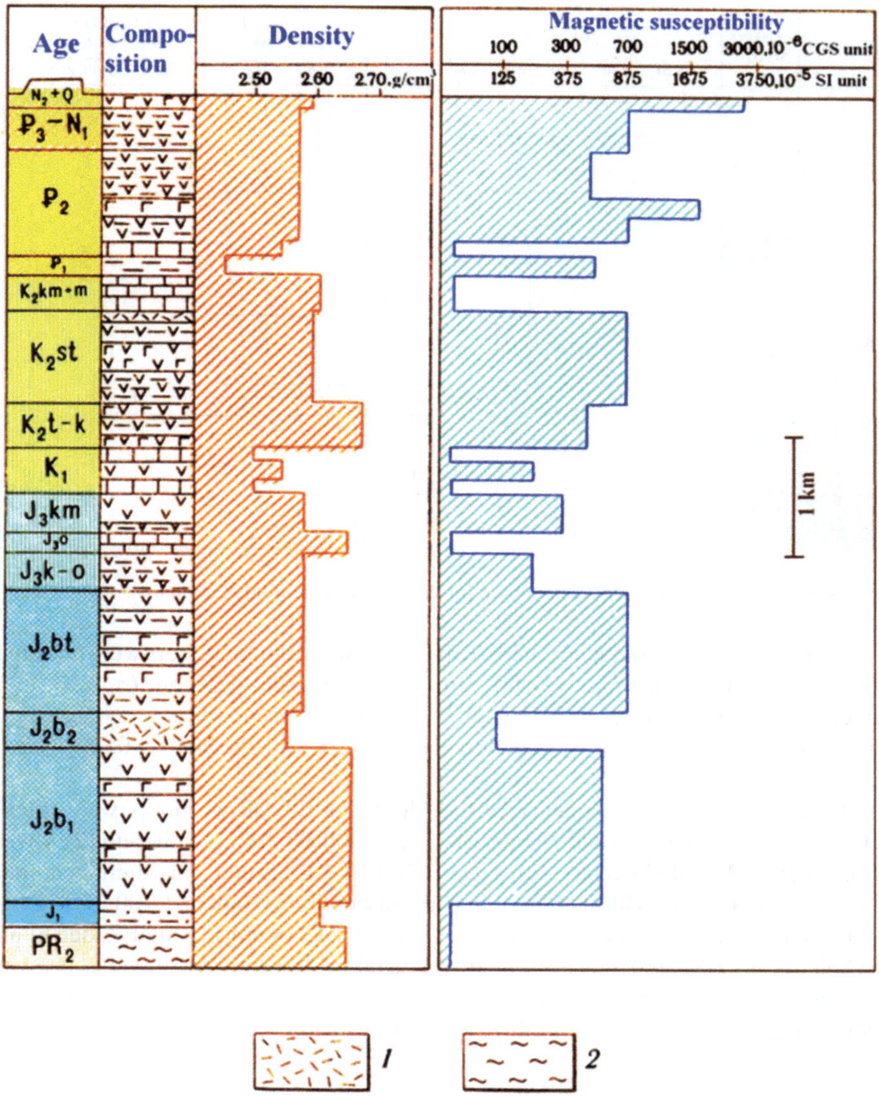

Fig. 4.7 Generalized petrophysical column for the Lesser Caucasus mega-anticlinorium. (*1*) liparites and plagioliparites; (*2*) metamorphized schists (other symbols are the same as in Fig. 4.5)

on these two petrodensity boundaries (top-down) are: +0.34 and 0.14 g/cm^3. Analysis of magnetic susceptibility behavior served to delineate five floors (Fig. 4.8): (1) Oligocene-Miocene ($\kappa_{aver} = 100{\cdot}10^{-6}$ CGS unit), (2) Eocene ($\kappa_{aver} = 1{,}500{\cdot}10^{-6}$ CGS unit), (3) Upper Cretaceous – Paleocene ($\kappa_{aver} = 50{\cdot}10^{-6}$ CGS unit), (4) Jurassic ($\kappa_{aver} = 1{,}500{\cdot}10^{-6}$ CGS unit), and (5) Paleozoic ($\kappa_{aver} = 10{\cdot}10^{-6}$ CGS unit).

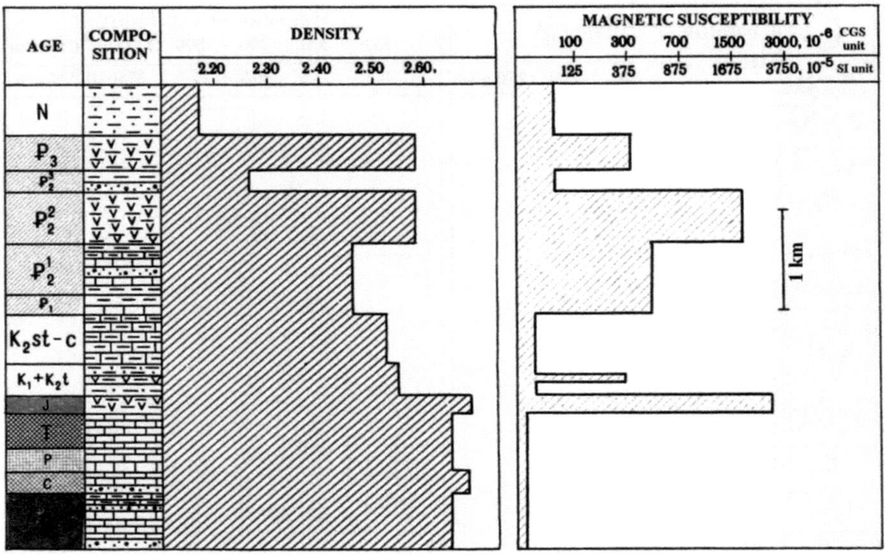

Fig. 4.8 Generalized petrophysical column for the Nakhichevan folded region

In the petrodensity sequence of the Talysh anticlinorium four floors were detected (Fig. 4.9): (1) Pliocene ($\sigma_{aver} = 2.00$ g/cm^3), (2) Oligocene – Miocene ($\sigma_{aver} = 2.25$ g/cm^3), (3) Eocene ($\sigma_{aver} = 2.53$ g/cm^3), and (4) Paleocene ($\sigma_{aver} = 2.45$ g/cm^3). The discontinuity jumps (top-down) are: +0.25, +0.28, and −0.08 g/cm^3, respectively.

In the petromagnetic sequence of the Mountain Talysh two floors were found (Fig. 4.9): (1) upper (Eocene-Neogene) $\kappa_{aver} = 300 \cdot 10^{-6}$ CGS unit, and (2) lower (Paleocene – Middle Eocene) with $\kappa_{aver} = 700 \cdot 10^{-6}$ CGS unit. It is significant that the Kosmalyan suite with its high magnetic susceptibility (about 1,700 10^{-6} CGS unit) and residual magnetization (about 1,900$\cdot 10^{-6}$ CGS unit) (Ismail-Zadeh et al. 1983b) is divided into intervals of direct (Middle Eocene) and reverse (Lower Eocene) magnetization. This fact confirms earlier conclusions about the nature of negative magnetic anomalies in this region (Khesin 1968).

4.1.4 Common Characteristics of Petrophysical Boundaries and Geological Associations

Density characteristics patterns, together with some common peculiarities can have both shared and distinct features (Fig. 4.10).

The shared feature of all these sequences is the presence (typical of all geostructures) of two petrodensity boundaries: (I) between the Quaternary (sometimes Neogene-Quaternary) deposits and the underlying denser associations ($\Delta\sigma_{aver} = 0.2$–$0.3$ g/cm^3)

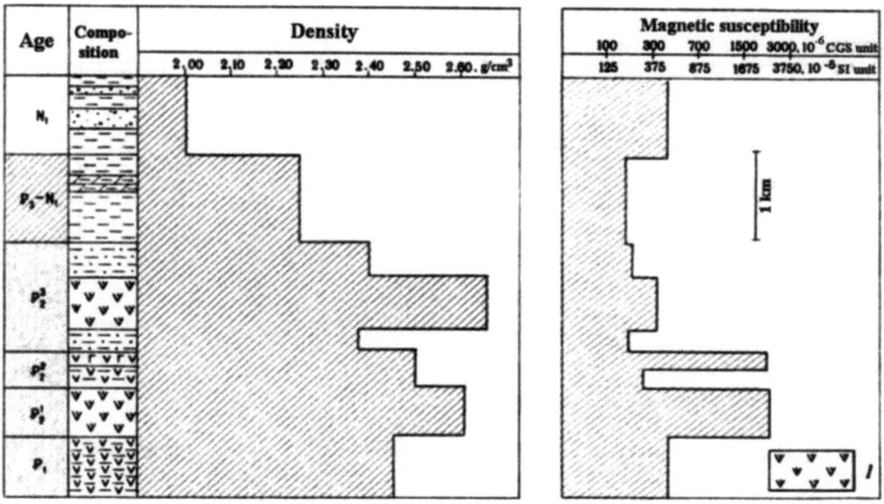

Fig. 4.9 Generalized petrophysical column for the Talysh anticlinorium. (*1*) trachyandesites and trachybasalts

and (II) the boundary between Cenozoic and Mesozoic associations ($\Delta\sigma_{aver}$ consists of 0.14–0.44 g/cm^3). However, the latter boundary in the sequence of the Lesser Caucasus mega-anticlinorium is poorly defined since $\Delta\sigma_{aver}$ reaches only 0.05 g/cm^3. Here the reference horizon of lower density and magnetization are liparites and plagio-liparites of the Upper Bajocian.

As a whole, the densest rocks are typical of intrusive formations in early- and late-geosynclinal stages. The rock density of intrusive formations in the early-geosynclinal stage ranges approximately from 2.80 to 2.95 g/cm^3 (excluding Bajocian plagio-granites whose chemical consistency leads to a density of 2.50–2.60 g/cm^3). The density of the intrusive formations of the late-geosynclinal stage ranges on average from 2.70 to 2.85 g/cm^3 and thus should be included in the class of heightened density.

The density of intrusive formations in the pre-orogenic stage has a clear tendency to decrease to the class of middle density (σ_{aver} = 2.65–2.75 g/cm^3). The intrusive formations of the orogenic stage have the lowest density (σ_{aver} = 2.55–2.60 g/cm^3).

The rocks composing effusive formations mainly belong to the class of heightened (σ_{aver} = 2.70–2.85 g/cm^3) and middle (σ_{aver} = 2.50–2.70 g/cm^3) density. Rocks of heightened density are typical of the effusive formations of the early-geosynclinal stage. The density of the later stages gradually decreases, and rocks of the orogenic stage can almost completely be assigned to the class of low density (σ_{aver} = 2.45–2.60 g/cm^3).

Among the background sedimentary deposits the densest (σ_{aver} = 2.60–2.75 g/cm^3) are the sand-shale associations of the mega-anticlinorium of the Greater Caucasus, as well as carbonate deposits of the Late Jurassic – Early Cretaceous. The tendency for decreased density from the early to late stages is also present here.

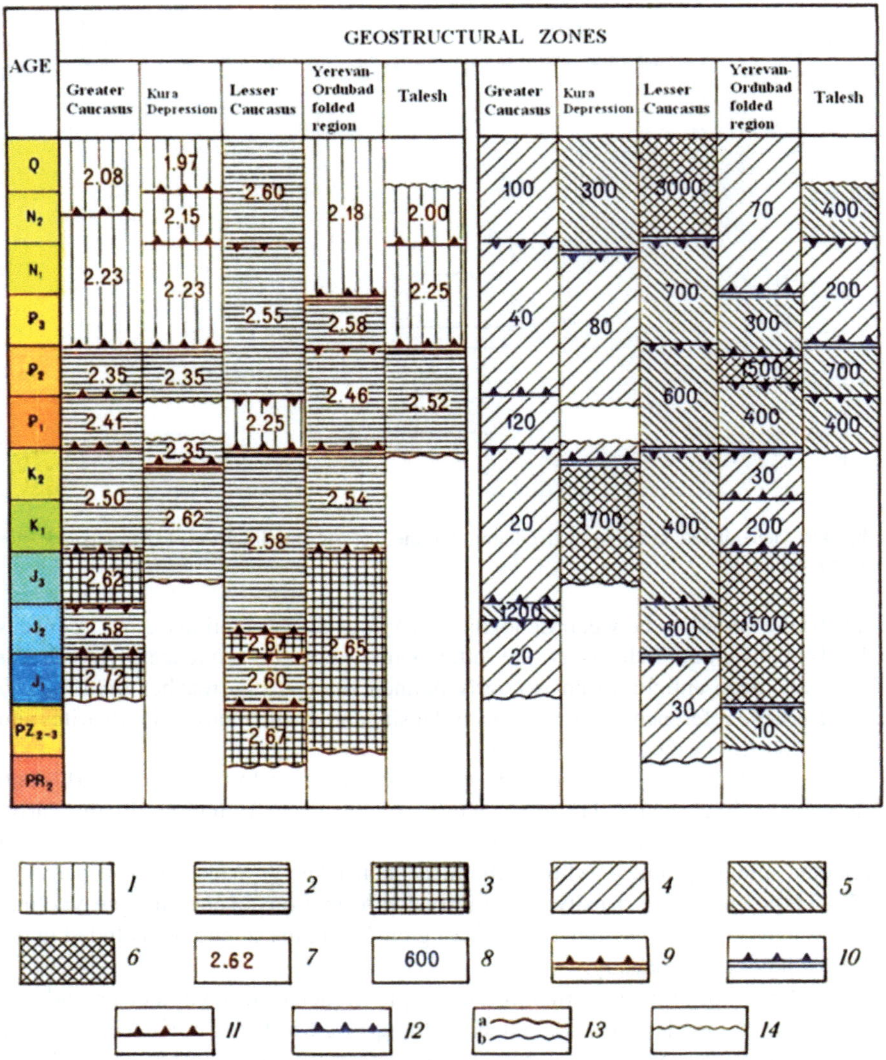

Fig. 4.10 Determining petrophysical floors (Ismail-Zadeh and Khesin 1989b). Intervals of average density values, g/cm^3: (*1*) 1.95–2.30; (*2*) 2.31–2.60; (*3*) 2.61–2.90; intervals of average magnetic susceptibility values, 10^{-6} CGS: (*4*) 0–250; (*5*) 251–700; (*6*) >700; (*7*) average value of density, g/cm^3; (*8*) average value of magnetic susceptibility, 10^{-6} CGS; clearer petrophysical boundaries: (*9*) density; (*10*) geomagnetic; less clear petrophysical boundaries: (*11*) density; (*12*) geomagnetic; (*13*) lower boundary of density (*a*) and magnetization, (*b*) investigation; (*14*) boundary of petrophysically unknown section. Note. The *arrows* in 9–12 are oriented in their direction of density or decreasing magnetization

The Pre-Alpine associations (metamorphic shales of the Upper Proterozoic and sub-platform terrigeneous-carbonate Paleozoic deposits) belong to the middle-density class.

As a first approximation, a geomagnetic sequence can be divided into three floors corresponding to the main tectonic stages. The lower floor (Pre-Jurassic), is represented mainly by terrigeneous-carbonate associations penetrated by acid intrusions, and is practically non-magnetic. One exception is the volcanogenic association revealed in the Shamkhor anticlinorium of the Lesser Caucasus. The middle floor (Mesozoic-Eocene), within which basic and middle consistency thick volcanites develop, is distinguished by its heightened magnetization. The upper floor (Post-Eocene) represented mainly by the terrigeneous deposits of the sedimentary cover, is characterized by a lower magnetization (although in the some areas Neogene-Quaternary lavas with high magnetization have developed).

It should be noted that in the intermediate floor, the volcanogenic associations of the Middle Jurassic have the highest average magnetic susceptibility (in the Lesser Caucasus their thickness attains approximately 4 km), and less thick vulcanites of the Upper Cretaceous. The residual magnetization of these rocks is parallel or antiparallel to the induced magnetization; in the zones of inverse magnetization typically there are relatively low-magnetic associations.

The Cenozoic volcanogenic associations of the Talysh and Pre-Talyshian as well as the lava covers in the central part of the Lesser Caucasus have a different characteristic. Here paleomagnetic zones of different polarities alternate. Computations indicate that the upward continued magnetic anomalies (associated with the gently sloping Cenozoic associations) were rapidly damped.

The lavas are characterized by the largest values of magnetization compared to piroclastic associations of the same consistency. Magnetization of hyperbasites is approximately proportional to their degree of serpentinization (at the same time the hyperbasite density decreases by degree of their serpentization).

Analysis of the sequence of the main Caucasian petrodensity and petromagnetic geostructures (Fig. 4.10) thus suggests that these floors are correlated in terms of age, but the average values of their physical characteristics and relations differ. Obviously, this is caused by the different histories of these geostructures' geological development. Nevertheless, the physical parameters can be assessed by creating 3D physical-geological models of the Earth's crust of the Caucasus with 3D gravity-magnetic modeling as described below.

4.2 Regional Geophysical Schemes

4.2.1 Quantitative Analysis and Regioning

Figure 4.11a exemplifies the use of the tangent method and the method of characteristic points, whereas Fig. 4.11b presents the method of characteristic areas (for the inclined thin bed model).

The Guton magnetic anomaly is situated in NW Azerbaijan (southern slope of the Greater Caucasus), near the border with Russia (Fig. 4.12). A detailed quantitative interpretation of this anomaly was carried out along 15 profiles crossing this

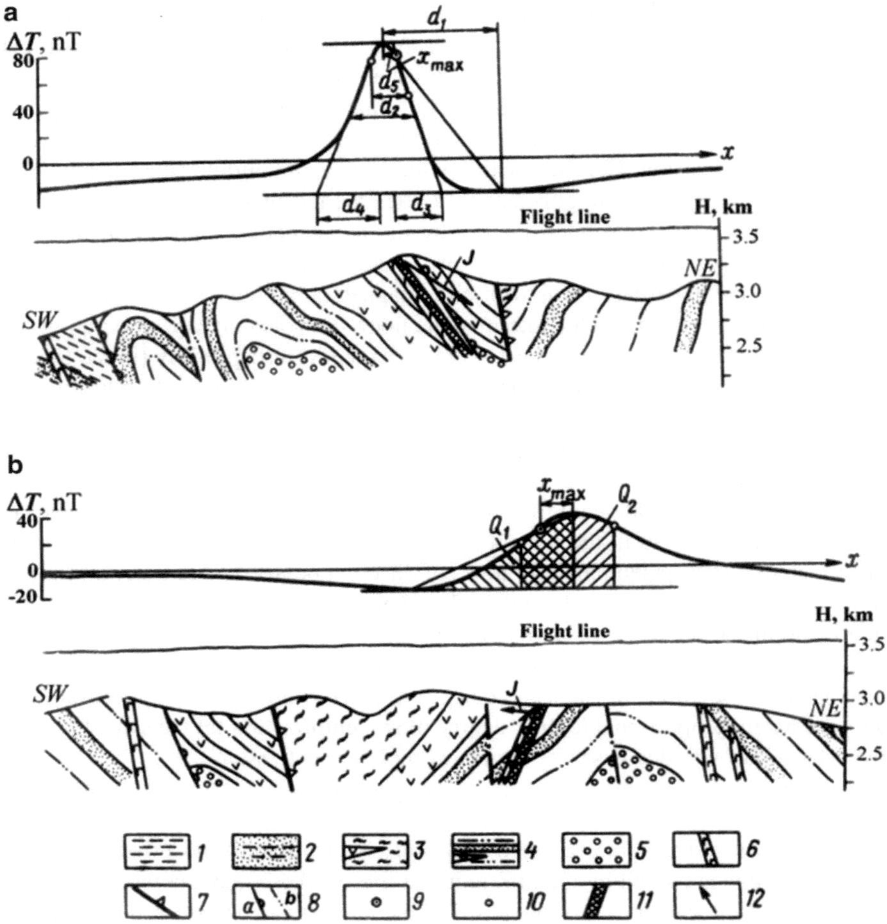

Fig. 4.11 Examples of quantitative interpretation of ΔT plots along profiles 171 (**a**) and 181 (**b**) in the area of the Big Somalit (southern slope of the Greater Caucasus). (*1*) Yalakhkam suite J_2aal_2; (*2*) Zainkam suite J_2aal_1; (*3*) Nagab suite J_1toa_3; (*4*) Tseilakhan suite J_1toa_3; (*5*) the Lower and Middle Toarcian suite J_1toa_{1-2}; (*6*) dikes of the gabbro-diabasic association; (*7*) the Major Caucasian upthrust-overthrust; (*8*) ore controlling (*a*) and ore distributing (*b*) upthrust-overthrusts; (*9*) Reford's point; (*10*) inflection points; (*11*) anomalous body according to the interpretation results; (*12*) obtained direction of the magnetization vector

anomaly. The results along one of these profiles (the anomalous body was approximated by a thick inclined bed) are presented in Fig. 4.13 (here, improved tangent, characteristic point and areal methods were applied). The methodology is described in detail in Khesin et al. (1996). The data indicate that the anomalous body is characterized by comparatively low magnetization (250 mA/m), considerable vertical thickness (about 30 km) and a steep dip of the lateral contacts. Analytical continuation and singular point methods applied on the same profiles

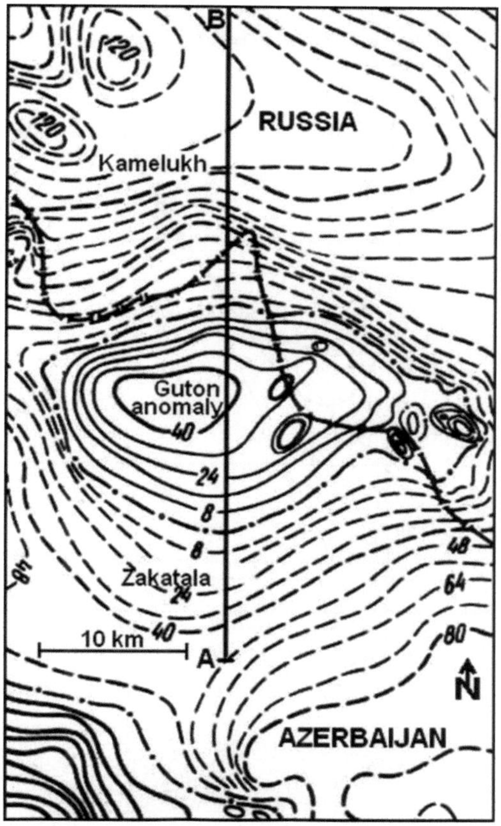

Fig. 4.12 Guton magnetic anomaly (results of an airborne survey) on the southern slope of the Greater Caucasus. The map shows the total magnetic field ΔT (isolines are given in mA/m) and the location of interpreting profile **A–B**

gave similar results (Fig. 4.14). The characteristics of this anomalous body testify to the intermediate-acid composition of this target (intrusion). The interpretation of the significant vertical thickness of this body agrees with the geothermic data on the depth of the Curie discontinuity in this area (about 30 km) (this value was calculated by A. N. Pilchin).

The outcroppings at the Earth's surface formed sub-volcanic and sub-intrusive bodies of different consistencies which are apparently fragments of this large magmatic massif penetrating the upper part of the section along the extended faults of the common Caucasian direction. This magmatic focus is associated with the rich pyrite-polymetallic deposits of the Belokan-Zakatala ore field and possibly other areas in the Greater Caucasus (Ismail-Zadeh and Khesin 1989b).

The calculation techniques should be decided upon when developing the initial model of the medium (see Sect. 4.1). After the field work is completed, the set of quantitative interpretation techniques can be corrected as a function of the particular types of anomalies that were observed. This set of calculations includes techniques to obtain the parameters of a certain class of model bodies typical to the area under survey. The techniques should meet the following criteria:

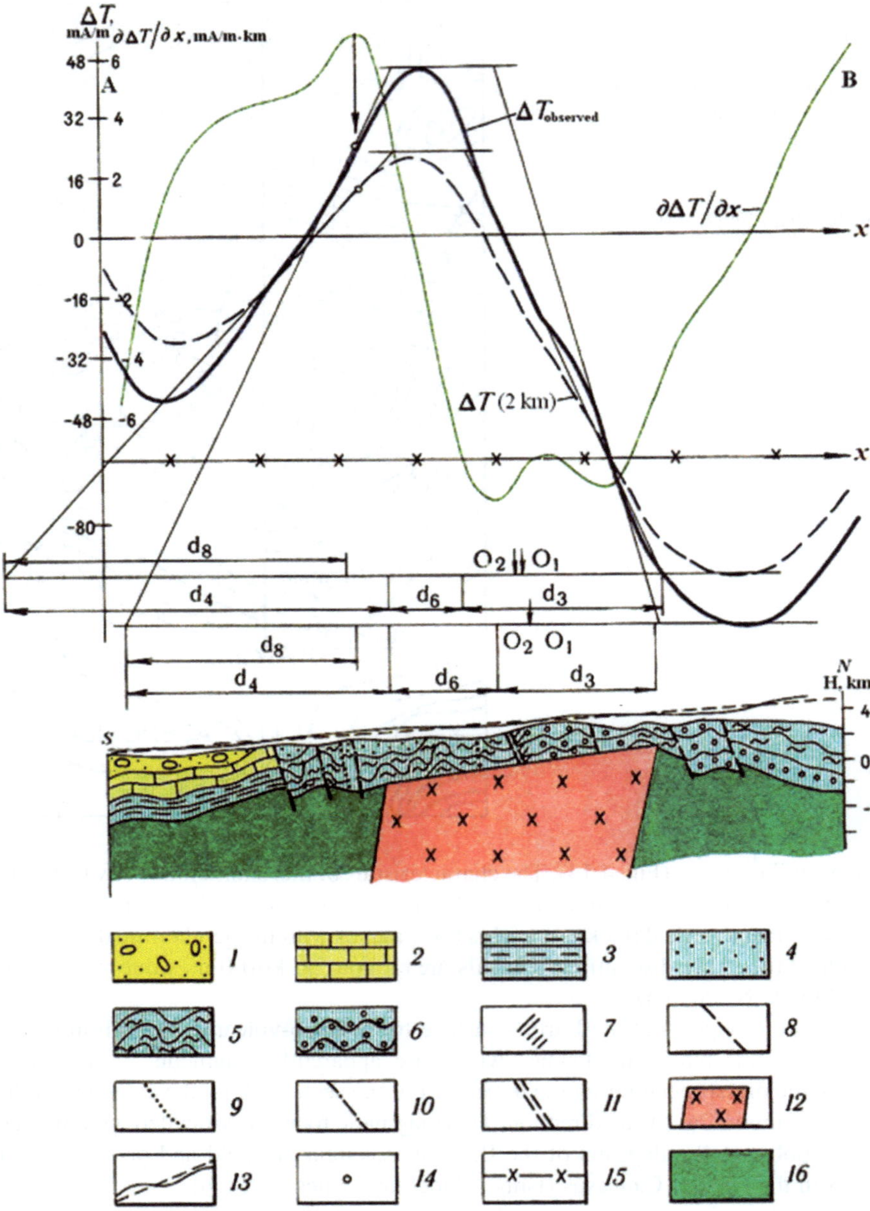

Fig. 4.13 Interpretation of ΔT graphs on two levels along profile 28 through the Guton anomaly (southern slope of the Greater Caucasus). (*1*) recent alluvial deposits; (*2*) limestones, tuff sandstones, clay shales (*K*); (*3*) mudstones, tuff sandstones (J_3); (*4*) monolith clay shales and coarse-grained tuff sandstones (J_2); (*5*) sandy-clay shales with horizons of sand flysch, metamorphosed clay shales and sandstones (J_2); (*6*) phyllitizated clay shales, sandstones, spilites (J_1); (*7*) dikes and sheet bodies of the gabbro-diabasic association (J_2); (*8*) regional upthrust-overthrusts; (*9*) upthrust-overthrusts separating the longitudinal tectonic steps of the second order;

applicability to the specific conditions of the geophysical investigation, in particular under the oblique polarization and rugged terrain relief characteristic of open areas in central and low latitudes, independence of the normal background level and the choice of the origin of coordinates. The optimal techniques are those which can successfully obtain the above- mentioned parameters. Of special interest is the geophysical field selection.[2]

The research and development briefly presented in Chap. 3, in Khesin et al. (1996) as well as in Eppelbaum and Gasanova (1990) and Eppelbaum (2010a) are worth recalling and are summarized in Table 4.4.

The parameters should be determined using a combination of techniques, so that independent data on the parameter values can be obtained. When the 2-D condition is satisfied, the calculations should ideally be carried out along a series of profiles in the mid-portion of the anomaly, where the field isolines are nearly parallel. The results should then be averaged. This will ensure higher validity in determining the parameter values.

To obtain the anomalous bodies parameters, the first stage involves interpolation selection methods (the method of characteristic points, the tangent method, the areas method, etc.). These make it possible to obtain a large number of parameters in a fairly simple way. The next stage involves the methods of singular points and analytical continuation, because by this time the researcher already has the necessary information about the quantitative characteristics of the object. Comparing the results obtained by the methods of interpolation selection to those obtained through the method of singular points will yield a more reliable and valid classification of singular points. The same applies to the method of analytical continuation. In this case, because the range of the upper edge occurrence depth is known, one can avoid the risk of field continuation into the vicinity of the sources.

To verify the calculations by the values of Z and X under oblique magnetization, it is advisable to calculate these components under vertical magnetization (Alexeyev and Khesin 1971) with the following interpretation of the obtained curves using simple techniques (Logachev and Zakharov 1979).

Specifically, having made the necessary corrections in the calculations (if the profile is non-horizontal and non-perpendicular to the body strike, etc.), we get a set of values for each parameter. Since the results of the calculations are probabilistic, it is useful to describe them not only by the average value of the parameter, but in

Fig. 4.13 (continued) (*10*) upthrust-overthrusts complicating the longitudinal tectonic steps; (*11*) transverse fractures; (*12*) magmatic intrusion of intermediate-acid composition according to the interpretation data (in non-segmented J_{1-2} complex); (*13*) the lines of flight and averaging inclined straight line; (*14*) inflection point of the plot ΔT nearest to the maximum on the *left*; (*15*) corrected zero line of the plots ΔT; O_1, O_2 are locations of the origin (middle of the anomalous body's *upper* edge) obtained from $x_{un.r}$ and $x_{un.l}$, respectively; (*16*) J_{1-2} complex

[2] The well-known term "selection" implies an approximation selection, if not stated otherwise techniques and other methods applicable under rugged terrain relief.

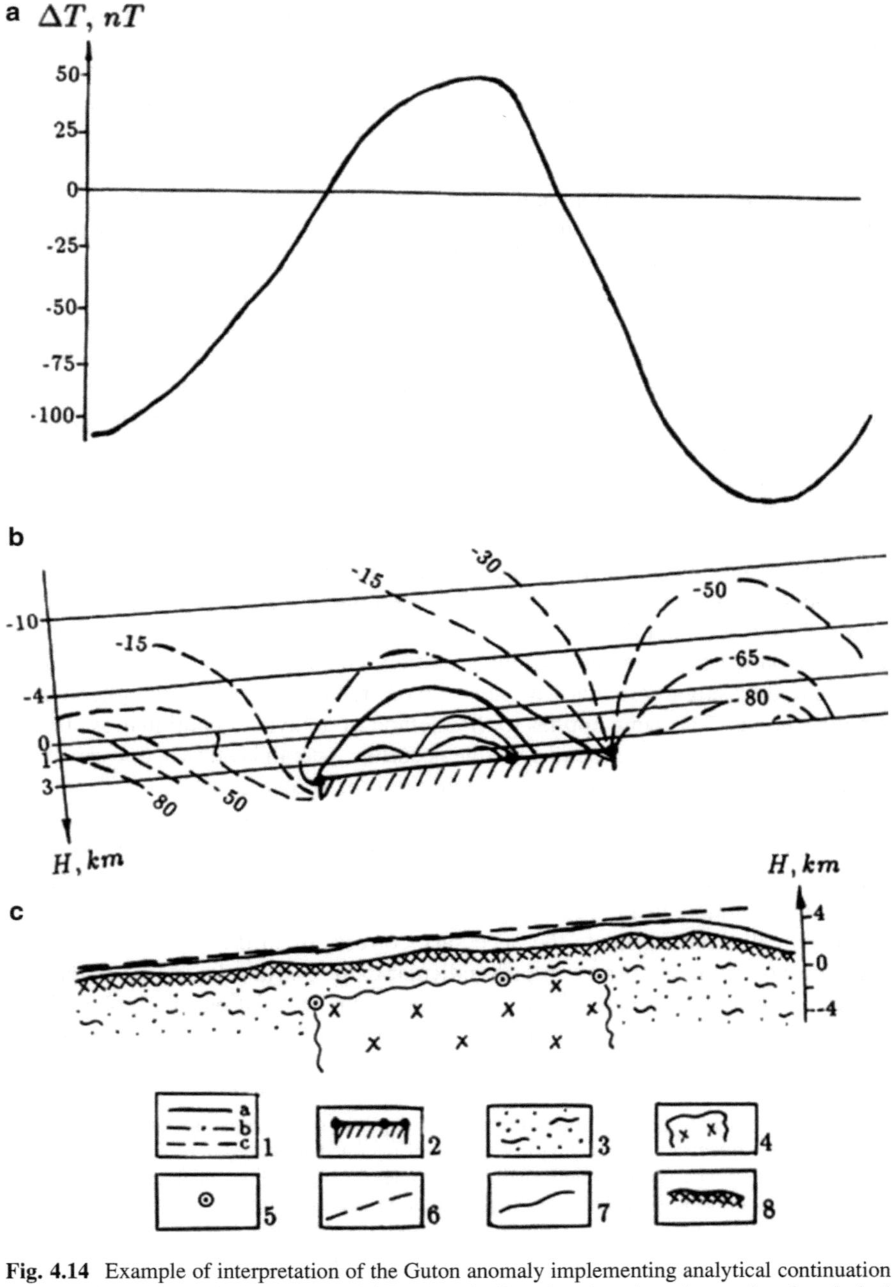

Fig. 4.14 Example of interpretation of the Guton anomaly implementing analytical continuation and singular point methods. (**a**) plot of ΔT; (**b**) an isoline chart based on the results of analytical continuation in an inclined half-space; (**c**) a schematic section of the anomalous object. (*1*) isolines, nT: (*a*) – positive, (*b*) – zero, (*c*) – negative; (*2*) magmatic intrusion of an intermediate-acid composition (according to the interpretation results); (*3*) the *upper* edge of the anomalous object, obtained by the method of analytical continuation (*black* points show logarithmic peculiarities of the field); (*4*) singular point; (*5*) inclined straight line approximating the observation profile

Table 4.4 Elements of a geophysical field processing and interpretation system under complex environments

FIELD	Time variation correction	Terrain correction using correlation method	Logico-Information algorithms for revealing targets	Inverse problem solution				3-D integrated modeling of fields
				Rugged relief	Arbitrary magnetization (polarization)	Approximation of anomalous object		
						1–3 models	*4–5 models*	
Magnetic	+ ⊙	⊙	+⊙	⊙	⊙	⊙	⊙	+ ⊙
Gravity	+	⊙		⊙	⊙	⊙	+	+ ⊙
Thermal	+ ⊙	⊙		⊙	⊙	⊙	–	⊗
DC	⊙	⊙		⊙	⊙	⊙	–	–
SP	+	+		⊙	⊙	⊙	–	–
VLF	+ ⊙	⊙		⊙	⊙	⊙	–	–
IP	⊖	⊙		⊙	⊙	⊙	–	–
Piezoelectric	⊖	⊙		⊙	⊙	⊙	–	–
Long wire	⊖	⊙		⊙	⊙	⊙	–	–

+ formal presence of procedure, – absence of procedure, ⊙ principally new or nonconventional procedure developed by the authors, ⊗ preparing theoretical basement for realization, ⊖ no need for calculation

terms of the standard deviation, as well. Therefore, for each of the anomalies the average value of parameters and the standard deviation should be calculated; the latter is an estimate of the inner convergence of the calculations obtained by different techniques.

The selection methods are used at the final stage of (Khesin et al. 1993b), when a sufficient amount of data on the object parameters has been acquired. These data enable important corrections and modifications in the initial medium model, thus forming an initial approximation to make the selection: the better the initial approximation, the faster the selection.

The initial approximation involves tracing the average parameters obtained by the above techniques onto the section (scheme) plotted when developing the initial model of the medium with previous data. The information on the sources' shapes and dimensions provides a rationale for selecting the type of approximating expression. The mean-square divergence in the parameters, with prior information and fieldwork experience taken into account, are used to impose constraints on the range of possible values of the parameters. With sufficient prior geological and geophysical data ensuring a valid initial approximation, there is no need for other methods of interpretation. By contrast, when prior information is inadequate, a physical simulation (Avdevich and Fokin 1978) can be useful for efficient preliminary rough estimates of the object parameters. This is followed by an interactive selection. Thus there is a successive application of relatively simple techniques (the methods of tangents, singular points, etc.) that lead to identification of the individual features of an object, followed by a more complicated selection, which determines the overall characteristics of the causative masses.

Under field conditions and at the very beginning of an interpretation calling for numerous calculations, it is advisable to estimate the change ranges for a number of values by some very simple calculations of the bounding parameters (Gladky 1967; Khesin 1969).

To enhance reliability when calculating the anomalous object parameters, techniques that measure different individual elements of the anomaly or the curve as a whole should be applied jointly. These include the methods of interpolation selection, singular points, analytical continuation and selection based on approximation optimization. The optimum combination is determined by the specific nature of the region under survey and the nature of the material. For example, the density of the observation network and the accuracy of the field measurements impose certain restrictions on the method of singular points. The results of this method also depend on the noise fields of the sources occurring above the singular points of the objects being localized.

The applicability of methods under rugged terrain relief is not, in most cases, a decisive one if the observed anomaly is initially converted into a horizontal or inclined plane. The need for anomaly reduction is determined by the chosen of set of techniques.

When interpreting anomalies quantitatively, it is essential to combine techniques that differ in their imposed constraints. For instance, while the limiting inequalities determine the largest possible depth, the method of singular points underestimates the depth of the upper edge of the object under survey. The latter method is characterized by a higher susceptibility to random errors and a lower sensitivity to background effects, whereas the opposite is true for the method of areas. The parameters determined by techniques with diverse constraints (and chosen by their informativity) make it possible to obtain the mean-square error of a parameter. In this case the results will be represented in the form of a confidence area rather than in the form of a point in the section. Note that the intersection of confidence areas for the results obtained by different geophysical methods helps choose the most reliable interpretation in terms of the geological data.

It is valuable to compare the average parameters obtained by a statistical approach to the results of a deterministic interpretation. The technique worked out by Klushin and Tolstikhin (1961) estimates the average occurrence depth of magnetized masses by computing the normalized autocorrelation function $R_n(\tau)$ of a field with sources represented by a set of thin vertical beds. The average depth of the upper edge of magnetized sources was calculated by the following formula:

$$\bar{h} = \frac{1}{\pi} \int_0^\infty R_n(\tau) d\tau = \frac{\tau_e}{\pi}, \tag{4.6}$$

where τ_e is the autocorrelation radous.

We obtained $\bar{h}$ value of 4.3 km which is close to the average depth of the magnetic field's singular points. This technique was applied to the meridional profile through Western Azerbaijan (Fig. 4.15) and confirmed Paffenholtz's (1959) geological concepts.

The very rugged relief in the Caucasian region means that various methods should be used to estimate different heights of geophysical field observations. The topographic mass attraction may be of no use, for example, for magnetic investigations directed at singling out a magnetized body among non-magnetic host rocks or for electric exploration to locate conducting ores in a high-resistance medium. In any case, however, the distorting effect of a non-horizontal observation line takes place when the object differs from the host medium by contrasting properties and causes an anomalous vertical gradient.

After applying all the corrections (for "free-air", plane-parallel intermediate layer and terrain relief with correct intermediate layer density (σ_{int})), the Bouguer anomaly corresponding to the local anomaly takes on small negative values (minima) in the direction of the relief depression, whereas the anomaly on the horizontal profile has no negative values. The ΔT anomaly takes a form on the inclined profile as though it corresponded to a smaller magnetization inclination, unlike the case of a horizontal profile. This shows a good fit to the theoretical results reported in (Khesin et al. 1983). Thus, applying topographic corrections does not eliminate the effect of different observation point heights with respect to the anomalous object. The same conclusion was drawn earlier by Nemtsov (1967).

As a general rule, to avoid the analytical continuation through the possible causative masses, these observation results need to be reduced to their common

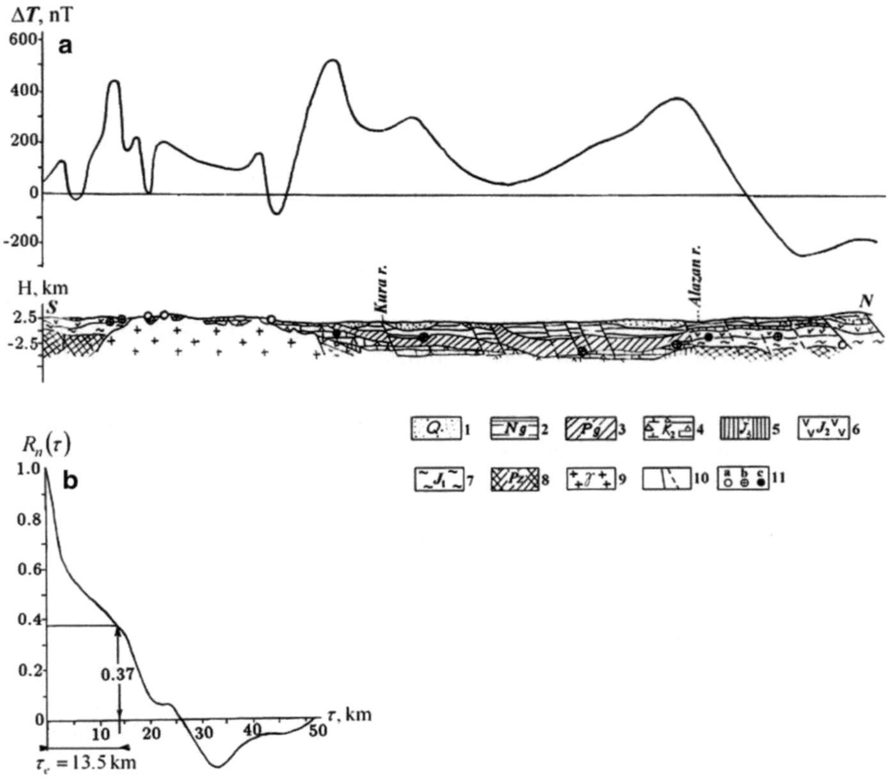

Fig. 4.15 Determination of the average occurrence depth of magnetized masses by the autocorrelation function of the magnetic field along the Gedabey-Belokan profile: (**a**) ΔT plot, (**b**) the plot of the autocorrelation function. Deposits: (*1*) Quaternary; (*2*) Neogene; (*3*) Paleogene; (*4*) Upper Cretaceous; (*5*) Upper Jurassic; (*6*) Middle Jurassic; (*7*) Lower Jurassic; (*8*) Paleozoic; (*9*) intrusive rocks of the granodiorite series; (*10*) disjunctive dislocations; (*11*) singular points of the magnetic field with the following magnetization values: (*a*) up to $0.5 \cdot 10^{-3}$ CGS, (*b*) from 0.5 to $1 \cdot 10^{-3}$ CGS, (*c*) larger than $1 \cdot 10^{-3}$ CGS

level, usually the highest point level (reduction to line). This means introducing a correction for an anomalous vertical field gradient; i.e., for the gradient component caused by anomalous bodies and the normal component of the vertical gradient.

Quite a number of methods for reducing to a horizontal level are known (Andreyev and Klushin 1962; Berezhnaya and Telepin 1966; Bhattacharyya and Chan 1977; Golizdra 1977; Khesin et al. 1983; Lukavchenko 1961; Roy 1970). The principle of the majority of them consists of using the successive approximation method to select such a field on the intermediate plane that coincides with the measured one when continued to the observation surface. The observed field is assumed to be a zero approximation; i.e., it is considered to have been recorded on the intermediate plane. Then, this field is continued upward to the observation line. The difference between the observed and computed fields is the correction at each point, which is added to the zero approximation (observed field). The first approximation is calculated in this way. Then, the first approximation field is continued again to the measurement line, and the correction is computed. Then the difference in the results between the first and the second upward continuations are calculated. This correction is added to the first approximation. After this manipulation, we obtain the second approximation, which is subjected to the same procedure until the correction approaches to the observation error. The final approximation is used for continuing to any horizontal plane. Among these techniques, Aronov's (1976) method is the most well-known.

When the above methods are carried out, this yields the upward continuation. As a result, both anomalies caused by relief forms as well as anomalies in the objects under investigation collapse or are attenuated (see, for example, Fig. 16.5 from (Dobrin 1976)). Therefore, the decision to use continuation should be evaluated with regard to the type of work, the problems and the available data.

It is well known that for the purpose of estimating the usefulness of reduction to the common level, the degree of anomaly distortion can be approximated from the formula for any component of the recorded field

$$\Delta U = \frac{\partial U}{\partial z} \Delta H, \tag{4.7}$$

where $\frac{\partial U}{\partial z}$ is the vertical derivative of the field, and ΔH is the relative elevation of the flight height (observation line), i.e. the terrain clearance of the aerial survey.

In aerial surveys along sinuous (in the vertical plane) routes and ground surveys, the sinuosity effect, i.e. reducing observations to horizontal or inclined planes is theoretically necessary. It substantiates the solution to certain problems associated with quantitative determination of anomalous body parameters in that neglecting the effect of variation in the observation point heights may introduce errors into the interpretation. Hence the reduction level should be as close to the acting masses as possible and, if required, several levels should be selected.

Within areas in which the surface profiles have close inclination angles, distortions of the field and its transforms due to the relief also occur. This means there is no need to carry out reduction to the plane if the preliminary regioning of the main forms has determined the dominant dips of the routes (see Fig. 3.14).

Subsequent analysis of the field and its transforms should be executed independently within the singled-out areas. Note that the results relate to fictitious bodies of identical forms, and the sizes are similar to those of the real anomalous bodies (see Chap. 3). The evidence shows that the positions of fictitious and real bodies in plane do not differ greatly (max. 0.4 of body depth for a relief dip angle up to 30°). This is not essential for shallow-seated objects. When the relief is very complex in certain plots it may be necessary to reduce the field to the plane, which should be inclined and as close as possible to the relief.

A general case of the complexity of the physical properties and the geophysical anomaly polygon distribution can be found in the Caucasus. Take for example the magnetic anomaly distribution in the Somkhet-Agdam tectonic-magmatic zone of

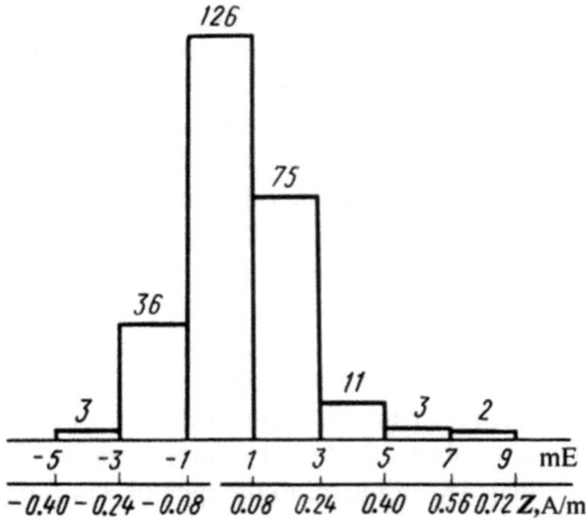

Fig. 4.16 Histogram of residual magnetic anomalies (*Z* component) distribution (after upward continuation to the level of 25 km) for the Somkhet-Agdam Zone of the Lesser Caucasus. Values over the histogram intervals show the frequencies of magnetic anomalies

the Lesser Caucasus (Fig. 4.16). Even within the limits of this zone the magnetic anomaly distribution does not correspond to a normal or lognormal distribution that fits either the Pearson criterion or the less rigid Kolmogorov criterion (Yaglom and Yaglom 1973). Interestingly, the observed magnetic field distribution the average values are close to the dispersion. This is specific to the Poisson distribution which represents the physical nature of anomalies as rare events.

The similarities in anomalies ΔT and Z_a for the Kura Depression can be explained by the northern ($\approx 45°$) dip of the disturbing bodies approximated by thin beds (Alexeyev 1976). This assumption is corroborated by the regional geological sections through the Kura mega-sinclinorium. For instance, in the upper part of the sedimentary cover there are large northern sharply dipping faults. In many cases these faults form a plateau whose depth curls to the north. Thus, the magnetic anomaly bodies occurring in the Kura Depression section at a depth (mainly in the Ganja region) were exposed by tectonic forces oriented to the north. Hence the magnetic data examination confirms the opinion of A. V. Peive and certain other investigators as to the significant role of tangential movement and the subduction of the Lesser Caucasus under the Greater Caucasus structure (e.g., Khain and Alizadeh 2005).

To delineate the ore mineralization associated with intrusive magmatism, the structures shown in Fig. 4.17 are especially instructive. They are presented as series of local anomalies corresponding mainly to the basic phases of granodiorite intrusive bodies.

There are several advantages of the direct use of topographic maps for geological purposes which are well-known in geomorphology. Topographic data (digital terrain models) are used to determine topographic corrections in geophysics. Furthermore, the data on the terrain relief (height field) can be a source of additional geological information in processing and interpreting geophysical observations.

Geological regularities, which manifest themselves in mountainous relief structures, can be highlighted by treating the height field of the area with the techniques employed in geophysical field analysis. For example, Borovko (1971)

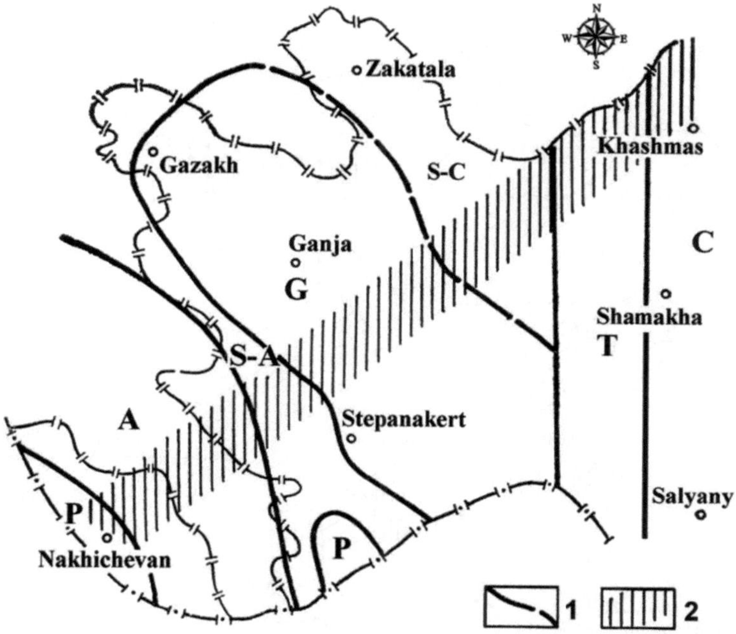

Fig. 4.17 Geophysical data of the Azerbaijan tectonic regioning (After Khesin 1976). (*1*) boundaries of tectonic units; (*2*) the largest zone of anti-Caucasus dislocations delineated in most geophysical maps

presented an effective application for ore prediction of the predominant strikes of relief isohypses. Later Khesin (1976) showed that this transformation can identify a very important indicator of endogenic deposits of various compositions and origins (Fig. 4.18). Large deposits tend to occur in the corners of blocks having a predominant strike of isohypses that differs from the Caucasus as the whole. This led to the recommendation to drill deep test wells in the mountainous regions of Azerbaijan.

The relief complexity, which was described above, governs the spatial distribution of topographic corrections, and can be estimated by a map of the specific sinuosity of height isolines. Khesin (1981) proposed applying the specific sinuosity of isolines (*SSI*) to characterize the complexity of a geophysical field.

It is calculated as follows: (1) the total length of isolines (L) for a given field is computed within a sliding area, (2) area S on the sliding cell is calculated and (3) the parameter

$$SSI = \frac{KL}{S}, \tag{4.8}$$

is computed, where K is the scale factor for converting L into kilometers, and S into square kilometers. The *SSI* is measured in km/km^2, and the value obtained refers to the center of this cell.

The map not only characterizes the complexity of the terrain relief and determines the requirements for eliminating its effect, but also enables geological

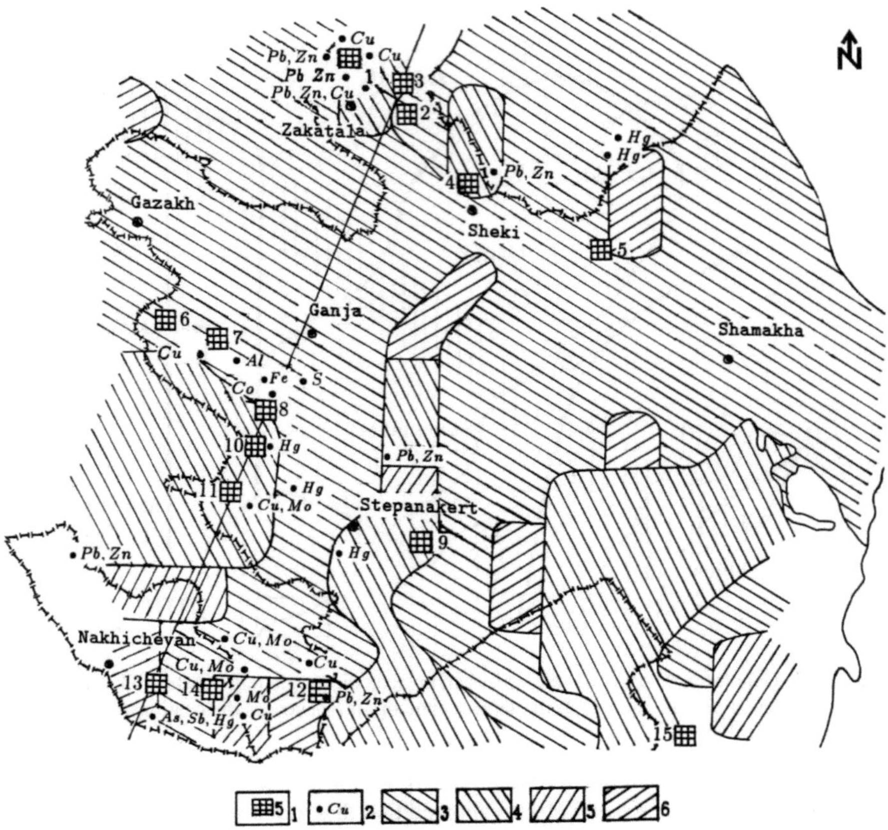

Fig. 4.18 Test well drilling site pattern in mountainous regions of Azerbaijan. (*1*) sites recommended for test well drilling and their numbers; (*2*) ore deposits; predominant courses of terrain relief isohypses: (*3*) W.N.W.; (*4*) N.N.W.; (*5*) N.N.E.; (*6*) E.N.E

interpretation. For example, when analyzing the large-scale data on the Mekhmana ore district (Lesser Caucasus), it was established that the oldest associations from the Lower Bathonian are characterized by a reduced specific sinuosity of horizontals, whereas in the areas of younger rocks a corresponding increase was observed. This parameter increases along disjunctive dislocations, and axes of synclines.

To summarize, the application of height field transformations enhances to the informativity of a set of geophysical investigations practically without any expenditure, since the terrain height data need to be calculated for topographic effects. These transformations of the digital terrain model can be executed together with the geophysical field conversion using programs developed for computing field entropy, and curvature or the standard deviation. Obviously, the spatial distribution of the entropy values will be close to the distribution of the *SSI* parameter.

4.2.2 *Field Differentiation into Regional and Local Isotropic Components*

Various techniques of averaging are used to single out regional isometric anomalies in the Caucasus. These include the method developed by Tikhonov and Bulanzhe (1945) who initiated the extensive implementation of transformations. Other techniques that are applicable are analytical upward continuation and smoothing with polynomials. However averaging of fields with asymmetric distributions using the conventional techniques may result in distorting the regional component by the intensive local anomaly. Therefore, it is recommended to apply the method of quantiles, in particular, the median method (Borovko 1971). The 50% quantile, the median, corresponds to the middle of the variation series composed of sequentially increasing field values within the transformation cell. To differentiate isometric anomalies by the median method, the radius r of the apparent graticule is chosen according to

$$r = \sqrt{r_1 r_2} \tag{4.9}$$

where r_1 and r_2 are the radii of field features to be separated (regional and local objects, respectively) using half of their amplitude.

Usually the transformation cell (sliding window) size should be larger than the local noise, less than the background features and close (or slightly higher) than the expected signal. Since there is always a correlation of results when overlapping the neighboring cells by a sliding window, it is not recommended to reduce the computation step to more than half a radius of the transformation cell.

Kunin (1968) proposed a relationship between the average area radius r and the continuation height H for the gravitational field, leading to a good fit of this transformation:

$$r \approx (2.5 \div 3)H. \tag{4.10}$$

The comparison of regional component charts of Azerbaijan obtained by averaging and upward continuation corroborates this conclusion regarding the magnetic field as well (Fig. 4.19).

Usually the choice of transformation technique is up to the researcher. A clear physical sense of the transformation by upward continuation suggested by Andreyev and Klushin (1962), and its comparability with aircraft observations testify to the advantages of this method. However, the uncontrollable influence of the discarded residual term in Poisson's integral often turns out to be very high.

In informal calculations, along with their fundamental advantages, productivity plays an important role in the selection of the type of conversion. The median method is attractive from this standpoint. Actually, the assumption of compensation of random deviations from the average within the limits of the transformation

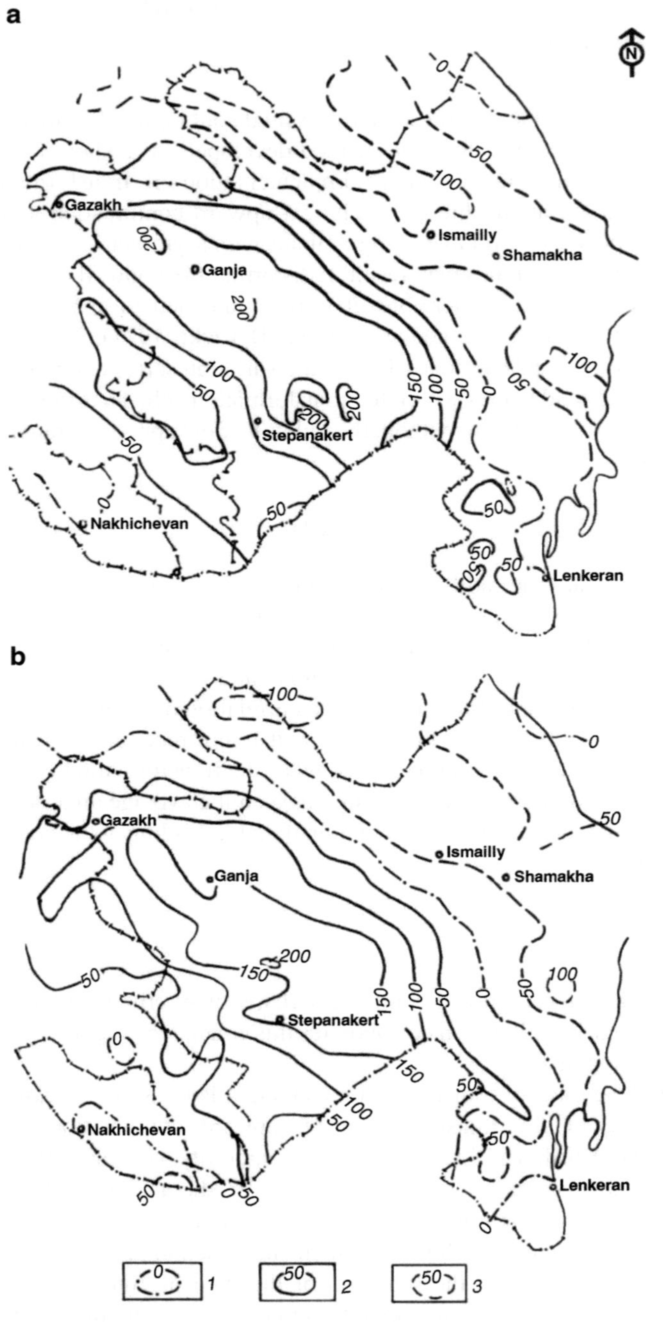

Fig. 4.19 Singling out the regional component of the magnetic field in Azerbaijan: (After Khesin 1981). (**a**) average within a radius of 60 km; (**b**) continued upward to a height of 25 km. Isolines in nanoTesla: (*1*) zero; (*2*) positive; (*3*) negative

cell (averaging method) may not hold true in smooth fields, whereas for sharply varying fields polynomial smoothing is completely inapplicable. The results of application of the sliding average method, as shown by Borovko et al. (1969), are more dependent on the anomaly amplitude than those of the median method. The latter exhibits very high productivity, which is confirmed by experience in Azerbaijan.

Polynomial and other types of smoothing serve to localize the targets in plane (for example, using the isoohm chart to single out the distribution of water-bearing associations). It should be noted that the same problem could also be solved in the vertical plane by vertical electric sounding data. This is done by separating the vertical section of the apparent resistivity ρ_a into its regional and local components.

High reliability can basically be attained in geophysical field regioning and reveal regional structures through areal autocorrelation analysis. However, the results of this method are highly dependent on the field description interval and the size of an apparent graticule. In addition this analysis is accompanied by a significant loss of the area on the map edges and only yields general results, which rarely or never have individual value.

In order to identify the general features of geophysical fields (and the height field as well) it is often advisable to calculate tertiary indicators; i.e., *SSI* by formula (4.8), the length of dislocations obtained by sliding window calculations (Khesin and Metaxas 1974).

The *SSI* distribution (Fig. 4.20) reflects certain geological features of the region. Interestingly, this chart correlates rather well with the map of seismic activity (Riznichenko et al. 1983).

The role and significance of various indicators in the indicator space is different when the goal is regioning. Here regions with different signs and intensity of the field are mainly singled out by maps of the initial field and the field on a number of levels of the upper semispace incorporating local and difference anomalies. Field region boundaries and other linear elements are traced by horizontal gradients and predominant strike isolines maps using the initial field and regional anomaly maps and employing local elongated anomalies from corresponding charts. The local anomalies are singled out by maps of local and difference anomalies on the basis of the horizontal gradient map. These together give an idea of the depth extension of the sources.

For example consider the regioning scheme of the gravitational field in Azerbaijan as compared to the regional steps of the magnetic field (Fig. 4.21). Regional peculiarities of the gravitational and magnetic fields determine the regional factors of both oil-and-gas and ore control. The Shamkhor-Gedabey-Dashkesan ore zone and the Mekhmana ore district (Lesser Caucasus), the Belokan-Zakatala ore field (Greater Caucasus) which were identified on the basis of systematic geophysical studies, as well as the Kutkashen-Ismailly district (Greater Caucasus), which has potential endogenic mineralization according to a number of indicators (Khesin 1976) are controlled by the positive difference field Δg_{8-20} (Fig. 4.22) that corresponds to zones 8 and 2, respectively, in Fig. 4.21.

The regional minima of the same component outline the prospective oil-and-gas regions. The lower density of the Cenozoic terrigeneous rock masses (2.0 $\div$ 2.3 g/cm^3) unambiguously defines the regions of lower density at depth (gravity

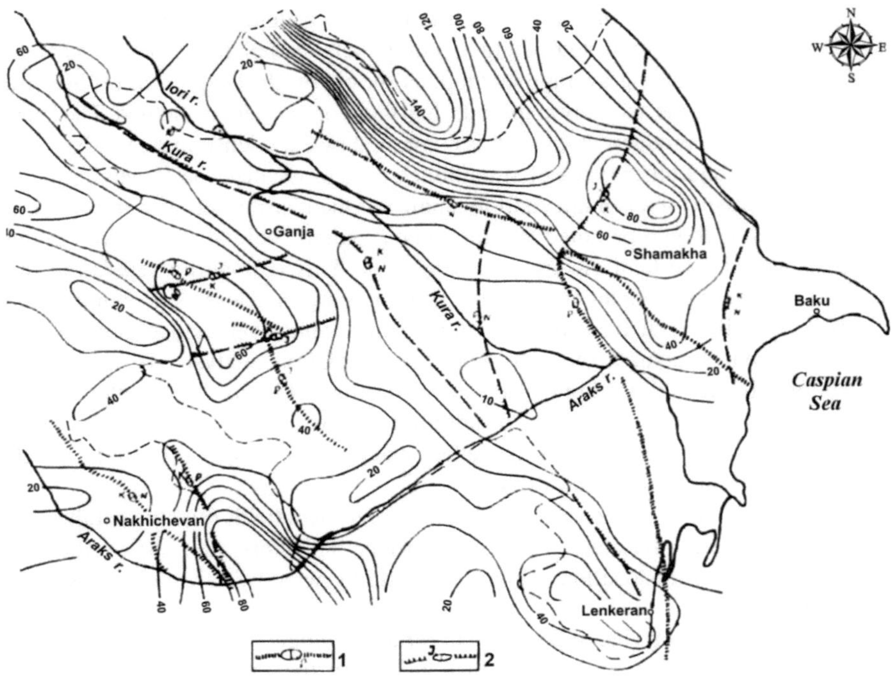

Fig. 4.20 Map of equal lengths of isohypses for the topography of Azerbaijan (After Khesin et al. 1996). (*1*) deep fault (*dots* show buried paths of the fracture); (*2*) flexures (on the surface as shown by *solid lines*, and buried, as shown by *dashed lines*); indices point to the expected section interval dissected by fractures

minima) as having the largest accumulation of sedimentary rock masses. The latter are promising as regards oil and gas resources in the Palaeogene-Miocene and Pliocene-Quaternary deposits. These regions are characterized by the conjunction of the Kuba-Khachmas and Iory-Agrichai Zones with the Kobustan-Lower Kura Zone in the east, and the Evlakh-Agdjabedy Zone (negative field regions 1, 5, 6, and 14, respectively, in Fig. 4.21).

Accumulated experience shows the importance of performing independent regioning using initial field and regional anomaly maps. Comparing the results in a unified scheme makes it possible to determine common features and differences in deep structure elements and in the upper portion of the section, and reveal the elements (primarily, disjunctive dislocations and blocks) that manifest themselves at the different structural stages.

The transforms reflecting the degree of heterogeneity of the field such as anomalous coefficients, dispersion, entropy, etc. are the most useful for singling out field areas in open regions. In different areas, these parameters can take on different geological meanings. They are also appropriate for solving more specific problems, in particular in the regioning of separate portions by the degree of

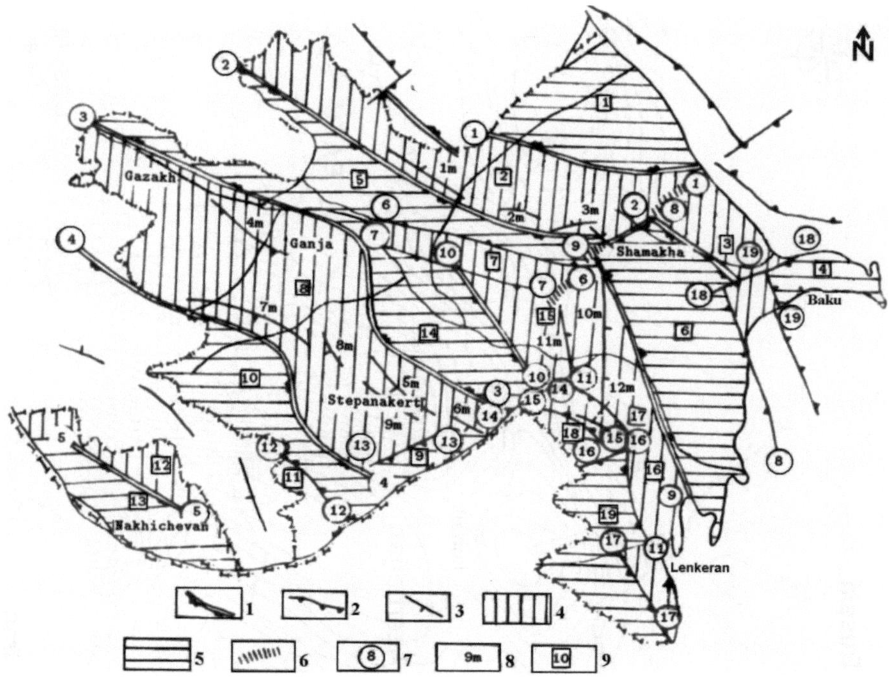

Fig. 4.21 Comparison of gravitational field regioning and regional steps of magnetic field mapping (After Khesin et al. 1996). (*1*) axes of intense extended gravity steps distinguished in the $\Delta g_{8\text{–}20}$ and $\Delta g_{4\text{–}10}$ difference fields and the horizontal gradient field obtained from Δg at a level of 2 km; (*2*) axes of marked gravitational steps distinguished in the same fields; (*3*) axes of regional magnetic steps distinguished in the magnetic field at a level of 10 km and in the field of its horizontal gradient at a level of 2 km; (*4*) $\Delta g_{8\text{–}20}$ positive gravity field blocks and zones; (*5*) $\Delta g_{8\text{–}20}$ negative gravity field blocks and zones; (*6*) weak boundaries of $\Delta g_{8\text{–}20}$ positive or negative gravity field areas; (*7*) gravitational step number (beginning and end of a step are indicated); (*8*) magnetic step number; (*9*) number of $\Delta g_{8\text{–}20}$ positive or negative gravity field zones and blocks. Note: The *arrows* on the symbols show the direction to the blocks of lower density or magnetization

heterogeneity of their lithological complexes. Here, it makes sense to compute the specific sinuosity of the isolines (see Sects. 4.2.1 and 4.2.2).

A map of predominant trends of magnetic anomaly distribution (Fig. 4.23) could be also used for regional tectonic-structural examination. Note that selected four trends (0°–45°, 45°–90°, 90°–135°, and 135°–180°) are sufficient for description of map of any complexity (see Sect. 3.9.5).

4.3 3-D Combined Modeling of Gravity and Magnetic Fields

The *GSFC* (Geological Space Field Calculation) program was developed to solve direct 3D gravity and magnetic prospecting problems under complex geological conditions (Khesin et al. 1996). This program was designed to calculate the field of

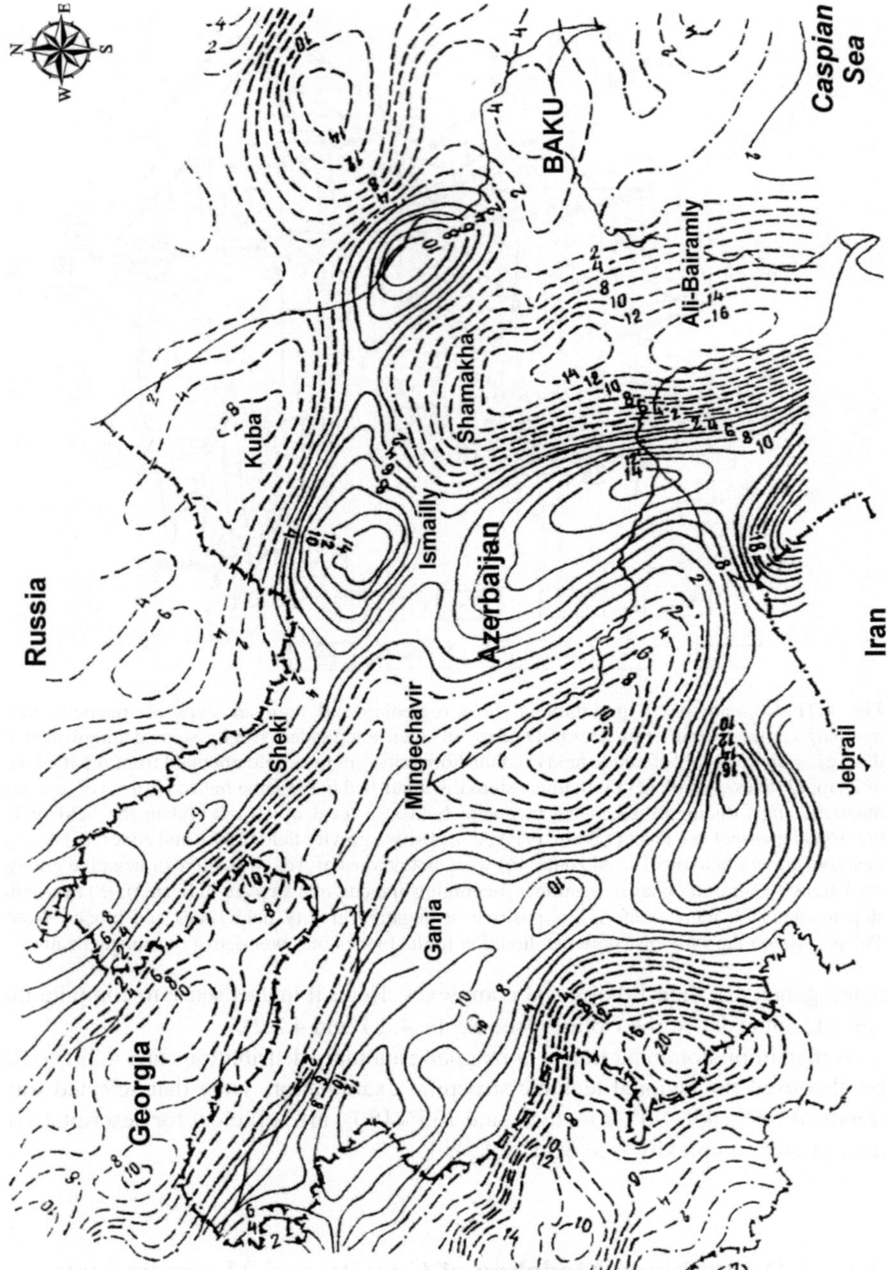

Fig. 4.22 Map of difference Δg_{8-20} anomalies computed as the difference between upward continued fields in the Bouguer reduction (terrain correction was computed with a radius of 200 km and the density of the intermediate layer was defined as 2.67 g/cm^3) at heights of 8 and 20, respectively (Khesin 1976; Gasanov 2001)

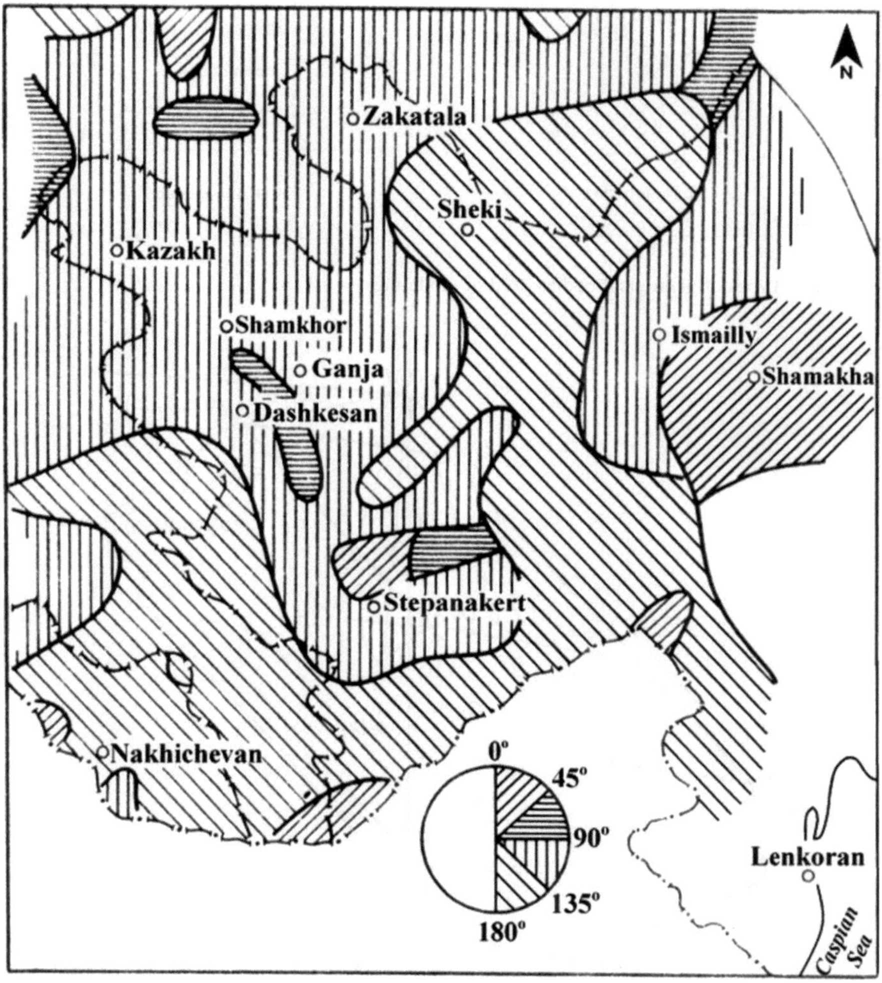

Fig. 4.23 Map of predominant trends of magnetic anomalies for the Eastern Transcaucasian (Khesin et al. 1983)

Δg (Bouguer, free-air or observed value anomalies), ΔZ, ΔX, ΔY, ΔT, as well as second derivatives of the gravitational potential under conditions of rugged relief and inclined magnetization. The geological space can be approximated by (1) three-dimensional, (2) semi-infinite bodies and (3) infinite along a closed strike, (L.H. non-closed, R.H. on-closed and open) (a simplified description is shown in Fig. 4.24). Geological bodies are approximated by horizontal polygonal prisms (Fig. 4.25).

The program has the following main advantages:(1) Simultaneous computing of gravity and magnetic fields; (2) Description of the terrain relief by irregularly

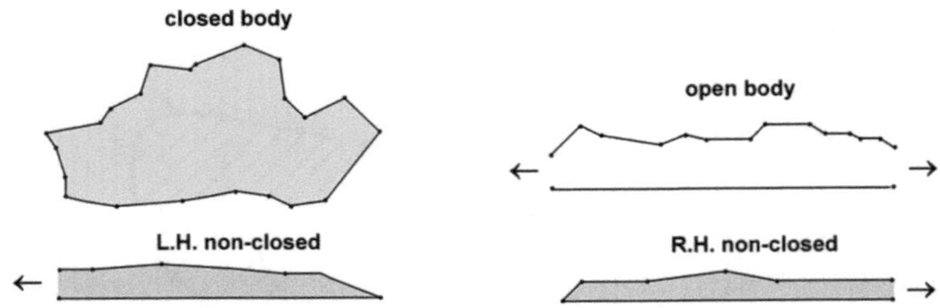

Fig. 4.24 Types of geological bodies used in modeling

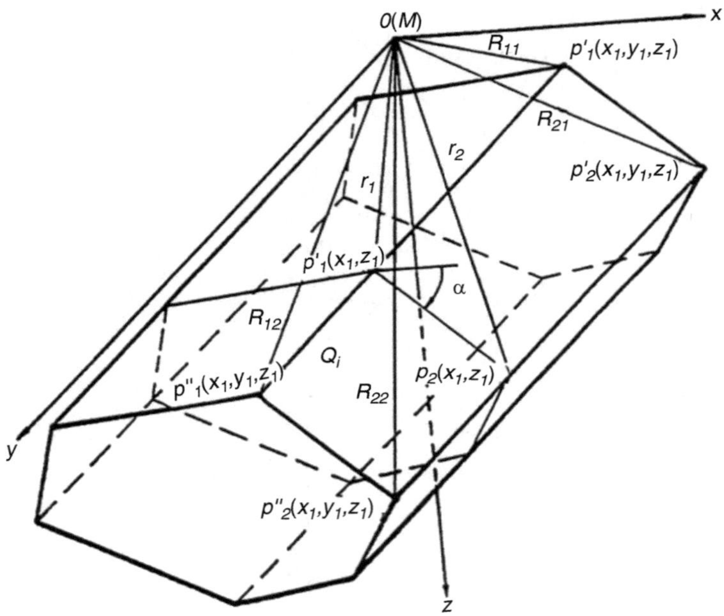

Fig. 4.25 A diagram for computing derivatives of gravity potential for a horizontal polygonal prism

placed characteristic points; (3) Computation of the effect of the earth-air boundary by the selection method directly during the process of interpretation; (4) Modeling of the selected profiles flowing over rugged relief or at various arbitrary levels (using characteristic points); (5) Simultaneous modeling of several profiles; (6) computation of a practically unlimited number of geological bodies and fragments. The basic algorithm implemented in the *GSFC* program is the solution to the direct 3-D problem of gravity and magnetic prospecting for horizontal polygonal prisms restricted to the strike direction (Fig. 4.25). In the algorithm, integration over a volume is realized on the surface limiting the anomalous body.

The analytical expression for the first vertical derivative of the gravity potential of the (*m*-1) angle horizontal prism (Fig. 4.25) is obtained by integrating a known analytical expression:

$$U_{z'} = -\int_S \frac{z}{(R+y)R} dxdz \Big|_{y_1}^{y_2}, \tag{4.11}$$

where $R = \sqrt{x^2 + y^2 + z^2}$, *S* is the area of normal section of the prism by the plane of *xOz*. A detailed description of analytical expressions of the first and second derivatives of gravity potential of the approximation model of the horizontal polygonal prism and their relationships to the magnetic field is presented in Khesin et al. (1996).

4.3.1 Computation of Gravity Field Reductions

The computation of the gravitational field with the Bouguer (**a**) and free-air (**b**) reduction and observed values (**c**) uses the following formulas:

$$\left.\begin{aligned} \Delta g_{\mathrm{B}} &= \Delta g_{\mathrm{M}} + \Delta g_{\mathrm{SR}}, \\ \Delta g_{\mathrm{f.a.}} &= \Delta g_{\mathrm{M}}, \\ \Delta g_{\mathrm{obs}} &= \Delta g_{\mathrm{M}} - 0.3086 H_{\mathrm{abs}}, \end{aligned}\right\} \tag{4.12}$$

where Δg_M is the field of the geological space model, Δg_{SR} is the negative effect of a homogeneous body with density σ_R (the top of the body coincides with terrain relief and the bottom is horizontal at sea level), H_{abs} is the absolute height of the selected point *M*, Δg_{obs} is the acceleration of gravity observed on the Earth's surface.

These formulas were obtained from conventional expressions of Bouguer reduction:

$$\Delta g_{\mathrm{B}} = g_{\mathrm{obs}} - \gamma_0 + 0.3086 H_{\mathrm{abs}} - 0.0419 \sigma_R H_{\mathrm{abs}} + g_{\mathrm{corr}}. \tag{4.13}$$

where Δg_{obs} is the observed value of gravity acceleration; γ_0 is the normal value of gravity acceleration on the Earth's spheroid, 0.3086 H_{abs} is the elevation correction, and 0.0419 σ_R H_{abs} is the correction for the attraction of the intermediate plane-parallel layer (infinite slab) with density σ_R and thickness H_{abs}. The value g_{corr} is the correction for the influence of the surrounding terrain relief.

From Fig. 4.26 it is obvious that

$$\Delta g_{\mathrm{SR}} = (0.0419 \sigma_R H_{\mathrm{abs}} - g_{\mathrm{corr}}). \tag{4.14}$$

The observed anomaly of gravitational field can be expressed as

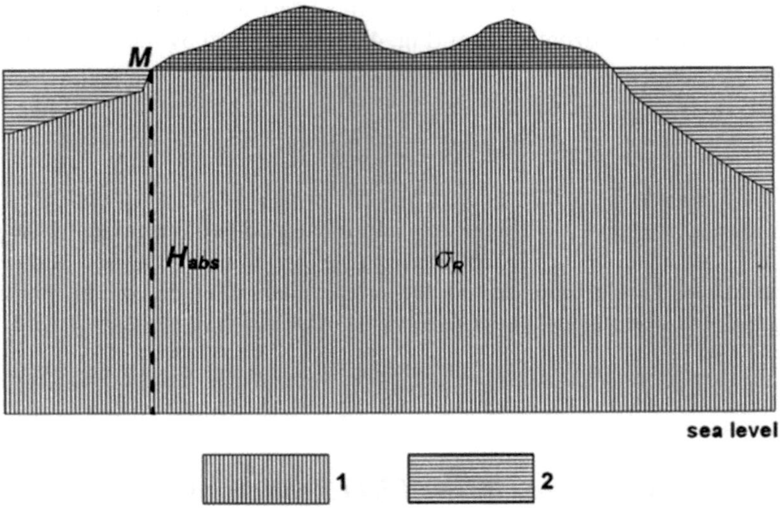

Fig. 4.26 Determining Δg_{B} (*1*) area where the terrain correction is computed; (*2*) homogeneous body with density σ_R (gravity effect at point *M* is equal to the effect from the plane-parallel layer with terrain correction)

$$\Delta g_{\mathrm{obs}} = g_{\mathrm{obs}} - \gamma_0. \tag{4.15}$$

Using expressions (4.14) and (4.15), expression (4.13) can be written in the following form:

$$\Delta g_{\mathrm{B}} = g_{\mathrm{obs}} - \gamma_0 + 0.3086 H_{\mathrm{abs}} + \Delta g_{\mathrm{SR}}. \tag{4.16}$$

If the plane-parallel layer and terrain correction are not introduced, we obtain the expression of free-air anomaly:

$$\Delta g_{\mathrm{f.a.}} = g_{\mathrm{obs}} + 0.3086 H_{\mathrm{abs}}. \tag{4.17}$$

This follows directly from Eq. 4.12.

This expression determines the field Δg_M from the geological space model with accuracy to a certain constant. It should be noted that in a geological model the effects of all bodies are computed to some deep horizontal level. Below this level, because of the absence of data, the density of rocks can be assumed constant. In addition, a constant effect of this deep layer can be considered as a part of the value γ_0, since the effect from the common model is calculated at terrain relief points. This is equivalent to introducing a free-air correction into the value Δg_{obs}.

4.3.2 Interactive Direct Problem Solution: Main Principles

The interpretation results obtained by rapid methods serve to make the necessary adjustments to the model of the medium before interpretation starts, and to form an initial approximation for selection-based interpretation. The following stage, although the most complicated one, plays a key role in analyzing the gravimetric and magnetic prospecting data (e.g., Brodsky and Strakhov 1987; Lyubimov and Lyubimov 1988; Strakhov et al. 1986). The electric (electromagnetic) anomaly sources represent rather sparse and isolated objects, whereas the gravity and magnetic sources with variable properties cover most of the geological medium under survey, and occupying the greater part if not the entire medium.

Complex media with a large number of anomalous bodies and parameters hinder the implementation of automated selection schemes. In these conditions it is more effective to use an interactive computer selection system, where an experienced interpreter solves the direct problems of magnetic and gravimetric prospecting, compares computed and observed curves, and records divergences and modifications of the petrophysical model. A *GSFC* program with graphical subroutines has been developed (Khesin et al. 1993b; Eppelbaum and Khesin 2004) for this purpose.

The advantage of the interactive computer selection system is that it eliminates the need for a great number of iterations, which are inevitable in automated selection algorithms that use successive parameter variation searches within a specified range.

Nevertheless, systems of automatic selection of geophysical fields (for the gravity field this approach was developed by Goldshmidt et al. (1981) and Bulakh et al. (1983)) have certain limitations. First the selection must be meaningful; e.g. changes in physical properties and 3-D geometrical parameters should be carried out in defined numerical and space intervals. These limitations stem primarily from the program. However, there are more serious issues. Take the example of a 3-D integrated gravity and magnetic selection over a comparatively simple section consisting of ten geological bodies. Each geological body has three petrophysical variables (density, value and inclination of the magnetization vector), geometric variables (left-hand (y_1) and right-hand (y_2) end faces of the body) and, finally, its geometric parameters in the plane of geological section. The number of points (variables) to describe bodies in the plane of section *a priori* is unknown. For simplicity and given that many of these points are calculated twice by the contouring objects, the number of these points can be assume to be ten. To calculate the number of possible combinations of all variables by combined 3-D modeling, we need to calculate the approximate ranges of the variables (Table 4.5). Obviously these ranges are relative and only estimate the number of combinations.

Thus, for only one body we have a number of combinations

$$C_{30}^1 \cdot C_{60}^1 \cdot C_{24}^1 \cdot C_{30}^1 \cdot C_{30}^1 \cdot C_{100}^1 \approx 4 \cdot 10^9$$

Similarly, for ten bodies we have $\approx 4 \cdot 10^{90}$ combinations (for a non-complex geological section). This number of combinations is prohibitive and computationally hard, even for modern computers.

Table 4.5 The number of possible combinations of variables

Variable	Interval of change	Range	Number of variants
Density, g/cm^3	2.3–2.6	0.01	30
Magnetization, mA/m	0–3,000	50	60
Inclination of magnetization, degree	0–360	15	24
Left-hand and face, y_1/x^*	0–20	Nonlinear	30
Right-hand and face, y_2/x^*	0–20	Nonlinear	30
Coordinates of body in the plane of geological section	Minimum 10 points	–	Minimum 10·10

$^*x$ is the maximum length of a selected profile

Moreover, it does not eliminate the burden of attaching a geological meaning to the distribution of physical sources. An interactive computer system enables the expert to modify a model by using geological categories expressed in the form of petrophysical sources.

A method of geological space approximation is crucial to the selection of a good algorithm. Sometimes, algorithms based on an approximation of the geological section by horizontal circular cylinders (HCC) are used for this purpose. However, this type of approximation introduces serious errors, which can lead to dramatic mistakes in geophysical field modeling. Decreasing the radii of the approximating HCC cannot reduce these errors. For instance, let a cylinder with horizontal radius R be inscribed into a geological body's cross-section which is approximated by a square with side $2R$. Clearly, the difference between the areas of the square and the HCC is $4R^2 - \pi R^2$, and the error value is above 20%. Now let us fill the area of the square with the HCC with radii R/n. Their total number is n^2. Thus their total area will also be πR^2. In other words, the error value will be the same for an arbitrary number of approximating HCC.

The *GSFC* program simultaneously computes the gravity and magnetic effects of an arbitrary number of boundaries and any complexity of the medium.

It should be emphasized that the development of quantitative models that do not take into account the 3-D nature of the geological medium can lead to serious mistakes. Figure 4.27 shows that the selection of the profile near the anomalous body can distort the amplitude of gravitational field and dramatically change its magnetic field behavior. This has led to the development and constant updating of integrated software for computing 3-D fields under conditions of complex geological media and rugged relief (Eppelbaum 2006).

4.3.3 Computation of Gravity Reductions and Magnetic Field as a Component-Wise Process

The terrain relief effect on gravitational and magnetic fields can be calculated in two stages: (1) calculation of that part of the field which is due to the known or "missing" topographic masses, and (2) conversion of the observed values to the

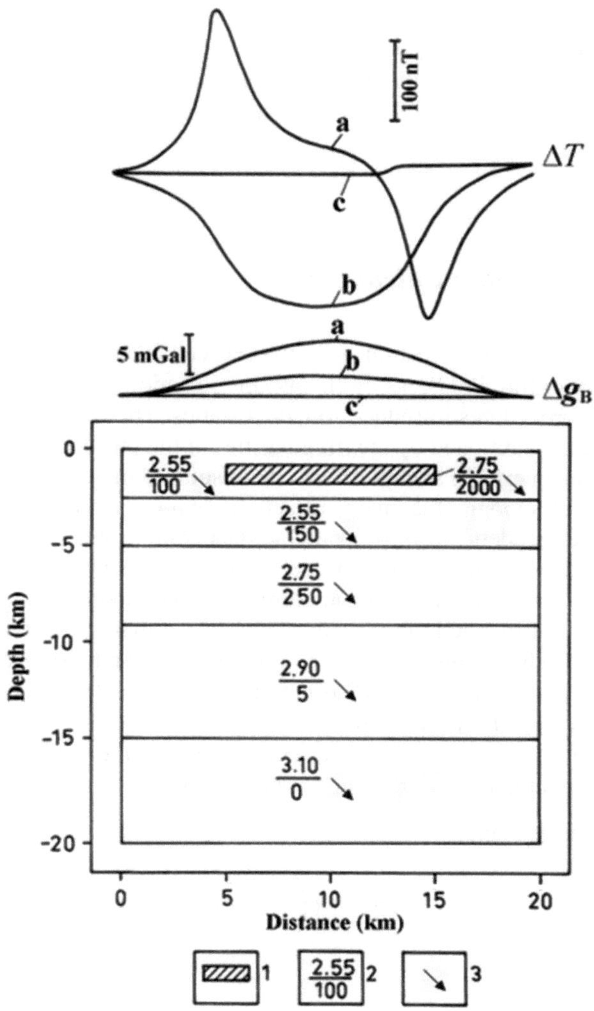

Fig. 4.27 Comparison of Δg and ΔT fields computed with various locations of end faces of the anomalous bed. Disposition of end faces, km: (**a**) $(-\infty, +\infty)$; (**b**) $(-0.5, 10)$; (**c**) $(5, 10)$. (*1*) anomalous body; (*2*) physical properties: numerator = density, g/cm^3, denominator = magnetization, mA/m; (*3*) direction of magnetization vector (additional data used in computing: azimuth of the selected profile is 50° and magnetic declination is 5°)

horizontal reference plane (reduction line correction). The attraction of morphostructures forming the terrain relief may considerably exceed the anomalies from the target, and reach or exceed 1,000 nT in magnetic prospecting and 10 or more mGal in gravity prospecting. At the same time large objects, which are most clearly detected by gravity survey in mountainous regions, such as acidic intrusions, produce a gravitational effect on the same order.

Figure 4.28 illustrates the field ratios for some situations involved (*GSFC* software was applied).

Apparently, in one case the Bouguer anomalies and in the other case ΔT anomalies caused by geologic boundaries are completely concealed by the relief form attraction.

4.3.4 Terrain Relief Calculation

The problem of terrain correction has received considerable attention in gravity prospecting (Gordin 1974) as well as in seismic prospecting, where the corresponding procedures introduce so-called statics. The first factor (topographic mass influence), conditioned by the irregularities of the physical (density, velocity, etc.) earth-to-air interface, always affects the results of these methods. Correcting for the effect of topographic mass attraction requires eithera large amount of topographic and geodetic work or available data from ground and aerial surveys, as well as the data on the physical properties of the upper portion of the section. These data are used for computing and eliminating the topographic mass attraction in gravimetric and magnetic prospecting, and by applying the statics in seismic prospecting, etc. Thus, a direct problem of geophysics can be solved.

The accuracy of various methods is determined by standard models; the attraction is calculated by analytical formulas.

To determine corrections in the central zone, the following analytical expressions of simple body attraction can be employed:

(a) For an inclined plane

$$g_r = 2\pi G\sigma_r R\left(1 - \cos\alpha\frac{K_\alpha}{K_o}\right), \tag{4.18}$$

(b) For a cone

$$g_r = 2\pi G\sigma_r R\tan\alpha(1 - \sin\alpha), \tag{4.19}$$

(c) For a conical hollow

$$g_r = 2\pi G\sigma_r R\tan\alpha(1 - \cos\alpha). \tag{4.20}$$

Here G is the gravitational constant, σ_r is the density of rocks forming the relief, R is the central zone radius, α is the dip angle of the inclined plane as in formula (4.18) and the cone generatrix dip angle as in Eqs. 4.19 and 4.20, K_α is the total elliptic integral of the first kind, $K_o = \pi/2 = 1.5708$ at $\alpha = 0$.

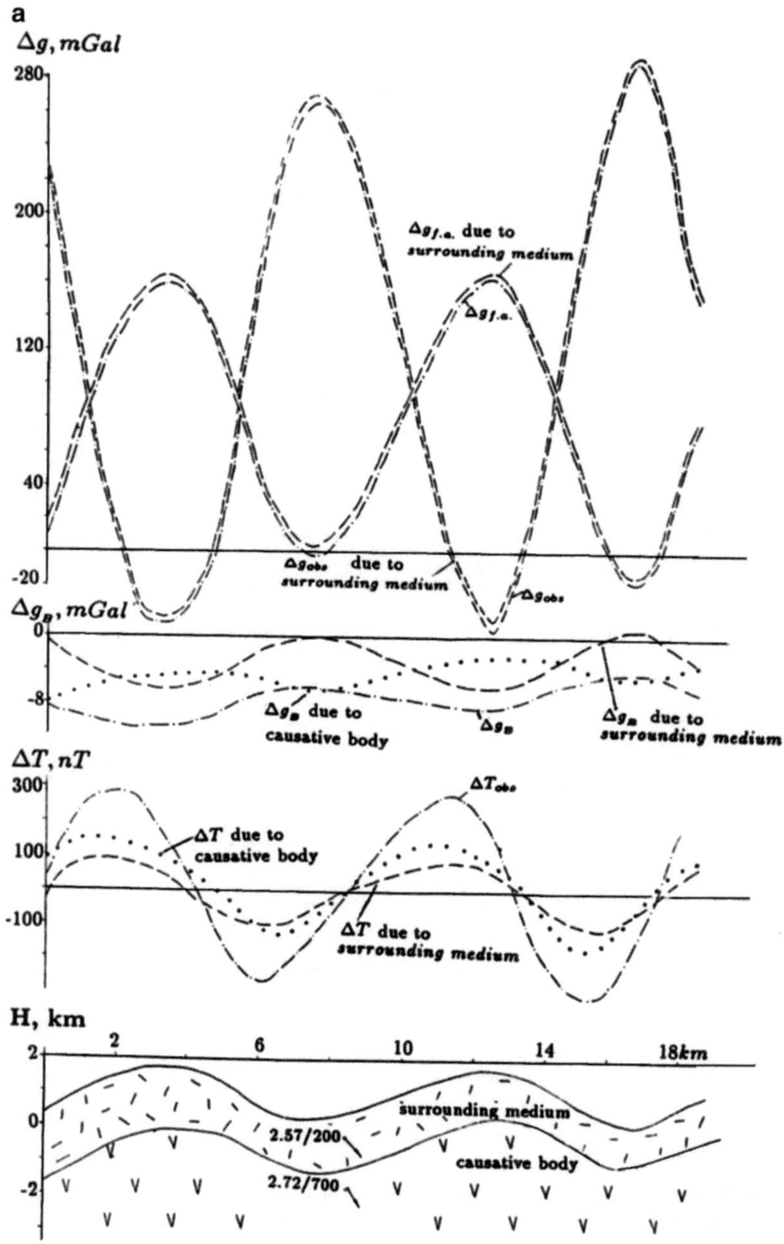

Fig. 4.28 (continued)

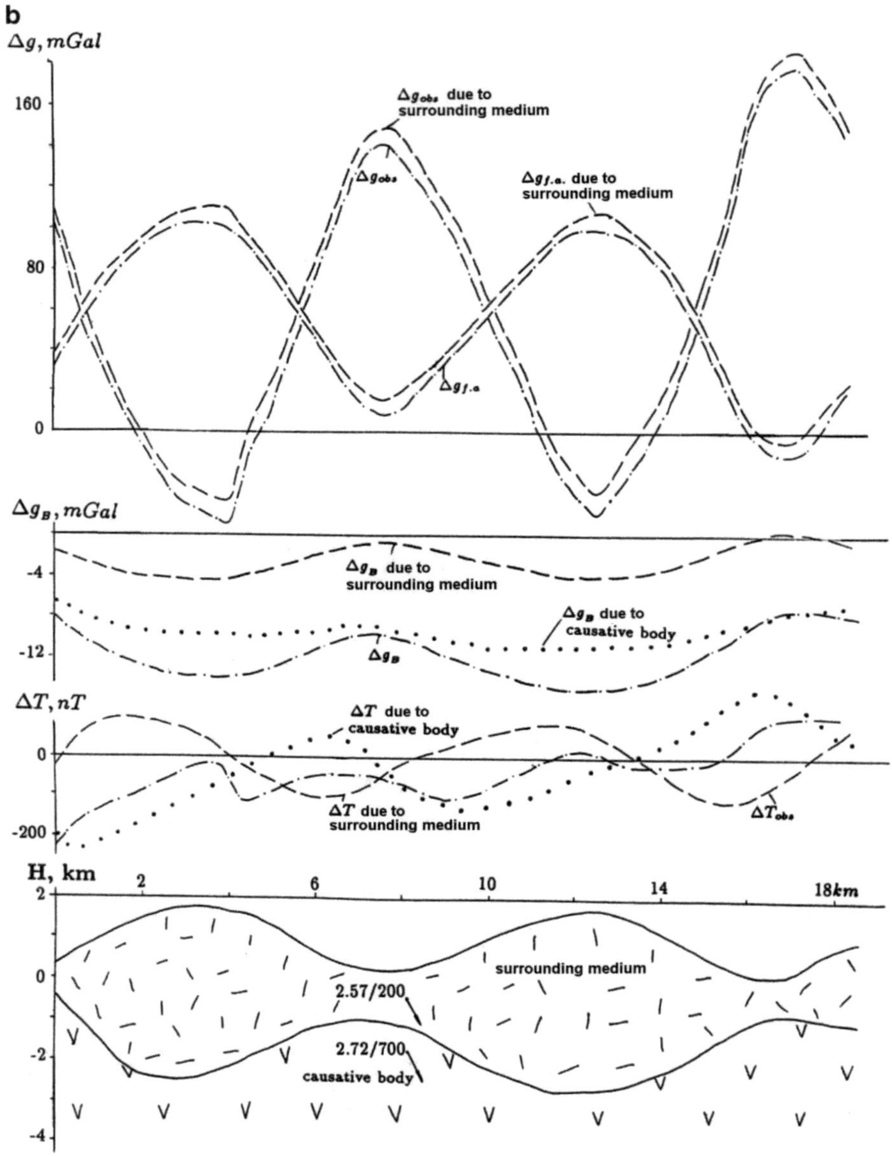

Fig. 4.28 Modeling of gravitational and magnetic fields at conformable (**a**) and discordant (**b**) depth boundaries with respect to the relief. (*1*) surrounding medium; (*2*) causative body; (*3*) physical properties (numerator = density, g/cm^3; denominator = magnetization, mA/m); (*4*) magnetization vector direction; (*5*) computed effects using the *GSFC* program: (*a*) physical interface, (*b*) medium, (*c*) total

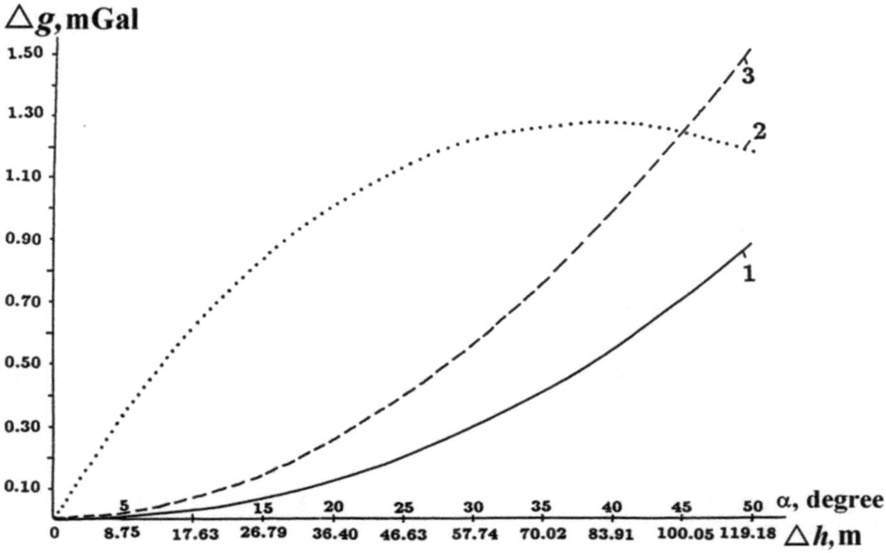

Fig. 4.29 Plots of correction for topographic mass attraction within a zone of 0–100 m computed from analytical formulas with an effective density of 1 g/cm^3. Terrain relief forms are approximated by inclined plane (*1*), cone (*2*) and conical hollow (*3*)

The gravitational effects of these bodies were computed by formulas (4.18–4.20) for $\sigma_r = 1$ g/cm^3 and R =0.1 km, provided α varied from 0° to 50° with a step of 5° (Fig. 4.29).

The plots in Fig. 4.29 were used to determine terrain corrections in the central zone with a radius of 0.1 km for 8 points located on the slopes and satisfying the model in formula (4.18), and for 2 points on the tops of rounded positive relief forms approximately satisfying the model describing by formula (4.19).

The standard deviation (SD) of corrections obtained from Eq. 4.18 as compared to the corrections determined for the same points by Berezkin's (1967) method was ±0.06 mGal, whereas the values ranged from Eq. 4.19 to ±0.09 mGal. However, in the former case a systematic discrepancy of 0.35 mGal was observed. This was caused, on the one hand, by the deflection of the real relief from the inclined plane, and on the other from the ineffective hyperbolic approximation assumed in Berezkin's method for an even slope. These data point to the need to carefully measure the terrain relief effect in the central zone during high-accuracy surveys.

Specifically a distinction should be made between its internal part where approximation by simple bodies is permissible, and its outer part. For the latter, terrain corrections can be computed using well-known programs, if the topographic evidence is available.

The combination of rapid methods of potential and quasi-potential field interpretation with selection based on the *GSFC* program may be treated as a system of quantitative interpretation under rugged terrain relief and oblique polarization of objects. Its last stage is a geological interpretation of the geological space model (Fig. 4.30).

This system can help optimize the selection of test wells and mines, thus cutting costs substantially.

4.4 Models of the Earth's Crust Along Regional Traverses

4.4.1 3D Combined Modeling of Gravity and Magnetic Fields

The bases for this type of modeling were described in the previous section. This section provides a detailed description of the order of execution of this process. It results in a 3D Physical-Geological Model (***PGM***) of some Caucasian regions using the results of observation of the gravitational and magnetic (ΔT (ΔZ)) fields under rugged terrain relief conditions and arbitrary magnetization of objects and the host medium.

4.4.1.1 Main Stages of 3D Combined Modeling of Gravity and Magnetic Fields

There are four major stages:

1. Selection of the interpretational profile location (azimuth and length),
2. Preparation of the initial model of the medium by inputting petrophysical data, all available geophysical, geological and drilling data, and using methods of rapid interpretation for magnetic and gravity anomalies developed for conditions of rugged terrain relief and oblique magnetization of objects,
3. Combined 3D gravity-magnetic modeling using the 3D *GSFC* program,
4. Development of 3-D ***PGM*** of the area under investigation.

The first stage involves choosing the location of the interpretational profiles. These are chosen so that they traverse the most significant geological structures (or their projections to the Earth's surface). At the same time it is essential to have some deep boreholes or profiles of deep seismic investigations in the vicinity of the projected profiles. Sometimes in the most important areas the profiles may cross, and obviously the geological sections should coincide at the points of intersection.

The second stage involves the following:

(a) A special geological section is formed where all the intrusive, effusive and other associations, as well as disjunctive dislocations and the surface of folded foundations are selected on the basis of geological data within a strip 15–20 km wide. The section is located in the middle of this strip. It characterizes the upper portion of the Earth's crust 2–3 to 5–8 km from the Earth's surface to the

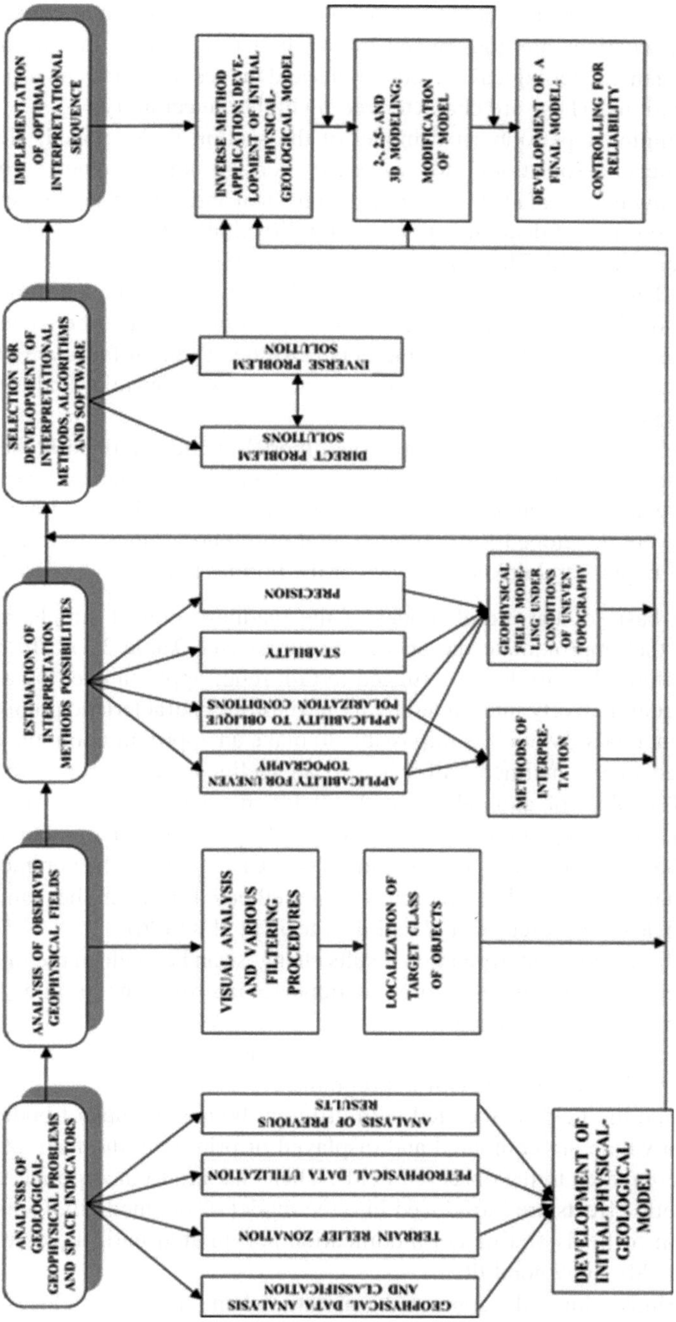

Fig. 4.30 Flow-chart of a system of geophysical fields quantitative interpretation for complicated terrain and geological conditions

Baikalian basement. The deeper parts of intrusion bodies and certain disjunctive dislocations are constructed by extrapolation, integrating the occurrence mode, general geological considerations and the results of previous geophysical explorations. These are characterized by higher uncertainty.

(b) A preliminary petrophysical model of the section is developed. Here all the geological associations are assigned density and magnetization values according to the data from earlier petrophysical and geophysical investigations. In the absence of data on magnetization direction, it is assumed parallel to the geomagnetic field. The magnetization direction can be refined over the course of mathematical modeling. The petrophysical model includes deep-seated layers of the Earth's crust: (1) "basaltic", (2) intermediate between the crust and the upper mantle and (3) the upper mantle. Their surfaces are plotted and physical properties are inputted from previous seismic and other deep investigations.

(c) The initial (preliminary) petrophysical model includes hidden bodies as well. Their location, thickness, depth and magnetization are obtained by quantitative interpretation of magnetic and gravitational anomalies by rapid methods. The latter were developed for non-horizontal observation lines, arbitrary polarization of objects and unknown levels of the normal field.

On the basis of the initial model of the medium, typical simple models are generated that approximate the anomaly sources (see Table 3.4). The anomalies observed along the profiles in a rugged terrain relief approximated by an inclined plane are quantitatively interpreted by more rigorous characteristic points, tangents and areal methods. When it is impossible to make an approximation, the anomalies are reduced in the inclined half-space to an inclined plane nearest to the relief surface. The plots of anomalies for the fields are constructed in the horizontal projection. The interpretation of such plots yields quantitative parameters of certain fictitious bodies. Then, applying the formulas in Chap. 3 gives the parameters of the real anomaly sources. The interpretation methods for anomalies attributed to different fields are presented in detail in Khesin et al. (1996).

On the basis of the interpretation results changes can be made in the initial model of the medium, and initial approximations to interpret the profiles using the selection method can be made.

The third stage selects gravitational and magnetic fields along the profiles using 3D software (in our case the *GSFC* program).

Each time the fields associated with different bodies, groups of bodies and the model as a whole are computed and displayed or printed in the form of plots, the latter are compared to the observed fields. Then changes to incorporate the gravity and magnetic effects are introduced into the model of the medium. Computations, comparison of fields and model modification are iterated until the computed and observed fields show good fit.

Each printed plot of the computed gravitational and magnetic fields is compared to the corresponding plot of observed fields. The comparison starts with the total effect of the bodies that formed the regional fields, and continues with a gradual introduction of effects of local bodies. On the basis of the comparison, a decision is made as to the necessity of introducing modifications into the configuration or physical properties of the bodies determining regional fields, local anomalies, the introduction of new local bodies or the exclusion of some of the available ones. These modifications are then included into the model and the computation is resumed.

First, a regional gravitational field is roughly selected. As a rule, the densities of deep-seated complexes are not changed and the modifications only concern the shape of their roof. Then, fields of local bodies are selected. If necessary, this is followed by the verification of the regional field and the field of local bodies.

In each step a separate analysis of gravitational and magnetic fields is performed. Coordinated variations verified at subsequent steps are then introduced into the model. This results in an integrated quantitative interpretation for the anomalous gravitational and magnetic fields. The selection is said to be complete when the computed gravitational and magnetic fields coincide, on average, with the observed ones with acceptable accuracy.

The fourth stage involves geological interpretation of geological space models selected by the profiles. A 3-D physical-geological model of the area under investigation is developed by incorporating the data from rapid interpretation and a qualitative interpretation. This results in the final geological plot that maps the geological objects, mineral controls and mineral deposits, including deep-seated ones in a much more complete fashion.

The geological interpretation of the complexes and local bodies of the final petrophysical model does not usually present difficulties, since in the implementation of the interactive selection system nearly all the bodies in the model acquire specific geological content. The geological nature of new sources introduced into the model during the selection and reflected either in the initial geological section, or in the initial physical-geological model, is determined as a function of the similarity of their physical properties, dimensions, and mode of occurrence with respect to the known objects. The age of the bodies is determined according to their interrelations with the host rocks.

This integrated interactive simulation of fields observed under rugged terrain relief, where geological objects can have an arbitrary direction of polarization (magnetization) vector is very rarely applied in geophysical practice. The advantages of this technique as compared to the precision achieved at the present time are as follows: (a) more complete consideration of real topographic and geological conditions of the region under investigation, (b) determination of up to six quantitative parameters of anomalous bodies with an admissible error of 10–20% for each field, (c) development of integrated 3-D physical-geological model of the region under investigation, (d) creation of final geological documents that are more complete, thorough and reliable in terms of their geological structure mapping.

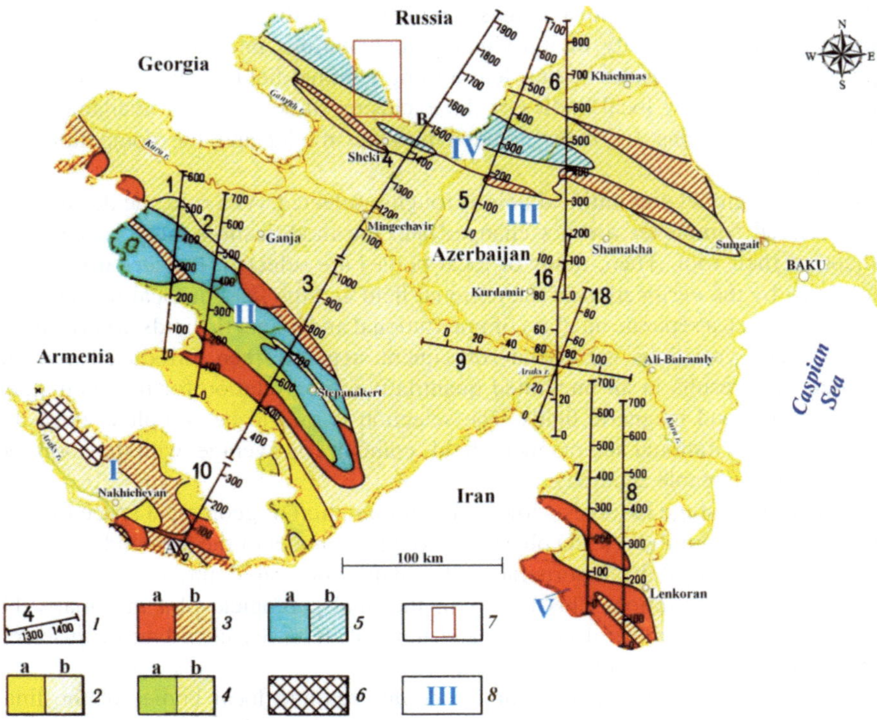

Fig. 4.31 Areal map of main profiles used for physical-geological modeling in Azerbaijan and adjacent regions of the Eastern Caucasus. (*1*) profiles and pickets; (*2*) Pg_3-Q: (*a*) orogenic magmatic associations, (*b*) background sedimentary deposits; (*3*) K_2-Pg_2: (*a*) pre-orogenic magmatic associations, (*b*) background sedimentary deposits; (*4*) J_3-K_1: (*a*) magmatic associations of the Late Alpine sub-stage, (*b*) background sedimentary deposits; (*5*) J_1-J_2: (*a*) magmatic associations of the Early Alpine sub-stage, (*b*) background sedimentary deposits; (*6*) *Pz* deposits; (*7*) contour of the Guton magnetic anomaly; (*8*) tectonical regions: (*I*) Nakhichevan folding region, (*II*) SE part of the Lesser Caucasus mega-anticlinorium, (*III*) central and SE parts of the Kura mega-synclinorium, (*IV*) SE part of the Greater Caucasus mega-anticlinorium, (*V*) Talysh anticlinorium

4.4.2 Examples of 3D Combined Modeling of Gravity and Magnetic Fields Along Interpretation Profiles

The location of interpretational profiles for 3D combined gravity-magnetic modeling is presented in Fig. 4.31.

4.4.2.1 Integrated Physical-Geological Model of the Saatly Super-Deep Borehole

For many years, the dominant point of view in Azerbaijan was that in the Kura Depression separating the mega-anticlinoria of the Greater and the Lesser

Caucasus, thick sedimentary deposits were present in the crystalline Pre-Alpine basement, and these structures were divided by subvertical deep faults. On the buried uplift of the basement, hypothesized on the basis of high densities and velocities of elastic waves, the Saatly superdeep borehole SD1 was designed in 1965. The drilling area was selected using the analysis of seismic profiles 9, 16 and 18 (Fig. 4.31) as well as regional gravity field analysis (Tzimelzon 1959, 1965).

However, analysis of the magnetic properties of rocks and the magnetic survey results showed that the basement was not magnetized, and a large part of the geological section of the Middle Kura Depression was occupied by Mesozoic magmatic associations of basic and intermediate composition with high magnetization (Khesin 1976). These mainly Jurassic associations are widely distributed in the north-eastern part of the Lesser Caucasus. They have a deep-seated gently sloping underthrust under the sand-shale thick series of the Greater Caucasus Jurassic.

The validity of the interpretation was fully confirmed by the results of the SD-1 drilling (Fig. 4.32). The borehole exposed Mesozoic volcanogenic rocks at a depth of 3.6 km and did not reachits bottom even at 8.2 km (Ismail-Zadeh and Khesin 1989b).

The geomagnetic model along the meridional profile 16 is significantly different from sublatitudinal profile 9 (Fig. 4.33). At the same time the gravity field along profile 16 is smoother. Obviously, these differences in potential geophysical fields reflect the specificities of the deep structure of the source of the Talysh-Vandam gravity maximum which is composed of a submeridional strike of deeply occurring massses with superfluous density and a sublatitudinal strike of younger magnetoactive associations.

The examination indicates that the area of the Talysh-Vandam gravity maxiumum is highly inhomogeneous in terms of geological structure, and individual elements of the gravity maxiumum – anomalies of second order – reflect a different genesis in areas of Earth's crust and suggest that they developed apart from one another (Khesin and Eppelbaum 2007).

Improved interpretation methods (tangents, characteristic point and the areal method) were applied to examine the gravity and magnetic anomalies along all profiles surrounding SD-1. Part of this interpretation along profile 18 is shown in Fig. 4.34. The behavior of magnetic ΔZ curve is striking, and graph $\frac{\partial \Delta g}{\partial x}$ is very similar, which testifies to the fact that these anomalies derive from the same geological objects. The quantitative analysis of the magnetic curve revealed two magnetic targets. The main target may be the source of the Talysh-Vandam anomaly (its upper edge coincides with the SD-1 drilling data). Interestingly, the results of $\frac{\partial \Delta g}{\partial x}$ graph examination fit well with the magnetic data analysis.

An examination of the gravity and magnetic fields was carried out for profiles 9, 16 and 18 (with seismic and thermal data) and were used to determine the main features of a spatial geological-geophysical model of the Earth's crust for the SD-1 area (Fig. 4.35). The diagram shows the main sources of gravity and magnetic anomalies in the eastern part of the area where SD-1 is located, and in the western part where the Zardob magnetic maximum was found. The magnetic and gravity field analysis (including 3D modeling of these fields) together with petrophysical data provided additional evidence to account for the Ganja regional magnetic maximum (Ismail-Zadeh and Khesin 1989b).

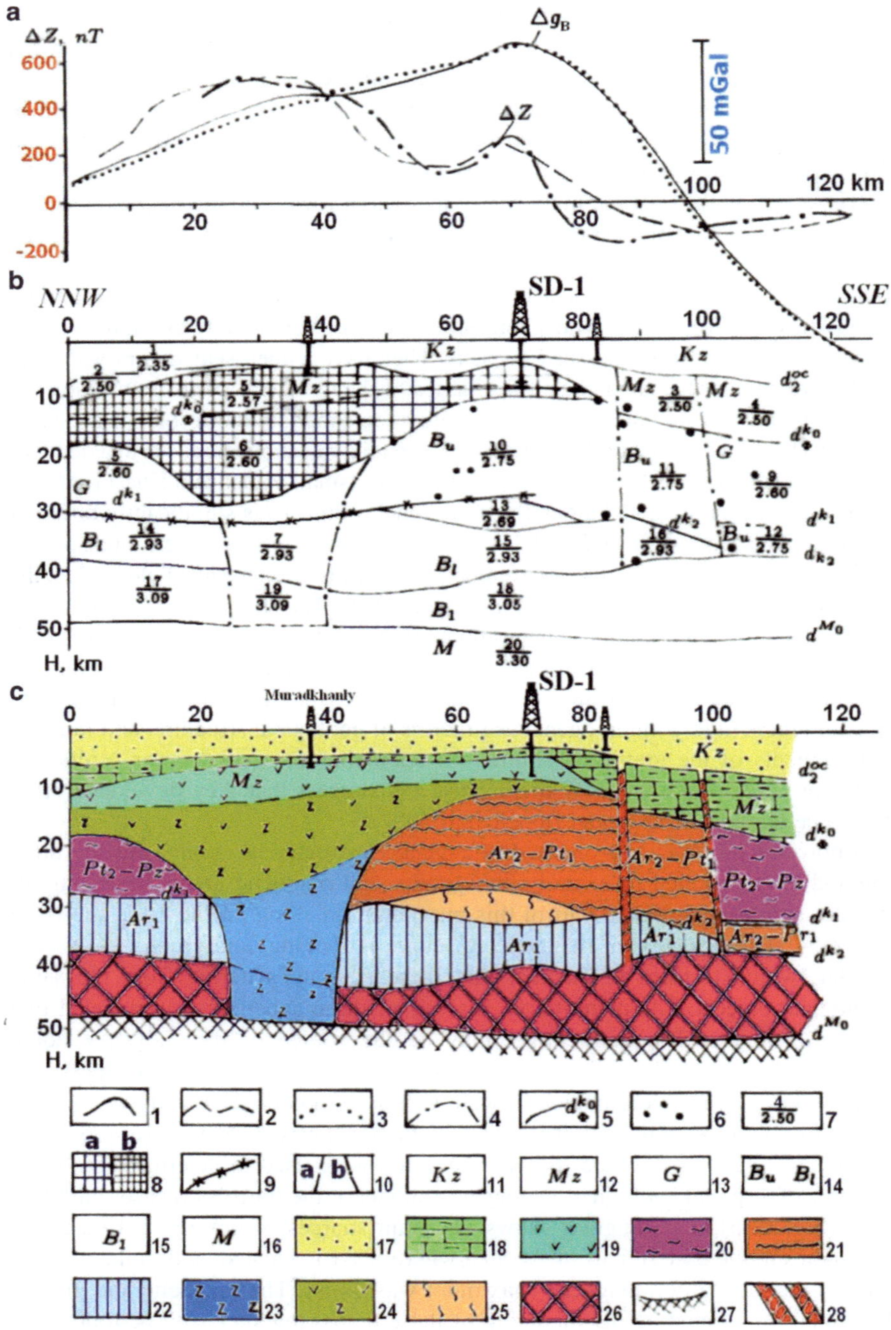

Fig. 4.32 Deep geological section of the Earth's crust in the SD-1 area (profile 9 in Fig. 4.30). (**a**) – gravitational and magnetic fields, observed and computed by the model **b**; (**b**) – petrophysical

The SD-1 was halted at a depth of 8.2 km. Given the location of the lower edge of the magnetized masses and by analogy of this geological section to ones in the Lesser Caucasian, it is likely that magmatic rocks occur down to 10 km (compatable with the Lower Bajocian rocks of the Lesser Caucasus).

Obviously, the main source of the Talysh-Vandam gravity maximum is associated with underlying high-density strongly metamorphized and initially chiefly sedimentary associations of the non-magnetic or low-magnetic Pre-Baikalian floor (this floor has a submeridional strike in the Russian and African platforms). The depth of the upper edge of these highly dense rocks is estimated at 9.5 km.

Thus, SD-1 has yet to discover (these drlling operations may or may not be continued) the source of the Talysh-Vandam regional gravity anomaly, but it has identified the origin of the Ganja magnetic maximum. This has enormous importance not only for the analysis of the tectonic-magmatic evolution of the Caucasus region, but also for evaluating the potential of ore- and oil&gas-bearing prospecting. For example, the geomagnetic models can extend prospecting of oil deposits in the Middle Kura Depression that are associated with zones of protrusions of Mezosoic magnetoactive associations. Many geophysical methods have mapped these dense Mezosoic associations, but only magnetic prospecting revealed basic and middle consistency magmatites (Eppelbaum and Khesin 2011).

4.4.2.2 Example of 3D Combined Interactive Modeling Along Profile 3–4

A visual example of 3D combined interactive modeling of gravity and magnetic fields along stakes 700–1,350 of profiles 3–4 (see scheme presented in Fig. 4.31) is shown in Fig. 4.35a–c. This profile crosses the Lesser-Caucasus, the Middle Kura Depression and comes to an end at the submountainous zone of the Greater Caucasus. An initial ***PGM*** presented in Fig. 4.35a, was constructed on the basis of drilling data, analysis of seismic data and petrophysical material as well as

Fig. 4.32 (continued) model; (**c**) – geological model. Observed curves: (*1*) Δg; (*2*) ΔZ; curves computed by the model **b**; (*3*) Δg; (*4*) ΔZ; (*5*) boundaries of the velocity and the density inhomogeneities and their indices; (*6*) diffraction points; (*7*) body number (numerator) and density value, g/cm^3 (denominator); (*8*) geological bodies with a magnetization of 2,500 mA/m (*a*) and 2,800 mA/m (*b*); (*9*) projection of Curie surface on the basis of geothermal data; (*10*) subvertical boundaries of bodies on the basis of magnetic (*a*) and gravitational (*b*) fields; (*11*) Cenozoic; (*12*) Mesozoic; (*13*) *G* complex (velocity analogue of the "granitic" layer); (*14*) B_u and B_l subcomplexes of *B* complex (complex *B* is the "basaltic" layer velocity analogue); (*15*) B_1 complex (presumably basite and eclogite composition); (*16*) *M* complex (presumed peridotite composition); (*17*) Cenozoic complex: mainly terrigeneous deposits; Mesozoic complex: (*18*) terrigeneous-carbonaceous formations; (*19*) mainly effusive associations of basic and intermediate composition; (*20*) mainly Baikalian complex (Pt_2-Pz): metamorphic (primarily terrigeneous) associations (the presence of younger deposits is possible in the *upper* part); (*21*) Pre-Baikalian complex (Ar_2-Pt_1): mainly gneisses and marbles; (*22*) ancient complex (Ar_1): gneisses and amphibolites; (*23*) root of the basic magmatism; (*24*) undivided effusive-intrusive complex; (*25*) rock complex of a low density (serpentinization zone ?); (*26*) complex of associations corresponding to crust-to-mantle transition; (*27*) *upper* mantle roof position; (*28*) large fault zones

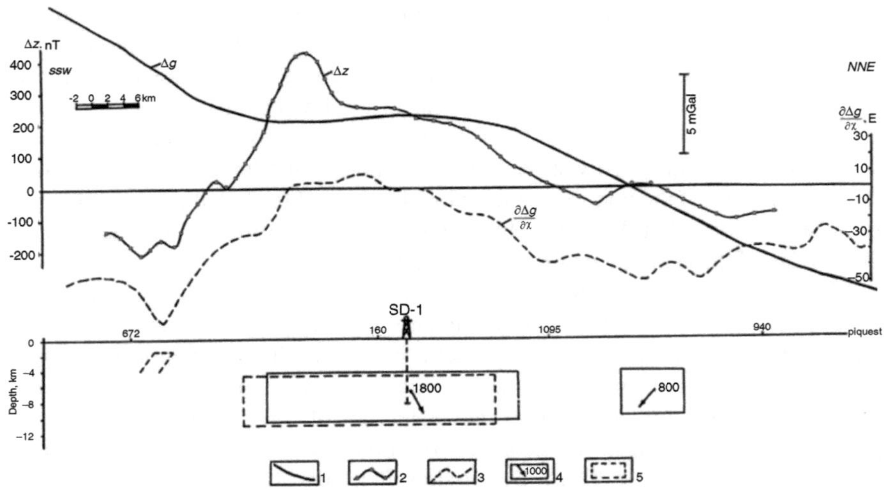

Fig. 4.33 Fragment of gravity and magnetic fields analysis along profile 18 (Eppelbaum and Khesin 2011). (*1*) gravity field Δg_B; (*2*) magnetic field ΔZ; (*3*) first horizontal derivative of the gravity field $\frac{\partial \Delta g}{\partial x}$; (*4*) contour of determined magnetized body and position of magnetization vector; (*5*) contour of body determined by analysis of $\frac{\partial \Delta g}{\partial x}$

advanced examination of magnetic and gravity anomalies (see Chap. 3). As shown in Fig. 4.35a, there are considerable discrepancies between the observed and computed gravity and magnetic fields. After a third iteration (Fig. 4.35b) this discrepancy was reduced and a model with satisfactory fit was achieved after the eleventh iteration (Fig. 4.35c).

4.4.2.3 Deep Structure of Typical Physical-Geological Models

Regional profile A–B (its location is shown in Fig. 4.31) is composed of three profiles (10, 3 and part of profile 4) and starts near the town of Djulfa (Nakhichevan autonomic region of Azerbaijan), crosses Armenian territory, and subsequently traverses the Azerbaijanian territory and ends near the town of Sheki (near the border with Georgia) (Fig. 4.36). It is the most important profile and this ***PGM*** will be described in detail.

In this ***PGM*** the decrease in the Earth's crust is observed in the Kura Depression. The surface of the « basaltic » layer displays the most uplift in the area of the core of the Lesser Caucasus mega-anticlinorium, where associations of the early-geosynclinal effusive formations are found.

The upper part of the geological section visually reflects the stages of geological formation in Azerbaijan. Orogenic granodiorit-porphyric and pre-orogenic gabbro-monzonite-dioritic intrusive formations developed in the beginning of the SW profile (Nakhichevan region); orogenic effusive dazite-andesite-basaltic formation

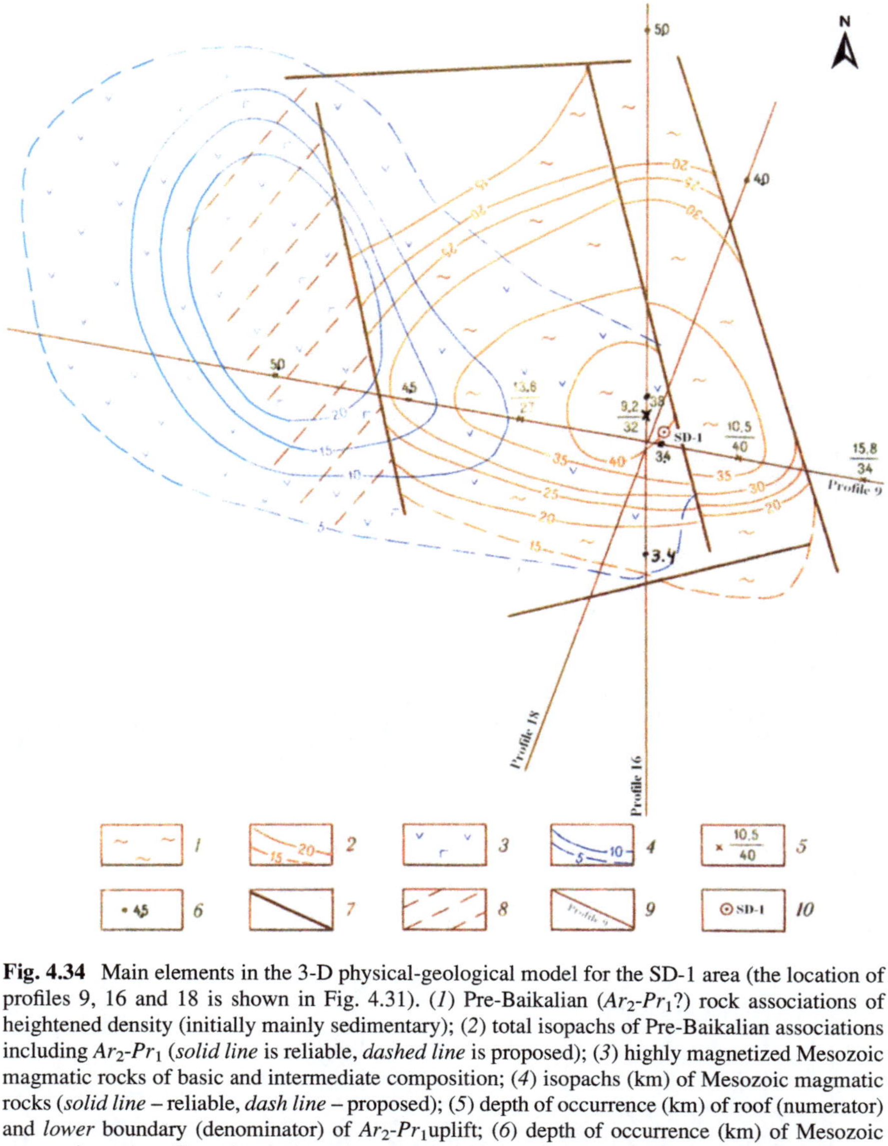

Fig. 4.34 Main elements in the 3-D physical-geological model for the SD-1 area (the location of profiles 9, 16 and 18 is shown in Fig. 4.31). (*1*) Pre-Baikalian (Ar_2-Pr_1?) rock associations of heightened density (initially mainly sedimentary); (2) total isopachs of Pre-Baikalian associations including Ar_2-Pr_1 (*solid line* is reliable, *dashed line* is proposed); (*3*) highly magnetized Mesozoic magmatic rocks of basic and intermediate composition; (*4*) isopachs (km) of Mesozoic magmatic rocks (*solid line* – reliable, *dash line* – proposed); (*5*) depth of occurrence (km) of roof (numerator) and *lower* boundary (denominator) of Ar_2-Pr_1uplift; (*6*) depth of occurrence (km) of Mesozoic magmatic rocks; (*7*) major faults – boundary of blocks; (*8*) projection of root zone of basic magmatism; (*9*) interpretation profiles; (*10*) SD-1 location

are found – SE of the town of Istisu, and orogenic sedimentary deposits (molassa) in the Kura Depression, to a thickness of 4–5 km.

Associations of pre-orogenic effusive formations are present in: (1) the SW part of this profile where they occur on sub-platform terrigeneous-carbonate deposits,

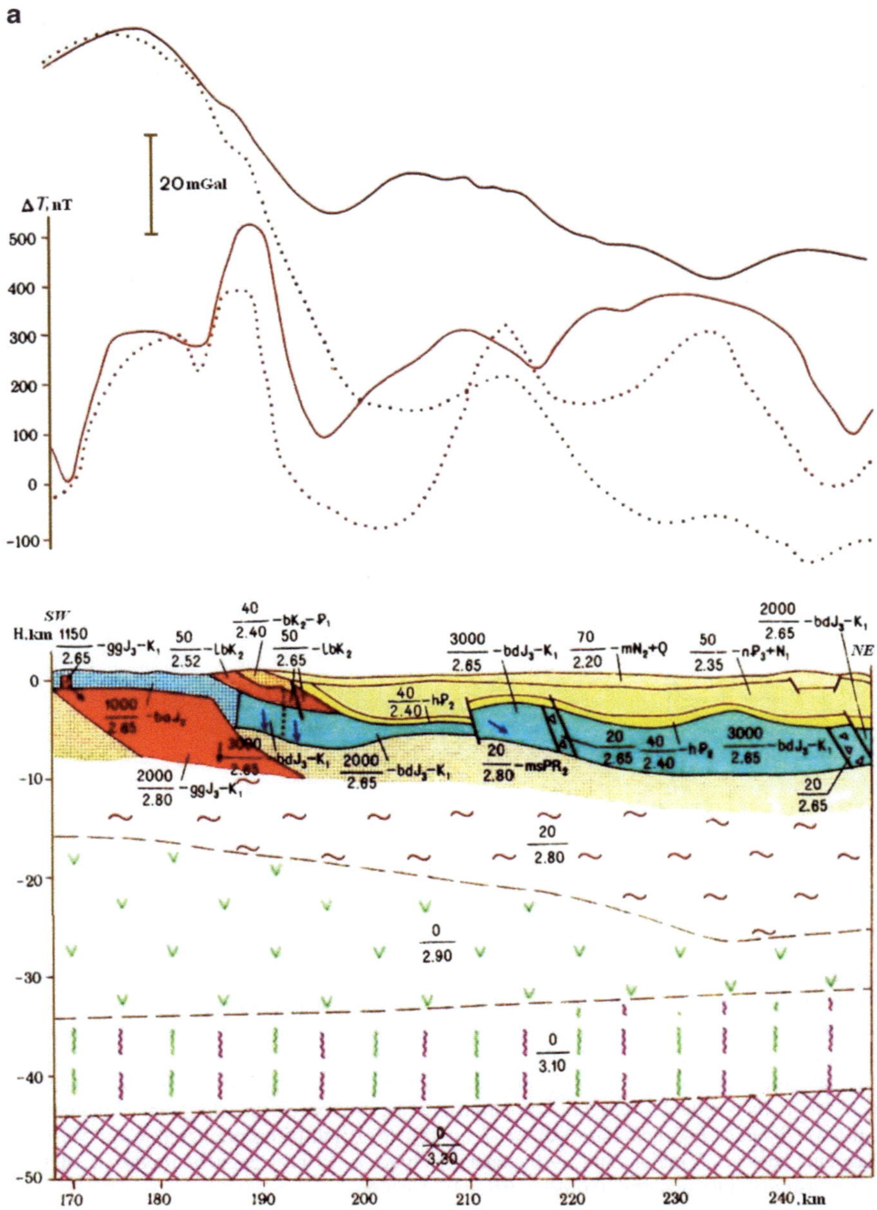

Fig. 4.35 (continued)

b

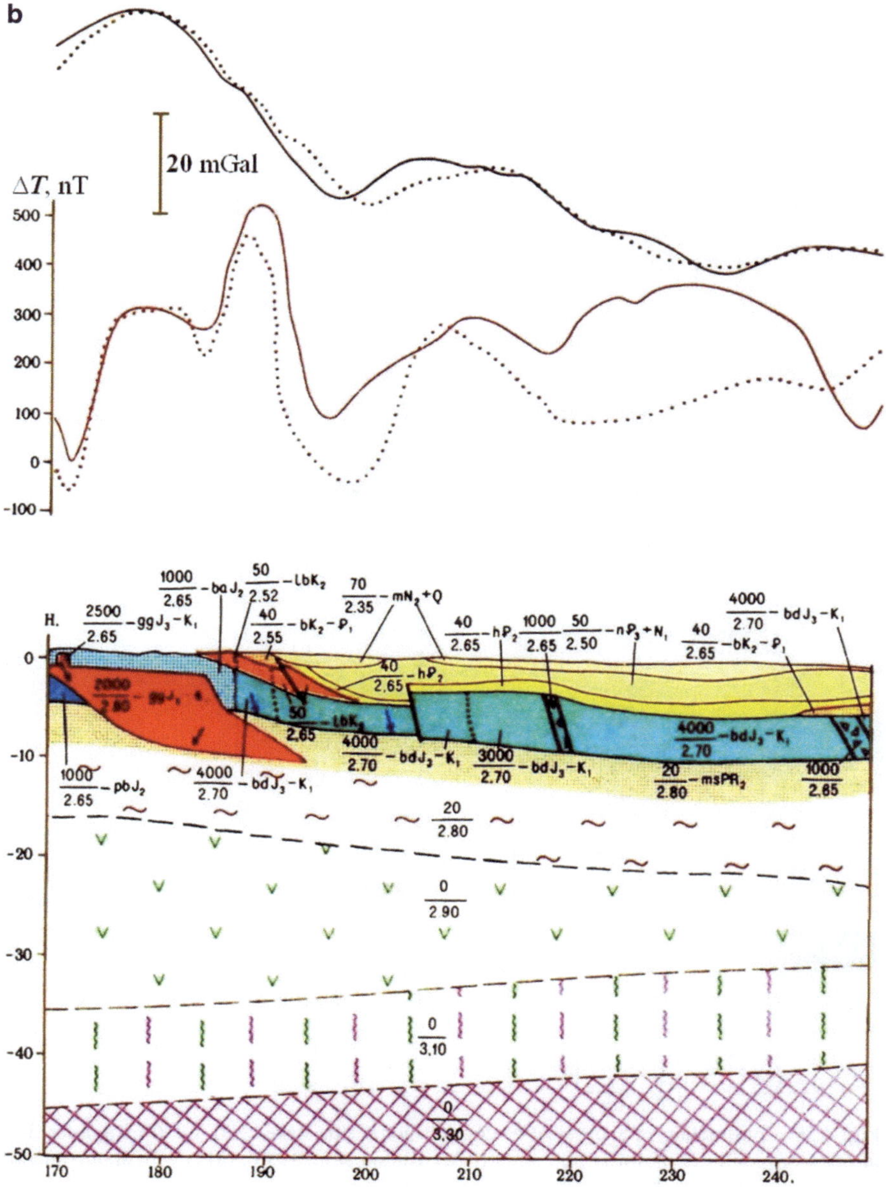

Fig. 4.35 (continued)

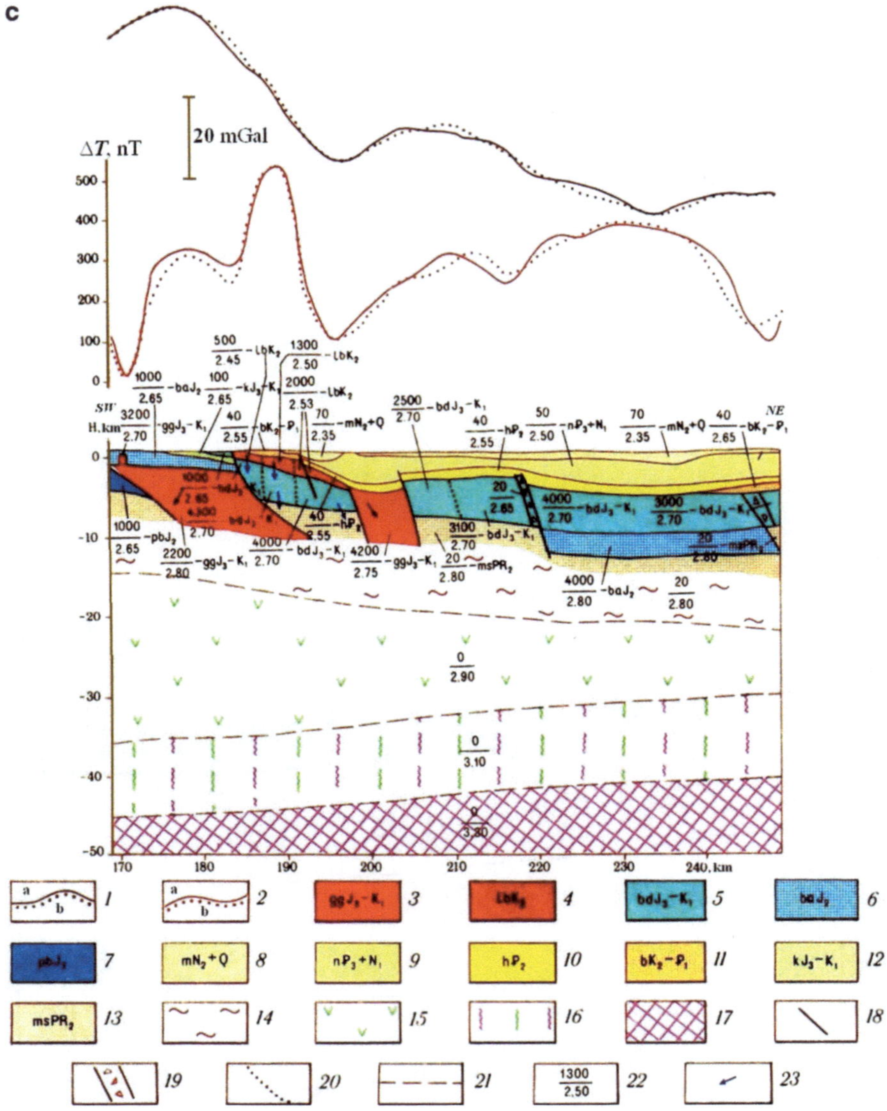

Fig. 4.35 Construction of a physical-geological model (for stakes 700–1,350 of profiles 3–4; location of profiles is shown in Fig. 4.31) (After Khesin et al. 1996). (**a**) A zero approximation model (initial) (**b**) The third approximation model (**c**) The eleventh approximation model (final). (*1*) intrusive gabbro-diorite-granodioritic association; (*2–5*) effusive associations: (*2*) liparite-basaltic, (*3*) basalt-andesite-dacitic, (*4*) basalt-andesitic, (*5*) basalt-andesite-plagioliparitic; (*6–11*) background sedimentary deposits: (*6*) upper molassic, (*7*) lower molassic, (*8*) terrigeneous, (*9*) terrigeneous-carbonaceous, in some places flyschoid, (*10*) carbonaceous-sandy, reef rocks, (*11*) metamorphic schists and other metamorphites; (*12–15*) deep-seated complexes: (*12*) granites and gneisses, in some places amphibolites, (*13*) basic rocks, (*14*) basic rocks – eclogites, (*15*) uppermantle peridotites; (*16*) faults, upthrusts; (*17*) crush zones; (*18*) boundaries of physical

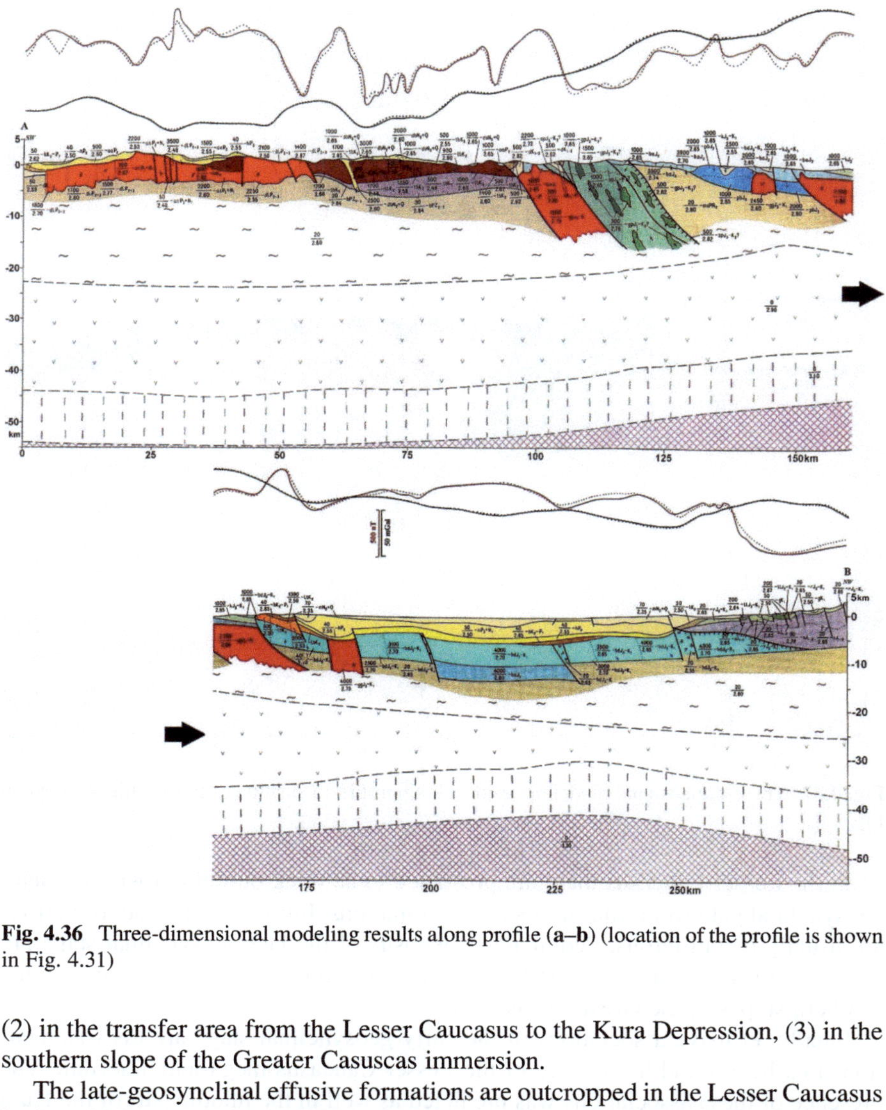

Fig. 4.36 Three-dimensional modeling results along profile (**a–b**) (location of the profile is shown in Fig. 4.31)

(2) in the transfer area from the Lesser Caucasus to the Kura Depression, (3) in the southern slope of the Greater Casuscas immersion.

The late-geosynclinal effusive formations are outcropped in the Lesser Caucasus in the middle part of profile. These formations create a so-called "ophiolitic zone". Physical-geological modeling indicates that these formations can occur under young sedimentary deposits in the Kura Depression and under Lower-Middle-Jurassic sand-shale associations in early-geosynclinal stage of the Greater Caucasus (where these formations were underthrusted as a result of tectonic processes).

Fig. 4.35 (continued) properties changing within the same association; (*19*) physical properties (numerator = magnetization, mA/m, denominator = density, g/cm^3); (*20*) direction of the magnetization vector differing from the geomagnetic field inclination

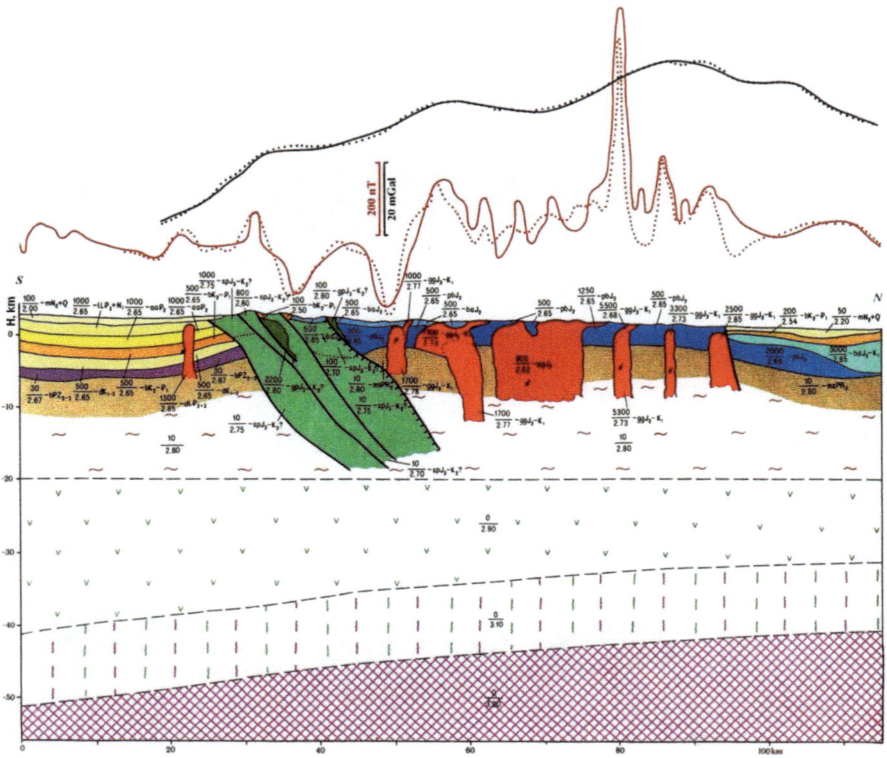

Fig. 4.37 Three-dimensional modeling results along profile 1 (location of this profile is shown in Fig. 4.31)

In the Lesser Caucasus there are presumed to be thick buried intrusives of late-geosynclinal gabbro-diorite-granodiorite formations. Intrusives of these formations are usually metalliferous. Sedimentary deposits of the late-geosynclinal stage are exposed in separate areas of the Lesser Caucasus and more widely present on the southern slopes of the Greater Caucasus.

Finally, effusive associations of the early-geosynclinal stage are (as was mentioned earlier) found in the core of the Lesser Caucasus mega-anticlinorium. The presence of these associations was predicted as well in the middle part of the Kura Depression below the late-geosynclinal effusives. Early-geosynclinal sedimentary deposits exist in the Greater Caucasus, where their thickness attains 7–8 km.

The Pre-Alpine basement is closest to the Earth's surface in the vicinity of the ophiolitic zone, where it depth is only 1–1.5 km. Other physical-geological models have been developed. Profile 1 stretches along a line Mez-Mazra – Gedabey – to Dzegam-Djirdakhan (Fig. 4.37), profile 2 – along a line Lake Karagel – Dashkesan – Beyuk-Kasik (Fig. 4.38), profile 5 – along a line Kutkashen – Mt. Shakhdag – village Chakh-Chakh (Fig. 4.39), profile 6 – Agsu – Lagich – Kusary (Fig. 4.40), profile 7 – along a line Kyalvazchai – Lerik – Djalilabad, and profile 8 – Mt. Shandankalasi – Masally.

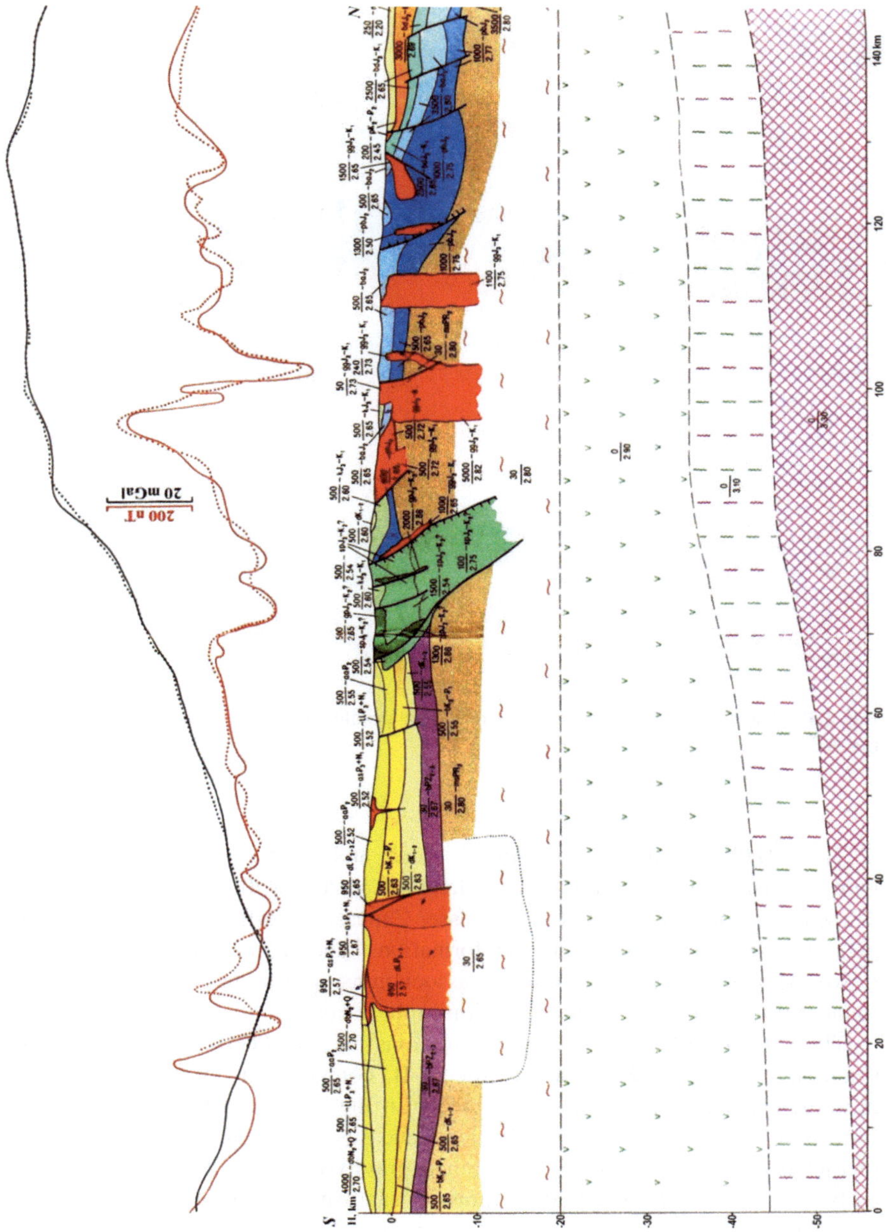

Fig. 4.38 Three-dimensional modeling results along profile 2 (location of the profile is shown in Fig. 4.31)

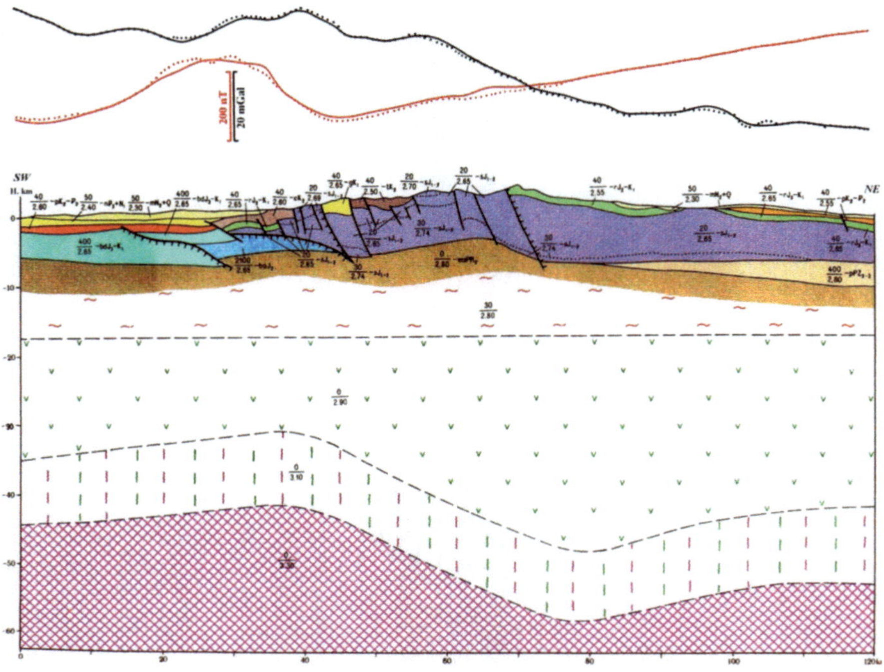

Fig. 4.39 Three-dimensional modeling results along profile 5 (location of the profile is shown in Fig. 4.31)

The ***PGM*** presented in models 1, 2, 5 and 6 show clearly fixed early-geosynclinal cores of the Lesser and Greater Caucasus with Jurassic associations. These associations are present in the Lesser Caucasus as effusive basaltic-andesite-plagioliparite and basaltic-andesite formations, and in the Greater Caucasus – Lower-Middle-Jurassic as sand-shale associations. The surface of the Pre-Alpine foundation occupies the highest position in these cores.

The late-geosynclinal effusive formations in the ***PGMs*** 1 and 2 (Lesser Caucasus), as in the ***PGM*** along line A–B, compose an ophiolitic zone. It is presumed that the same formations occur in the NE submersion of the Lesser Caucasus. In ***PGMs*** 5 and 6 (Greater Caucasus) their wide evolution can be predicted by the results of 3D combined gravity-magnetic modeling under the early-geosynclinal Lower-Middle-Jurassic sand-shale deposits of the Greater Caucasus (occupying this position as a result of overthrust), as well as under young deposits of the Kura Depression.

Pre-orogenic and orogenic intrusive and effusive formations are located in the southern parts of ***PGMs*** 1 and 2. Sedimentary deposits of these stages developed in the northern parts of profiles 1 and 2, and in southern and northern parts of profiles 5 and 6.

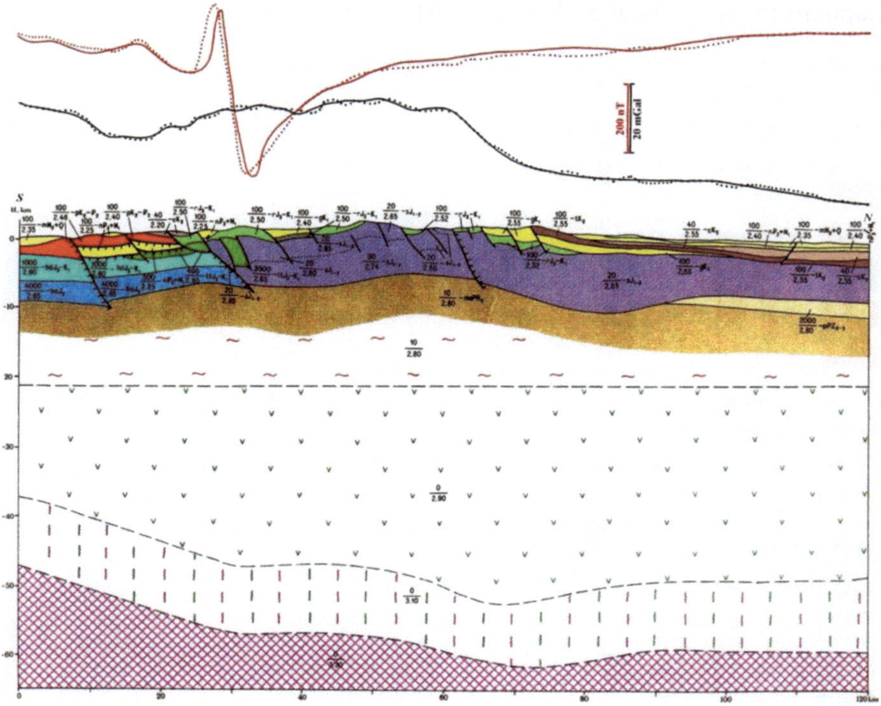

Fig. 4.40 Three-dimensional modeling results along profile 6 (location of the profile is shown in Fig. 4.31)

In the northern parts of profiles 5 and 6 at a depth of 9–10 km are associations of significant magnetization, presumably related to tuff-terrigeneous sub-platform associations of the Middle-Late-Paleozoic.

The behavior of the Moho discontinuity in the aforementioned ***PGMs*** on the whole agrees with the ***PGM*** along direction A–B. In ***PGMs*** 1 and 2 a smooth uplift of the Moho discontinuity can be observed from south to north at the depth of −52–54 km up to −42 km. In the ***PGMs*** 5 and 6 its behavior is more complex: in ***PGM*** 5 (from SE to NW) there is an uplift from −44 km to −41 km (under the area of the abovementioned overthrust), then submersion to a depth of −58 km (below the area of conjunction of anticlinoria of the Main Ridge and Side Ridge) and then an uplift to a value of −52 km; in the ***PGM*** 6 (from north to south) there is a stepped submersion from a depth of −47 km to −60 km (under the same conjunction area) and then a smooth uplift to a depth of −58 km.

An interesting peculiarity of the southern part of ***PGM*** 2 is the presence of a zone of strong rock decompaction in the root parts of the Dalidag intrusive with values from −4 to −17 km and a horizontal thickness of 30 km. This decompaction zone produces the greatest Kelbadzhar-Dalidag gravity minimum. This large "granite room" apparently was a source of pre-orogenic and orogenic granitoid magmatism.

Captions for the physical-geological models

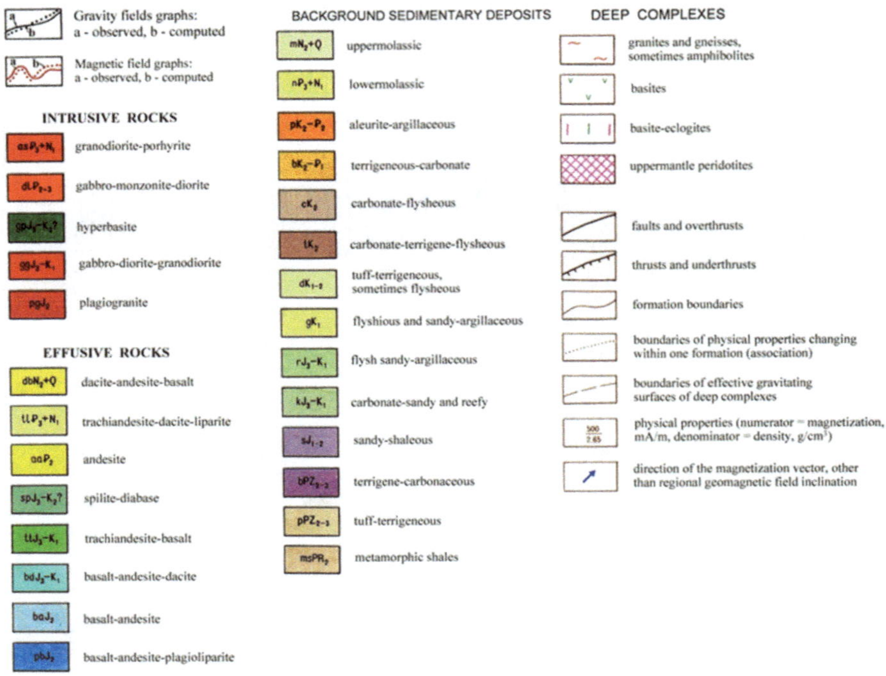

PGM 7 and 8 are located in the Talysh geostructural zone (see scheme in Fig. 4.31). There is a Moho uplift from south to north (from −52 km to −42 km and from −44 km to −31 km, respectively) and as well as to the East (on the side of the Caspian Sea).

Within the Talysh Zone, buried intrusive bodies from ultrabasic to acid consistency were delineated. Applying 3D gravity-magnetic modeling, surface trachite-trachiandesite-trachibasalt effusive formations that are widely present at the surface were divided into blocks of differing magnetization and (or) density. These blocks, apparently, have tectonic contacts. The bodies of gabbro-scientific and gabbro-monzonite-dioritic formation were characterized quantitatively.

4.5 Deep Structure Maps and Their Prognostic Importance

The complex geological structure of the Caucasian region governs its highly intricate gravity and magnetic fields which reflect the effects of outcropped and deeply buried bodies and structures. Thus to identify anomalies arising from a range of geological sources both the observed geophysical fields and their various transformations must be used. Therefore, the legend of such a deep structure map must include a classification of spatial features (regional and local anomalies), morphological features (isometric anomalies, elongated anomalies or ledges), and the signs and intensities of geophysical fields.

A description of gravity anomalies was based on: (1) difference fields ("ring" anomalies) $\Delta g_{B(8-20)}$ and $\Delta g_{B(4-10)}$ computed as the difference between upward continued fields in the Bouguer reduction (terrain correction was computed with a radius of 200 km and the density of the intermediate layer was defined as 2.67 g/cm^3) at heights of 8 and 20, 4 and 10 km, respectively, (2) the residual field $\Delta g_{B(0-4)}$ (local anomalies were calculated as the difference between the initial land Bouguer gravity and this field continued upward to a height of 4 km), (3) horizontal gravity gradient field at a height of 2 km ($\Delta g_{x(2)}$).

These features were analyzed on the basis of the following descriptors.

1. The residual field $\Delta g_{B(0-4)}$ in the mountain-folded regions reflects the influence of near-surface Alpine structures composed of volcanogenic and volcanogenic-sedimentary associations, and intrusions (both outcropping to the Earth's surface and near-surface), and in the Kura Depression as regards the distribution of dense inhomogeneties of the upper part of the sedimentary floor.
2. The difference field ("ring" anomalies) $\Delta g_{B(8-20)}$ which is present to a significant extent in the mountain-folded regions reflects the influence of the internal structure and inhomogeneties of the consistency of the metamorphic basement (Baikalian folded basement) and deeply occurring (and buried) parts of intrusions in the Kura Depression, in general the influence of Mesozoic associations.
3. Significant anomalies of $\Delta g_{x(2)}$ should be taking into consideration by delineation and examination of regional gravity anomalies.

These descriptors were determined on the basis of analysis of petrophysical parameters (see Sect. 4.1.2) as well as data on density of metamorphic foundation outcropped in Armenian territory.

According to Nikolsky et al. (1975), basement outcrops in Armenia are characterized by the following average density values: Aparan (2.80 g/cm^3), Arzakan (2.76 g/cm^3), Lok (2.76 g/cm^3), Eranos (2.87 g/cm^3), Nyuvadi (2.80 g/cm^3), and Akhum (2.63 g/cm^3). It should be noted that the Shamkhor (Beyukkishlak) protrusion, located in Azerbaijanian territory 20 km east of Akhum, is also characterized by an average density of 2.63 g/cm^3. At the same time, the density of separate types of rocks composing these protrusions may vary considerably. For instance, muscovite and sericite-quartz shales have a density of 2.66 (Aparan) to 2.79 g/cm^3 (Lok), quartzite from 2.60 (Lok) to 2.67 g/cm^3 (Nyuvadi), phyllite from 2.63 (Akhum) to 2.84 g/cm^3 (Arzakan), greenstone rocks (plagioclase porhyrites, tuff-sandstone, etc.) – from 2.71 g/cm^3 (Lok) to 2.83 g/cm^3 (Aparan), gneisses – from 2.60 g/cm^3 (Arzakan) to 2.94 g/cm^3 (Nyuvadi), limestone and marble – from 2.66 g/cm^3 (Aparan) to 2.78 g/cm^3 (Arzakan), amphibole shales – from 3.00 g/cm^3 (Aparan) to 3.03 g/cm^3 (Eranos).

In the Shamkhor (Beyukkishlak) cericite-muscovite-graphite outcrop, shales were found with comparatively low density (which is typical for the average density of the Baikalian series). An analogical situation was identified on the Akhum outcrop where at the Earth's surface phyllite clay shales representing the less metamorphosed part of the geological section were identified (Nikolsky et al. 1975). Thus, despite this variability in density, the average density of the rocks composing the Pre-Alpine foundation in SW Azerbaijan can be estimated as 2.80 ± 0.05 g/cm^3.

For the Armenian territory, the regional gravity map $\Delta g_{B(reg)}$ was constructed by a methodology that included using points of outcropped metamorphic basement (6 points), boreholes that crossed the basement (7 points), and other data. This map as a whole coincides with the $\Delta g_{B(8)}$ – map obtained by analytical continuation of the surface gravity field to a height of 8 km (the absolute levels of these maps are different) (Sirotkina and Nikolsky 1971). Presumably that analogical ratio is correct and for the Lesser Caucasian part of Azerbaijan territory. Thus the $\Delta g_{B(8)}$ map mainly reflects the variations of the upper surface and inhomogeneties of the consistency in the Pre-Alpine metamorphic floor and deeper layers of the Earth's crust; the effects of the Alpine floor are weakly present in this map. Therefore, the difference field $\Delta g_{B(8-20)}$ ("ring anomalies") primarily reflects the influence of the internal structure and inhomogeneties of the consistency of Baikalian and possibly the Pre-Baikalian floor. In addition as regards the transition from the Lesser Caucasus to depression regions, the difference field mainly represents the structure and consistency of the Mesozoic associations which act here as a dense basement.

Thus, the anomalies in the $\Delta g_{B(0-4)}$ map indicate local (close to the Earth's surface) objects; anomalies recognized in the $\Delta g_{B(8-20)}$ map represent some regional (deep) objects. The $\Delta g_{B(4-10)}$ map has a pattern similar to $\Delta g_{B(8-20)}$, but with a more differentiated field.

To examine regional magnetic anomalies the following measures were used: field ΔT (in the Middle Kura Depression – field Z_a) at 6 and 10 km, and the $\Delta T(Z_a)$ horizontal gradient field at 2 km. Attempts to reveal local magnetic anomalies as a residual field of $\Delta T_{(0-6)}$ or $\Delta T_{(0-10)}$ were unsuccessful because of the considerable difference in the observed magnetic field in the Lesser Caucasus. Therefore local magnetic anomalies were selected directly in the initial field, and magnetic maxima were singled out irrespective of whether they were found in a positive or negative field. Magnetic minima were identified in a similar fashion. Some of these minima were conjugated with the corresponding maxima in relation to the oblique magnetization (see Chap. 3) typical of the Caucasian region. Other minima corresponded to objects with a low magnetization or reversely magnetized targets. The map of gravity-magnetic anomalies (Fig. 4.41) shows both these minima since the problem of their nature can be solved at the stage of anomaly examination (Ismail-Zadeh and Khesin 1989b).

In the $\Delta g_{B(8-20)}$ map intensive (first tens of mGals) and extended (first hundreds of km) gravity ledges were identified that split the areas of positive and negative field. Similar ledges were identified in the $\Delta g_{B(4-10)}$ map. The intensive linear maxima or minima correspond to them in the horizontal gravity gradient map. Axes of these ledges are shown in the map of gravity-magnetic anomalies (Fig. 4.41) as two bold colored lines: brown from the side of the positive field and green from the

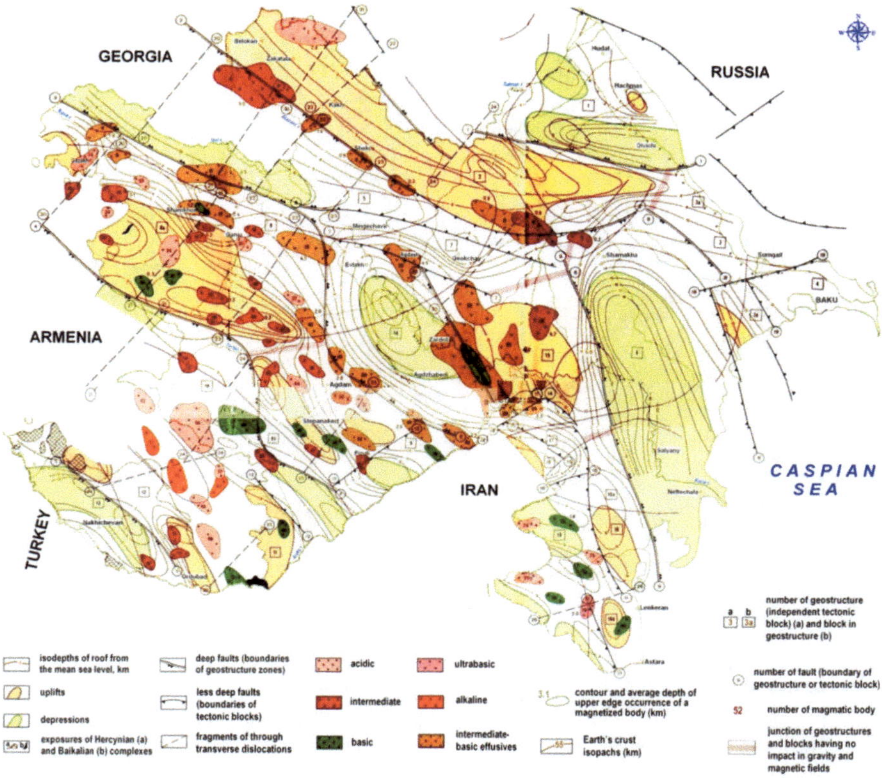

Fig. 4.41 Map of the deep structure of Azerbaijan with its adjacent regions according to gravity and magnetic data (After Ismail-Zadeh and Khesin 1989a; Khesin et al. 1996)

side of negative one; arrows in the green line are oriented toward the block of lesser density. Less intensive (several mGals) and less extended gravity ledges are represented by an analogous symbol with lines of less density.

Regional magnetic steps were identified by analysis of $\Delta T_{(10)}$ and $Z_{a(10)}$ maps by taking into consideration the map of the horizontal magnetic gradient $\Delta T(Z_a)$ at 2 km. Axes of these ledges are depicted in this map (Fig. 4.41) as two bold colored lines: blue – is from the side of positive field and red from the side of negative field; arrows on the red line are oriented toward the side of the block with lesser magnetization (Ismail-Zadeh and Khesin 1989b).

Regional gravity maxima and minima (with an intensity up to 20 mGals) as revealed in the $\Delta g_{B(8-20)}$ map, are contoured by the points with values approximately equal to half the extremal values. In the map (Fig. 4.41) they are shown as solid bold brown (maxima) and green (minima) lines, respectively with arrows oriented to the side of decreasing field. Within certain contours are located several (1 to 3–4) anomalies of $\Delta g_{B(4-10)}$ but are not shown to make the map pattern more legible.

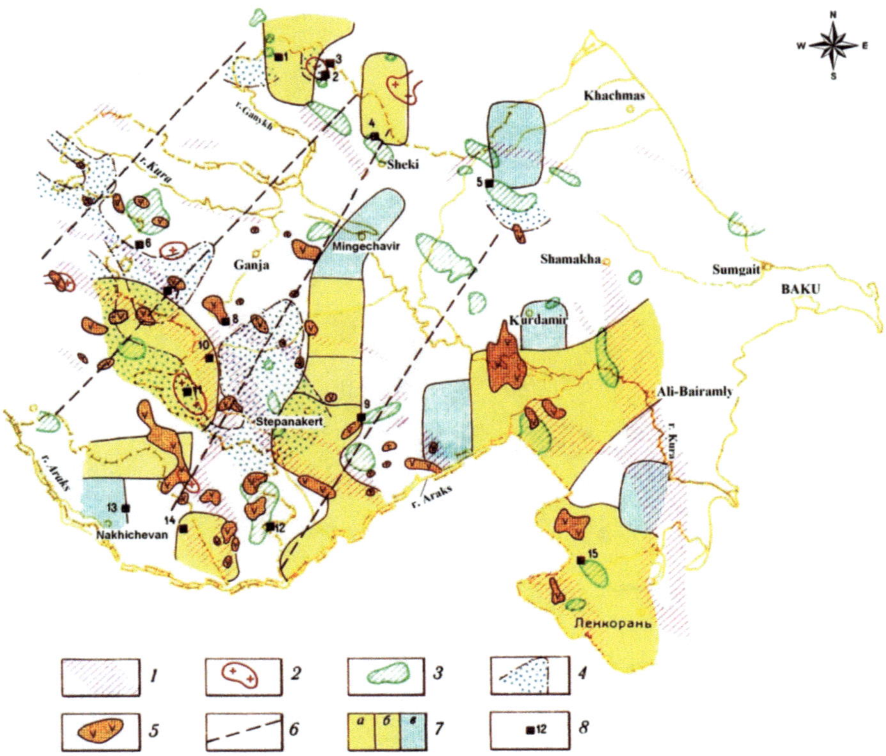

Fig. 4.42 Location of areas for deep drilling derived from integrated examination of geophysical fields. (*1*) zones of large faults detected by the use of the highest gradients of the gravity field in the Bouguer reduction; (*2*) acid intrusions revealed by negative local gravity anomalies; (*3*) dome-shaped structures or areas of basic rock development revealed by positive local gravity anomalies; (*4*) uplifts of basic and middle magnetized generations found after determination of singular points in the reference aeromagnetic net; (*5*) areas of basic rock development corresponding to local magnetic anomalies; (*6*) transversal faults marked by local magnetic anomalies; (*7*) neotectonic blocks corresponding to areas with anomalous (different from common Caucasian) strikes of relief isolines: (*a*) NNW, (*b*) NNE, and (*c*) – NEE; (*8*) areas of recommended reference boreholes and their numbers

Local gravity and magnetic maxima indicated respectively in maps $\Delta g_{B(0-4)}$ and ΔT (Z_a) are represented by points with the half-maximum of the corresponding extremum. The maxima and minima are divided into intensive (exceeding the absolute value of 3–4 mGals and 300–400 nT) and less intensive (intensity of 1–2 mGals and 100–200 nT).

In the map (Fig. 4.41) intensive local gravity maxima are depicted by the dense dark shading surrounded by a brown line, and intensive minima are indicated by the dark green line. Less intensive local gravity maxima in a positive field are shown by light brown shading contoured by a brown line; the same maxima in a negative field are represented by light brown shading surrounded by a green line. Similarly less intensive gravity minima are shown by light green shading surrounded by a green line in a negative field and a brown line in a positive field.

A similar system of symbols was applied to depict local magnetic anomalies: intensive maxima are shown by dark blue shading surrounded by a blue line in a positive field, and red in a negative field. The less intensive maxima are shown by light blue shading surrounded by blue or red lines depending on the sign of field in which the maximum appears. The local magnetic minima are shown as dark or light red shading depending on field intensity, and a red or blue contour line depending on the sign of surrounding field.

The points of high intensive magnetic maxima and minima (exceeding 2,000 nT in absolute value) are presented respectively by a blue cross and a bold red dot (imitating the tail and head of an arrow – magnetization vector).

The areas or intersections of local gravity and magnetic anomalies are shown by combined symbols: strips of corresponding color and shading intensity oriented in a NE direction at an angle of 45°.

On the basis of a combined analysis of gravity and magnetic fields selected areas for deep drilling were charted (Fig. 4.42).

Chapter 5
Mining Geophysics

The specific nature of mining geophysics, models of media, the details and the sequence of the interpretation process, including petrophysical studies have been amply covered in books (Khesin 1969,1976; Khesin et al. 1983,1996) and Eppelbaum's (1989) Ph.D. dissertation. A separate chapter (Khesin et al. 1988) in (Borisovich and Eppelbaum 1988) deals with the optimization of interpretation for detailed prospecting under mountainous conditions. These works form the basis for the Caucasian mining geophysics presented in this and previous chapters.

The Caucasus Mountains are a typical Alpine-type mountainous region. The most specific conditions for geophysical mining studies are described below. In these regions, solid heterogeneous associations usually outcrop at the Earth's surface. These associations are multiply folded, with intensive rupture tectonics (including thrust tectonics). Rocks of various origins and compositions with a broad range of physical properties rapidly change along both the vertical and lateral directions. This predetermines the complexity of the images of geophysical fields. Due to the curvature of the earth-to-air interface and the rugged observation surface, the relief exerts a pronounced effect on observations. The dissected relief, complex geology, difficulties in transporting and observation – are all factors that affect the investigation procedure. They restrict efficiency because they require more intricate and cumbersome equipment and survey systems.

At the same time, it would be a mistake to view mountainous conditions solely as an obstacle. Deep erosional truncation and a lack or low thickness of loose deposits encourages visual geological methods. These make it possible to obtain extensive geological evidence on the nature of anomaly sources, and to correlate them with geophysical data, as well as to study physical properties of rocks and ores in natural and artificial exposures. Rugged relief can be used to calculate the real physical parameters of a section exposed to erosion based on measurements of the corresponding fields on an uneven surface. It also facilitates geological application of topography data. Outcroppings of mountainous regions promote the integrated application of conventional geophysical methods not only along with visual and geological methods, but also geophysical, geochemical, petrophysical and

L.V. Eppelbaum and B.E. Khesin, *Geophysical Studies in the Caucasus*,
Lecture Notes in Earth System Sciences, DOI 10.1007/978-3-540-76619-3_5,

physical-chemical investigations (including areal study), which can dramatically lessen the ambiguity of geophysical interpretation.

According to Solovov (1985), open areas where ore-bearing rocks outcrop onto the surface or are covered with eluvial and deluvial products of their weathering can be divided into two groups. The first which has a severe topography, is considered unfavorable for geophysical prospecting. This opinion as regards typical orogens (mountainous folded areas) of the open (one-stage) type can be easily explained since complex geophysical equipment are difficult to transport and use in highly inaccessible mountainous regions, and anomalies caused by the relief are pronounced and difficult to take into account.

However, mountainous regions objectively call for a systematic application of geophysical methods of investigation. Moreover, only these methods can ensure a sequential, deep and sufficiently rapid study of the endogenic mineralization distribution and its relations to geological structure. Prospecting for large hidden deposits is conducted with these issues in mind. Many valuable deposits of different types are located in mountainous regions above 2,000 m, which make up about 12% of the total continental area, and in areas over 1,000 m, which account for as much as 30% of the total land. A large portion of this area consists of mountainous structures formed or rejuvenated during the Alpine epoch (Khain 1984).

The requirements for geophysical techniques in mountainous conditions are less stringent in both regional and detailed prospecting. In regional prospecting, airborne surveys may be tied in with comparatively stable elements of the relief. In ground surveys, observations are linked to routes or riverbeds when making geophysical measurements with cumbersome equipment. In this way key features are intersected and are then used as references for large-scale surveys of the area.

Geophysical mining in the Caucasus can be divided into three parts: the Northern Caucasus, the southern slope of the Greater Caucasus, and the Lesser Caucasus.

5.1 Petrophysical Examination of Ore Areas

Petrophysical studies in ore areas are an important component of mining geophysics. They have been significantly impacted by the development of physical-geological models. A few examples are presented briefly below.

Brodskaya and Gogua (1988) examined the formation conditions of several Adjarian intrusives using petromagnetic data. As a result, the origin of many intrusives (including the known Uchambo intrusive body) was modified. Sobolev and Starostin (1993) conducted detailed analysis of the ore-bearing Eldzhurtinsky massif (Northern Caucasus) for density, porosity, saturation and elastic parameters, ultrasound velocities, Poisson coefficients, shear and Young's moduli and piezoelectric characteristics. Karkoshkin (1972) investigated the petrophysical properties of the Gedabey mining district (Lesser Caucasus). An important analysis of the palaeomagnetic characteristics of Middle Jurassic deposits in the Lesser Caucasus was carried out by Karkoshkin et al. (1974). Ginzburg et al. (1986) described the most important physical characteristics of rocks and ores in the southern slope of the Greater Caucasus.

A detailed examination of petrodensity (Gadjiev et al. 1984) and petromagnetic (Ismail-Zadeh et al. 1983b) characteristics of geological associations of Azerbaijan covers numerous ore areas in the Lesser Caucasus and the southern slopes of the Greater Caucasus.

An example of a study on the physical properties in the Kyzyl-Bulakh gold-pyrite deposit is presented in Table 5.1. In this table, density and magnetic susceptibility were assessed by Eppelbaum (1989), the resistivity data are from Novikova et al. (1983) and measurements of thermal conductivity were performed under the supervision of S.N. Ginzburg in TzNIGRI (Moscow).

5.2 Borehole Logging

Borehole logging in ore deposits can reveal enough physical characteristics to develop physical-geological models enabling electric, electromagnetic and radiometric application methods. The results of borehole logging are found mainly in industrial geophysical reports.

Khesin and Sattarov (1967) successfully applied a gamma-ray log in the Dashkesan and Filizchay ore fields. In the Dashkesan ore field, gamma-ray analysis reliably detected intervals with tuffs and limestone; in the Filizchay boreholes new prospective ore intervals were detected. Muradkhanov (1971) carried out comprehensive statistical processing of geophysical data from boreholes drilled in the Filizchay deposit. The results of borehole η_a measurements in the Karadagh copper-porphyry deposit (Lesser Caucasus) are shown in Fig. 2.18.

Radio-frequency crosshole investigations (RFCI) on frequencies $f = 21$, 32, 128, and 512 kHz were employed in several deposits in the southern slope of the Greater Caucasus. Figure 5.1 shows the results of RFCI in the Filizchay deposit in boreholes 670 and 675 (these boreholes were located 550 m apart).

To conduct the RFCI, well 670 was drilled deeper than was projected by 100 m. In the inter-well space two anomalous areas were detected. The first (located at a depth of 600 m) was caused by a small ore body and totally contoured. This ore target pinched out at a distance of about 100 m from well 675. The second anomalous area caused by the main ore body began at a depth of 709 m in well 670 and continued uninterruptedly to the boundary of the RFCI domain (in a line to well 675) for 260 m.

Other methods include the Self-Potential (Spontaneous Polarization), the Electric Method of Sliding Contacts, and the Method of Electrode Potentials. Abdullaeyev and Bagirov (1996), and Muradkhanov and Magerramov (1996) give a brief review of these mine-borehole and borehole logging methods applied to Caucasian ore deposits.

5.3 Northern Caucasus

5.3.1 *Gravity*

In the Northern Caucasus several skarn deposits were discovered and exploited (e.g., Tyrnyauz, Upper Chegem, etc.). Unfortunately, during the Soviet era most of the geophysical data on these deposits were prohibited from publication.

Table 5.1 Physical properties of rocks and ores in the Kyzyl-Bulakh gold-pyrite deposit

Age	Lithology	σ, g/cm^3		κ, 10^{-5} SI		ρ, Ohm·m		λ, W/m°C	
		Average	Range	Average	Range	Average	Range	Average	Range
Q_3	Alluvial soil, clay and loam	–	–	50	13–125	–	–	–	–
Q_1	Deluvium conglomerates, pebbles, brecciated rocks	–	–	500	350–1,080	–	–	–	–
J_3k–J_3o	Calcareous sandstone	2.64	2.62–2.66	260	39–760	–	–	–	–
	Silicified calcareous sandstone	2.66	2.63–2.67	690	480–1,030	–	–	–	–
	Limestone	2.56	2.53–2.59	10	1–20	–	–	–	–
J_2bt	Andesite-dacitic porphyrite	2.47	2.27–2.59	1,400	115–4,800	–	–	2.80	2.44–3.0
	Andesitic porhyrite	2.59	2.48–2.73	3,260	810–69,000	700	140–1,200	–	–
	Intensely silicified andesitic porhyrite	2.45	2.36–2.56	210	30–660	–	–	2.87	2.72–3.0
J_2b	Liparite-dacitic porphyrite	2.48	2.38–2.55	590	210–1,750	450	400–500	2.84	2.73–2.94
	Tuff breccia	2.52	2.42–2.60	1,630	580–4,130	–	–	–	–
	Silicified brecciated lava	2.30	2.20–2.56	24	0–100	–	–	–	–
J_3	Dike of andesite-basalt composition	2.72	2.64–2.90	1,050	450–2,130	–	–	3.20	3.03–3.50
	Malachitized rock with rare inclusion of magnetite	2.55	2.46–2.73	8,750	1,880–26,300	320	140–500	–	–
	Veinlet-disseminated pyrite-chalcopyrite ore	3.13	2.80–3.54	14	2–31	8	1–20	3.97	3.50–4.19
	Dense-impregnated chalcopyrite ore	3.20	2.90–3.74	19	6–29	–	–	4.5	4.30–4.84
	Massive chalcopyrite ore	3.50	3.10–4.0	15	13–19	1	0.1–5	5.7	5.14–7.30

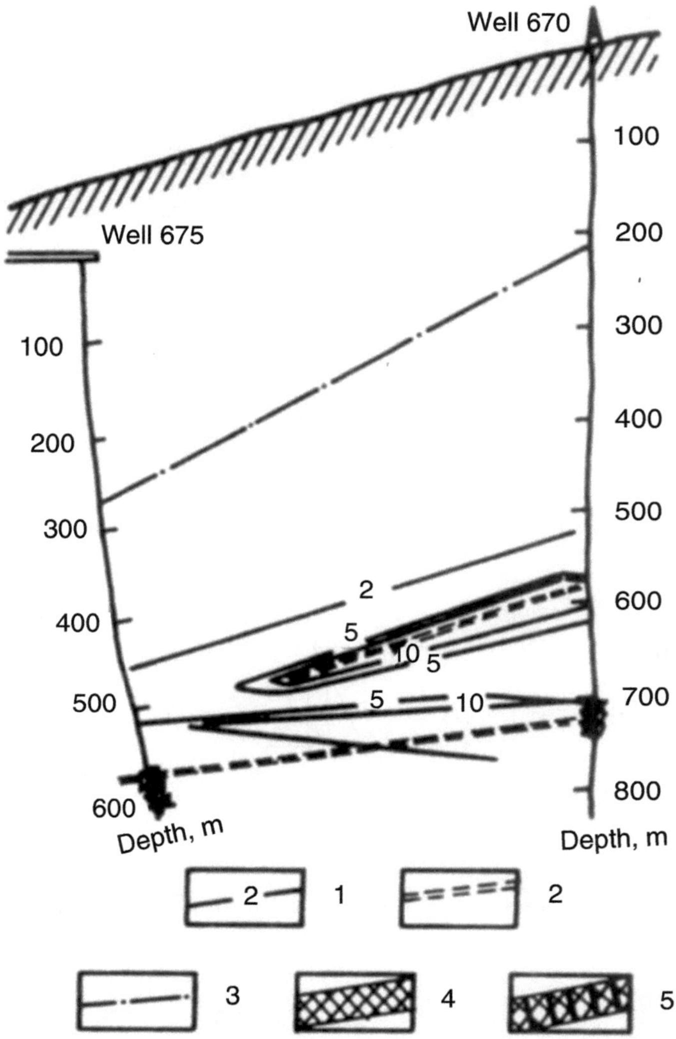

Fig. 5.1 Results of radio-frequency crosshole investigation (RFCI) between wells 675 and 670 ($f = 21$ kHz) in the Filizchay deposit (southern slope of the Greater Caucasus) (Borisovich et al. 1988). (1) isolines of shielding coefficients, (2) axis of shield, (3) boundary of the RFCI, (4) massive ore, (5) streaky ore

A generalized physical-geological model of the skarn deposit is presented in Fig. 5.2.

The examples presented in Fig. 5.2 illustrate the rapid interpretation of gravity anomalies $\left(\Delta g_{\mathrm{Bouguer}}\right)$ made on the rugged terrain relief by the interpretation methods presented in Sect. 3.4. The horizontal circular cylinder (HCC) represents the approximation of the ore body. Quantitative analysis used the tangent and

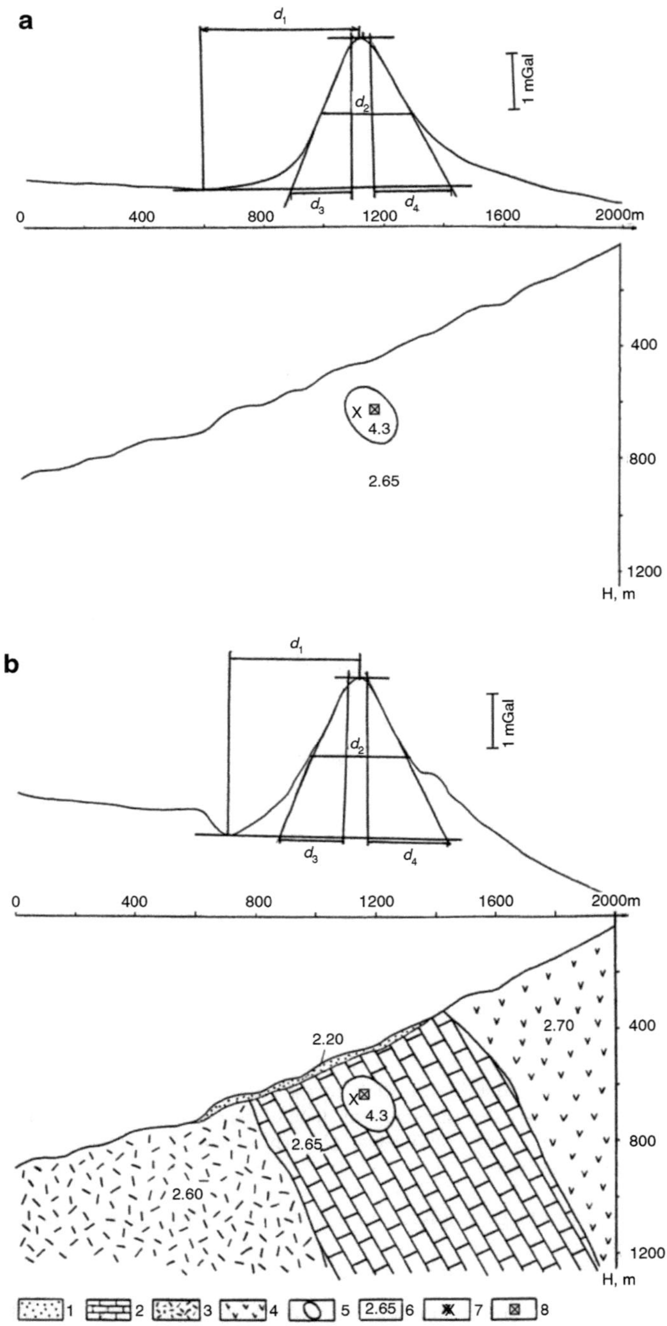

Fig. 5.2 Rapid interpretation of Δg anomaly over the model of the North Caucasian skarn deposit occurring in a homogeneous (**a**) and heterogeneous (**b**) medium. (1) loose deposits; (2) limestone; (3) acid volcanites; (4) andesites; (5) contour of the skarn deposit; (6) density, g/cm^3; location of the *HCC* center according to the results of Δg interpretation: (7) fictitious, (8) real

characteristic point methods. Interpretation accuracy drops when passing from the homogeneous host medium (Fig. 5.2a) to the heterogeneous one (Fig. 5.2b). The location of the "fictitious" body was determined before calculating the disturbing influence of terrain relief, and the location of "real" body was found after application of formulas (3.45). It should be noted that even in the case of a heterogeneous medium, the accuracy of interpretation is still about 15%. For mountainous conditions this level of accuracy is entirely sufficient for rapid quantitative interpretation.

5.3.2 VLF

VLF is one of the fastest and mobile geophysical methods. The main limitation of this method is its depth of investigation, which is usually 0.6–0.8 of the skin-layer (Gordeyev et al. 1981).

Figure 5.3 illustrates an application of the correlation method (see paragraph 3.2.4.4) in the area of the Eastern Kur-Kol ore zone (Northern Caucasus). Linearity of morphostructures is typical of Caucasian Alpine structures. Here a distant VLF transmitter (Rugby, England) was used that emitted at a frequency of 16.0 kHz. In the geological section (Fig. 5.3c) pyrite zones occur in effusive rocks of the middle and upper Devonian. The coefficient of correlation r (the anomalous zone was not calculated) was 0.99. The linear coefficients were used to calculate the correction for relief as $H_{x\ \mathrm{corr}}$ (Fig. 5.3b). As is clear from the figure, $H_{x\ \mathrm{corr}}$ is acceptable for a qualitative analysis.

Figure 5.4 gives an example of a quantitative interpretation of the VLF curve over a copper-pyrite ore body in the Northern Caucasus. It should be noted that the measured position of the upper edge can be slightly shifted from the real upper edge downward along the bed dip. This is because linear currents are focused in the upper portion of the conductive object, and this portion may be situated below the upper edge of the anomalous body.

5.3.3 Electromagnetic Methods

The method of charge with measurement of magnetic field was successfully employed for shallow pyrite deposits of the Northern Caucasus (Rogachev 1965). As a result, anomalous axes coinciding with ore body strikes were constructed. Rogachev (1965) used a model of a linear non-equipotential conductor. He determined the location of this pinching out, and estimated the influence of the overburden of finite conductivity.

The transient-field anomalies along about a 1,000 m profile across the ore are shown in Fig. 5.5. There is a clear increase in the secondary field over the ore body.

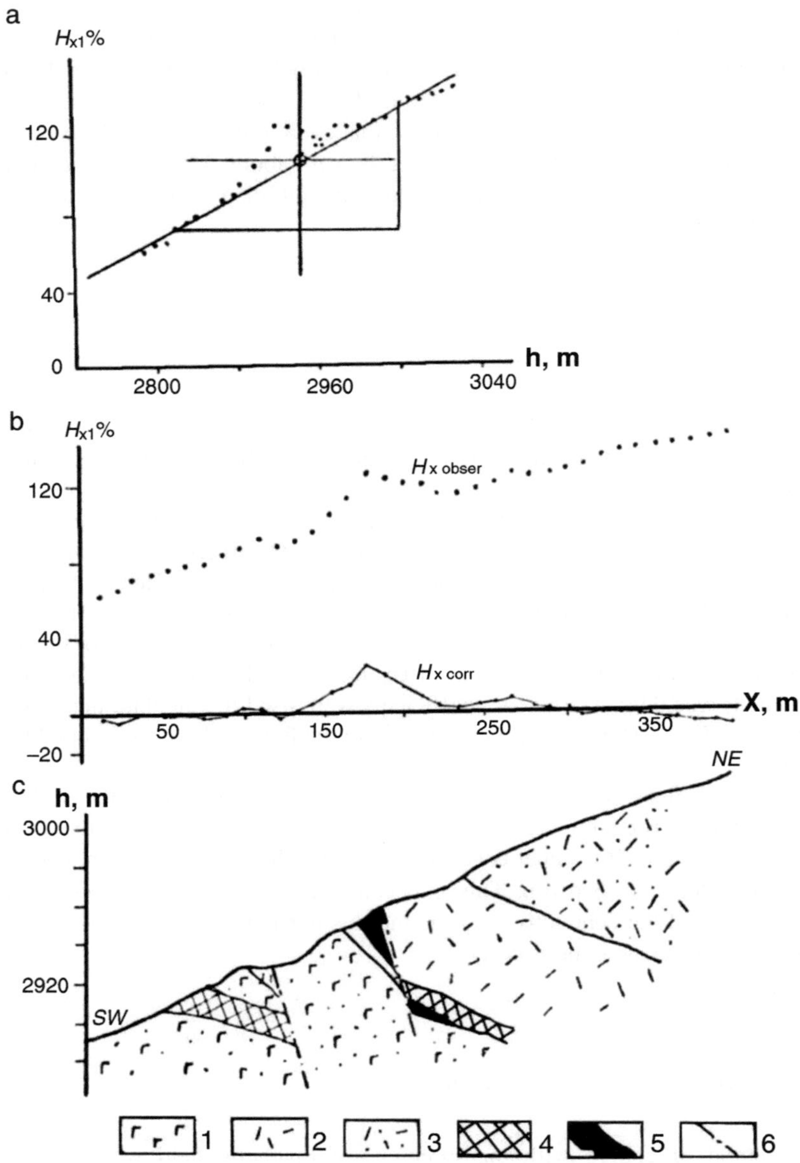

Fig. 5.3 Correlation technique for reducing terrain relief in a district of the Eastern Kur-Kol zone (Northern Caucasus). (**a**) correlation field, (**b**) observed and corrected H_x graphs, (**c**) geological section (observed field H_x and geological section after Gordeyev (1970), interpretation after Eppelbaum (1984b)). (1) tuffs, (2) tuffites, (3) tuff-sands, (4) and (5) poor and rich polymetallic ores, respectively, (6) faults

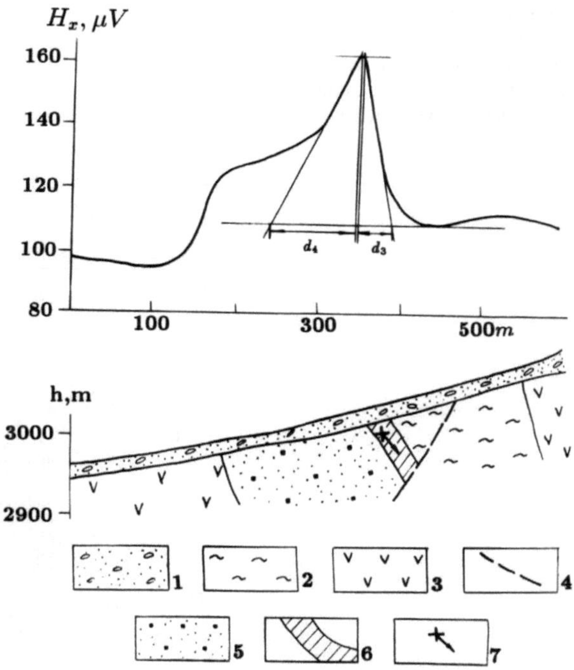

Fig. 5.4 Quantitative interpretation of the H_x anomaly in the area of a copper-pyrite deposit of the Northern Caucasus (initial data from Gordeyev et al. (1981)). (1) loose deposits; (2) spilites; (3) andesitic porphyrites; (4) disjunctive dislocation; (5) mineralized zone; (6) massive ore; (7) location of the upper edge of the conductive object and direction of its dip according to quantitative interpretation results

The resistivity of the glacial overburden was reported to be 100 Ω m and its thickness above the ore is as much as 40 m.

Detection of copper-sulphide ores in the Northern Caucasus by the non-grounding loop method is shown in Fig. 2.12.

Gezin (1964) and Komarov (1980) note several successful cases of IP method on polymetallic deposits in the Northern Caucasus.

5.4 Southern Slope of the Greater Caucasus

5.4.1 Physical-Geological Models of Ore Deposits

Physical-geological models (***PGMs***) are the easiest to grasp because they are displayed both in graphic and table form. ***PGMs*** make it possible to resolve the dual problem of interpretation: (1) enhancing the informativity and reliability of geophysical data interpretation, (2) substantiating the selection of a set for further geophysical investigations and plan a strategy.

Several examples of physical-geological modeling for barite deposits in Western Georgia can be found in Pirzhalava et al. (1992).

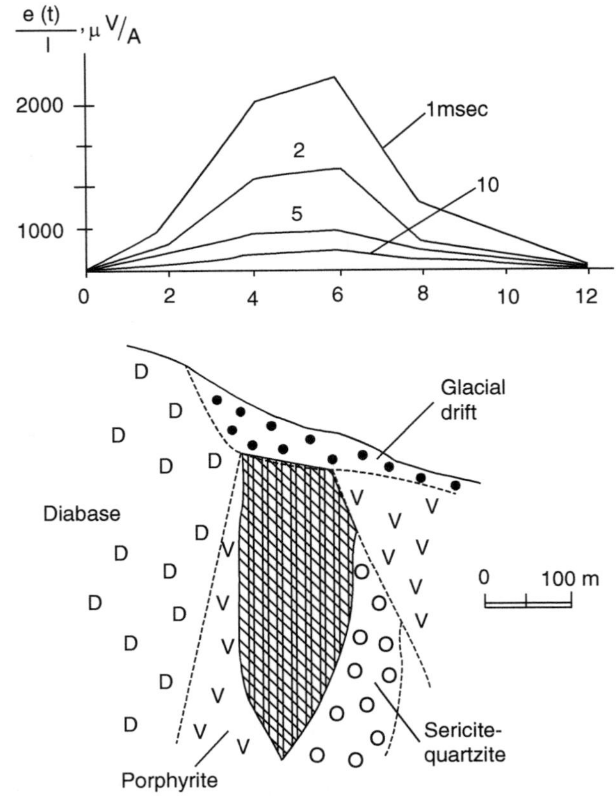

Fig. 5.5 Transient electromagnetic field profile across an orebody (Cu ore) in the northern Caucasus. Loop 200 × 200 m (After Fokin 1971)

Ginzburg et al. (1986) proposed using a formalized Ostrovsky (1980) method to develop 3D models of ore deposits on the basis of surface and underground geophysical measurements.

5.4.2 Physical-Geological Model of a Pyrite-Polymetallic Deposit of the Filizchay Type

This subsection discusses a ***PGM*** of pyrite-polymetallic deposit of the Filizchay type (this deposit is one of the largest polymetallic deposits in the world). This ***PGM*** represents a large group of pyrite-polymetallic deposits concentrated in the pyrite-bearing provinces with a predominance of carbonaceous-terrigeneous sediments (Borodaevskaya et al. 1977). Characteristic pyrite-polymetallic deposits of the Filizchay type are Sullivan (British Columbia, Canada), Coeur d'Alène

(Idaho, USA), Mount-Isa (Australia), Rammelsberg (Germany), Ozernoye and Kholodninskoye (Baikal region, Russia), Filizchay and other deposits (Katsdag, Katekh, Djikhih, etc.) on the southern slope of the Greater Caucasus (Gorzhevsky et al. 1987; Tvalchrelidze 1978). These deposits on the southern slope of the Greater Caucasus are situated in a severely rugged relief and occur in sandy-shale rock masses of the Jurassic. The ***PGM*** for this type of deposit can be represented in the form of steeply dipping massive sulfide sheet-like deposits, which differ from the host medium on a number of contrasting properties such as excess density ($1.3 \div 1.8$ g/cm^3), higher (by a factor of 10^2–10^3) conductivity, thermal conductivity (two to threefold) and polarization (up to tenfold) and in some cases magnetization (10–50-fold) (Muradkhanov 1971).

The development of numerous disjunctive dislocations impeding geophysical data interpretation is typical of these deposits.

In the fields predicted by a typical ***PGM*** of the Filizchay type (Fig. 5.6) deposits were computed using the *GSFC* program for different depths of occurrence of the upper orebody edge h (Table 5.2). For SP, VLF, IP and thermal prospecting methods, the determination of these effects took into account the results of physical and mathematical modeling, and field investigations.

The resulting physical-geological model substantiated the interpretation criteria and optimal geophysical sets for prospecting and estimation. For deposits of the Filizchay type this set comprised gravimetric and magnetic prospecting and the SP method. The IP method can be also a significant tool (see, for instance, Fig. 5.10). If ore objects differ only slightly in terms of their magnetic properties from the host sandy-shale deposits, it is worth using the VLF technique in the set instead of magnetic prospecting. Near-surface thermal prospecting can be also employed (Eppelbaum and Khesin 1988).

When comparing the two forms of ***PGM*** presentation, it should be noted that graphic ***PGM*** models (which are primarily quantitative) are easier to grasp. However, whereas the development of quantitative graphical ***PGM*** using computer-aided gravity and magnetic field computations presents no technical problems, quantitative computation of the temperature and self-potential fields using the same ***PGM***, as well as fields due to distant *VLF* transmitters leads to many mathematical and computational difficulties. This is why the combined application of these two approaches supplementing each other seems more promising (Eppelbaum and Khesin 1988).

5.4.3 Gravity

The gravity method can be illustrated by a comprehensive gravity field examination in the Katekh deposit.

The Katekh pyrite-polymetallic deposit is situated in the southern slope of the Greater Caucasus (northern Azerbaijan) under conditions of severe rugged topography. According to the data of the "Azerbaijangeologiya" Association, the

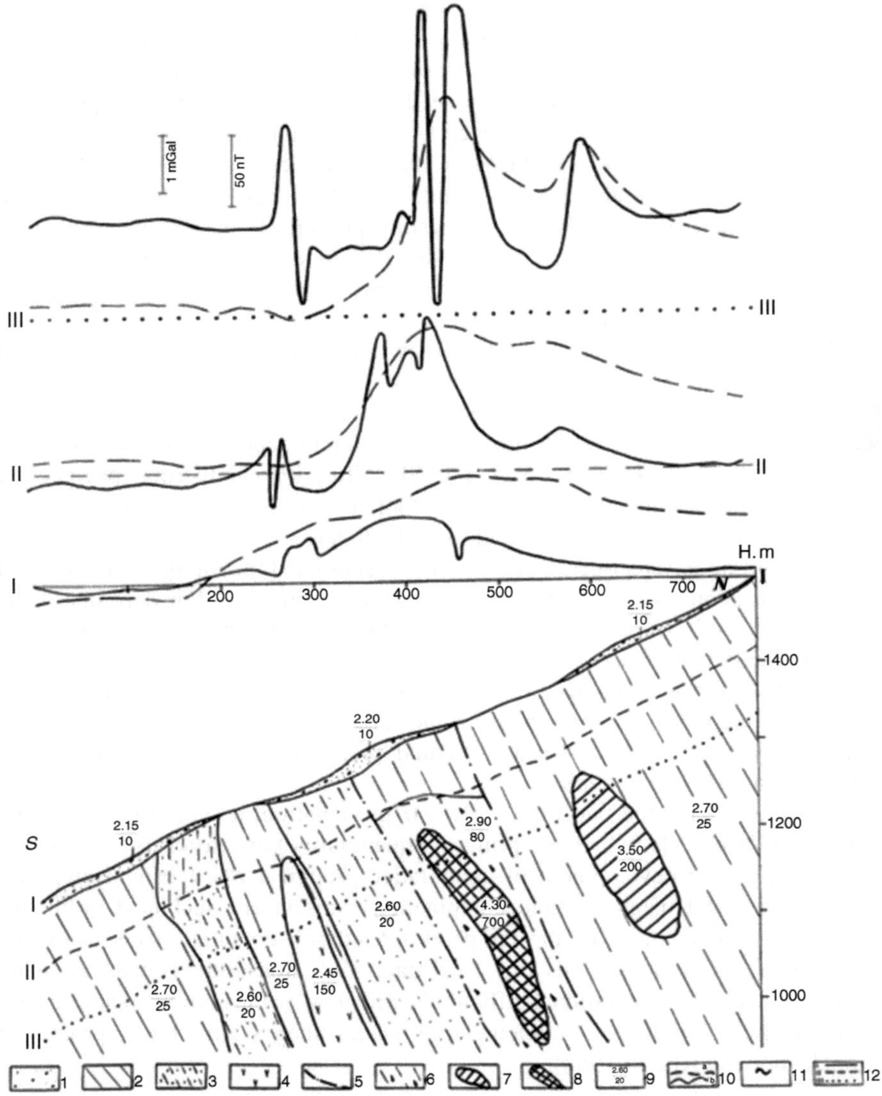

Fig. 5.6 A generalized physical-geological model of the pyrite-polymetallic deposit of the Filizchay type. (1) loose deposits; (2) clay shales; (3) clay shales with sandstone bands; (4)liparite-dacites; (5) disjunctive dislocations; (6) ore zone; pyrite-polymetallic ores: (7) impregnated-veined, (8) massive; (9) physical properties: numerator = density, g/cm^3, denominator = magnetization, mA/m; (10) computed values: (a) Δg, (b) ΔT; (11) boundary line for the anomaly plot overstepping the limits of the drawing; (12) levels of the erosional truncation: (I) – earth's surface, (II) and (III) depths of 80 and 160 m from the earth's surface, respectively

Table 5.2 Geophysical effects for a ***PGM*** of the Filizchay type

h value, m	Anomaly					
	Gravity, mGal	Magnetic, nT	*SP*, mV	*VLF*,%	*IP*,%	Temperature, °C
20	1.0 ÷ 2.2	150 ÷ 200	−(120 ÷ 300)	25 ÷ 50	7 ÷ 15	0.7 ÷ 2.0
60	0.6 ÷ 1.2	50 ÷ 70	−(60 ÷ 100)	10 ÷ 15	5 ÷ 9	0.2 ÷ 0.7
100	0.2 ÷ 0.6	25 ÷ 30	−(20 ÷ 40)	2 ÷ 4	2 ÷ 7	0.05 ÷ 0.10

geological section of the area is composed mainly of interstratifications of sandy-clay associations of the Upper Aalenian. Two sub-parallel stratified sheet-like bodies make up the Katekh deposit; tectonic faults control the space dislocation of all known economic ore bodies.

The ore bodies in this deposit are primarily characterized morphologically as lenticular. However, a combination of latitudinal and longitudinal faults complicates this type and they acquire the form of steam-chest beds. The Katekh deposit was investigated by mining and drilling up to a depth of 500 m. However, some experts note that due to the extremely complicated tectonics these operations failed to delineate the ore bodies completely.

The following types of texture ores were identified in the Katekh deposit: (1) massive, (2) veiny-clastic and (3) spotty-disseminated. The main ore minerals of the Katekh deposit are pyrite, sphalerite, chalcopyrite and galena. The secondary ore minerals are represented by hepatic pyrite, wurtzite, arsenopyrite and melnikovite; rare minerals are silver and gold (Mekhtiev et al. 1976; Zaitseva et al. 1988).

A 3-D combined modelling of the field $\Delta g_{\text{Bouguer}}$ (gravity field in the Bouguer reduction) and the magnetic field ΔZ (vertical component of the total magnetic field) was performed using the following procedure. A detailed physical-geological model of the Katekh deposit with a length of 800 m and depth of 400 m was constructed from the Mekhtiev et al. (1976) and Zaitseva et al. (1988) generalized data. Then, all the available data for this area on density (Gadjiev et al. 1984) and magnetic susceptibility (Ismail-Zadeh et al. 1983a,b) were utilized. For enhanced calculation of the surrounding terrain topography a digital terrain relief model was created. The expressed SW-NE regional topography trend in the area of the Katekh deposit led to the selection of a rectangular digital terrain relief model (DTRM) with a length of 20 km and a width of 600 m (the physical-geological profile with a length of 800 m was located in the geometric center of the DTRM) (Fig. 5.7). On the whole, 1,000 characteristic points describing the DTRM (with these points located close to the center of the DTRM and more rarely on these margins) were utilized.

The results of the first iteration of the gravity and magnetic fields modeling are presented in Fig. 5.8a. As indicated in this figure, the plots of ΔZ and ΔU_{SP} (self-potential anomalies) oscillate around zero and cannot provide useful information about the buried targets. This is due to the peculiarities of the mineralogical composition of ores in the Katekh deposit. The almost total absence of magnetic mineral pyrrhotite causes the virtually non-magnetic nature of the ores. On other

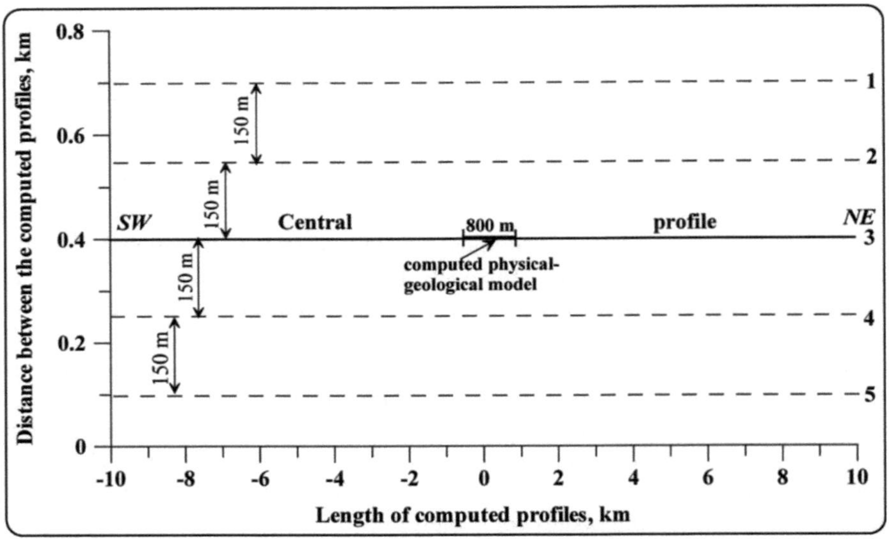

Fig. 5.7 Location of computed physical-geological model and profiles used for calculation of the surrounding terrain relief influence

hand, the fairly large lead content impedes the normal course of oxidation-reduction reactions needed to trigger of intense SP anomalies. Thus, essential geophysical information can only be derived from the $\Delta g_{\mathrm{Bouguer}}$ curve.

The analysis of the observed and computed gravity fields (Fig. 5.8a) shows that the initial physical-geological model has a certain deficit of anomalous masses. 3-D modeling of gravity field was carried out using about 25 sequential iterations. It yielded following results (Fig. 5.8b). Two ore bodies of massive composition (not reflected in the previous geological constructions) were identified in the southwestern and northeastern segments of the deposit. It should be noted that the conclusion as to the presence of a hidden ore object in the southwestern portion of the profile is consistent with the results of independent investigations; namely, underground geothermal observations and a ground geochemical survey. A temperature anomaly of 0.5–0.8°C was recorded in adit 8 during the underground geothermal investigations at 250–300 m; the surface zone containing a large amount of lead and zinc was revealed at 150–200 m (according to Ginzburg et al. (1981) and data from the "Azerbaijangeologiya" Association).

5.4.4 Induced Polarization

The quantitative interpretation of electric (electromagnetic) survey data is very complicated without reference to gently sloping discontinuities. However, qualitative techniques in the IP method often provide valuable information about the

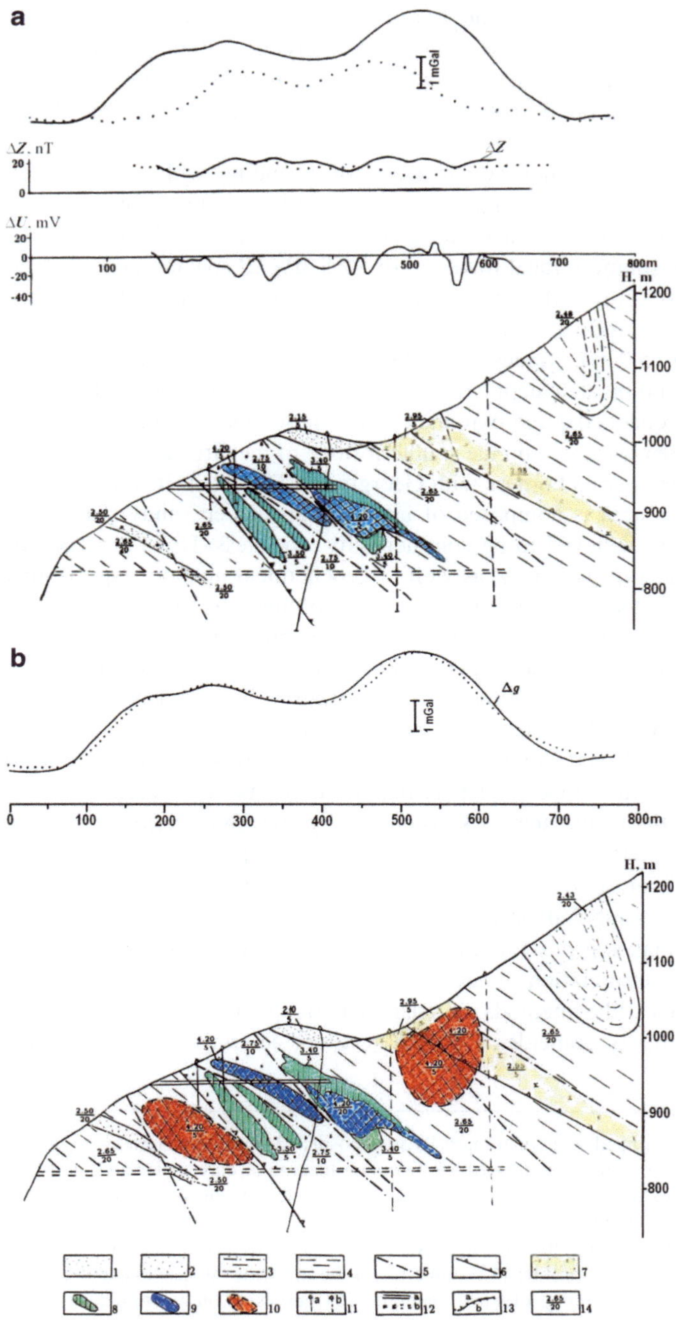

Fig. 5.8 (**a**) Computation of a geophysical effect based on a known geological section in the Katekh pyrite-polymetallic deposit (southern slope of the Greater Caucasus) (Eppelbaum and Khesin 2004). (**b**) Revised geological section model after a gravity field analysis (Eppelbaum and Khesin 2004).

geological section in question. For instance, visual analysis of η_a sections combined with ρ_a sections are often used while studying sulfide deposits in Canada and France. Such an approach is appropriate for the assessment of poorly accessible regions, where rapid interpretation results are necessary for planning further surveys.

The VES-IP method (vertical electric sounding with Schlumberger array) is widely used at various stages of prospecting in the almost inaccessible regions of the Greater and Lesser Caucasus (Khesin 1969; Alexeyev 1970; Khesin et al. 1997). Reference submeridional traverses of VES-IP (*AB* spacing up to 2–4 km, observation step 0.5–1.0 km) were performed from the foothills to the main watershed of the Greater Caucasus Ridge (Fig. 5.9).

In Fig. 5.9 the values of apparent polarizability η_{a} are shown under each sounding point at a depth of half the AB spacing. The pseudosection η_{a} thus obtained reflects the approximate change in the polarizability with depth.

This area is located on the southern slope of the Greater Caucasus, in the NW of Azerbaijan near its borders with Georgia and Dagestan. Its relief is highly complicated. This area is composed of intensely dislocated sandy-shale deposits which were earlier believed to be commercially worthless. However, combined electric and magnetic prospecting, as well as geochemical methods (spectrometallometry) led to the discovery of the Filizchay polymetallic deposit, the largest in the Caucasus. The main metals in this and several smaller ore deposits found in the same area (Katsdag, Katekh, Djikhikh, Katzmala, etc.) are copper, zinc, lead among others (Khesin et al. 1996). All these deposits formed a new ore province, confirmed by VES-IP reference crossings. In Fig. 5.9 and in other traverses, extended zones (I, II, III and IV) of high polarizability (tens of%) are evident. These zones include several copper and polymetallic deposits discovered later. The essential polarizability of the host rocks is caused by their pyritization and graphitization.

Figure 5.10 presents the interpretation results for the observed anomaly η_a obtained using a gradient array. The polymetallic body of complex composition occurs in a sand-shale stratum of the Upper Aalenian. The shape of the observed anomaly makes it possible to apply both inclined thin bed and horizontal circular cylinder interpretation models. The upper edge depth values for this body, computed by the methods of characteristic points and tangents for a thin bed model virtually coincide. The HCC centers determined by the same techniques

Fig. 5.8 (continued) (1) Quaternary loose deposits; (2–4) Middle Jurassic deposits: (2) massive fine and meso-grained sandstones, (3) interstratification of clay shales and sandstones, (4) rhythmical alternation of aleurolites and clay shales; (5) disjunctive dislocations; (6) upthrust-overthrusts; (7–9) pyrite-polymetallic ores: (7) spotty, (8) stockwork-veiny, (9) massive; (10) contour of orebodies introduced during selection; (11) prospecting boreholes: (a) on the profile, (b) projected on the profile; (12) adits: (a) in the plane of the geological section, (b) projected onto the plane of the geological section; (13) curves of gravitational and magnetic fields: (a) observed, (b) selected; (14) physical properties: numerator = density, g/cm^3, denominator = magnetization, mA/m (1–9 and 11–12 according to mining and drilling data)

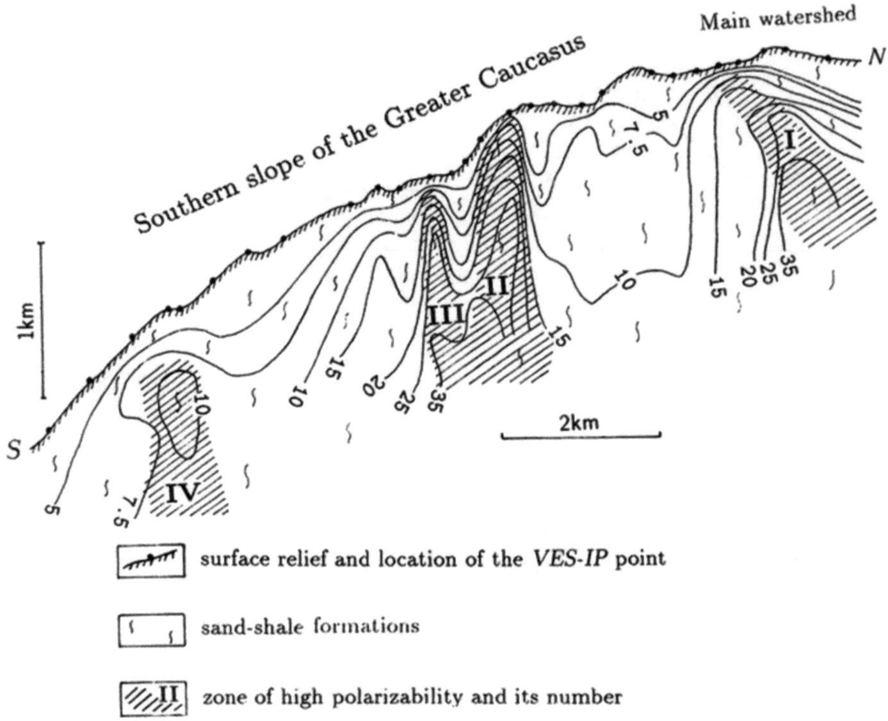

Fig. 5.9 Pseudosection of η_a (southern slope of the Greater Caucasus). Isolines are shown in % as a function of AB/2. The points of VES-IP are denoted by black dots (electrode spacing is oriented across the profile strike). The zones of high apparent polarizability in the Jurassic formations are shaded (Khesin et al. 1997)

(assuming that a horizontal circular cylinder may be inscribed into the upper portion of the orebody), are slightly shifted with respect to each other. From the analysis of the geological section (see Fig. 5.10) and both the thin bed and HCC models (see Sect. 3.6.2) it was shown that both models are applicable. Thus, such results provide higher reliability for the location of an orebody. As shown in Fig. 5.10, the inclination of relief and polarization vector cause a displacement of the anomaly maximum relative to the projection of the upper edge of the body onto the earth's surface. As a result, without the quantitative interpretation, a testing well drilled on the maximum of η_a curve can miss the target.

5.4.5 VLF

Figure 5.11 illustrates the correlation between the topographic relief and the total horizontal component of the VLF magnetic field in the area of the Katsdag pyrite-polymetallic deposit. The relief slope is about 40° and the ore bodies occur in Middle Jurassic deposits. The frequency was 19.6 kHz with the transmitting station

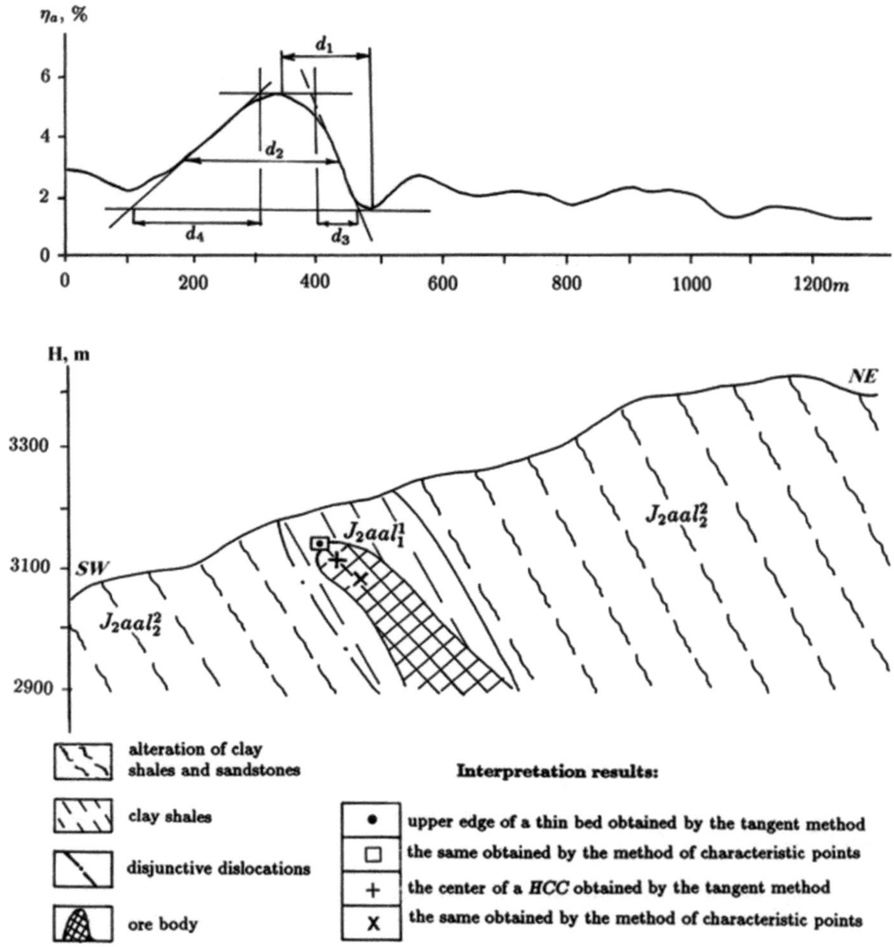

Fig. 5.10 Interpretation of the η_a anomaly (gradient array) in the Bazardyuzyu ore area (southern slope of the Greater Caucasus) by the developed techniques (Khesin et al. 1997)

located near London and the profile azimuth was 80°. The different geological units had the following resistivities: pyrite-polymetallic ore bodies (occurring in the bottom portion of the geological section) – several Ohm·m; mineralized zone (outcropping in the upper portion of the section) – 10–30 Ω m: sandstones – 1,200–1,500 Ω m; clay shales – 700–1,000 Ω m; and liparite-dacites – 400–700 Ω m. These values indicate a relatively uniform background medium as far as resistivity is concerned. However, a distorting topographic effect tends to impair visual detection of an anomaly from a concealed ore body having the shape of a tilted plate. The correlation coefficient r was 0.97 (the broad anomaly from a mineralized zone was neglected, as can be seen from Fig. 5.11b).

When the topographic effect is removed, a weak anomaly due to the ore body and a strong field from the thick mineralized zone appear more clearly on the $H_{\varphi\mathrm{corr}}$ plot (Fig. 5.11a).

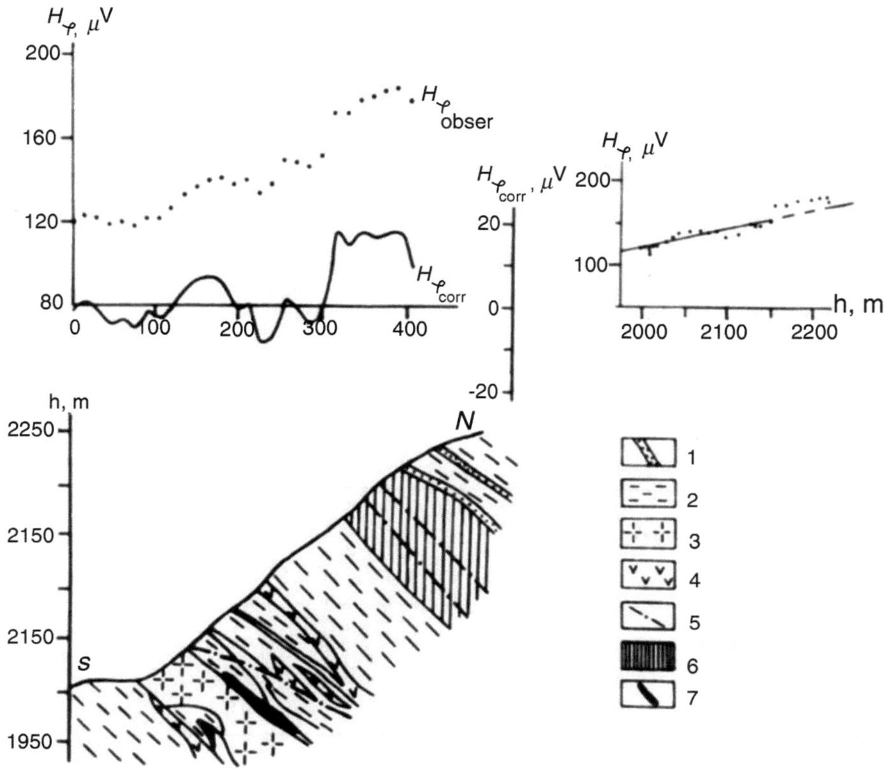

Fig. 5.11 Correlation technique for reducing the terrain relief effect on a portion of the Katsdag pyrite-polymetallic deposit (southern slope of the Greater Caucasus, Azerbaijan). (**a**) plots of observed and corrected H_φ values, (**b**) correlation, (**c**) geological section (After Eppelbaum 1991). (1) bands of sandstone, (2) clay shales, (3) liparite-dacites, (4) dioritic porphyrites, (5) dislocations with breaks in continuity, (6) zone of intense ore mineralization, (7) massive ore

The next and geologically more complicated example illustrates the application of the correlation technique using the VLF method on a portion of the Katekh pyrite-polymetallic deposit (Fig. 5.12) in the same region. A frequency of 16.0 kHz was used in the investigation (the transmitter was in Rugby, Great Britain), and the profile azimuth was 60°. The ribbon-like band of the pyrite-polymetallic ore was located at a depth of about 60–80 m in the sandy argillaceous series of the Upper Aalenian. The average resistivity of the nearly uniform surrounding medium was in the range of 700–900 Ω m; the resistivity of the massive ore body was a fraction of 1 Ω m. The skin depth in the surrounding medium was about 110 m. Since a topographic anomaly was superimposed on a small signal from a relatively deep-seated anomalous object, it was difficult to detect the latter. Two different correlations were calculated for the southwestern and the northeastern slope,

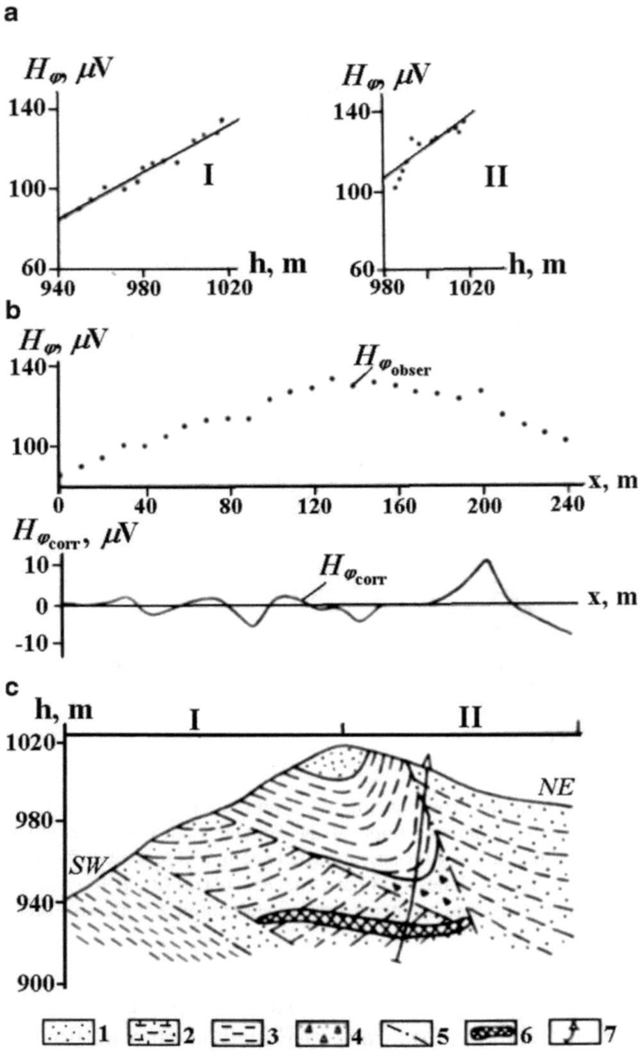

Fig. 5.12 Correlation technique for reducing the terrain relief effect on a portion of the Katekh pyrite-polymetallic deposit. (**a**) correlation, (**b**) graphs of observed and corrected H_φ values, (**c**) geological section. (1) fine- to medium-grained sandstones, (2) alternating sandstone and clay shale strata, (3) clay shales, (4) brecciated zone, (5) fractures, (6) ore body, (7) borehole

yielding correlation coefficients (between the relief heights and VLF intensity) of $r_1 = 0.988$ and $r_2 = 0.85$. The results for the southwestern slope were

$$H_{\varphi\,\mathrm{appr}} = 84 + (0.6 \pm 0.09)h,$$

and the northeastern slope

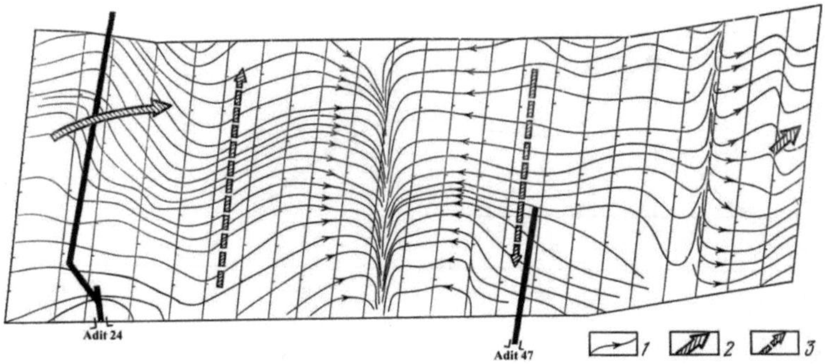

Fig. 5.13 Results of the VLF azimuth survey at the Adangea polymetallic deposit (southern slope of the Greater Caucasus) (Ginzburg 1982). (1) isolines of the anomalous total magnetic component $H_{\phi(\mathrm{an})}$ of the electromagnetic VLF field; electric axes: (2) local, (3) regional

$$H_{\varphi\,\mathrm{appr}} = 105 + (0.79 \pm 0.49)h.$$

After removing the relief effect, a positive anomaly could clearly be seen on the northeastern slope in the $H_{\varphi\,\mathrm{corr}}$ plot. This anomaly may have been due to the edge effect of a deep ore body, whereas the minimum on the southwestern slope could correspond to the opposite edge of the subhorizontal tabular ore body, since the anomaly was not very large in this case (Eppelbaum 1991).

Figure 5.13 illustrates the use of the VLF method in the Adangea polymetallic deposit (Georgia). VLF magnetic field $H_{\phi(\mathrm{an})}$ represents three large whirls disposed in a vertical plane (Ginzburg 1982).The distribution of the field $H_{\phi(\mathrm{an})}$ traces the position of geological blocks and the faults between them. Regional axes in this figure represent the anomaly field from the block structure of the whole ore zone, and local axes show the anomaly effects from ore bodies. The degree of regional field deformation depends mainly on the depth of massive polymetallic ore occurrence.

5.4.6 Near-Surface Temperature Survey

Figure 5.14 presents the results of the interpretation of a temperature anomaly applying the techniques described above (see Sect. 3.5.2). In the figure, the results are given for a district with the steeply inclined relief of the Katekh deposit (southern slope of the Greater Caucasus). The temperature was measured in 1.0 m deep blastholes. The temperature plot shows two anomalies, one of which is due to the subhorizontal ore deposit and is less pronounced. The thermal conductivities for the host rock and thick pyrite-polymetallic ore were 1.45 ± 0.35 and 3.87 ± 0.57 W/m °C, respectively (Zverev et al. 1982); i.e., their ratio exceeded 2.5. The upper

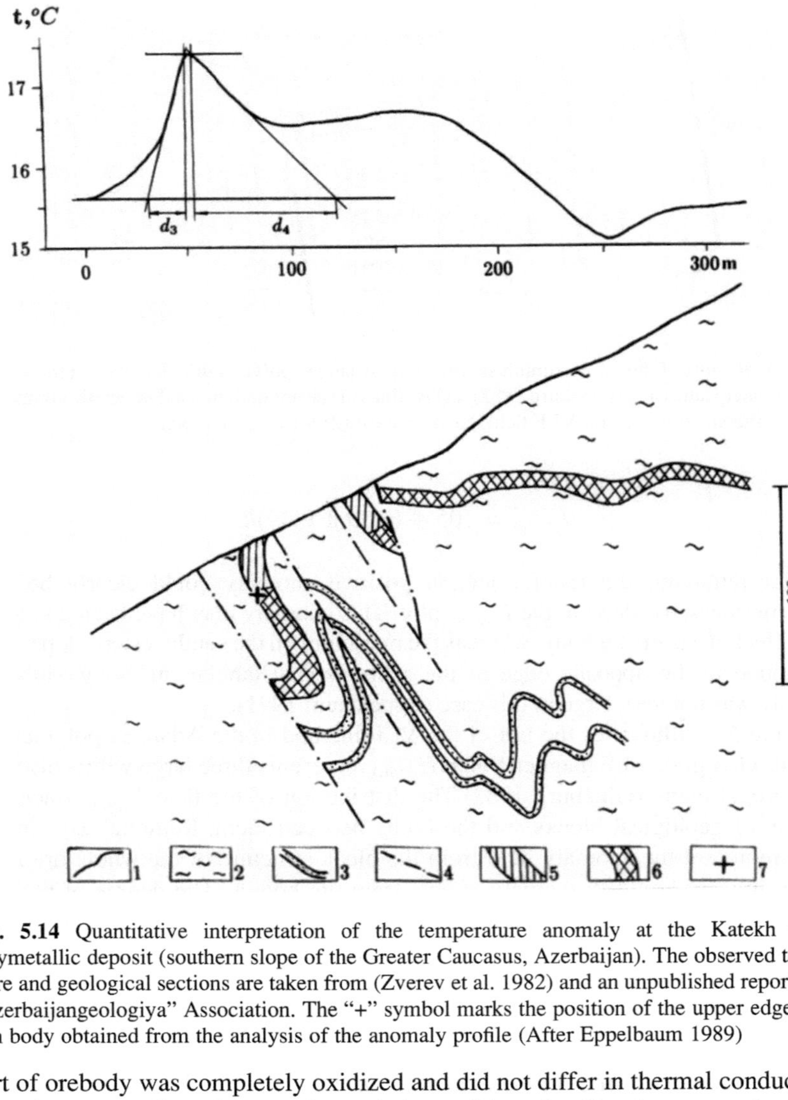

Fig. 5.14 Quantitative interpretation of the temperature anomaly at the Katekh pyrite-polymetallic deposit (southern slope of the Greater Caucasus, Azerbaijan). The observed temperature and geological sections are taken from (Zverev et al. 1982) and an unpublished report of the "Azerbaijangeologiya" Association. The "+" symbol marks the position of the upper edge of the thin body obtained from the analysis of the anomaly profile (After Eppelbaum 1989)

part of orebody was completely oxidized and did not differ in thermal conductivity from the host medium. The results made it possible to localize the upper edge of the subvertical orebody within an acceptable error.

Figure 5.15 illustrates the temperature anomaly observed by the geophysical team under the supervision of S. Ginzburg (Central Research Institute of Non-Ferrous and Precious Metals, Moscow) along the profile across the Kvaisa pyrite-polymetallic deposit (the Greater Caucasus). The anomaly amplitude exceeds 2°C, which is traceable to the additional effect of the fracture located at the edge of the orebody.

The temperature was measured in 1.0 m deep blastholes. The ore was pyrite-sphalerite in composition and there was a subvertical occurrence in volcaniclastic

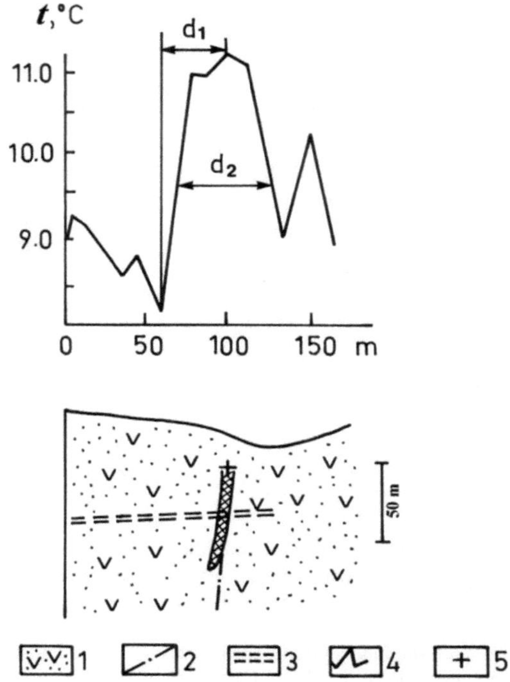

Fig. 5.15 Quantitative interpretation of temperature anomalies in the area of the Kvaisa pyrite-polymetallic deposit (southern slope of the Greater Caucasus, Georgia) (observed data and geological section after Ginzburg et al. 1974). The massive sulfide orebody is shaded. The "+" symbol marks the position of the upper edge of the thin body, as obtained from the analysis of the anomaly profile

host rocks. Thermal conductivities for sandy-argillaceous host rocks and thick pyrite-polymetallic ore were 2.0 ± 0.5 and 5.0 ± 1.0 W/m °C, respectively, differing by a factor of 2.5.

Quantitative interpretation was carried out using the characteristic point method. This served to locate the upper edge of the ore deposit occurring within unconsolidated volcaniclastic deposits from the Middle Jurassic. The results were confirmed by mining.

5.4.7 *Self-Potential Survey*

SP surveys were frequently applied on the pyrite-polymetallic deposits of the southern slope of the Greater Caucasus.

Four cases of *SP* interpretation are presented in Fig. 5.16a, b. A very intensive SP anomaly (about 600 mV) was observed in the Filizchay copper-polymetallic field (Fig. 5.16a). Use of the approach described above (see Sect. 3.5.3) determined the position of upper edge of the ore body in conditions of sharply inclined relief. Three SP anomalies were successfully interpreted in the Katsdag copper-polymetallic deposit (Fig. 5.16b).

Quantitative examination of an intensive (about 300 mV), wide *SP* anomaly (Fig. 5.17) in the Uchambo ore field (Georgia) delineated the position of the *HCC* center, which clearly locates the undrilled edge of a flat-lying orebody.

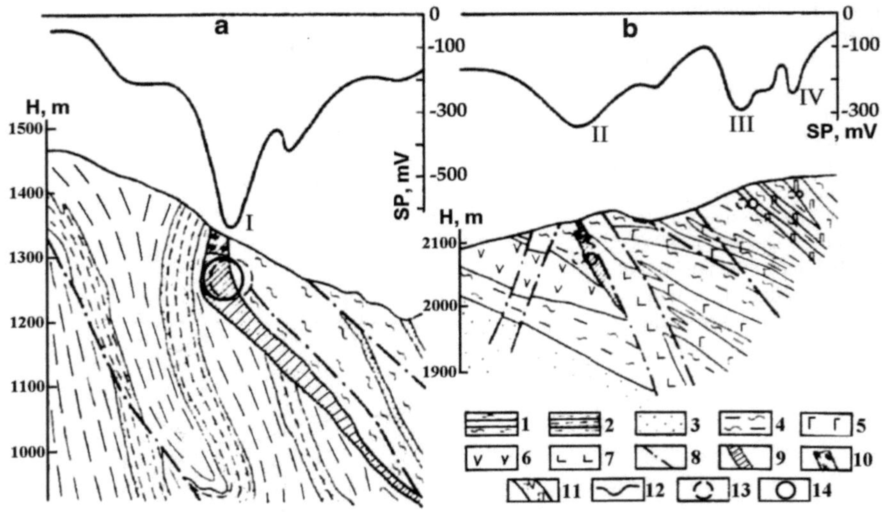

Fig. 5.16 Quantitative interpretation of *SP* anomalies at the Filizchay (**a**) and Katsdag (**b**) copper-polymetallic deposits on the southern slope of the Greater Caucasus. I – IV: numbers of interpreted *SP* anomalies. (1) interbedding of sands and clay schists, (2) clay schists with the flysh packages, (3) clay sandstone; (4) sand-clay schists; (5) diabases, gabbro-diabases and diabasic porphyrites; (6) andesites and andesite-porphyrites; (7) dacitic porphyrites; (8) faults; (9) massive ore of pyrite-polymetallic composition; (10) oxidized ore; (11) zones of brecciation, crush and boudinage with lean pyrite-polymetallic ore; (12) *SP* curves; location of anomalous source (13) without calculation of inclined relief influence, (14) after introducing correction for relief

5.4.8 Magnetic Survey

Average results of helicopter magnetic surveys, the distribution of ore and prospective sites in the Belokan-Zakatala ore region are shown in Fig. 2.13.

Two examples of vertical components of magnetic field Z_a examination in the Filizchay ore field are shown in Fig. 5.18. Figure 5.18a illustrates the application of Logachev's method (Logachev and Zakharov 1979). As shown in Fig. 5.17a, reducing Z_a by the use of Logachev's method and interpretation of the obtained graphs Z_a and X_a by vector methodology defined the upper edge of ore body above the Earth's surface. Thus, this method is unacceptable in complex geological conditions (oblique magnetization, inclined rugged relief and an unknown level of the normal magnetic field).

Application of the methodology as described in Khesin et al. (1983, 1996) determined the position of the buried ore body with high accuracy (Fig. 5.18b).

5.4.9 Electromagnetic Methods

Among the electromagnetic methods used in mining geophysics the most important are methods of partial extraction of metals (PEM), the transient electromagnetic method (TEM) and the contact method of polarization curves (CMPC).

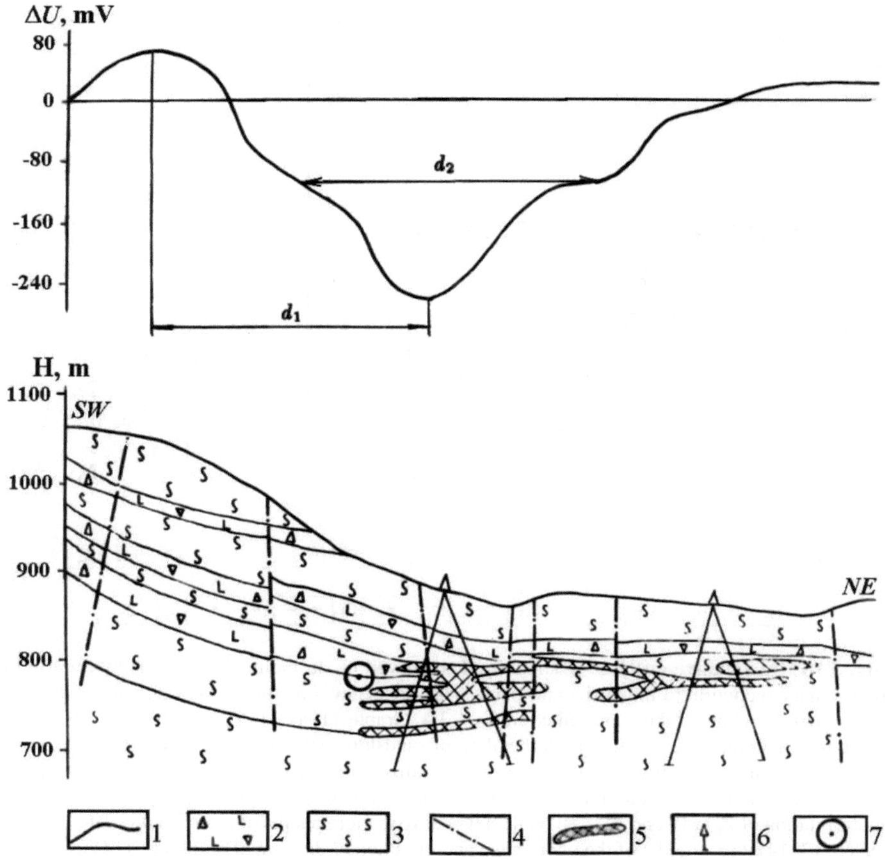

Fig. 5.17 Interpretation of SP anomaly by the characteristic point method in the area of the Uchambo ore field of the Adjar group of copper-polymetallic deposits (Greater Caucasus, Georgia). (1) observed *SP* values; (2) heteroclastic tuff breccia and their tuffs; (3) cover trachyandesite-basalts with pyroclastic interbedding; (4) disjunctive dislocations; (5) zones of increased mineralization; (6) drilled wells; (7) location of HCC center according to the interpretation results ((1–6) from Bukhnikashvili et al. (1974), (7) after Eppelbaum (1989))

The CMPC is based on consecutive excitation of electrochemical reactions at the boundary of electron conductive minerals with moist rocks, and the recording of electrochemical processes such as polarization curves. The CMPC was initially thought to be capable of solving the main problems of mining geophysics including quantitative calculation of metal components in each ore body. Field testing of this method in the southern slope of the Greater Caucasus unfortunately failed to confirm its effectiveness.

The PEM method (some scientists believe that this method is more geochemical than geophysical) was applied in the southern Caucasus to assess copper, zinc, lead and gold content. In general, this method correlates well with other geophysical, geochemical and geological methods.

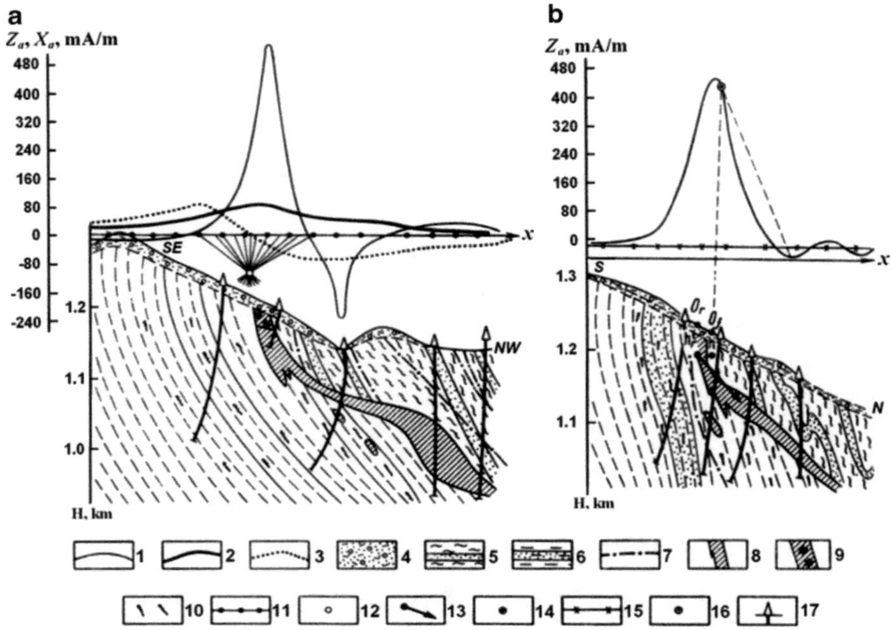

Fig. 5.18 Quantitative analysis of anomaly Z_a observed on an inclined relief. (**a**) by the use of preliminary reduction to the horizontal plane with Logachev's method (Logachev and Zakharov 1979), (**b**) using the methods presented in Sect. 3.4. Graphs (1–4): (1) Z_a (x,z), (2) Z_a $(x,0)$, (3) X_a $(x,0)$, (4) Quaternary deposits; (5) interbedding of sandstone, siltstone and shales; (6) shales with flysch; (7) faults; (8) pyrite-polymetallic ore; (9) oxidized ore, (10) veiny ore, (11) horizontal level of magnetic anomaly reduction, (12) location of the middle of the upper edge of the ore body by interpretation of Z_a $(x,0)$ and X_a $(x,0)$ on the level of reduction, results of interpretation of observed curve (13–14): (13) middle of the upper edge and direction of dipping of real body, (14) middle of the upper edge of fictitious body, (15) corrected line of the normal background of curve Z_a (x,z), (16) Reford's point (Reford and Sumner 1964), (17) boreholes. 0_f and 0_r are the calculated origins of the coordinates for fictitious and real bodies, respectively (Khesin et al. 1983)

Two variants of the TEM method have been applied: the one loop method and the frame-loop method; the size of the loops was 400 × 400, 200 × 200, and 100 × 100 m. This method covers large depths of investigation. It was proved to successfully calculate rugged terrain relief influence by the use of correlation methodology (Khesin 1981).

5.4.10 Integrated Analysis

The discovery of the Filizchay ore deposit by the integration of SP, IP and magnetic fields is shown in Fig. 2.12.

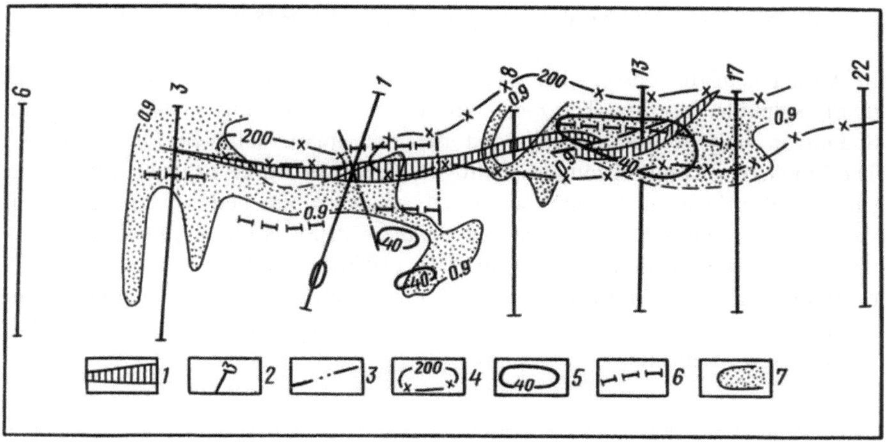

Fig. 5.19 Inverse probability and cross-correlation method testing on the Filizchay polymetallic deposit (southern slope of the Greater Caucasus). (1) upper edge projection of the deposit on the earth's surface by geological data; (2) profiles; (3) disjunctive dislocations; (4) isopotential lines of *SP* (mV); (5) isolines ΔZ (nT); (6) ΔZ anomaly axes obtained bycross-correlation; (7) areas the probability of detecting the targets exceeded 0.9 after applying the inverse probability method

The inverse probability method (IPM) was used on the Filizchay pyrite-polymetallic deposit revealed by geophysical methods. Application of IPM with the purpose of singling out weak magnetic anomalies showed that the ore deposit could be detected using this method (Fig. 5.19) even in that part of the area where it was not located in the observed field.

5.5 Lesser Caucasus

5.5.1 Self-Potential Survey

5.5.1.1 SP-Survey in Azerbaijan

In the 1930s, a SP survey was successfully carried out to study Chiragidzor pyrite stocks (M. L. Ozerskaya, S. N. Kondrashev, A. G. Surikov). A new Toganaly pyrite deposit was discovered on the contour of a SP anomaly in 1938 (Khesin 1962a). SP observations were used together with other methods in the Kyzyl-Bulakh gold-pyrite deposit for geological-geophysical mapping (Mekhmana ore district of Nagorny Karabakh). Altogether, SP investigations were applied at a few dozens of ore targets.

5.5.1.2 SP-Geological Models of Several Ore Deposits in Armenia

In Armenia, in the Lesser Caucasus, there are known copper, polymetallic and quartz-sulphide deposits. The most typical are the Shamlug copper deposit, the Shaumyan polymetallic deposit, and the Zod quartz-sulphide (gold) deposit. The lithological composition and hydrogeological conditions of these deposits are similar. The host media are composed of dense cracked volcanogenic and volcanogenic-sedimentary associations. In these deposits the ore/host media contrast can exist 8–20 times. Gamoyan et al. (1986) carried out a large volume of theoretical and modeling investigations to calculate the expected anomalous and space distribution of SP effects from both from single bodies and group of bodies for several ore deposits.

5.5.2 Physical-Geological Model of the Copper-Pyrite Deposit of the Lesser Caucasian Type

The ***PGM*** of the Lesser-Caucasian type represents copper-pyrite deposits occurring in pyrite-bearing provinces with a predominance of volcanogenic rocks (Borodaevskaya et al. 1977). As reported in (Tvalchrelidze 1978), the following provinces are classified as the Phanerozoic pyrite provinces of the Lesser-Caucasian type: the Caledonian (Tuva, Salair, Western Sayan), the Hercynian (Central Kazakhstan, Altai Mountains), the Cimmerian (Canadian Cordillera, Sierra-Nevada, Lesser Caucasus) and the Alpine (Eastern Serbia, Bulgarian Middle Upland, Anatolian Pont, Lesser Caucasus, Northern Iran, Japan, Taiwan, Philippines) geological structures. The pyrite deposits of this type, which are also termed South-Uralian or Kuroko, are probed in detail in the Lesser Caucasus.

Copper-pyrite deposits of the Lesser-Caucasian type in the Somkhet-Karabakh Zone (e.g., Alaverdy, Gedabey and Kyzyl-Bulakh) are located at middle-height relief and localized in volcanogenic and volcanogeno-sedimentary associations of the Jurassian and Cretaceous. Constructing a physical-geological model of the Lesser-Caucasian type deposits is complicated by the enormous variability of rocks and ores and distinct tectonics. A ***PGM*** can be schematically represented as a set of small flat-dipping bodies of lens-like form, which differ from the host medium by certain contrasting properties. These include excess density (from 0.4 to 0.9 g/cm^3), higher conductivity (by a factor of $5{\cdot}10$ to $5{\cdot}10^2$), polarizability (several-fold) and thermal conductivity (1.5–2.5-fold) and lower magnetization (by a factor of $5{\cdot}10$ to $2{\cdot}10^2$).

The fields predicted by the typical ***PGM*** of deposits of the Lesser-Caucasian type (Fig. 5.20) are compiled in Table 5.3.

A combination of gravity and magnetic prospecting, the VLF method and near-surface thermal prospecting has proved to be efficient for determining deposits of the Lesser-Caucasian type (Eppelbaum and Khesin 1988).

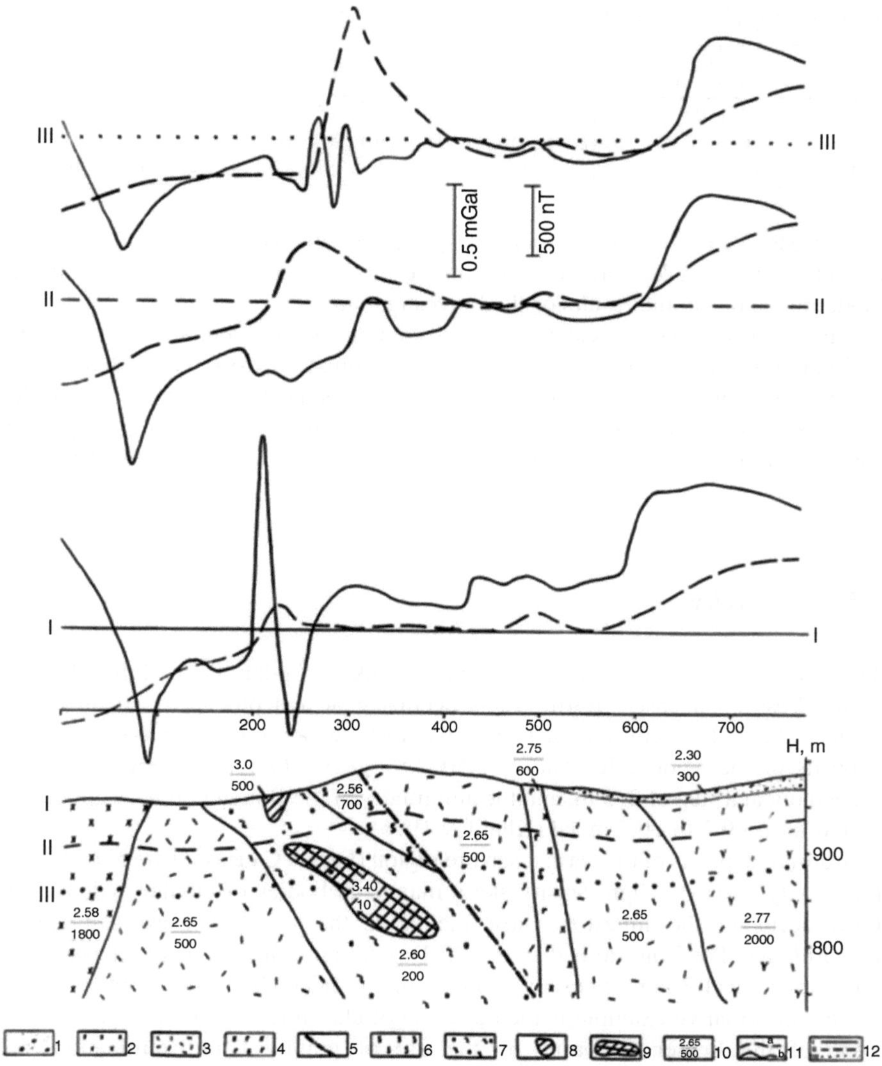

Fig. 5.20 A generalized physical-geological model of copper-pyrite deposit of the Lesser-Caucasian type (Eppelbaum and Khesin 1988). (1) deluvial deposits; (2) andesitic porphyrites; (3) tuffs of liparite-dacites; (4) lavas of dacitic porphyrites; (5) dike of andesite-basalts; (6) disjunctive dislocations; (7) zone of brecciation and crush with weak traces of pyritization; (8) zone of breccia-tion, crush and boudinage, with lean ore; (9) oxidized orebody; (10) massive pyrite-chalcopyrite ore; (11) physical properties (numerator = density, g/cm^3, denominator = magnetization, mA/m); (12) computed geophysical fields: (a) Δg, (b) ΔT; (13) levels of the erosional truncation: (I) earth's surface; (II) and (III) depths of 50 and 100 m from the earth's surface, respectively

Table 5.3 Geophysical effects for the ***PGM*** of the Lesser-Caucasian type

h value, m	Anomaly					
	Gravity, mGal	Magnetic, nT	*SP*, mV	*VLF*,%	*IP*,%	Temperature, °C
20	0.3 ÷ 0.5	−(70 ÷ 100)	−(20 ÷ 50)	15 ÷ 30	4 ÷ 12	0.5 ÷ 1.2
60	0.08 ÷ 0.15	−(20 ÷ 30)	−(15 ÷ 30)	5 ÷ 10	3 ÷ 6	0.1 ÷ 0.3
100	0.05 ÷ 0.08	−(5 ÷ 10)	−(5 ÷ 10)	1 ÷ 3	1 ÷ 4	0.03 ÷ 0.05

For SP, VLF, IP and thermal prospecting methods, the determinations of expected effects take into account the results of physical and mathematical modeling, and also the results of field investigations.

These physical-geological models substantiate the interpretation criteria and optimal geophysical sets for prospecting and estimation work. A combination of gravity and magnetic prospecting and VLF methods has proved to be efficient for deposits of the Lesser-Caucasian type. IP, SP and near-surface thermal prospecting can be employed as subsidiary methods (Eppelbaum 1989; Khesin et al. 1996).

5.5.3 Gravity

The Chovdar barite deposit is located in the Somkhet-Agdam Zone of the NE slope of the Lesser Caucasus. Barite veins occurring at a depth up to 20 m have a density of 3.7–4.5 g/cm^3 and density of the host media (volcanogenic-sedimentary deposits of the Middle Jurassic) – 2.50–2.87 g/cm^3. Gravity observations were carried out at a step of 2–5 m, and the amplitudes of the Bouguer gravity anomalies varied from +0.2 to +0.7 mGal (Khalilov 1996).

Iron-ore deposits of the Dashkesan ore region (Lesser Caucasus) are good targets for gravity surveys. High density skarn-magnetite bodies of tabular or lenticular form occur in the limestone host media of the Upper Oxford and Lower Kimeridgian. The iron-ore bed in the SE Dashkesan skarn-magnetite deposit was traced by a local Bouguer gravity anomaly of 1.6 mGal (Malakhova 1996).

One informative example is the detection of chromite ores in the Sevan-Akera Zone of the Lesser Caucasus by the use of gravity field examination and metallometry as shown in Fig. 2.10.

The Kyzyl-Bulakh gold-pyrite deposit situated in Mekhmana ore district of Nagorny Karabakh, Azerbaijan (Lesser Caucasus) has been investigated in detail by mining and drilling operations and can thus be used as a reference for testing geophysical method results. More than 2,500 observation points were set up in the deposit over a network of 10 × 25 m and the density and magnetization of 620 samples from the surface and drill cores were studied (Eppelbaum 1989; Khesin et al. 1993a). The accuracy of measurement was +0.03 mGal. The heights of the observation points and their horizontal projections were determined to an accuracy of ±0.08 and ±0.8 m, respectively.

Table 5.4 Terrain correction zones in conventional gravity prospecting techniques

Zone	Radius of zone, km		Author of the methodology, SD (mGal)	Scale of topographic materials	Step of digital relief model (km)
	Inner	Outer			
Central	0.0	0.1	Berezkin (1967), ±0.06	1: 2,000	–
Near	0.1	2.0	Lomtadze (1982), ±0.05	1: 25,000	0.1
Middle	2.0	10.0	Lomtadze (1982), ±0.05	1: 50,000	0.3
Distant	10.0	50.0	Lukavchenko (1951), ±0.05	1: 200,000	–

Small ore objects and their low excess densities are expected to produce gravity anomalies of 0.2–0.3 mGal. The errors due to neglecting the effect of very rugged topography are, therefore equivalent to the expected anomalies. Hence the gravity results were processed by two methods: a conventional technique (Berezkin 1967; Nemtsov 1967; Veselov 1986) and by using a special processing scheme.

When faced with several problems it seems worthwhile to consider eliminating the topographic effect step in 3D modeling of the whole medium. By excluding a separate procedure for terrain correction, calculation of the total error of processing and interpretation can take much less time. An example of such an approach was obtained in an area of the Kyzyl-Bulakh deposit (Khesin et al. 1993a).

An isoline chart of $\Delta g_{B(0\text{–}50)}$, i.e. Bouguer anomalies including the total correction within a zone of 0–50 km (the heights of the surrounding relief elements varied from 0 to 4 km within a radius of 50 km) was compiled using the conventional method of terrain correction (Table 5.4).

Variations over the whole area ranged from 4.2 to 7.4 mGal, with maxima in the brook and crests and minima on the slopes. The Δg_B chart incorporating these corrections is characterized by a standard deviation amounting to ±0.11 mGal and is inconsistent with the geological evidence obtained by mining and drilling (Fig. 5.21a, b).

Therefore, a special scheme for obtaining the gravitational field of Bouguer anomalies was developed. It is based on determining the difference between free-air anomalies $\Delta g_{f.a.}$ and the gravitational field computed by a 3-D digital model of the homogeneous medium with $\sigma = 2.67$ g/cm^3 and real topography.

In the new method, the terrain relief effect was accounted during the interpretation. The terrain relief was simulated using six selected profiles 2c–7c in the central portion of the Kyzyl-Bulakh area, crossing the deposit of the same name (800 m in length) (Fig. 5.22). Experimental calculations were carried out using the *GSFC* program to investigate the relief digital description range. These calculations were made along profile 6c for both the 2-D and 3-D models of the medium. For the 2-D model, relief description intervals of 0.8 and 10 km were used, and for a 3-D model, 0.8, 10 and 80 km.

For the 3-D model the relief was described for all design profiles and, additionally, for four other profiles, two of which were to the left of profile 2c and the other two located to the right of profile 7c. The border profiles were selected so that they coincided with those observed in the Kyzyl-Bulakh area. In this case, the Δg values

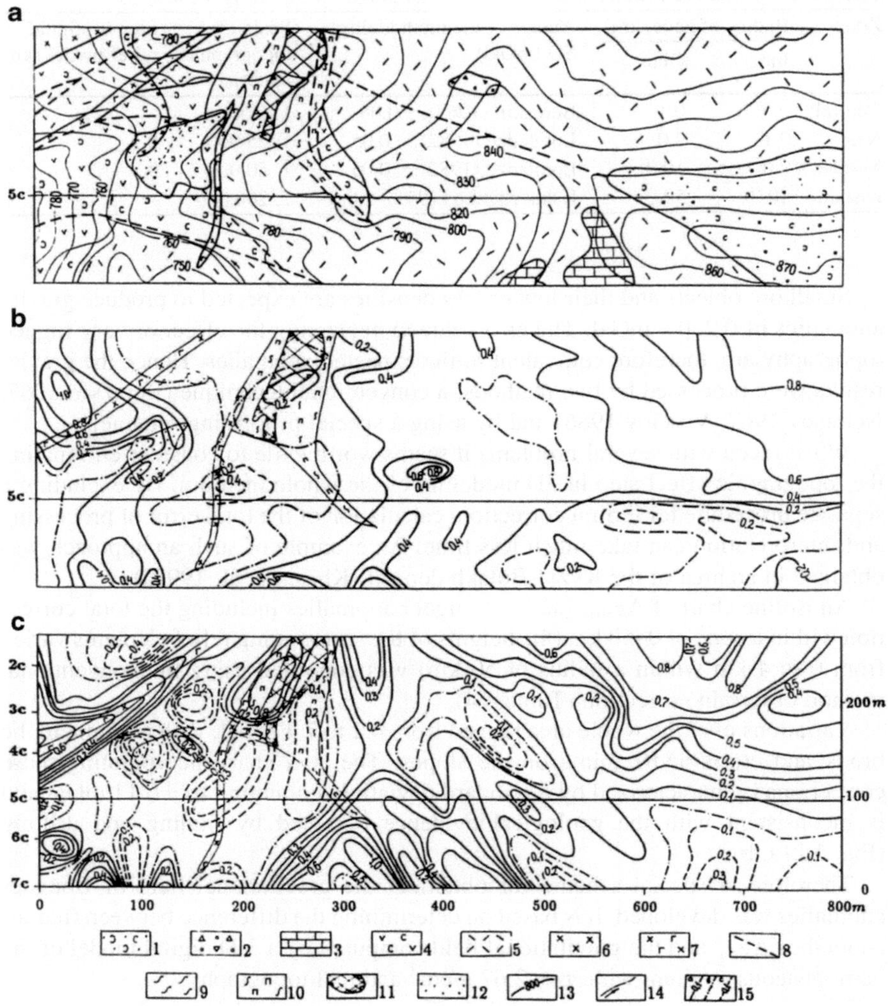

Fig. 5.21 Detailed gravimetric prospecting: A new method for $\Delta g_{\mathrm{Bouguer}}$ computing (Eppelbaum, 1989). (**a**) geological map of the central part of the Kyzyl-Bulakh gold-pyrite deposit; (**b**) and (**c**) fragments of Δg field charts obtained by conventional and special techniques respectively. (1) Quaternary deluvial deposits, (2) lens of tuffaceous conglomerate from the Upper Jurassic; (3) lenses of bone chert; (4) tuffs and lavas of andesitic porphyrites; (5) tuffs of liparite-dacitic porphyrites from the Upper Bajocian; (6) subvolcanic body of andesitic porphyrites; (7) dykes of andesite-basalts from the Upper Jurassic; (8) disjunctive dislocations; (9) zones of brecciation and crush with weak pyrite-chalcopyrite mineralization; (10) zones of brecciation lean to pyrite-chalcopyrite ore content; (11)outcrop of orebody represented by oxidized pyrite-chalcopyrite ore; (12) dead rock and ore component pile; (13) terrain relief isolines; (14) brook bed; (15) isoanomals, mGal: (a) positive, (b) zero, (c) negative

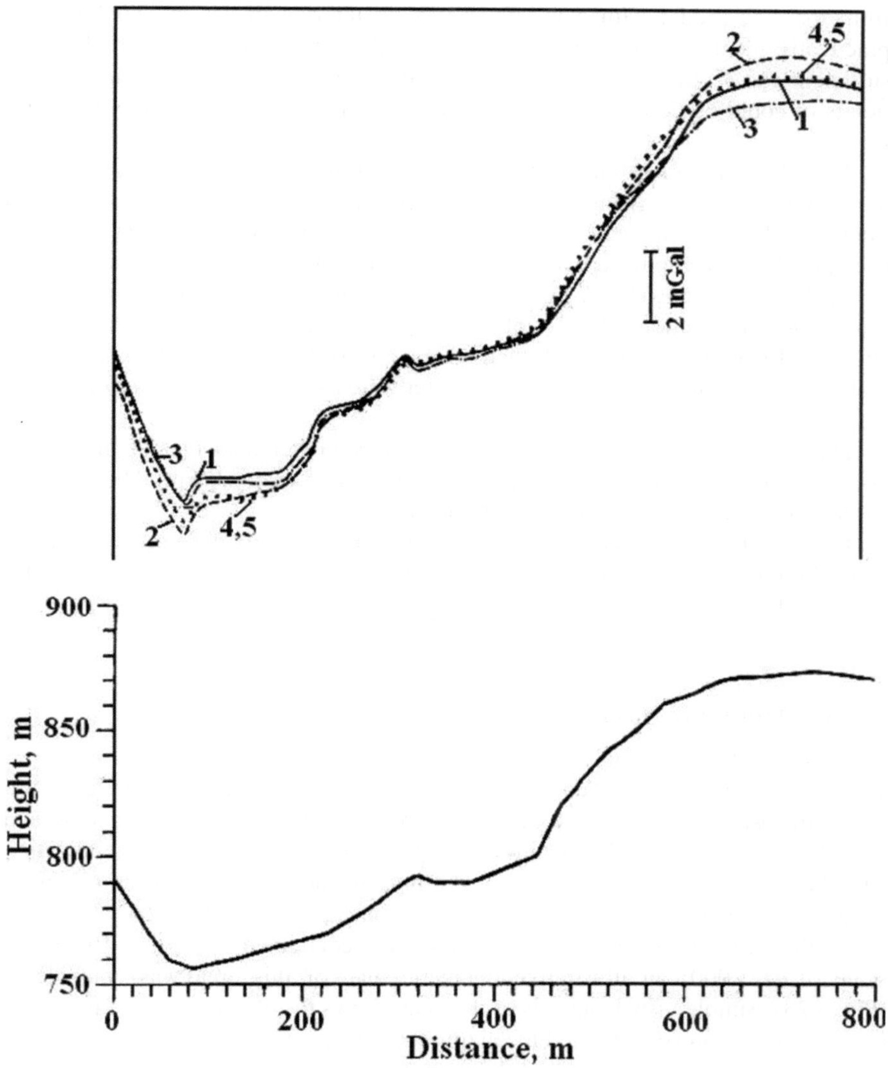

Fig. 5.22 Comparison of topographic mass attraction plots for different terrain relief descriptions. (**a**) gravity plots computed along profile 6c of the Kyzyl-Bulakh area, (**b**) topographic profile along profile 6c. (1) 2-D model with a description interval of 0.8 km, (2) 2-D model with a description interval of 10 km, (3) 3-D model with a description interval of 0.8 km, (4) 3-D model with a description interval of 10 km, (5) 3-D model with a description interval of 80 km (Khesin et al. 1993a)

were computed along profiles 2c–7c with a unified digital description of the terrain relief. The analysis showed that the difference in the results of 2-D and 3-D modeling was appreciable (up to ± 1.5 mGal). At the same time, the curves obtained from the 3-D model with a relief description interval of 10 and 80 km were practically identical. This suggested that the description interval of 10 km was sufficient for 3-D modeling. When the terrain relief receded from the selected points, it could be described more schematically, by singling out solely its major forms.

After this experiment subsequent processing of material from the Kyzyl-Bulakh area involved the following steps. A 3-D terrain relief model with a description interval of 10 km, in the points of profiles 2c–7c with a step of 8.46 m (95 points per each profile) was employed to compute the gravitational effect of the medium with $\sigma = 2.67$ g/cm^3 using the *GSFC* program. This effect was equal to an incomplete topographic correction (Δg_{ITC}) of the opposite sign. These corrections were subtracted from (i.e. added with their own sign) to the values of $\Delta g_{\mathrm{f.a}}$ and Δg_{Bm} (subscript "m" stands for "model"). The latter were used to construct an isoanomaly map.

The accuracy of this Δg_{ITC} was considerably higher than that of the terrain correction by conventional techniques within a zone of 0–50 km. The standard deviation of Δg_{Bm} amounted to ± 0.06 mGal.

The chart of Δg_{Bm} isoanomals (Fig. 5.21c) is much more differentiated than the Δg_{B} chart generated by conventional techniques (Fig. 5.21b) and is in better agreement with the available geological data (Fig. 5.21a).

5.5.4 Magnetic Survey

The Dashkesan iron-ore area of the Lesser Caucasus contains a group of magnetite-cobalt deposits. These massive magnetite ores contain of 79–90% magnetite and 30–10% non-ore minerals. Scarn-magnetite ores contain cobalt mineralization (Azizbekov 1972). Obviously, one of the first magnetic surveys in the ore areas of the Caucasus was carried out in the Dashkesan deposit (see Fig. 2.5).

On the basis of a magnetic helicopter survey (conducted in the mid-1960s at a scale of 1:25,000) an illustrative magnetic map of this area was constructed (Fig. 5.23). Simple visual analysis of this map shows the effectiveness of magnetic prospecting for prospecting and localizing these magnetite-containing ores.

A comparison of this map and the map of a surface magnetic survey with the results of large-scale geological mapping indicated that within the anomalous zones bordering the Dashkesan intrusive from north and south, together with iron-ore deposits, gabbroids of various composition were outcropped. The presence of maghemite testifies to the shallowness (up to 2 km) of the intrusive generation (Nagatha 1961). Magnetic anomalies from magnetite bodies are weakly present in the observed field (both ΔZ and ΔT) due to their deep occurrence and the subhorizontal location of the ore bodies (Khesin et al. 1983). Therefore, to determine the sharp contacts of gabbroids (with which may be associated with ore bodies) and localize the effects from a strongly magnetized lens, a map of the horizontal gradients of ΔT was constructed (Khesin 1976). It showed that the

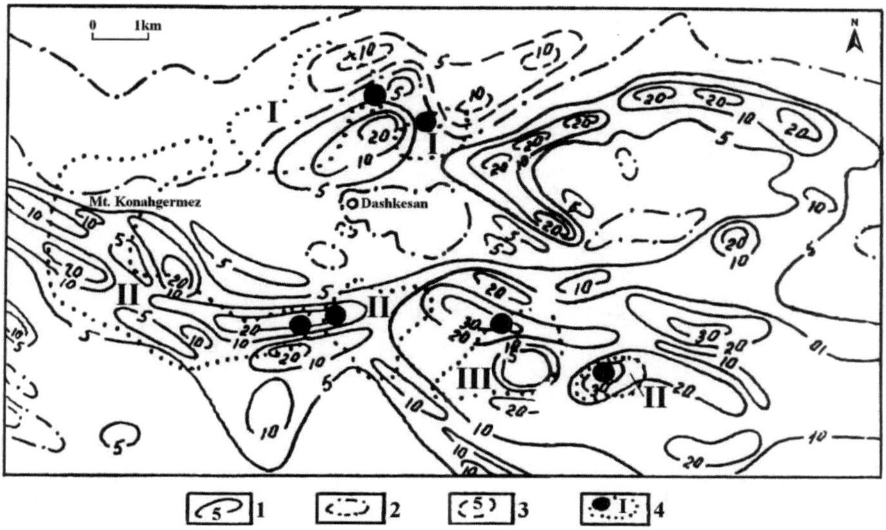

Fig. 5.23 Map of ΔZ isolines (in milliOersted) in the area of the Dashkesan group of iron-ore deposits. ΔZ isolines: (1) positive, (2) zero, (3) negative; (4) contours of skarn-magnetite deposits: I – Dashkesan, II – south Dashkesan and its eastern portion, III – Damir

maximal ΔT gradients coincided with the contours of the predicted ore reserves. Thus, the map of ΔT gradients could be used to discover new magnetite bodies in this area. A scheme of magnetic mass distribution at a depth was constructed by analyzing surface and airborne magnetic surveys, and the map of ΔT gradients incorporating several geological sections where quantitative computations were carried out. It showed that all the known iron-ore deposits were in the zones where gabbroids occur at comparatively low depths. This factor was an important search criterion.

To reduce the influence of surface inhomogeneties and better identify the ore bodies the relief disturbing effect was calculated differently to include a correlation method (see Fig. 3.13).

By applying Eq. 3.77, a few nodes of hidden structure crossing were revealed (Fig. 5.24). This parameter does not guarantee absolute effectiveness, but the relation of the displayed nodes to the known ore mineralization was confirmed.

5.5.5 *Induced Polarization*

The IP method was one of the most frequently used geophysical methods for searching and prospecting ore deposits in the Lesser Caucasus.

Semi-quantitative interpretation of the VES-IP curves can serve to localize subhorizontal bodies with a high polarizability. This was applied to localize a skarn-magnetite deposit in the Dashkesan ore field in the Lesser Caucasus (Khesin 1969).

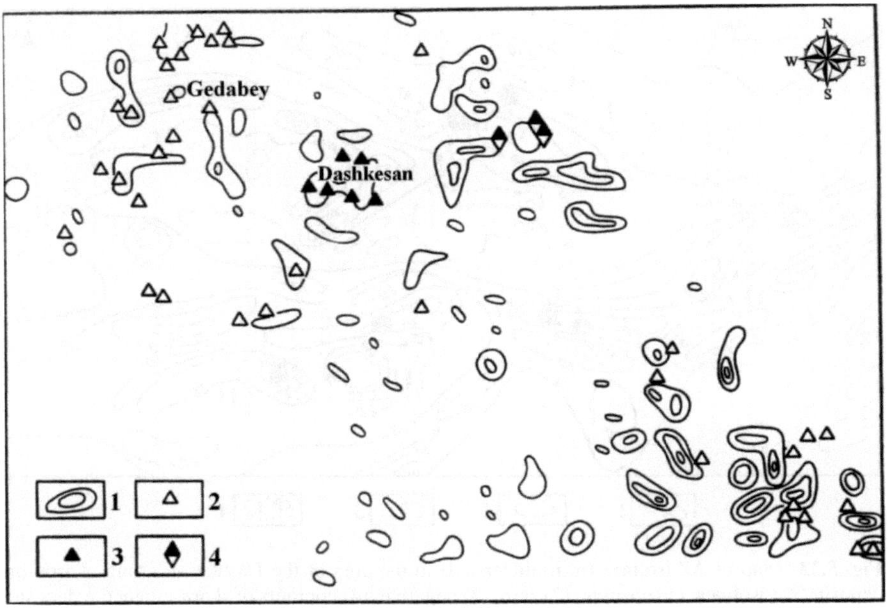

Fig. 5.24 Comparison of the nodes of hidden crossings of structures delineated by magnetic data with the distribution of ore mineralization in the north-east slope of the Lesser Caucasus. (1) isolines of parameters, ore deposits and manifestations: (2) copper-pyrite, (3) skarn-magnetite, (4) iron-manganese

The results of borehole IP measurements in the Karadagh copper-porphyry deposit are shown in Fig. 2.18.

An IP survey in the Lesser Caucasus revealed a smoother relief and not very complex tectonics. However, there were a wide range of magmatic and other rocks. This caused significant variability in the physical properties, both lateral and vertical, which made it difficult to interpret the geophysical data. The application of an approximate IP interpretation method successfully determined the depth of the upper edge of the magnetite body. It was characterized by an abrupt rise in the η_a value in the VES curve (Fig. 5.25).

Figure 5.26 illustrates the application of IP and resistivity methods in the chrome deposit of Shorzhinskoe in Armenia (Gazaryan 1965). Ore chrome bodies and certain dunite bodies contain 10–25% magnetite which produce significant IP anomalies. The electric conductivity of chrome bodies is not high and negative ρ_a anomaly in the center of profile is caused mainly by a thick ore zone (dunites enriched by magnetite) in the center of geological section. The 17% IP anomaly over the ore zone was about 60 m in width. On the basis of the IP curve analysis a contour of the ore zone was drawn (Gazaryan 1965). Interpretation of the IP anomaly by the characteristic point and tangent methods (see Sect. 3.6.2) detected the middle of the upper edge of this ore zone.

The Cretaceous Madneuli barite-gold-copper-pollymetallic deposit is a major deposit of the Georgian Bolnisi mining district, located in the Artvin-Bolnisi Zone

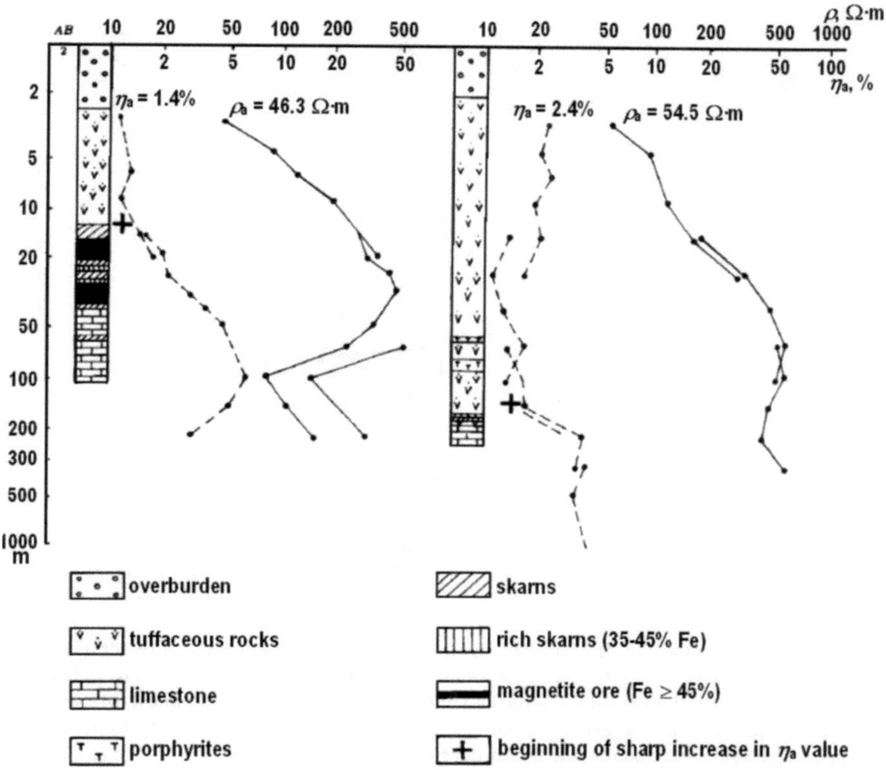

Fig. 5.25 Semi-quantitative interpretation of η_a curves in the Dashkesan iron deposit (Lesser Caucasus) (Khesin et al. 1997)

(Popkhadze et al. 2009). In this deposit self-potential, magnetic, gravity resistivity and induced polarization methods were applied. Figure 5.27 presents the results of resistivity and IP method observations. The resistivity curve behavior reflects the position of the left and right ends of the orebody (which occupies and intermediate position between the thick bed and thin horizontal plate interpretation models); a some small resistivity anomaly is observed over ore zone in the right portion of the profile. The IP anomaly over the orebody reached 18% and 8% IP anomaly was recorded over the ore zone. The interpretation of IP anomaly by an areal method developed in magnetic prospecting (Khesin et al. 1983; Eppelbaum and Khesin 2002) identified the middle of the upper part of the orebody.

5.5.6 VLF

Figure 5.28 illustrates the interpretation results for the H_φ curve along the profile across the Kyzyl-Bulakh gold-pyrite deposit in the Mekhmana ore district. The anomaly over the ore object exposed by prospecting boreholes is shown in the

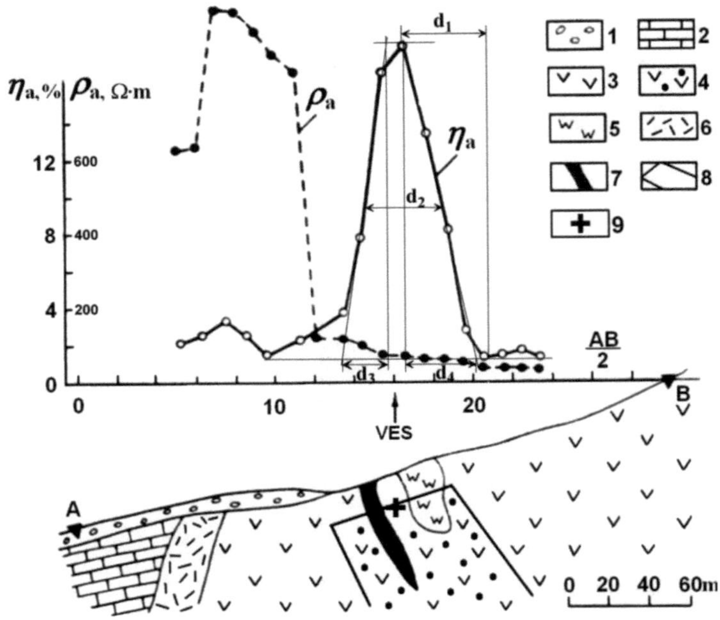

Fig. 5.26 Interpretation of IP anomaly in chrome deposit Shorzhinskoe (Lesser Caucasus, Armenia) (initial data (1–8) after Gazaryan 1965, with modifications). (1) loose deposits, (2) limestone, (3) dunites, (4) dunites with significant volume of magnetite, (5) peridodites, (6) listvenites, (7) chrome ore body, (8) suggested contour of ore zone by IP data analysis, (9) determined center of upper edge of the ore zone by the use of developed methods

central portion of the profile. In the south-western portion the anomaly over an anticipated object is marked.

In both cases the approximation model was represented by an inclined thin bed. As shown in Sect. 2.1.2, the magnetic field ΔZ was an analogue of H_x and H_φ components in the VLF method for the case of an inclined thin bed. A model magnetic field ΔZ_m due to the a host medium and a near-surface ore body was computed for the central part of the profile. For modeling purposes the following parameters were used: magnetization value: 300 mA/m for the host medium, and 1,000 mA/m for the anomalous (ore) body; the vector of magnetization was assumed to be vertical for the host medium and along the dipping for the orebody. The azimuth of the selected profile was assumed to be 70°, which corresponds to the angle between the incoming *VLF* field and the real azimuth of the profile. It is clear from the figure that the H_φ and ΔZ_m curves are in good agreement. This provided additional proof of the similar nature of these fields.

5.5.7 Near-Surface Temperature Survey

Figure 5.29 illustrates the application of the correlation technique for eliminating the topographic effect in the district of the Kyzyl-Bulakh gold-pyrite deposit.

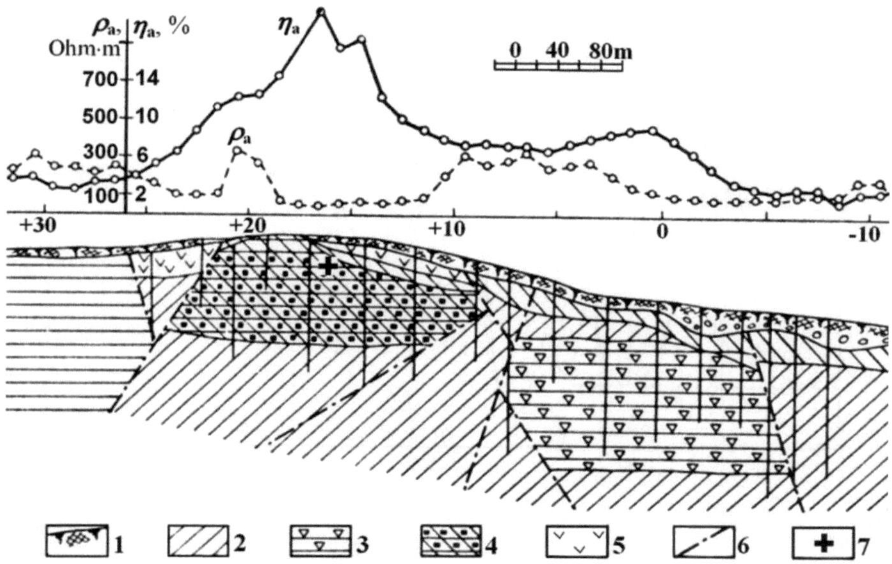

Fig. 5.27 IP data analysis in the Madneuli copper-pyrite deposit (Lesser Caucasus). (1) soil and deluvium, (2) silicificated acid effusives, tuffs and tufogenic sandstone, (3) secondary quartzites with disseminated copper-pyrite ore (ore zone), (4) secondary quartzites with barite-polymetallic ore (orebody), (5) andesites, (6) faults, (7) determined center of upper edge of the ore zone by the use of developed methods ((1)–(6) after Komarov (1980) and Popkhadze et al. (2009))

The equation for the terrain relief correction takes the form $T_{appr} = 15.6 - 0.07H$, where the correlation coefficient between the temperature T and the height H is -0.8. Once the inclined relief effect has been eliminated, the anomalies from disturbing objects (massive ore in the southwestern part of the profile, and a vast zone of disseminated ore in the northeastern part) are more pronounced.

5.5.8 Electric and Electromagnetic Methods

Petrovsky and Skaryatin (1929) successfully applied a radiowave method to the translucent Chiragidzor pyrite stocks in Somkhet-Karabakh Zone (Azerbaijan). Balavadze (1944) delineated cobalt-bearing veins in the Dashkesan deposit using electric resistivity profiling. The identification of cobalt-bearing fragmentation zones by combined resistivity profiling and an emanation survey in the Dashkesan ore district (Lesser Caucasus) is shown in Fig. 2.9.

The same electromagnetic methods (mentioned in Sect. 5.4.9): namely PEM, TEM and CMPC were applied to ore deposits of the Lesser Caucasus. However, the effectiveness of these methods in this region (due to greater differentiation of

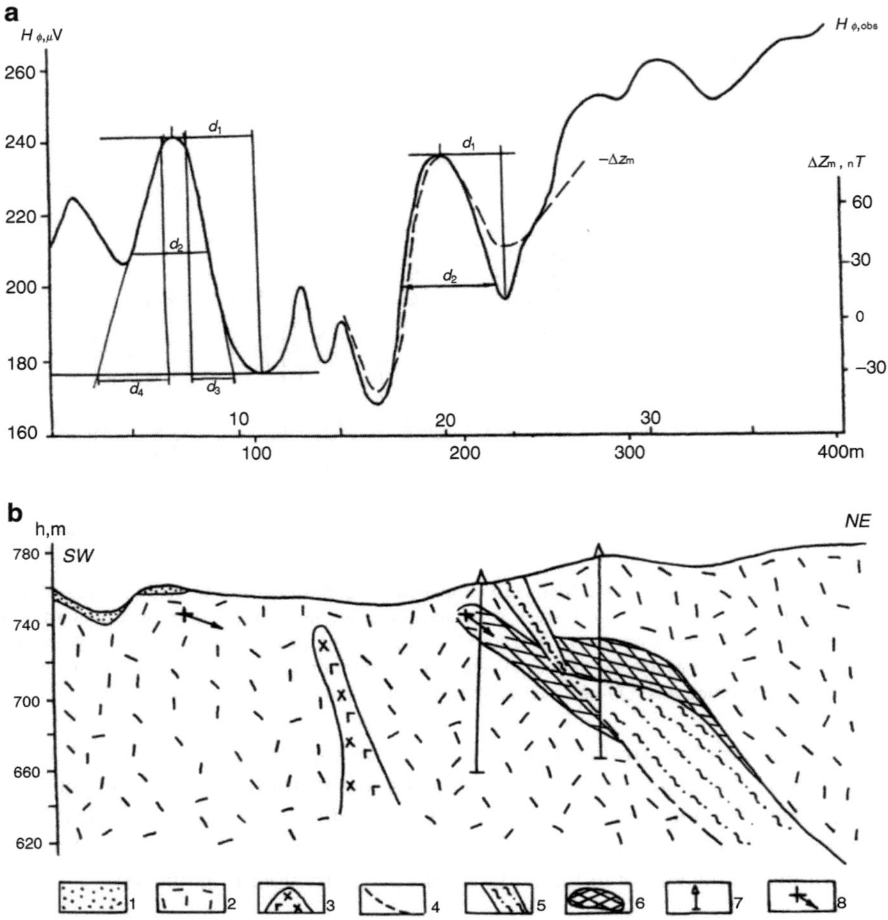

Fig. 5.28 Quantitative interpretation of the H_ϕ field in the area of the Kyzyl-Bulakh gold-pyrite deposit (Lesser Caucasus): (**a**) the plots of H_ϕ and model magnetic field ΔZ_m, (**b**) geological section (Eppelbaum 1989). (1) loose deposits; (2) tuffs of liparite-dacitic porphyrites; (3) dike of andesite-basalts; (4) disjunctive dislocation; (5) ore body; (6) zone of boudinage; (7) prospecting wells; (8) location of the conductive bodies' upper edge and direction of their dip by quantitative interpretation results

electric and electromagnetic properties and the smaller size of the ore targets) was lower compared to the Greater Caucasus deposits.

An airborne long wire electromagnetic method was employed for geological-geophysical mapping of a stretched uranium field of the hydrothermal type in the Lesser Caucasus (Armenia) (Baryshnikova et al. 2011).

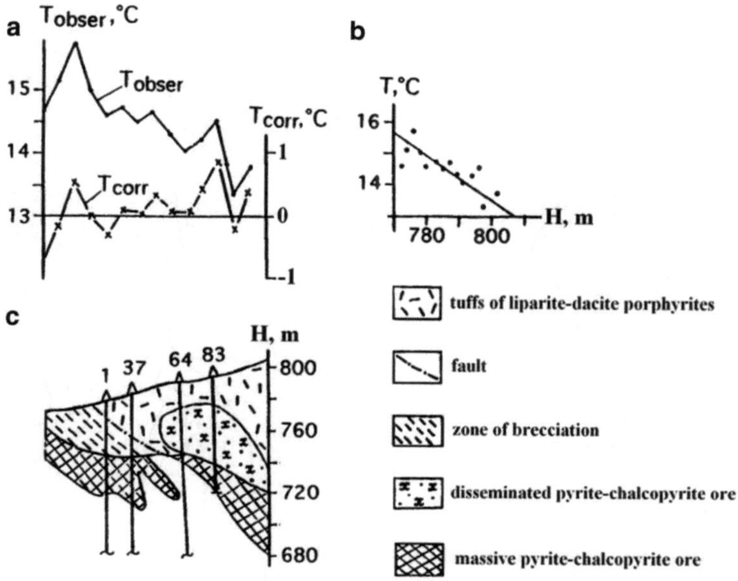

Fig. 5.29 Correlation technique for reducing the terrain relief effect in near-surface thermal prospecting in a district of the Kyzyl-Bulakh gold-pyrite deposit (the Lesser Caucasus, Nagorny Karabakh): (a) plots of observed and corrected temperature values, (b) correlation, (c) geological section. Drill holes 1, 37 and 64 continue below the section. Drill hole 83 terminates as shown (Eppelbaum 1989)

5.5.9 Simple Integrated Analysis

Visual analysis of the field is often enough to fulfill the geological objective, especially when a set of geophysical methods is employed. The mapping of intrusive rocks on the basis of data obtained by magnetic and emanation surveys on a portion of the Lesser Caucasus revealed an intrusive body which intruded the limestone with sufficient precision, and even made it possible to classify its composition (Fig. 5.30).

5.5.10 Integrated Analysis of the Basis of PGM

The techniques of processing and interpretation (presented mainly in Chaps. 2–4) form a continuous sequence of procedures aimed at a final geological product – a map, a section, etc., in other words, a plot representing the final physical-geological model (***PGM***) of the medium.

For example, by summing up the amounts of information contained in the Δg and ΔT fields, and in the vertical component of VLF magnetic field H_z, we get the

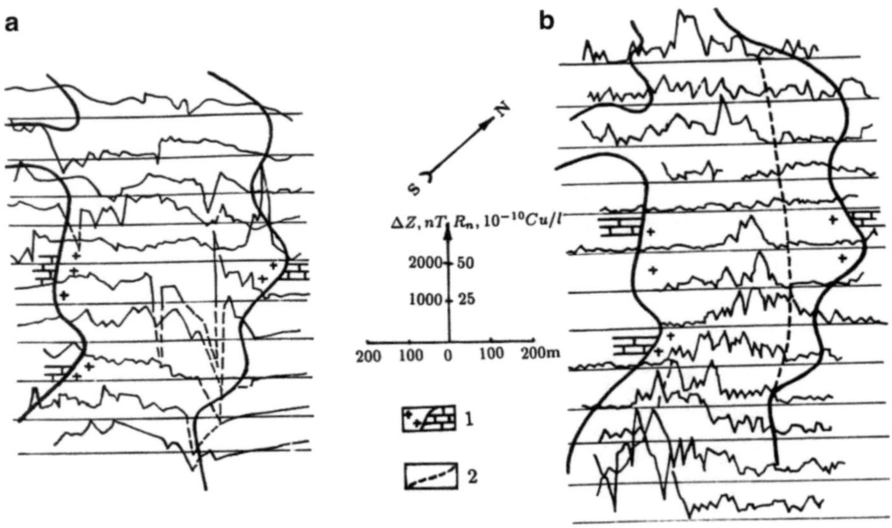

Fig. 5.30 Magnetic (**a**) and emanation (**b**) survey data employed for intrusive rock mapping (dashed line shows negative values of magnetic field graphs) (Lesser Caucasus). (1) intrusion/limestone contact; (2) interface of alkaline rocks and gabbro-diabases by the emanation survey data

anomaly of the parameter $\frac{1}{3}\sum_{i=1}^{3} J_i$, which is more marked and reveals a gently sloping ore object on profile 5c through the Kyzyl-Bulakh gold-pyrite deposit in the Lesser Caucasus (Fig. 5.31a).

Because of the limited sampling, Eq. 3.94 was used. For this calculation ΔU_i was taken as the double standard deviation of U_i. The zero line of the ΔT field was selected at the level of the anomaly maximum. The values of $U_i(\Delta T)$ were calculated from this level in the direction of the ΔT field decrease (since pyrite deposits are characterized by lower magnetization against to strongly magnetized volcanogenic host rocks).

The rapid interpretation of Δg, ΔT, H_z and H_φ plots (Fig. 5.31b) confirmed the available geological predictions and formed an initial approximation of the model of the medium which was further refined by interactive physical-geological modeling of the Δg and ΔT fields (Fig. 5.31c).

5.6 Underground Geophysics

Underground geophysics plays an important role in the investigation of ore deposits occurring in mountainous conditions. In the Caucasus, underground geophysics is employed in all three regions: the Northern and Southern Caucasus, and the Lesser Caucasus. Below the geophysical methods that are best suited for underground geophysics are discussed.

5.6.1 Gravity

The methodology of underground gravity surveys is determined by the nature of the problem, as well as the physical-geological and environmental mine-technical conditions.

The following conditions are favorable for searching ore bodies: (a) presence of contrast density and a sufficient volume of ore target causing a gravity effect of no less than 0.05–0.08 mGal, (b) comparatively small distance between the adit and

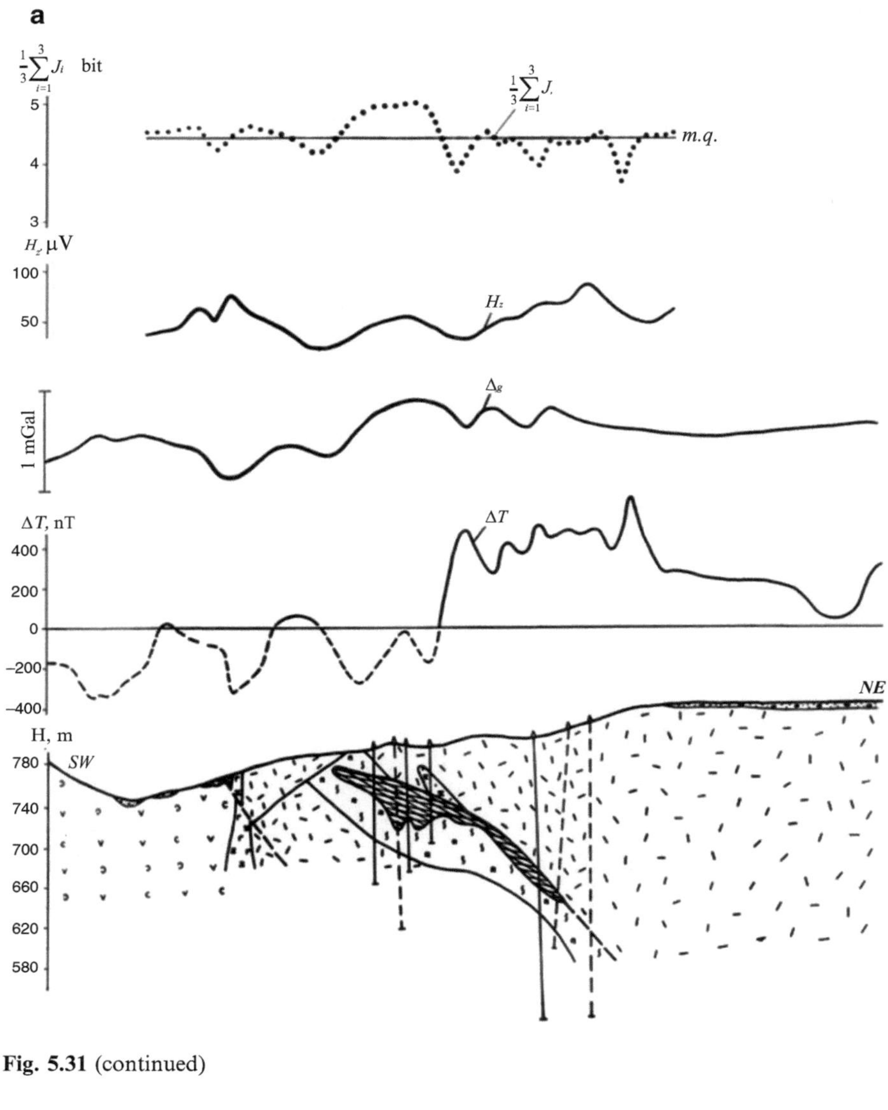

Fig. 5.31 (continued)

b

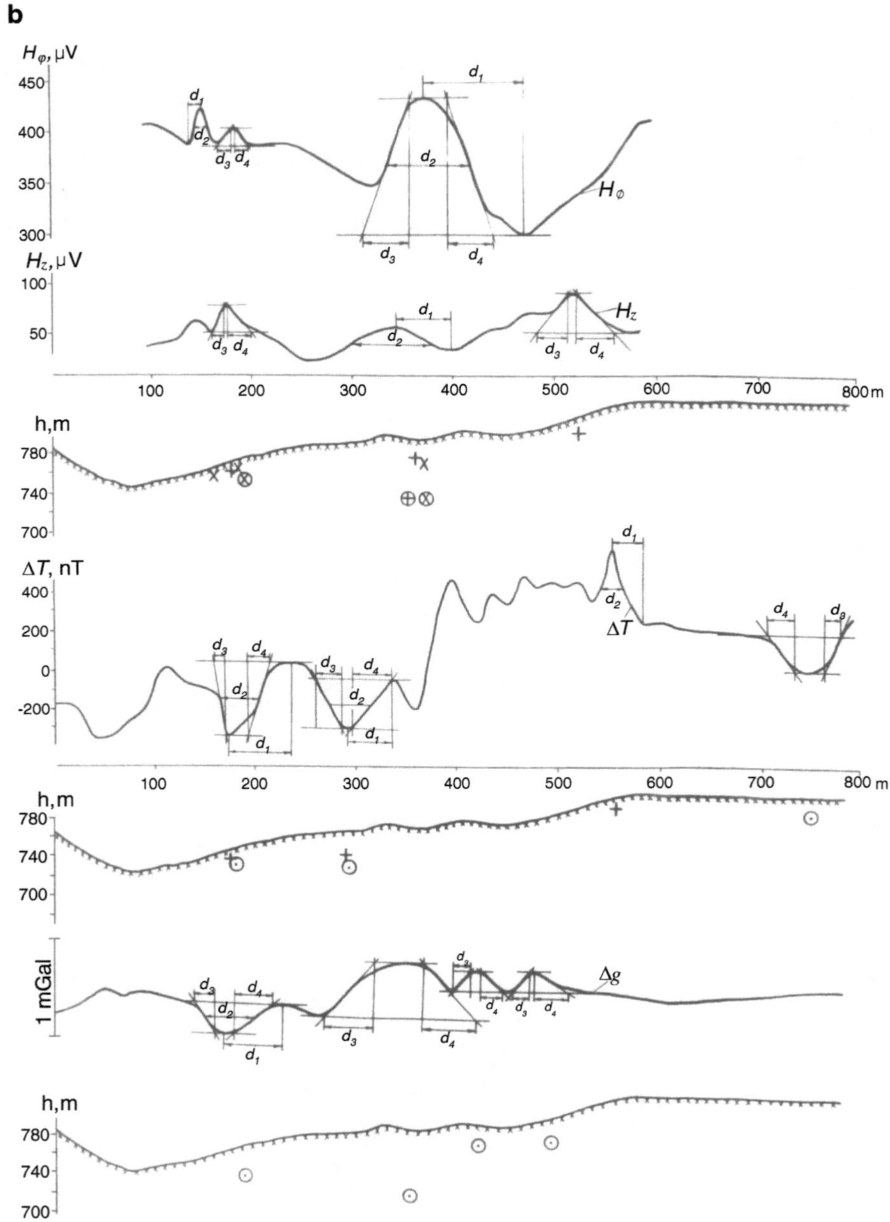

Fig. 5.31 (continued)

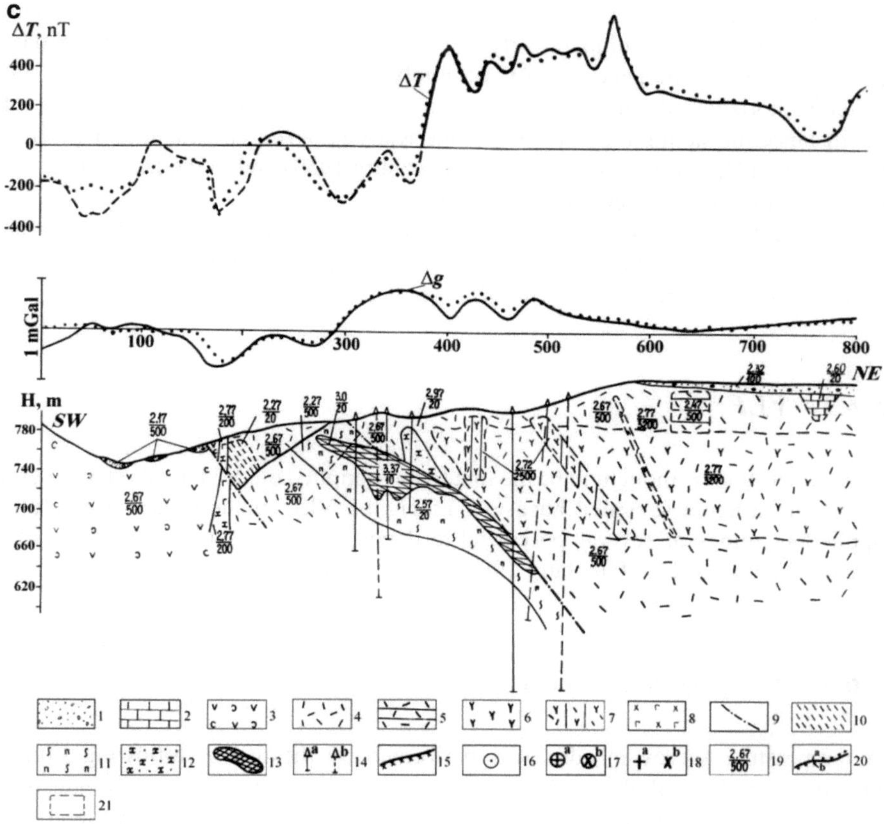

Fig. 5.31 Stages in geophysical data interpretation at the Kyzylbulakh gold-pyrite deposit (Lesser Caucasus) (Eppelbaum 1989; Khesin et al. 1993a). (**a**) Identifyingthe target object (ore deposit) by summing up the information obtained by different geophysical methods. (**b**) Quantitative rapid interpretation of ΔT, Δg, H_z and H_φ fields. (**c**) Selection of a geological section on the basis of gravity and magnetic data. (1) Quaternary deluvial deposits; (2–8) Middle and Upper Jurassic rocks: (2) silicificated limestone lens, (3) tuffs and lavas of andesitic porphyrites, (4) tuffs of liparite-dacitic porphyrites, (5) deconsolidated tuffs of liparite-dacitic porphyrites, (6) lavas of dacitic porphyrites, (7) consolidated lavas of dacite-porphyrites, (8) dikes of andesite-basalts; (9) disjunctive dislocations; (10) zone of brecciation and crush; (11) zones of brecciation, crush and boudinage with lean pyrite-chalcopyrite ore; (12) zone of brecciation, crush and boudinage with rich impregnating mineralization; (13) massive pyrite-chalcopyrite ore; (14) drilled wells: (a) on the profile, (b) projected on the profile; (15) terrain relief in the profile (in Fig. 5.31b only); results of rapid interpretation; (16) location of the center of HCC from the interpretation of Δg and ΔT plots, (17) location of the *HCC* center from the interpretation of the H_z (**a**) and H_φ (**b**) plots, (18) location of the upper edge of the thin bed from the interpretation of H_z and ΔT (**a**) and H_φ (**b**) plots; (19) physical properties (numerator = density, g/cm^3); denominator = magnetization, mA/m; (20) gravitational and magnetic fields: (**a**) observed, (**b**) selected; (21) body contours assumed in the process of 3D modeling

ore target and its asymmetric position with respect to the adit (especially by dipping close to vertical), (c) comparatively uniform host medium (or presence of data about density inhomogeneties), (d) desirably dense net of operational mines at different horizons. An underground gravity survey was successfully carried out for instance, in the mining openings of the Katsdag pyrite polymetallic deposit in the Southern Caucasus (Poltoratsky and Ginzburg 1989) and in the Bazun iron-ore deposit in the Lesser Caucasus (Fidanyan 1986).

Interesting results by calculating the statistical reduction in an underground gravity survey in one of the Lesser Caucasian ore deposits (Armenia) was presented by F.M. Fidanyan (Khesin et al. 1988).

5.6.2 VLF

Gazaryan (1968) is thought to be the first to have applied the VLF method in several mines in Armenia (Lesser Caucasus). Later, the VLF method was successfully applied in mines at several polymetallic deposits on the southern slope of the Greater Caucasus (e.g., Ginzburg 1982; Brodovoi et al. 1989).

5.6.3 Temperature Survey

A temperature survey in adits was performed by the use of following simple methodology. First of all, for this survey non-working adits without any industrial activity (primarily, without air supply) were selected. As a rule, positive temperature anomalies (or thermal flow) and heightened values of geothermal gradients are measured over ore bodies. Hence, inside the target of heightened thermal conductivity, the geothermal gradient will be below the normal value. The width of the anomalous temperature interval gives definite information about the size of anomalous object, and the temperature gradient is indicative of the distance to the observation line. The distribution of the temperature graph can in many cases be used to draw conclusions about the location of the anomalous target.

Temperature measurements in adit 10 of the Katsdag deposit were made in glasses filled with water at an observation step of 10 m (Fig. 5.32). Analysis of the temperature field indicated that the temperature anomaly observed in the right portion of this geological section in general corresponded to the anomalous body. At the same time, the temperature anomaly observed in the left portion of this section testified to the presence of some additional ore bodies not detected by geological means.

An example of the influence of rugged terrain relief on adit temperature measurements is presented in Fig. 5.33. As is known, surface relief influences the temperature field at a distance of 4–5 amplitudes of the relief's maximal amplitude (e.g., Cheremensky 1972). A temperature survey was performed in blast-holes of

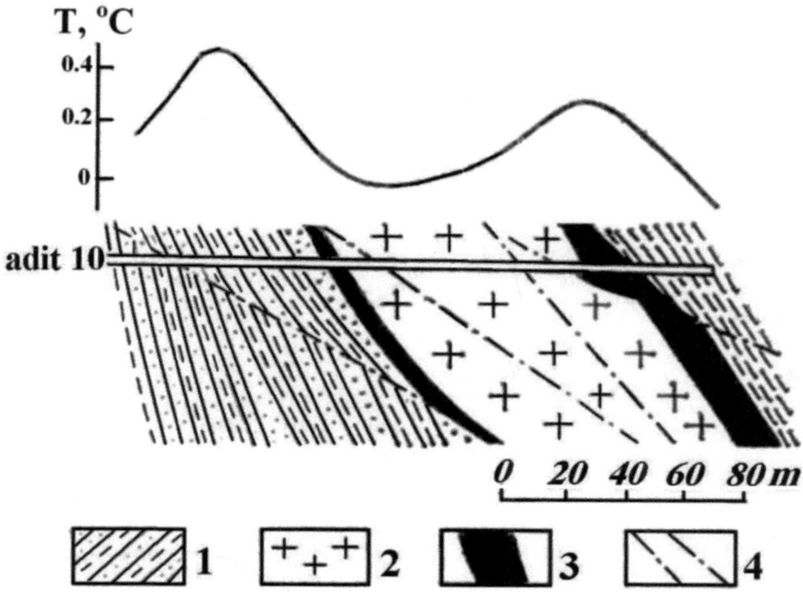

Fig. 5.32 Graph of the anomalous temperature distribution along the main opening of adit 10 in the Katsdag polymetallic deposit, southern slope of the Greater Caucasus (this survey was carried out by L. Eppelbaum in 1978 under the supervision of S. Ginzburg (Ginzburg et al. 1981)). (1) sandstone, (2) liparite dacites, (3) massive copper-pyrite ore, (4) faults

1.5 m in the adit walls. The search targets in this case; namely, massive pyrite-polymetallic ore bodies, occurred in the sandy-argillaceous medium that provided a high thermophysical contrast. The observed temperature curve (Fig. 5.33b) was strongly distorted by the rugged relief influence. After application of the correlation approach (coefficients of linear regression were obtained by the least-square method), the corrected temperature graph was constructed (Fig. 5.33a). Quantitative analysis of the graph led to calculation of the position of the center of the anomalous target (the quantitative interpretation method is presented in Sect. 3.5.2).

In the Katsdag pyrite-polymetallic deposit, an areal temperature survey was performed (L. Eppelbaum, as a student in the Geophysical Dept. of the Azerbaijan Univ. of Oil & Chemistry, took part in these field works) in the underground drifts (Fig. 5.34a). The temperature map for the normal field (which incorporates the influence of the surface rugged relief, the depth of attenuation of annual temperature variations propagating from the earth's surface and other factors) is shown in Fig. 5.34b. A residual anomalous temperature map is depicted in Fig. 5.34c. Clearly, the final anomalous map (Fig. 5.34c) differs considerably from the initial observations (Fig. 5.34a).

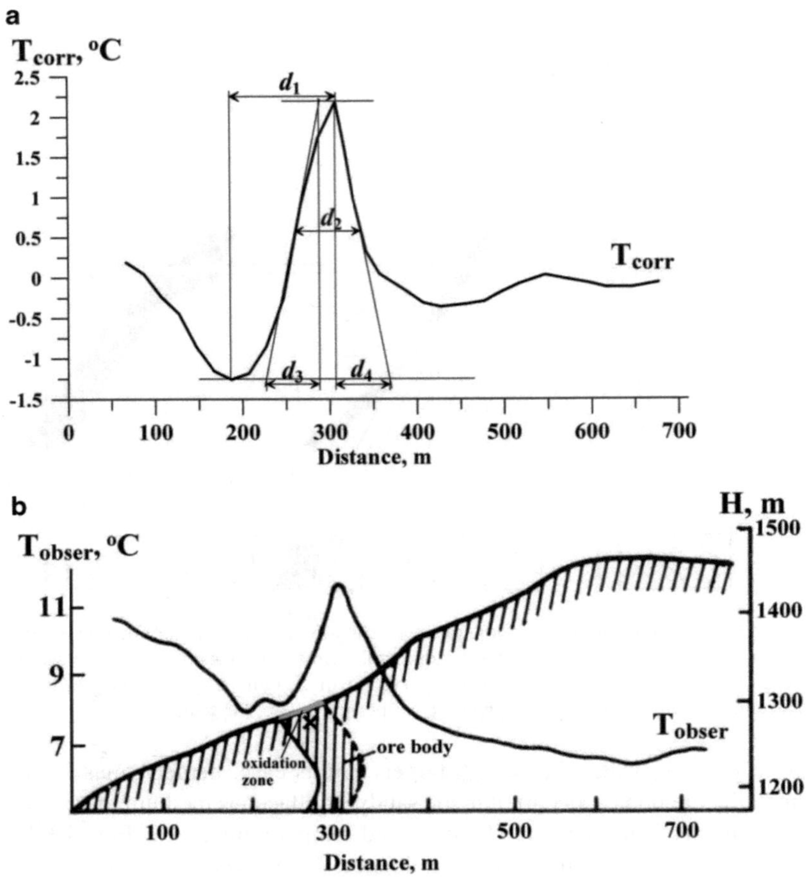

Fig. 5.33 Quantitative analysis of near-surface thermal investigations carried out in an adit of the Filizchay pyrite-polymetallic deposit (southern slope of the Greater Caucasus). Observed temperature field and geological section are adapted from (Borisovich et al. 1988). (**a**) quantitative analysis of corrected temperature graph, (**b**) observed temperature field and geological section. The "+" symbol designates the position of the middle of the upper edge of the anomalous body

5.6.4 Self-Potential Survey

An example of SP field changing at the various stages of underground mining in the Chiragidzor sulfur deposit (Lesser Caucasus) is shown in Fig. 2.6.

A classic example of a SP field distribution along a highly electrical conductive orebody is presented in Fig. 5.35. In the upper part of the massive orebody a value of (−600) mV was observed, and in the lower part the value was (+400) mV (obviously, the absence of SP measurement in the outermost lower points precludes observing the equivalent value of +600 mV).

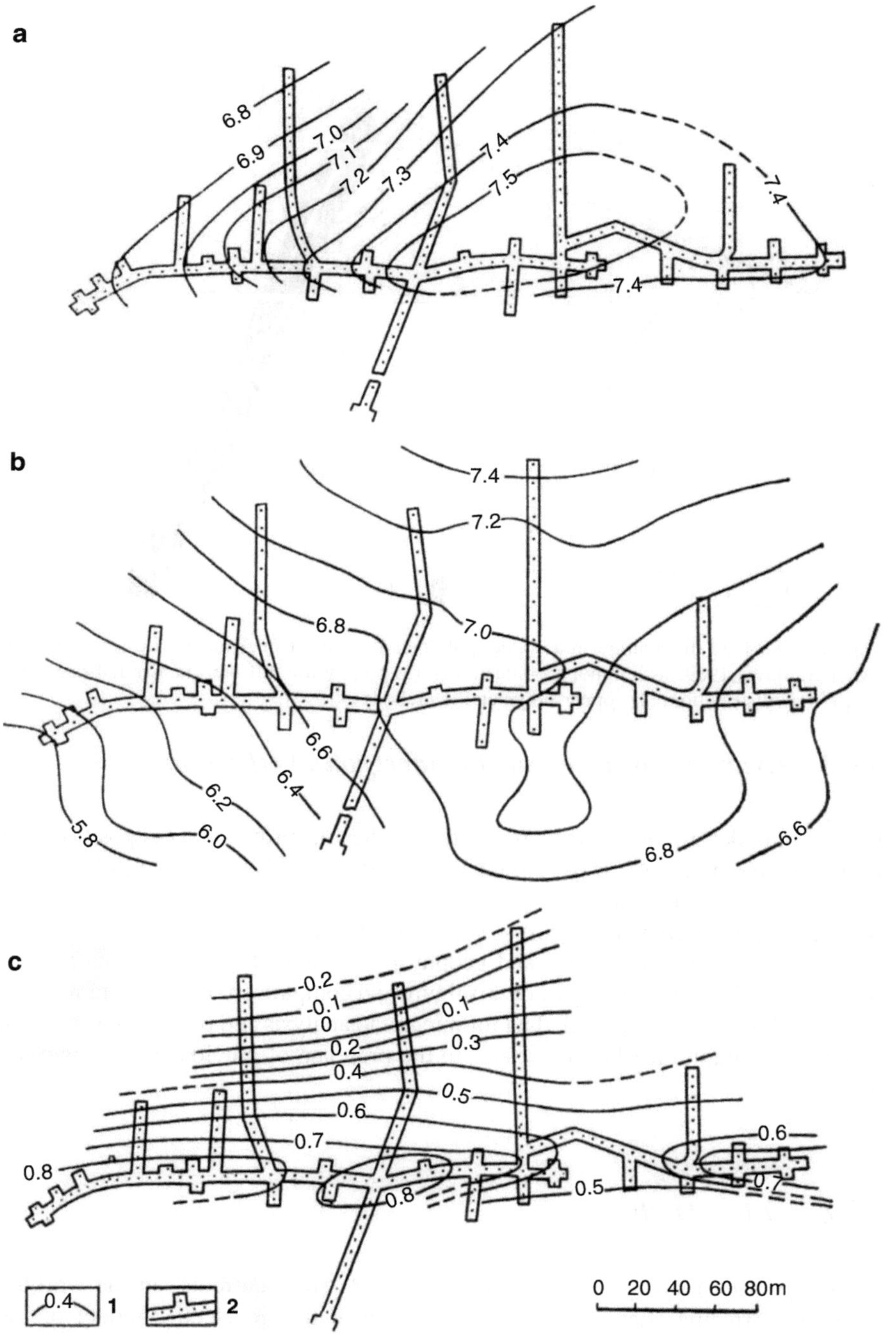

Fig. 5.34 Results of mine thermal prospecting in the Katsdag pyrite-polymetallic deposit (southern slope of the Greater Caucasus) (After Ginzburg and Maslennikov, with minor modifications). Temperature maps: (**a**) observed values, (**b**) normal field, (**c**) anomalous values. (1) temperature isolines, (2) underground shifts

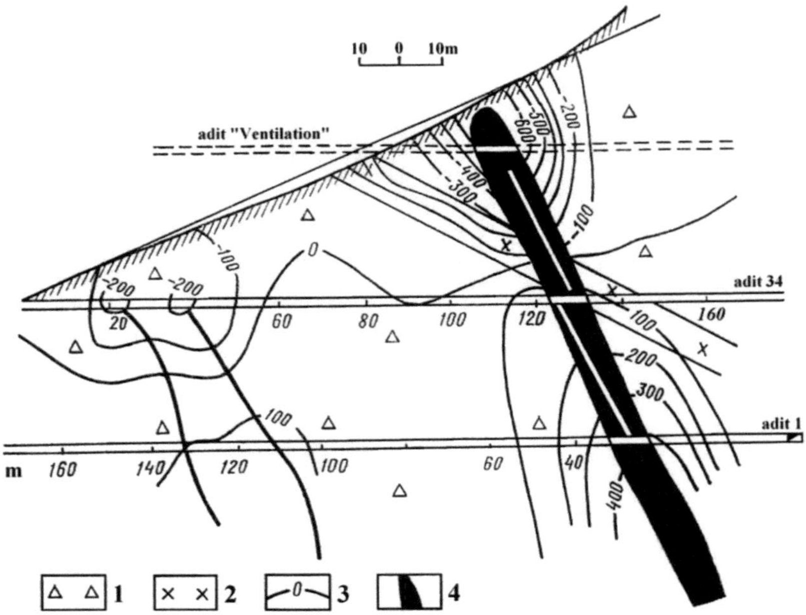

Fig. 5.35 SP field distribution in a copper-polymetallic deposit in Armenia (Lesser Caucasus) (After Brodovoi 1989, with minor modifications). (1) porphyritic tuff breccia, (2) diabasic dyke, (3) SP field distribution, (4) copper-polymetallic orebody

5.6.5 *Examples of Integrated Underground Observations*

Results of underground electric prospecting in the Alaverdy copper deposit (Lesser Caucasus) by the use of stray currents, IP and SP methods are shown in Fig. 2.17.

An impressive integration of gravity, temperature and VLF observations in the Katsdag deposit (Greater Caucasus) is presented in Fig. 5.36 (L. Eppelbaum took part in these field observations and geophysical data analyses). Analysis of geophysical graphs in the lower part of this figure and comparison of these data with the geological section provide rich data for integrated analysis and prospection. Certain geophysical graph distributions testify to the presence of previously unknown ore bodies.

5.6.6 *Other Methods*

Gamoyan (1986) suggested employing so-called "stray currents" in the working mines of Armenia (Lesser Caucasus) to enhance the geoelectric model of the surrounding media. The results of underground electric prospecting in the Alaverdy copper deposit (Lesser Caucasus) are shown in Fig. 2.16. The electric correlation

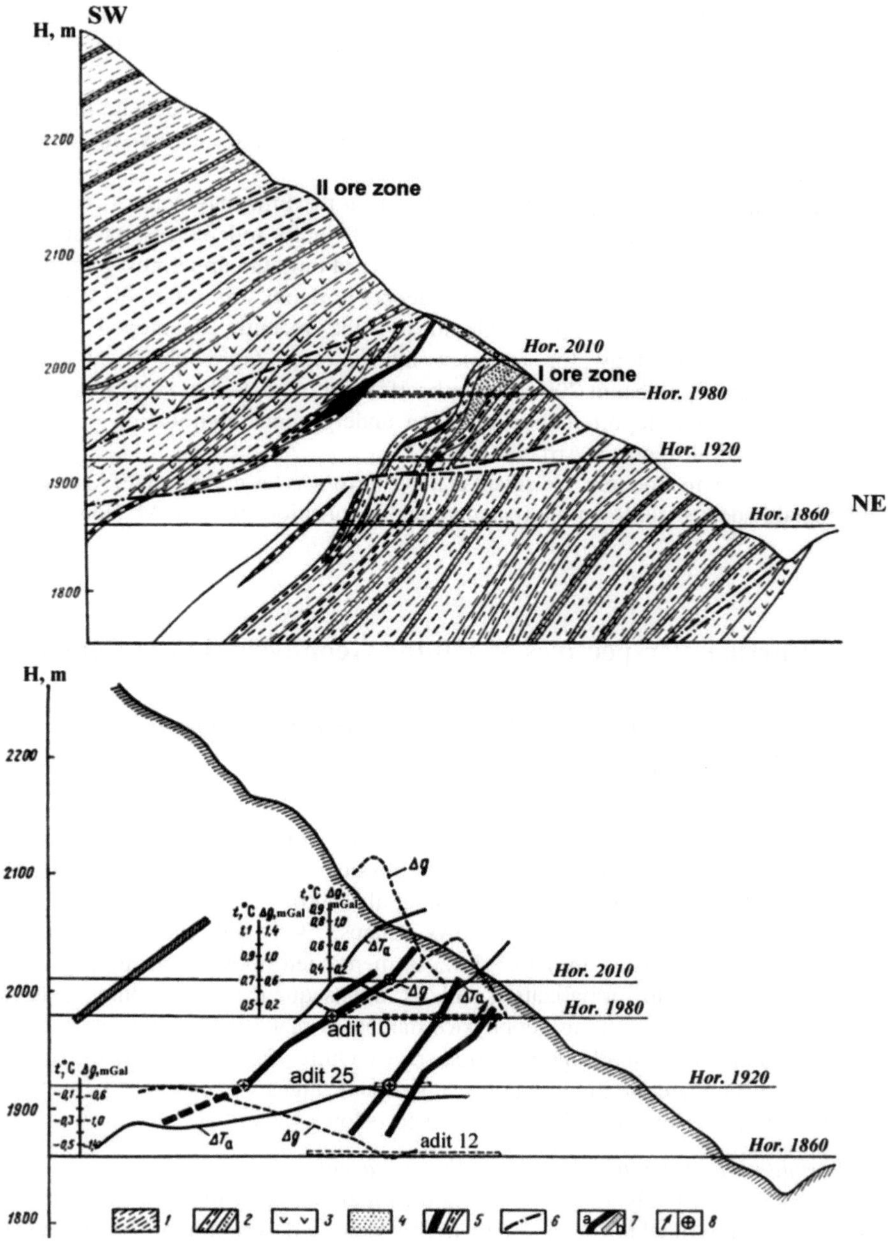

Fig. 5.36 Integrated geophysical investigations in mines of the Katsdag deposit (southern slope of the Greater Caucasus) (Poltoratsky and Ginzburg 1989). (1) shales, (2) interbedding of shales, sandstones and siltstones, (3) dioritic porphyry, (4) oxidizing zone, (5) polymetallic ores of I and II zones, (6) faults, (7) apparent beds of redundant: density (**a**) and thermal conductivity (**b**), (8) current axes revealed by the VLF method

methods for the linkage of ore undercutting between various horizons and surface observations were carried out in the Armanis and Shaumyan polymetallic deposits (Arutyunyan 1986).

According to Kulkov et al. (1991), underground recordings of cosmic rays was successfully applied in the Tyrnyauz deposit (Northern Caucasus) to identify geological blocks with different densities.

The seismic "central ray method" employed in the Filizchay deposit (southern slope of the Greater Caucasus) did not yield sufficiently reliable results since the area of high gradients of seismic velocities covered only the uppermost part of the geological section and information on the deeper horizons was extremely distorted (Karaev and Rabinovich 2000). Obviously, the recently developed seismic multifocusing technology for complex media (e.g., Gurevich et al. 2002; Berkovitch and Eppelbaum 2009) might be effectively applied in such areas.

Badalyan (1968) reported the use of an underground modification of the IP method in several mines in Armenia.

Piezoelectric method (e.g. Neishtadt et al. (2006)) was tested in mid-1970 in some polymetallic deposits of Georgia (personal communications with N. M. Neishtadt).

5.7 Further Perspectives of Mining Geophysics in the Caucasus

5.7.1 Development of the Caucasian Mining Geophysics Databases

The first priority in mining geophysics is the development of a comprehensive computer database covering all types of geophysical surveys (e.g., Khesin 1975a), the observed data and the findings (Khesin et al. 1996).

The interpretation sequence follows Chaps. 3–4 and the interpretation procedures (stages) listed in Table 5.5. As mentioned above, not all stages are equally susceptible to formalization and, as a consequence, to computer-assisted processing. However, mining geophysics databases make it possible to store the input and output data of all interpretation stages. Quite often in this case the output data of one stage are the input data for subsequent stages.

This interpretation technology is reviewed below.

1. *Formulation of the initial model of the ore target*
   This stage basically employs the non-formal methods of prior information synthesis described in Chaps. 3 and 4. This yields a description of the geological space of the investigated area represented by a set of geological bodies (including ore objects) belonging to a particular association. These bodies may be either diffuse or localized, and characterized by certain physical properties (for instance, density and magnetization). The coordinate representation of these objects is obtained according to description rules for each type of body. These

Table 5.5 Construction of mining geophysics databases: Main principles

No.	Interpretation procedures (stages)	Operations
1	Forming an initial ***PGM*** of ore target (on the basis of geological, drilling and mining data as well as preliminary geophysical data examination including satellite imaging) and modifications	Geological, geophysical and satellite data preparation and generalization, primary loading and DB modification
2	Forming an indicator space of ore deposits according to the observation results	Computation of direct effects due to the relief forms and other known bodies. Excluding the computed effects from the observed fields
2.1	Eliminating complications due to the terrain relief effect and other known factors	Conversion of the upper half-space to different levels, averaging with different radii, computing gradients, calculating ruggedness measures, etc
2.2	Computing secondary indicators by the initial field(s)	
3	Revealing and localizing; identifying the targets	Classification (taxonomy) of the investigation area by a specified (variable) set of indicators, including ruggedness. Correlation of linear zones and identifying ring zones. Determination of axes and contours of objects
4	Determination of the quantitative parameters of anomalous bodies	Parameter determination by application of quantitative analysis. Non-formalized simulation (modeling). Formalized simulation (optimization)
5	Integrated interpretation and physical-geological modeling; construction of a composite picture of the of ore deposit	Computing appropriate measures (probabilities, amount of information, etc.) by a set of fields of different kinds. Combined modeling of various fields
6	Graphic representation of the findings	Computer graphics operations (output of charts, schemes, plots, maps, 2D and 3D physical-geological sections)

data are loaded into the database together with the height information obtained over the area of investigation as a whole. In addition, gravity and magnetic fields as well as other measured fields are also loaded into the database.

2. *Formulation of an indicator space*

   Drawing on the analysis of the observed fields and based on the objectives of the investigation, the interpreter makes decisions regarding the type of secondary indicators, ways to obtain them and the transformation parameters. If terrain corrections are called for, or correction of the fields due to large distant objects, appropriate data are loaded into the database as described above.

3. *Revealing and localizing targets*
   Secondary indicators computed during the preceding stage are extracted from the database using special programs. The isoline maps are plotted for corresponding transforms. Their analysis and the display of the findings follow the rules given in Chap. 3. Data for initial modifications of the model of the geological target (ore manifestation, deposit or field) are obtained at this stage. They are introduced into the database.
4. *Determination of quantitative parameters of anomalous body* (*inverse problem solution*)
   The substage involving choice of parameters for determination and preparation of the anomaly (numerical values or graphs) for computation precedes this stage (see Chap. 3). At the end of this stage additional data for model modification are obtained, and these data are introduced into the DB.
5. *Integrated interpretation and physical-geological modeling*
   At the beginning of this stage the interpreter has a sufficiently complete idea about the majority of sources of the anomaly field including deep-seated hidden sources. Therefore, the interpreter can distinguish objects (classes of objects) with expected properties. Physical-geological modeling on the basis of gravity, magnetic, resistivity and other fields is done with software in an interactive mode. At each modeling stage the fields caused by the entire model are computed over a given profile (or a group of profiles). These fields are compared with the observed ones, and a decision is made regarding the necessary changes to the model and the nature of these changes. The modifications may be related to the physical properties of individual bodies or to the form of these bodies' surfaces (or both); certain bodies can be eliminated or, on the contrary, added. The changes are made on the basis of the analysis of both fields, which characterizes the integrated nature of physical-geological modeling. The modifications are loaded into the DB with the help of a special subroutine. Field computations, comparison of the computed fields with the observed ones and modifications of the model are iterated until the fields fall within the defined limits of observation accuracy.
6. *Developing a final model*
   Since modeling applies to geological objects rather than to physical sources, the final model has a geological content. The assignment of new bodies introduced during the selection to a certain classification is determined by analogy with bodies with close physical properties, proceeding from general geological notions in the form of graphs, charts, maps, 3D ***PGM***, etc.

5.7.2 *ROV Geophysical Surveys for the Delineation of New Caucasian Ore Deposits*

A long-term perspective in mining geophysics in the Caucasus, as in other regions of the world, involves the broad scale application of geophysical measurements by

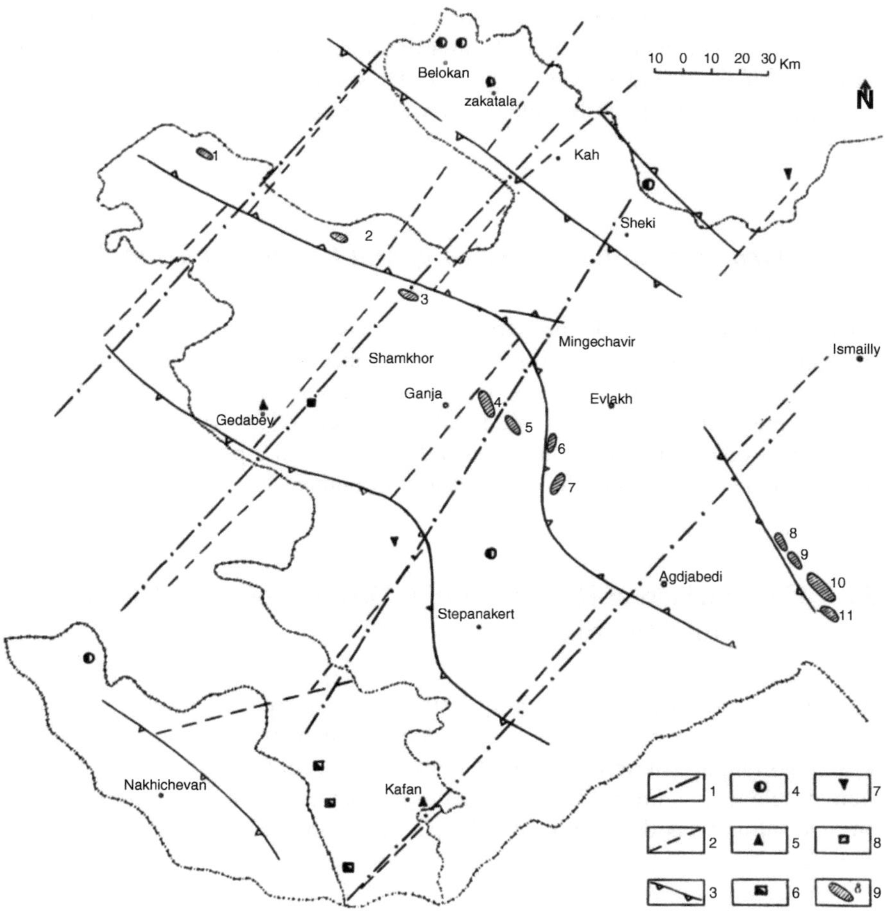

Fig. 5.37 Comparison of regional rupture dislocations and distribution of economic deposits in the Eastern Caucasus (Khesin et al. 1996). (1) generalized transverse dislocations; (2) revised position through transverse dislocations from (Ismail-Zadeh and Khesin 1989); (3) deep faults as geostructural zone boundaries from (Ismail-Zadeh and Khesin 1989a); typical ore deposits: (4) polymetallic ore (lead, zinc, copper), (5) copper, (6) copper and molybdenum, (7) mercury, (8) aluminum; (9) oil and gas deposits from (Mamedov et al. 1985): 1 – Takhtatope, 2 – Gyurzundakh, 3 – Tarsdallar, 4 – Kazanbulakh, 5 – Adjidere, 6 – Naftalan, 7 – Mir-Bashir, 8 – Zardob, 9 – Shykhbagy, 10 – Muradkhanly, 11 – Jafarly

the use of Remote Operated Vehicles (ROV) (e.g., Eppelbaum and Mishne 2011). The main advantages of ROVs are the absence of an aircraft crew (GPS controlled unmanned surveys can be carried out even at very low altitudes and in risky conditions). In addition, most methods of geophysical field analysis require knowledge of the field distribution at different levels over the Earth's surface. The ROV facilitates the collection of these data without having to use transformation methods. Finally, unmanned air geophysical surveys are extremely cheap (many tens of times less than conventional aircraft surveys, and in many cases are less expensive than a land survey). These ROV geophysical surveys (first of all, magnetic and VLF) must be integrated with ground geophysical investigations and satellite imaging.

5.7.3 Geophysical Examination of Old Caucasian Mine Spoils

Some geophysical methods (primarily, electric and electromagnetic) may be used to examine old Caucasian mine spoils (for instance, such "old" mines in today's terms are located in the Gedabey area of the Lesser Caucasus). The old spoils, according to Samykina et al. (2005) and Surkov et al. (2008) may contain large amounts of economically valuable minerals – in fact on a par with the volume of minerals mined throughout the entire history of humanity.

5.8 Comparison of Regional Fault Dislocations and Distribution of Useful Minerals

Intersections of faulted zones, highly dislocated crust areas and geologically non-trivial areas are key indicators of large endogenic deposits of various types (Khesin 1987). To identify these indicators, the special procedures described in the book were developed. Figure 5.37 shows some of the factors common to controlling both ore and oil-and-gas deposits in Azerbaijan and both ore and underground water deposits.

Chapter 6
The Kura Depression and Adjacent Basins

6.1 The Kura Depression

It was initially believed that the Kura Depression (Basin) (megasynclinorium) (see Fig. 2.4) was composed of thick sedimentary rocks deposited on the crystalline Pre-Alpine basement, and that these megastructures were separated by subvertical deep faults. In 1965, the Saatly superdeep borehole (SD-1) was drilled on the buried uplift of the basement of the Middle Kura Depression (see Sect. 4.4.2.1).

The SD-1 borehole (8,324 m) were discovered Premolassic formations beginning Middle Jurassic which (as Upper Jurassic and Lower Cretaceous) composed by vulcanites. In the SD-1 section more than 5 km thickness vulcanites was found. Khain (2000) noted that these rocks have expressed island arc nature with prevalence of basalts and andesites (with the exception of the Upper Jurassic part). Carbonaceous deposits of Upper Senonian and marls of Lower Paleogene overlie these vulcanites. All these generations compose large gentle folds of north-west strike.

Numerous deep wells have been drilled in this area. They confirm this model of the deep structure of Kura Depression and the occurrence of hidden stocks of Mesozoic magmatic rocks (see Sect. 4.4.2.1). The Inan et al. (1997) findings also mainly confirm this model.

Obviously, the oil and gas potential of the Kura Depression may be enormous (Kerimov 1996). The most recent investigations (e.g., Vincent et al. 2010) indicate that sandstones in the southern part of the Kura Depression have good and very good reservoir petrophysical properties.

L.V. Eppelbaum and B.E. Khesin, *Geophysical Studies in the Caucasus*,
Lecture Notes in Earth System Sciences, DOI 10.1007/978-3-540-76619-3_6,

Results of geophysical observations in boreholes are described in many publications (e.g., Kerimov et al. 1989; Akselrod et al. 1991; Kerimov 1996). Seismic survey results are compiled mainly in Kerimov (1996) and Alizadeh (2005).

6.1.1 Magnetic Survey

Magnetic surveys were used to classify a Mesozoic section into magmatic (magnetic) and carbonaceous (non-magnetic) complexes (for example, Fig. 6.1) according to their composition. As a result, oil and gas traps of a previously unknown type were revealed in zones of carbonaceous rock pinch-outs near the stocks of igneous rocks and in eroded roofs of these stocks (Khesin et al. 1983).

6.1.1.1 Magnetic Field Transformations

It is well-known that hydrocarbon deposits typically manifest in a magnetic field in the form of low amplitudes where a weak negative field over a pool is bordered by ring of positive values (e.g., Donovan et al. 1979). This pattern is in line with the notion of a secondary generation of magnetic minerals caused by hydrocarbons migrating along subvertical zones (methane and hydrogen sulphide) confined to peripheral parts of the deposit whereas in the central part the hydrocarbon migration is less intensive.

This superimposition of the observed magnetic anomalies often makes it difficult to differentiate these circular structures from subsurface inhomogeneties. Therefore, in such situations special methodologies are used (transformations, filtering, upward continuation, computing difference anomalies, etc.) to separate the useful signals from the noise background (e.g., Khesin 1976; Khesin et al. 1988; Eppelbaum and Mishne 1995; Eppelbaum 2007b; Eppelbaum et al. 2008).

A high-precision differential geomagnetic survey was carried out in the North Caspian to locate buried hydrocarbon deposits (Brusilovsky et al. 2009). The authors noted that after transformations (including "reduction to the pole") fragments of circular structures associated with the targets were reliably detected. This type of methodology could be effectively applied to hydrocarbon search in the Middle Kura Depression.

6.1.2 Gravity Survey

One of the most effective gravity surveys in the Middle Kura Depression was carried out by Tzimelzon (1965). His gravity data analysis delineated the Muradkhanly anticline structure (Fig. 6.2) where later an oil deposit was

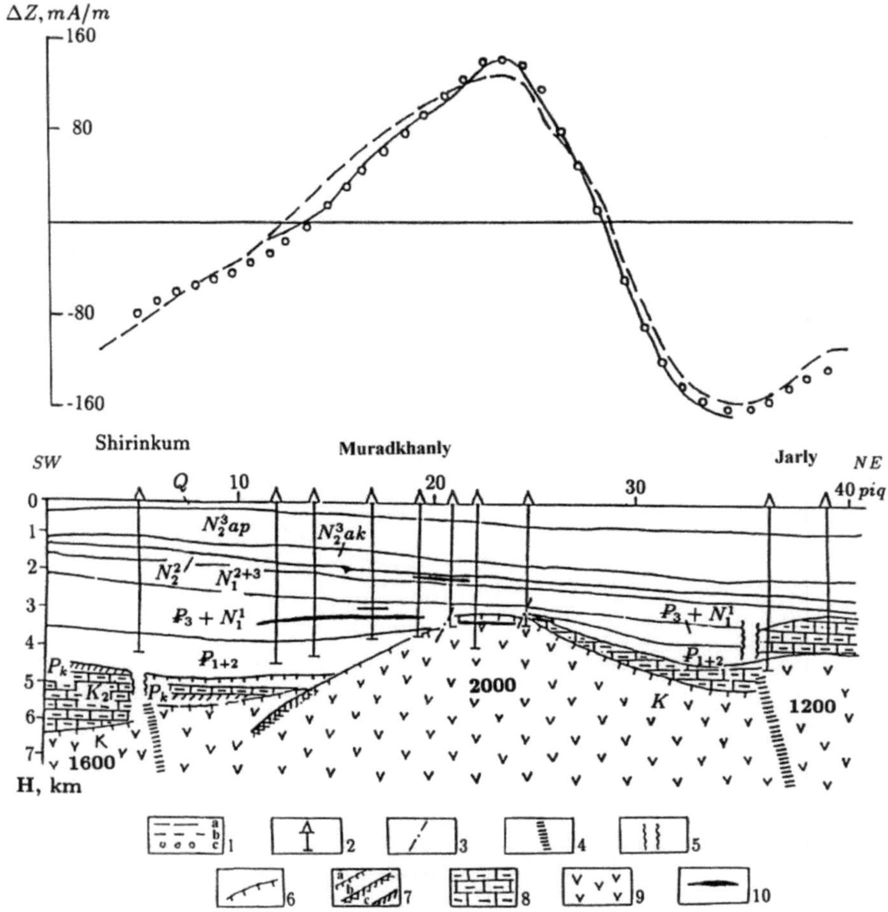

Fig. 6.1 Geological-geophysical section of a profile across the Shirinkum – Muradkhanly – Jarly areas (Middle Kura Depression) (Khesin and Eppelbaum 1997). (*1*) ΔZ_a curves: (*a*) and (*b*) observed, (*c*) calculated; (*2*) deep boreholes; faults revealed by the data from: (*3*) drilling and seismic prospecting; (*4*) gravimetric and magnetic prospecting; (*5*) zones of complicated seismic recording; (*6*) location of magmatic rocks roof determined by modeling; (*7*) conventional seismic horizons: (*a*) *Mz* roof, (*b*) *Mz* volcanogenic rocks roof, (*c*) *Mz* carbonaceous rocks roof; (*8*) carbonaceous-terrigeneous rocks; (*9*) magnetized magmatic rocks (in the figure the magnetization ***J*** is given in mA/m); (*10*) oil-bearing layers

discovered. More recently, Gadirov (2009) presented a more detailed physical-geological model of the Myradkhanly deposit on the basis of gravity (Δg_B) and magnetic (ΔZ) modeling. Amiraslanov (1990) identified several dozen local gravity maxima in the central part of the Middle Kura Depression joined to 20 anticline belts. According to this author, these gravity maxima correspond to uplifts and outcrops of Mesozoic (mainly Jurassic) rocks.

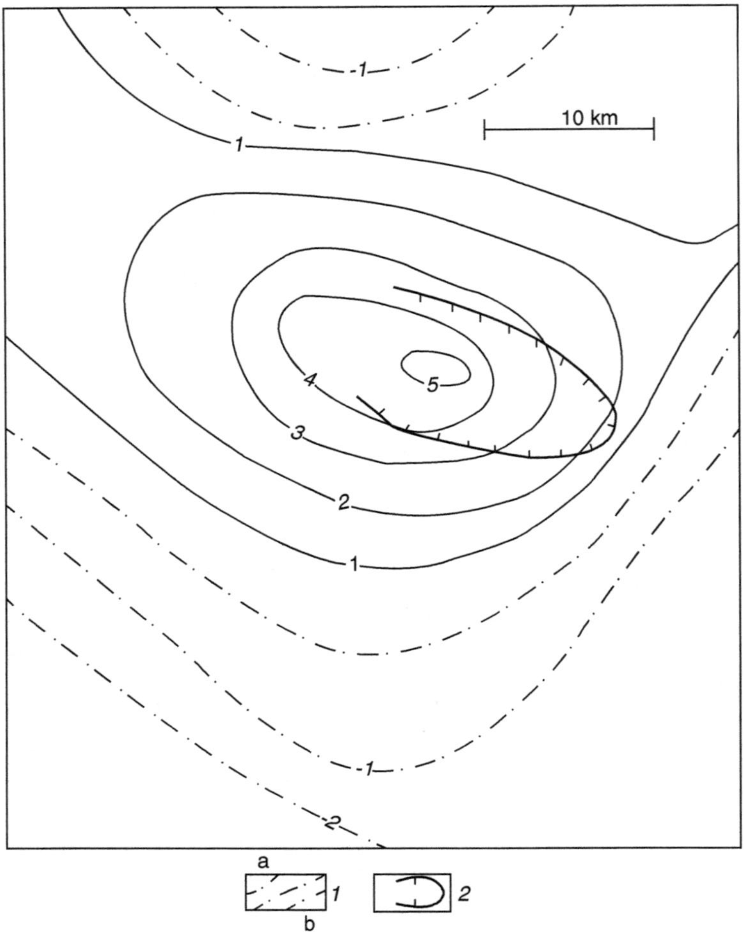

Fig. 6.2 Gravity method as a successful tool for delineation of the buried Muradkhanly structure (After Tzimelzon 1965). (*1*) difference gravity field Δg_{0-10}, mGal: (**a**) positive, (**b**) negative; (*2*) contour of Muradkhanly uplift according to seismic prospecting data and drilling

6.1.2.1 Gravity Field Transformations

Regional gravity field transformations (upward continuation) and calculating difference anomalies for Azerbaijan and adjacent areas are analyzed in Chap. 4 and in Eppelbaum and Khesin (2011).

Observed and analytically continued at different levels (1, 2.5, 4, 6, 8 and 20 km) the gravity field was employed as for direct analysis, and then for computation of difference gravity anomalies ($\Delta g_0 - \Delta g_6$, $\Delta g_0 - \Delta g_8$, etc.) in the Middle Kura Depression (e.g., Nasruyev et al. 1975; Amiraslanov 1987). As a result, several previously unknown local gravity anomalies were detected and geologically interpreted.

Kadirov (2000) applied the Hartley transform to separate the gravity field in the Shamakha-Gobystan and Absheron regions. The author's analysis of the gravity field power spectrum indicates that the average depth of the targets ranged from 24.5 (crystalline basement surface) to 3.2 km (surface within the Cenozoic deposits). According to Kadirov (2000), most of the regional anomalies are caused by the topography of the Pre-Mesozoic crystalline basement.

The regional gravity component was removed using correlation methods and maps of local gravity anomalies were constructed for the northern Naftalan-Gedakboz (Middle Kura Depression) and Byandovan (Lower Kura Depression) areas (Gadirov 2009). These local gravity anomalies (≈ -0.2 mGal) in the SW part of the Byandovan area may reflect early unknown hydrocarbon deposits.

6.1.3 Thermal Data Analysis

Batsevich (1881) was apparently the first to carry out scientific thermal (and applied geophysical as a whole) measurements in the Caucasus. He made temperature measurements at different depths in oil wells of the Sabunchy area (Absheron Peninsula). These investigations were successfully continued by Stopnevich (1913) and Golubyatnikov (1916).

6.1.3.1 Near-Surface Temperature Survey

This method was applied in the Muradkhanly oil deposit in Central Azerbaijan (Fig. 6.3), where the temperature was measured in 3 m deep wells and the data thus obtained were smoothed by the use of a sliding interval average of three points. A fault was discovered by a field crew from the "YuzhVNIIGeofizika" (Baku) using gravity and magnetic methods, which was confirmed by drilling data. Thermal observations were also conducted by the "YuzhVNIIGeofizuka" (Sudzhadinov and Kosmodemyansky 1986). The position of the upper edge of the fault was calculated using the tangents and characteristic points method (Khesin and Eppelbaum 1994); its projection in plane coincided exactly with the fault position indicated by the independent geophysical and geological data.

6.1.3.2 Regional Temperature Analysis

On the basis of analysis of numerous temperature data in boreholes of Azerbaijan Ovnatanov and Tamrazyan (1970) predicted that deep tectonic signatures could be obtained by examining thermal data in wells with depths of 1–2 km. This would make it possible to localize hydrocarbon deposits at a depth of 3.5–4.5 km (and more) and estimate the occurrence of buried uplifts in the lowermost strata.

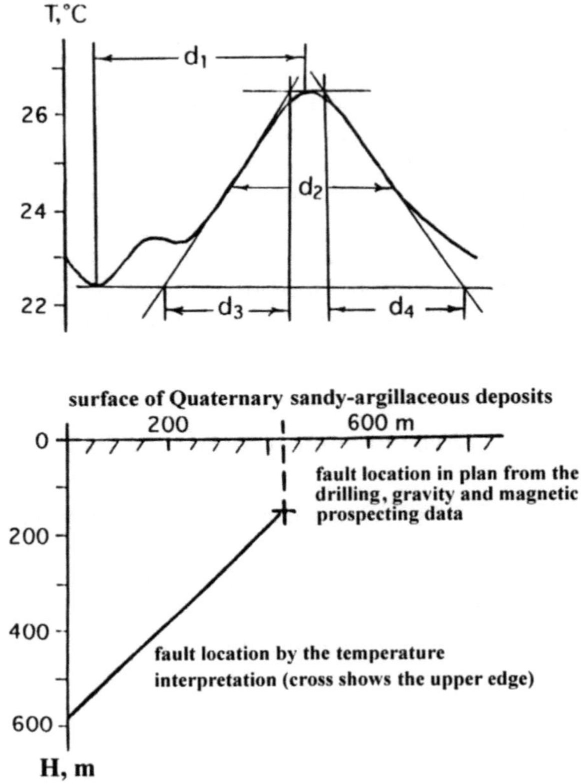

Fig. 6.3 Interpretation of the temperature anomaly in the district of the Muradkhanly oil field in the Middle Kura depression (Khesin and Eppelbaum 1994)

Values of the vertical geothermal gradient in the Middle Kura Depression (Azerbaijan) change at depth intervals of 0–2,000 m and 4,000–6,000 m from 25–43°C/km to 20–30°C/km (e.g., Kerimov et al. 1989). Using geothermal data for the Southern Caucasus, the depths of isotherm 200°C (H_1), 400°C (H_2) and the Curie temperature (H_c) for the Middle Kura Depression were determined. The findings are presented in Table 6.1. Values of H_m indicate the depth of the magnetized bodies' lower edges on the basis of magnetic data (Pilchin and Khesin 1981).

Table 6.1 shows that depth H_1 for isotherm 200°C increases from 5.5 km in the west to 7.5 km east of the Middle Kura Depression. Depth H_2 of isotherm 400°C also increases from 13 to 14 km in the west to 21–22 km in the east of the Middle Kura Depression. Therefore, in the Middle Kura Depression the most probable depth interval for the surface of wustite (iron suboxide) decomposition is the interval from 5.5÷7.5 km to 13÷22 km.

It can be seen from Table 6.1 that this interval is consistent with the depth of the magnetized bodies' lower edges calculated using magnetic data. It is obvious that this correlation is not accidental. Furthermore, the depth of the magnetized bodies' lower edges (H_m) only coincides with the depth of the Curie surface (H_c) in the area

Table 6.1 Depths of the Curie point isotherm (200 and 400°C) and lower edges of magnetized bodies in the Middle Kura depression, Azerbaijan (After Pilchin and Khesin 1981; Pilchin and Eppelbaum 1997)

Name of magnetic anomaly	H_1 (200°C), km	H_2 (400°C), km	H_C, km	H_m, km
Ismailly	5.5–6.5	14	25	14
Shamkhor	5.0–6.0	12	18	20
Sarkyar	5.5	13	22	18
Borsunly	5.5–6.5	14	24	16
Lyaky	5.5–6.5	15	30	22
Karajaly	5.5–6.5	15	30	14
Sor-Sor	6.0–7.9	16	30	13
Ragimly	6.0–7.0	16	32	12
Imishly	7.5	20	44	11
Comushly	7.5	19	42	14
Saatly	7.5	18	40	9
Levonarch	5.5–6.6	14	24	16
Bashkarvand	5.5–6.5	15	26	10
Gindarch	5.5–6.5	15	25	15
Beilagan	6.5–7.0	15	25	26
Pervomai (old name)	6.5–7.5	19	42	7

of three magnetic anomalies (Shamkhor, Beilagan and Zardob). These anomalies are located in the zone of large faults, where conditions are favorable for the transformation of iron suboxide into iron oxide (Pilchin and Eppelbaum 2006). Oxidation conditions in such zones may be present at great depths. The depth of the magnetized bodies' lower edge for 14 other magnetic anomalies reflects the depth of the interval by the following eutectic reaction (Pilchin and Eppelbaum 1997):

$$4\mathrm{Feo} \leftrightarrows \mathrm{Fe} + \mathrm{Fe_3O_4}.$$

Heat flow and geothermal gradients for some areas of the Kura Depression and the South Caspian depression are listed in Table 6.2.

Pilchin (1983) calculated values of geothermal gradient for the Middle Kura Depression as a function of depth. For instance, for the depths intervals 0–2,000, 2,000–4,000, and 4,000–6,000 m these values consist of 25–43, 20–35, and 20–30°C/km, respectively.

6.1.4 Radiometric Survey

Hydrocarbon deposits within the Absheron Peninsula and Kura Depression are reflected in the gamma field by typical oval shaped negative anomalies with a smooth drop of gamma radiation intensity (Aliyev and Zolotovtskaya 1996). In the Middle Kura Depression several geological structures with the fluent fall of 1–1.5 μR/h were delineated after reducing the gamma background. Within the

Table 6.2 Values of heat flow and geothermal gradient for some regions of Azerbaijan (After Pilchin 1983; Eppelbaum and Pilchin 2006)

Area	Heat flow, (mW/m^2)	Geothermal gradient, ($^{\circ}C/km$)
Middle-Kura depression, Azerbaijan (western part)	45–104	30–43
Middle-Kura depression, Azerbaijan (eastern part)	33–50	20–30
Lower-Kura depression, Azerbaijan	17–42	10–25
South-Caspian depression, Azerbaijan	17–42	10–20

Muradkhanly structure, the gamma field consists of 4.5–7 μR/h. It was suggested that gamma surveys in geophysical integration increase the likelihood of detecting hydrocarbon deposits (Aliyev 1994).

6.1.5 Integrated Analysis

Seismic and gravimetric prospecting methods have revealed Mesozoic associations with high velocities of elastic waves and densities under the Cenozoic terrigeneous cover. Similarly magnetic prospecting divides the Mesozoic associations into magmatic and carbonaceous according to their composition. As a result, oil-and-gas traps of a previously unknown type there were revealed in the zones of carbonaceous rock pinch-outs near the highs of magmatic associations and in eroded roofs of these highs (Khesin et al. 1983). A respective scheme on the basis of Kh. Metaxas' data (Khesin et al. 1996) is shown in Fig. 6.4.

Figure 6.5 presents a generalized distribution of Mesozoic magmatic highs in the zones of the most active faults (Gasanov 2001).

Figure 6.6 illustrates the relation between hydrocarbon accumulations in the Kura Basin and regional faults. The map shows many faults with different orientations and styles of movement (sublatitudinal, Caucasian, Anticaucasian, submeridian, etc.) and depths of occurrence (from pre-Mesozoic to the younger Cenozoic). These findings were used to examine the fault systems and oil-and-gas content in eastern Azerbaijan (Zeinalov 2000).

6.2 South Caspian Basin

Investigating the South Caspian Basin to study deep structure and localize commercial oil-and-gas deposits is of great interest to geoscientists (e.g., Bagirov et al. (1998), Abdullayev (2000), Katz et al. (2000), Buryakovsky et al. (2001), Diaconescu et al. (2001), Jackson et al. (2002), Allen et al. (2003), Brunet et al.

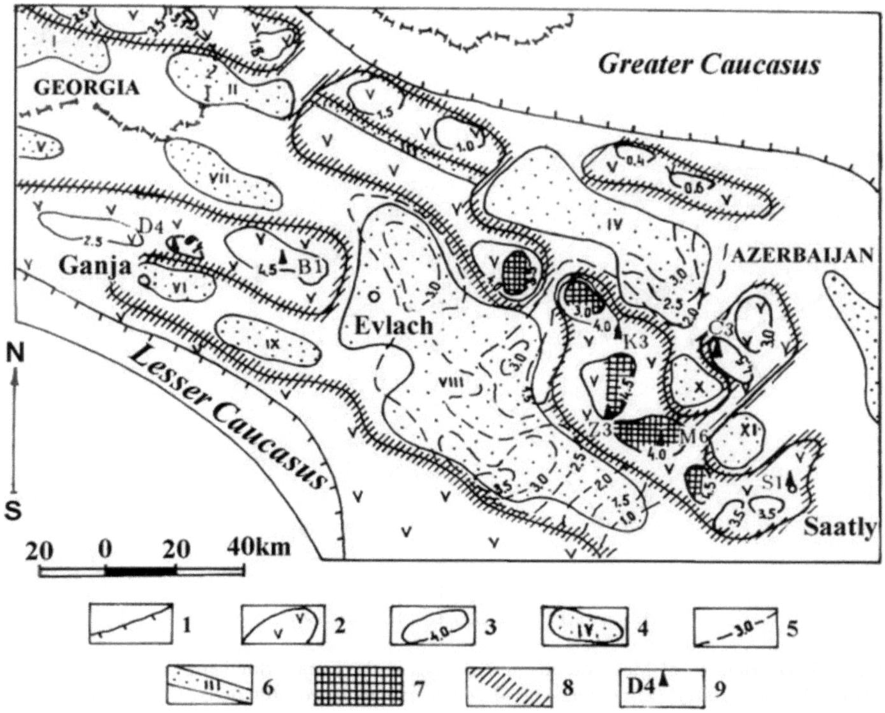

Fig. 6.4 Schematic chart of oil-and-gas prognosis for the Middle Kura depression using the interpretation of magnetic and other geophysical data (Khesin et al. 1996). (*1*) contour of mountain structures; (*2*) contour of Mesozoic magmatic associations of basic and intermediate composition according to magnetic prospecting data; (*3*) isodepths of magnetized (mainly, effusive) Mesozoic rocks according to magnetic prospecting data; (*4*) basins of normal sedimentary Meso-Cenozoic rocks according to a set of geological and geophysical data; (*5*) isopachs for an Upper-Jurassic-Cretaceous complex of normal sedimentary deposits according to seismic and magnetic prospecting data; (*6*) part of the North-Kura fault zone with the most intensive movements (according to gravimetric and magnetic data); (*7*) most promising areas for detecting oil deposits in eroded magmatic rocks in arch structures; (*8*) most promising areas for detecting oil-and-gas deposits mainly in traps of a non-structural type (on the various pinched-out rock associations); (*9*) wells for deep drilling

(2003), Knapp et al. (2004), Manley et al. (2005), Artyushkov (2007), Granath et al. (2007), Egan et al. (2009), Green et al. (2009)).

In the Caspian Sea area, adjacent to Azerbaijan from the north-east and east, large areas of the Middle Caspian relating mainly to the epi-Hercynian platform and the South Caspian Basin (SCB) associated with the Alpine-Himalayan orogenic belt (Khain 1984) have been pinpointed. The SCB appears to behave like a relatively rigid aseismic block within the otherwise deforming Alpine–Himalayan orogenic belt. The comprehensive seismic data suggests there is a fundamental compositional difference between the crust of the south Caspian basin and that of the surrounding region (Mangino and Priestley 1998).

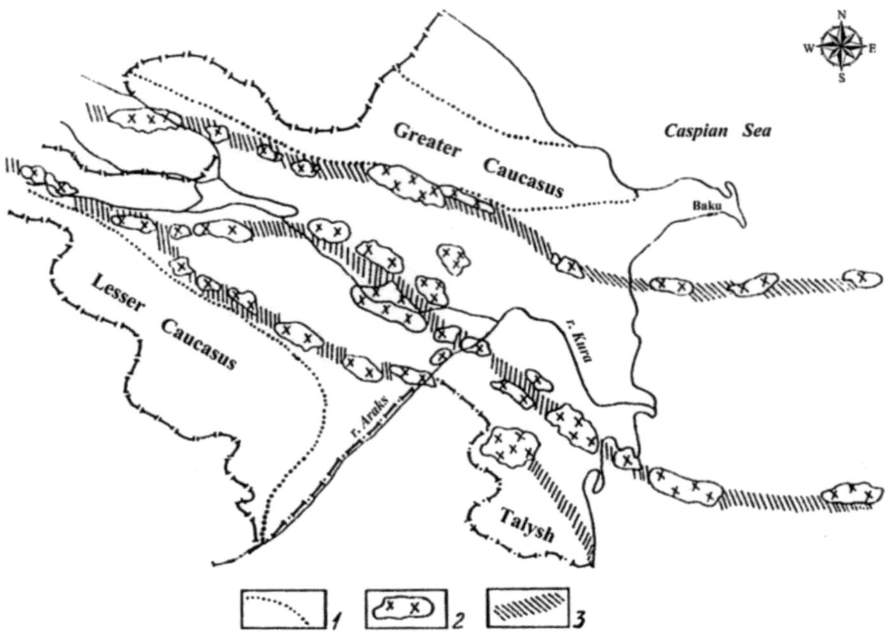

Fig. 6.5 Distribution of Mesozoic magmatic highs in the Kura – Southern Caspian Depression (After Gasanov 2001). (*1*) contours of Mesozoic mountain constructions; (*2*) magmatic rocks highs; (*3*) zones of deep faults along which magma substances erupted in the Mesozoic

The central part of the Caspian Sea is characterized by a mosaic magnetic field; within the uplifted Kara-Bogaz-Middle Caspian block a close association of negative and positive magnetic anomalies is found. This is typical in particular of the more elevated Kara-Bogaz arch where Permian-Triassic and Jurassic rocks are absent and terrigeneous low-magnetic rocks occur directly on the Pre-Permian basement.

The SCB is one of the deepest basins in the world with a depth to basement of over 20 km in places (Shikhalibeily and Grigoriants 1980; Brunet et al. 2003). Although it is widely accepted that the SCB was initiated by a Mesozoic back-arc extension related to the subduction of the Tethys Plate (e.g., Zonenshain and Le Pichon 1986), more than half of the 20 km subsidence presently observed occurred within the tectonic framework of the evolution of the Alpine–Himalayan orogenic belt (Egan et al. 2009). The SCB is characterized by a smoothed positive magnetic field intensity which decreases to the south along the common abrupt subsidence of the surface of crustal basement and a sharp increase in sedimentary Mesozoic-Cenozoic complexes. In the western part of the SCB (area of the Baku Archipelago) there are a large number of local positive and negative anomalies of small intensity (up to 100 nT) reflecting geological peculiarities of the sedimentary association structure.

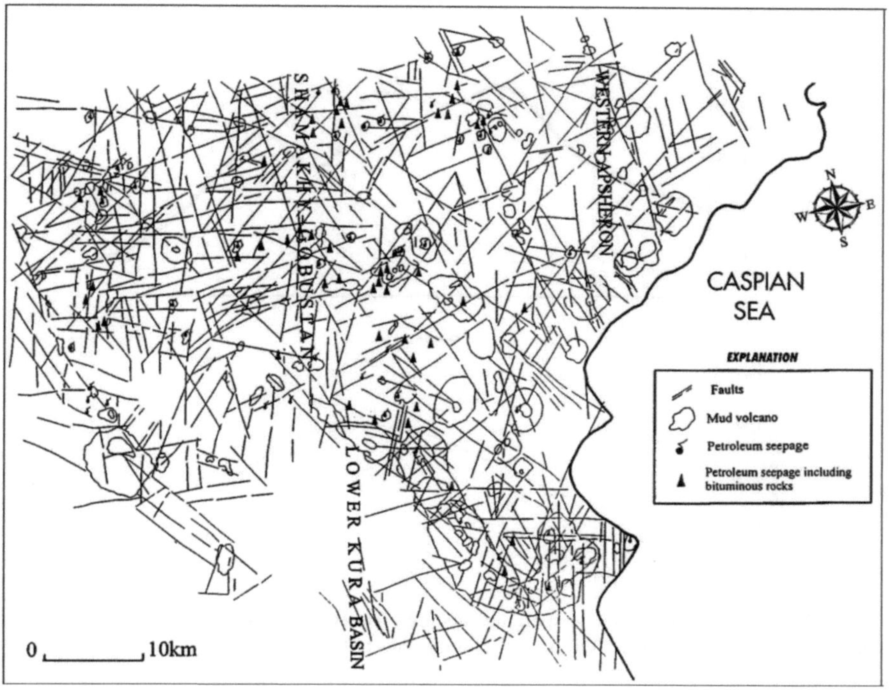

Fig. 6.6 Map of faults from analysis of Landsat data for central–eastern Azerbaijan (After Zeinalov 2000, with minor modifications)

Interestingly, up to now the boundary between the Epi-Hercynian platform and Alpine-Himalayan orogenic belt in the Caspian Sea has not been completely determined.

Sub-meridional zones in the SCB were identified by the combined geological, seismic and potential geophysical field analysis. The western part of the SCB is fixed by clear meridional trends of geological structures; the eastern part of the SCB is divided into two zones: the western zone is characterized by comparatively smoothed tectonics, and the western part by more intensive tectonics.

The geological nature of the regional magnetic maximum in the central part of the SCB is still being debated. According to Dzabayev (1969), the depth to the upper surface of the anomalous body is 17–19 km, which coincides with the depth to the crystalline basement found by deep seismic sounding. Dzabayev (1969) connected this anomaly to the influence of a complex of magmatic rocks of basic consistency occurring in the body of the basement. This point of view was confirmed by the behavior of the magnetic field continued upward to a level of 4 and 10 km: parameters of magnetic maximum change weakly with the level of upward continuation. At the same time, a careful analysis of this anomaly indicates that it is composed of several different anomalies. It was clearly detected from a combined interpretation of residual (0–10 km) anomalies and the regional magnetic field at

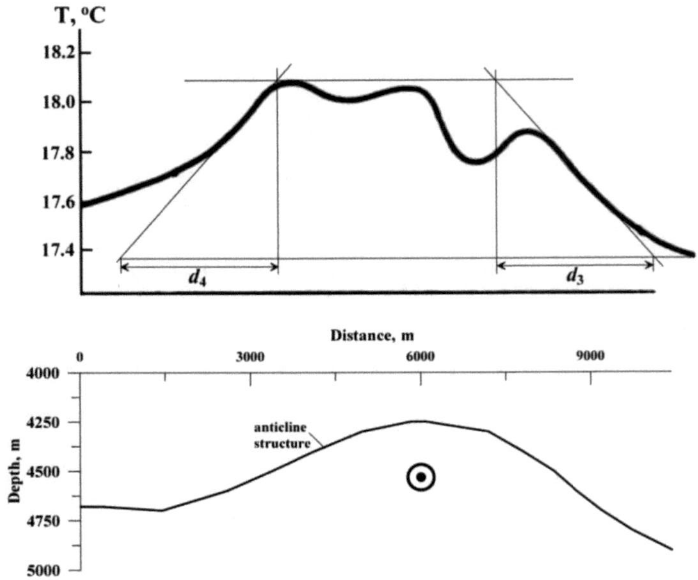

Fig. 6.7 Interpretation of the temperature anomaly observed in near-bottom sediments over the oil&gas Bakhar deposit (Azerbaijan) (Initial data after Artemenko and Malovitsky 1977). The "⊙" *symbol marks* the position of the *center* of anticline structure

a level of 10 km. It should be noted that the maximum of the magnetic field is localized in the northern area of a large gravity maximum corresponding to the western part of the SCB. A similar interrelation of gravity and magnetic anomalies was studied in the Middle Kura Depression and was explained by the physical-geological model when younger (Mesozoic) magmatic rocks were introduced into a weakened zone on the peripheral part of the uplift of the Pre-Alpine basement.

The Pliocene-Anthropogenic deposits in the SCB are the thickest. The paleomagnetic characteristics of these deposits on the eastern and western sides of the SCB are identical. This enables interpretation by applying the principles of directly magnetized deposits (Baku, a small interval of Absheron and the lower part of Ahchagyl) and inversely magnetized deposits (Absheron, upper part of Ahchagyl and productive red bed for Western Turkmenistan) (Ismail-Zadeh 1983 and his unpublished reports (1972–1980)). The average values of magnetic susceptibility usually range from 1,000–2,000 (up to 5,000)$\cdot 10^{-6}$ SI in Eastern Azerbaijan and 1,000$\cdot 10^{-6}$ SI in Western Turkmenistan.

It is interesting to note that the Western Caspian Sea Basin and the Dead Sea Basin have several common geothermal and tectonic features (Eppelbaum and Pilchin 1998) that can lead to joint geological-geophysical data examination.

One example of quantitative examination of the near-bottom temperatures observed in the SCB over the oil&gas Bakhar deposit (Fig. 6.7) is discussed below. The amplitude of the anticline structure where this deposit occurs is more than 600 m. The accuracy of the near-bottom single temperature measurement was

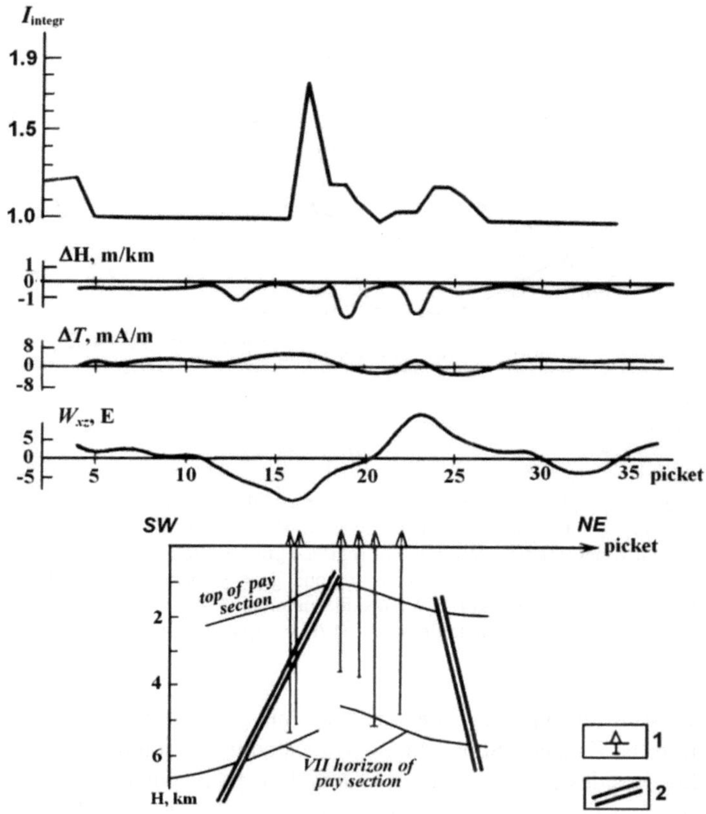

Fig. 6.8 Qualitative delineation of faults by computing parameter I_{integr} at the area of Bulla (Baku Archipelago, Caspian Sea). (*1*) deep wells; (*2*) faults delineated by combined analysis of geological and geophysical data

within 0.01°C (Artemenko and Malovitsky 1977). As a result of quantitative interpretation (a model of a horizontal circular cylinder was used; see Sect. 3.4.2) the center of the anticline structure was identified. It is possible that the small negative temperature anomalies in the central part of the profile are caused by the influence of oil&gas deposits.

At times, integrated geophysical field analysis is extremely useful even at the qualitative level. To calculate the parameter I_{integr} (see Eq. 3.95) in the area of Bulla-Sea (Bay of Baku) three different fields were employed: local magnetic anomalies ΔT (marine survey data), the second horizontal derivative of the gravity potential W_{xz} (data from the bottom gravity survey were utilized), and ΔH/H (relative changes of the sea bottom topography) (Fig. 6.8). As can be seen from the graph of I_{integr}, both faults are reflected. The large amplitude of the SW anomaly of the I_{integr} parameter compared to NE may be explained by the proximity of the SW fault to the surface of the sea bottom.

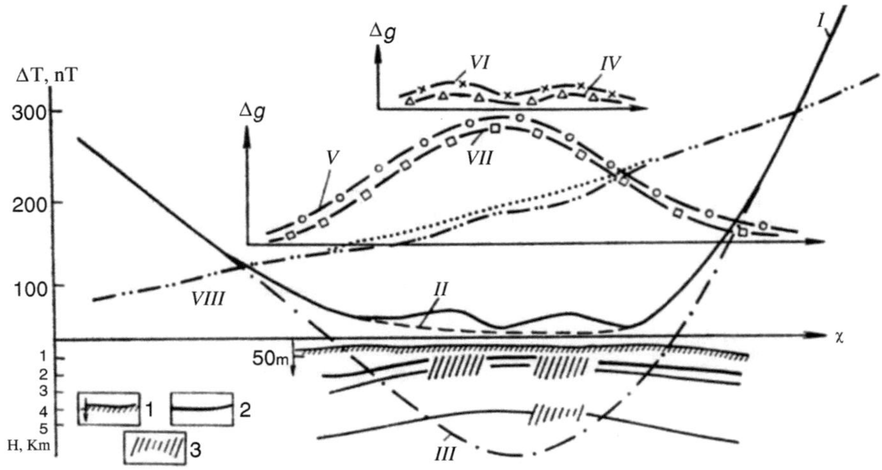

Fig. 6.9 Integrated geophysical-geological section across the Zhdanov Shoal (Malovitsky et al. 1977). (*1*) sea *bottom* (50 m only *marks* the position of the sea *bottom*), (*2*) reflective seismic horizons, (*3*) fault zones. (*I*) Δg_B, (*II*) graph of regional gravity background (parabolic interpolation with average radius of 3 km), (*III*) graph of regional gravity background (parabolic interpolation with average radius of 4.5 km), (*IV*) residual anomaly I–II, (*V*) residual anomaly *I–III*, (*VI*) computed gravity effect from fault zones, (*VII*) computed gravity effect from Cheleken-Livanov high, (*VIII*) graph ΔT

An interesting example of integrated seismic-gravity-magnetic analysis was carried out in the Zhdanov Shoal (Fig. 6.9) located in the south-eastern part of the Caspian Sea. Besides conventional seismic and magnetic surveys, detailed bottom gravimetric investigations were performed (Malovitsky et al. 1977). Quantitative analysis of the bottom gravimetric data revealed a significant local gravity maximum against the regional gravity minimum. It was assumed that this maximum was caused by a gravitational effect of the Cheleken-Livanov high, and other small local anomalies associated with the fault zones (Fig. 6.9). Later in this area the Djigalybek oil deposit was discovered.

The level of the Caspian Sea has fluctuated greatly in the last century with a drop of 3.25 m from 1900 to 1977 and a rise of 42 m from 1977 to 1994. The causes are still poorly understood but seem to have a cyclicity. The Kara-Bogaz Gol (KaBo) bay (literally the “black throat lake”), is connected by a narrow strait to the east of the Caspian Sea and, according to some researchers, significantly decreased its level. level. The KaBo water level closely follows that of the Caspian Sea, but due to high evaporation rates its level remains lower. In addition, the level of the KaBo was artificially altered by a dam built in 1980, which led to an unintentional catastrophic drying out (Leroy et al. 2006). It is known that composition of KaBo water is very close to that of the Dead Sea. Therefore, to investigate KaBo stratification and hydrodynamic patterns a method developed for Dead Sea water flow geodynamics (Eppelbaum and Yakubov 2010) might be applied.

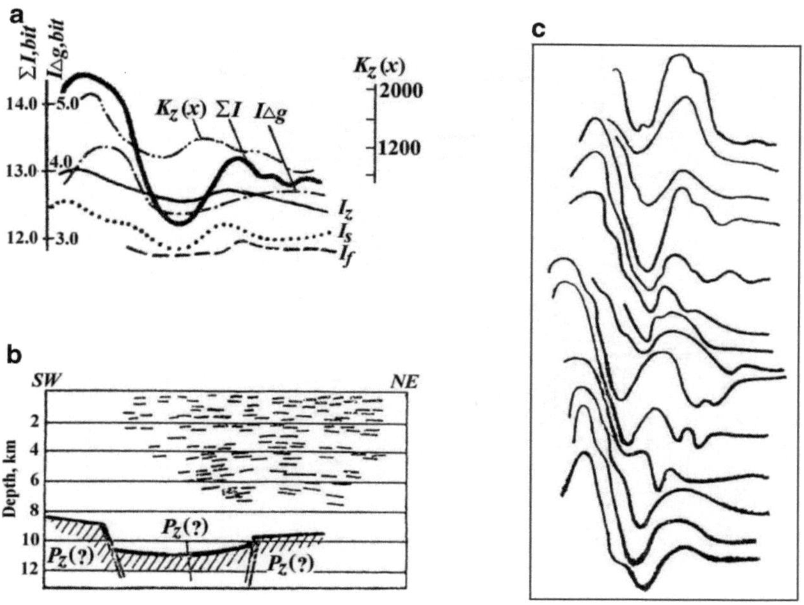

Fig. 6.10 Integrated interpretation of geophysical data using information units for the delineation of hidden fault zones under thick sedimentary cover in the Precaspian-Kuba oil-and-gas area (Kusar-Divitchi Basin) (After Pototsky and Khesin 1971). (**a**) graphs of amount of specific information derived from the results of various geophysical methods and total information $\Sigma I = I_{\Delta g} + I_z + I_S + I_f (\Delta g)$ is the gravity field, Z is the vertical component of the magnetic field, S is the series admittance according to VES data analysis, and f is the observed frequency of reflections (visible frequency of reflected waves) through comparison with the cross-correlation function $K_z(x)$, (**b**) distribution of reflected surfaces along seismic profile No. 85-10-03 and diagram of the proposed basement structure, (**c**) graphs of the total amount of information along 14 profiles including profile No. 85-10-03

6.3 Other Basins

6.3.1 Kusar-Divitchi Basin

The Kusar-Divitchi Basin is located in eastern Azerbaijan (see Fig. 2.4). Integrated analyses of gravity, magnetic, VES and seismic data are shown in Fig. 6.10 (the applied methodology is presented in Sect. 3.8.7). Note that parameter ΣI clearly reflects the position of the predicted basement structure P_z.

6.3.2 Ossetia Depression

A comprehensive investigation of the deep structure of the Ossetia Depression (northern Caucasus) and perspectives for hydrocarbon discovery were presented in

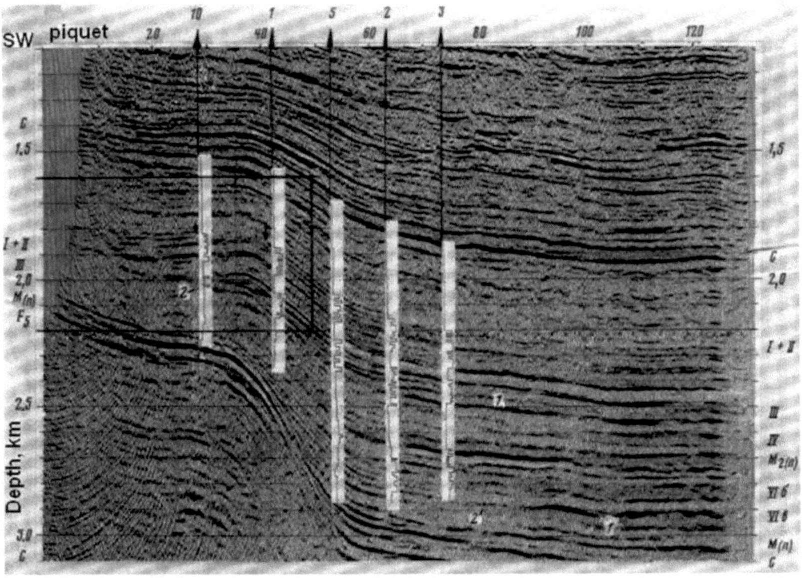

Fig. 6.11 Revealing facies replacement in the West-Kuban depression: Time section with curves of elastic impedance (Profile № 028111). Numbers indicate zones of sandstone development and replacement (Binkin et al. 1986)

Shempelev and Nevsky (2008). Based on numerous results from seismic, magnetotelluric, magnetic and gravity studies the authors developed several combined physical-geological models. For instance, these authors indicated that under the Ossetia Depression which is composed of Paleogene-Quaternary sediments, there is a horizontal low-resistivity layer (< 10 Ω·m) was detected at a depth of 6–13 km. It was presumed that this layer was composed of interbedding of anhydrides, halite, and gypsum, and hydrocarbons were thought to be associated with the underlying deposits. At the same time, hydrocarbon pools may occur in various near-fault zones, some anticline areas, and structural-tectonic blocks (Shempelev and Nevsky 2008).

6.3.3 *Taman and Kuban Basins*

The Taman Peninsula is located under a complex tectonic pattern that involves the coupling of the Caucasus and Crimea mountains, and the West-Kubansky and Sorokin regional troughs (the seismic profile across the Taman Peninsula is shown in Fig. 2.8).

A new interpretative approach was applied to obtain a precise gravity dataset for the peninsula (Chernov 2008). On the basis of the horizontal derivatives, strip filtration, and other parameters, a model of the isoprobabilty of hydrocarbon deposits was constructed. Geophysical methods are actively used in the Taman Peninsula to investigate mud volcanism (e.g., Sobisevich et al. 2008) which directly correlates with hydrocarbon accumulation.

The Kuban Basin occupies almost half of the eastern part of the Azov Basin coastal plain (see Fig. 2.4). The Azov-Kuban trough is the western part of the North Caucasian foredeep. This area is characterized by a relatively simple asymmetrical structure with a gentle and wide northern flank, and a very narrow and steep southern one that corresponds to the location of the main back thrust system (Mikhailov and Smolyaninova 2002).

Geophysical investigations to reveal oil-and-gas structures have been conducted in this area for many years. An example of seismic data processing in the sand complexes of the West-Kuban depression is shown in Fig. 6.11. To delineate the facial replacement of the Maikop sand horizons (deposited in conditions of underwater hydrodynamic flows) a combined seismostratigraphic, seismofacial, kinematic and dynamic analysis was used (Binkin et al. 1986).

The Kuban Basin is estimated at present to be a high-perspective area for discovering new commercial hydrocarbon deposits (e.g., Khain et al. 2005; Klett et al. 2011).

Chapter 7
Geophysics in Hydrology

7.1 Main Specifities of Geophysical Prospecting in Hydrogeology

7.1.1 Methodological Principles of Geophysical Prospecting for Underground Waters

The Transcaucasian and areas of the Greater Caucasus within the territories of Azerbaijan and Georgia alone incorporate more than 53,000 rivers, lakes, water reservoirs, glaciers and swamps with total water reserves of 168 km^3 (Svanidze and Tzomaya 1988). It is obvious that the total Caucasian water reserves consist of much larger volumes.

Hydrogeological studies require a large volume of drilling and other expensive operations, similar to petroleum exploration. Actually, revealing water flow ways, water collectors and impermeable strata, as well as direct water detection is similar to indirect and direct hydrocarbon prospecting. The use of geophysical methods cuts down drilling time, increases the uniformity of the medium, and reduces total expenses and environmental damage (Shakhnazaryan and Khesin, 1967; Melkanovitsky et al. 1982; Khesin and Shakhnazaryan 1986; Melikadze et al. 2005). Hydrogeophysics methodology is well developed for platform conditions with large structural forms, stable petrophysical sections and favorable near-surface environments. The geological-geophysical characteristics of sub-mountain and mountain regions including the Caucasus are more complicated, and the hydrogeophysics methodology is specific to it.

The development of new oil production and transportation techniques in the Caucasus prompted the need to provide reliable groundwater protection (e.g., Aliev and Askerov 2002). The most expedient is to use modern wavelet procedures for various geophysical method integration (e.g., Eppelbaum et al. 2011a, 2011b).

Underground fresh and/or low mineralization waters in mountain and sub-mountain regions are mainly distributed within alluvial fans and artesian basins. The upper part of these multilayer systems is the most valuable (zone of active

L.V. Eppelbaum and B.E. Khesin, *Geophysical Studies in the Caucasus*,
Lecture Notes in Earth System Sciences, DOI 10.1007/978-3-540-76619-3_7,

water exchange) and is studied to find solutions to water supply, land reclamation and irrigation. The key features of these systems are heterogeneity and variability. Water-bearing deposits in the foothills are mainly represented by boulder pebbles with thin clay bands (upper molasses formation). As it moves off the mountain frame, the content of thin-grain matter (clay particles) in the water-bearing deposits increases. These lithological variations are reflected in the differentiation of petrophysical characteristics, and first of all, by rock resistivity (Fig. 7.1).

It is clear that boulder-pebble deposits have the highest resistivity. Clayey deposits are characterized by low resistivity. In general, the electrical properties of rocks are the most sensitive to their lithological composition, water content and its mineralization. The increase in mineralization decreases water-bearing collector resistivity, whereas their polarizability (chargeability) intensifies the IP effects over pyrite-rich disseminated ores. Self-potential anomalies indicate sites of water filtration (e.g., Parasnis 1986). These features explain the widespread application of electric prospecting methods especially since similar applications of seismic prospecting are complicated by hard near-surface seismical-geological conditions and considerably higher cost. Quantitative interpretation of vertical electric sounding (VES) gives the thickness and true specific resistivity (ρ) of each geoelectric layer. Comparison of these data with drilling results showed the dependence between the concentration of gravels and sands in the studied layers (C) and their ρ in terms of the log-normal statistical distribution for gravel resistivity and alluvial suite thickness (Khesin and Shakhnazaryan 1986):

$$C = 8.67\rho^{0.4}. \tag{7.1}$$

where $C \geq 8.67\%$, and ρ varies from 1 up to 450 Ω·m.

At the same time, other geophysical methods can supply important information on the deep structure that often controls the spatial distribution of aquifers and aquicludes (Fig. 7.2). In some cases, these methods facilitate the lithological separation of water-bearing sections. For example, boulder gravels and clays are differentiated by their low and high velocity of compression (longitudinal) waves, respectively (Fig. 7.2).

Shear (transverse) waves do not pass through liquids and are quickly attenuated in water-rich rocks. The velocity of compression waves increases sharply at the boundary between dry and water-rich rocks. Such effects facilitate detailed seismic prospecting for detection of the groundwater level. Seismic methods are known to be highly useful for the study of deep structures. Gravity prospecting is also effective for the structure characterization of the dense basement underlying relatively loose water-bearing cover. Magnetic prospecting is usually required to reveal igneous rocks of basic composition that are sometimes water collectors or barriers to water flows; they often fill fault zones within the crystalline basement. Basement rocks can be an additional source of fresh water, particularly, in areas of carbonate rocks and fault zones.

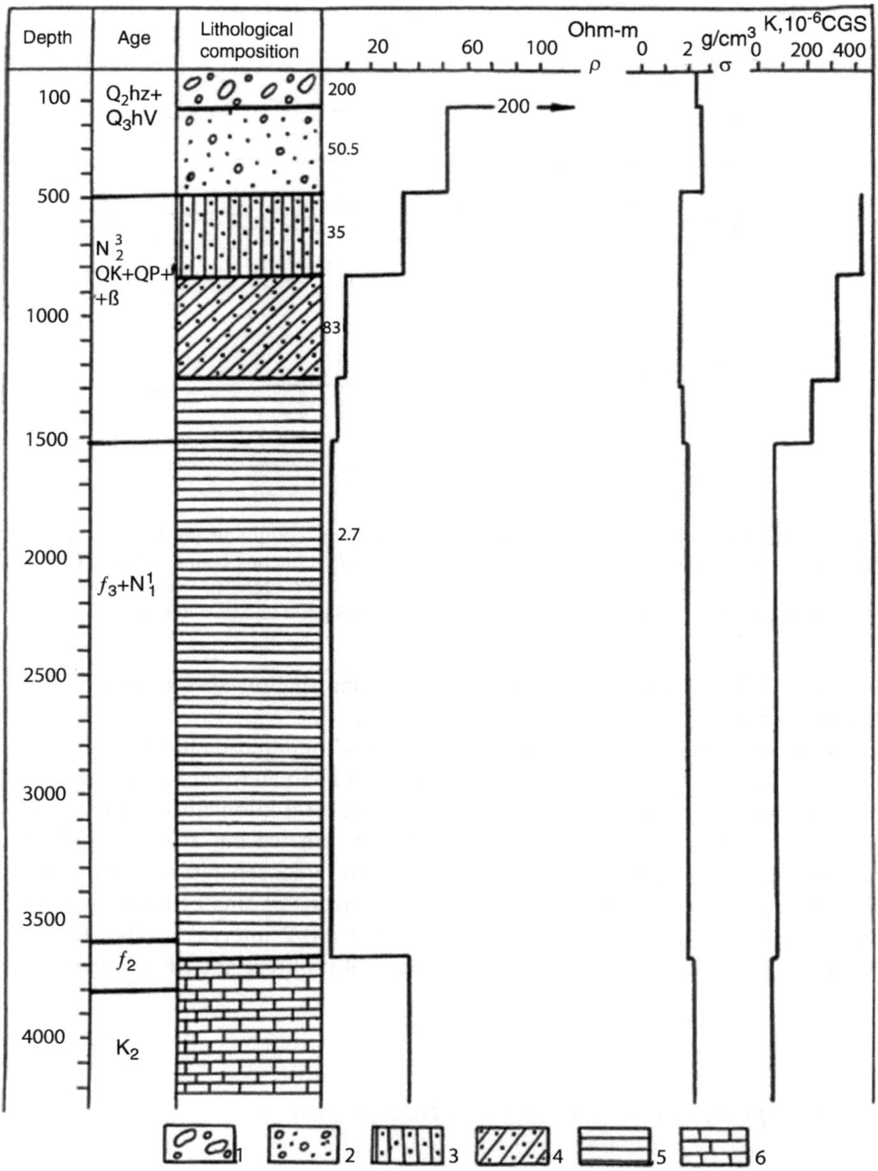

Fig. 7.1 Generalized petrophysical column of the Mesozoic-Cenozoic deposits in the north-eastern immersion of the Lesser Caucasus (Ganja region). (*1*) boulder pebbles; (*2*) sand pebbles; (*3*) chiefly sand deposits; (*4*) clayey-sand deposits; (*5*) sandy-clay deposits; (*6*) chiefly carbonate deposits

Deep fault zones and basement highs often contain thermal waters. Near-surface thermal prospecting can detect descending or rising water filtration by negative or positive anomaly, respectively. Thus, the integration of geophysical methods is

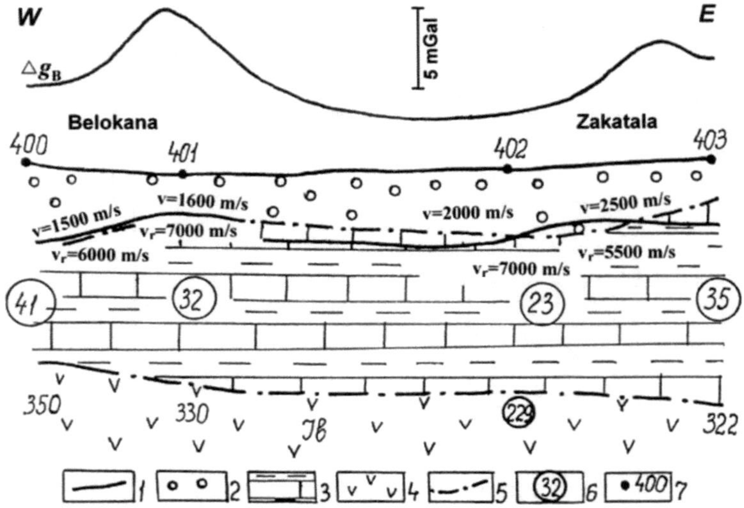

Fig. 7.2 Geological-geophysical section along the southern foothills of the Greater Caucasus. Note: the vertical scale of the section is multiplied by 10. (*1*) refracting boundary; (*2*) Quaternary boulder pebbles; (*3*) Cretaceous carbonate-clay deposits; (*4*) Jurassic volcanogenic rocks; (*5*) geoelectric layer boundary; (*6*) ρ, Ohm·meters; (*7*) VES point and its number

worthwhile for an exhaustive study of the deep and shallow specificities of water-bearing sections.

Water exploration can reach a depth of several hundred meters (usually up to 300 m); the extent of alluvial fans (foothills of the Greater and Lesser Caucasus) and artesian basins (Kura Depression) go down several kilometers or more. Thus, integrated geophysical studies based on VES are carried out at larger and more regional scales (usually, 1:100,000–1:200,000), and VES separations often attained 5–10 km. For the detection of water-bearing horizons of small thicknesses, additional conventional separations are added in the AB/2 interval of 500–1,500 m. Sometimes, the VES are combined with electrical profiling and SP methods.

7.1.2 Specificities of VES Data Interpretation

The key types of analysis of highly variable geoelectric sections are the following: (1) statistical analysis of geoelectric layer parameters, quantitative criteria for section differentiation and horizon correlation; (2) enhancement of qualitative techniques to identify water controls; (3) determination of statistically significant relations between geophysical (resistivity, polarizability, velocity of seismic wives) and hydrogeological (lithology, filtration properties, water mineralization, groundwater level) parameters (Khesin and Shakhnazaryan 1986).

The geoelectric layer is a rock association with similar lithology and mineralization of the waters it contains. Their characteristics are obtained using quantitative interpretation of the VES (first of all performed at the reference drilling points) and well- logging with the construction of model VES curves. The main geoelectric layers in the foothills of the Greater and Lesser Caucasus are the following (Khesin and Shakhnazaryan 1965, 1986):

1. The modern alluvial, proluvial-deluvial sandy loam, loam and sandy-pebble accumulations (sometimes marine deposits) usually have thickness up to 50 m; their ρ varies from several fractions of Ohm·m for salinized sand-clayey deposits up to several hundred and more Ohm·m for dry pebbles.
2. Sandy-pebble and boulder-pebble freshwater-bearing deposits of alluvial fans usually have thickness from several tens up to 400–500 m. This layer is usually represented by a combination of interstratified lens-like horizons of varied resistivity, mainly in the range of 50–300 Ω·m. In the Alazani-Agrichai valley (southern foothills of the Greater Caucasus) the layer resistivity reaches 700–800 Ω m and its thickness attains 1,000–1,400 m. Deposits of low resistivity are characteristic of inter-fan zones. The second layer is sometimes subdivided into several geoelectric layers.
3. The aquiclude layer of marine clayey deposits with ρ in a constant range of 2–5 Ω·m.
4. Carbonate and volcanogenic Mesozoic-Cenozoic rocks of high resistivity underlie the clayey aquiclude. It is fixed by the right rising branch of the VES curve that is inclined on 45° to the x-axis. The top of this layer is traced to a depth of 800–4,000 m.

From a water-content point of view, characteristic values of resistivity are 1–5 Ω·m, on average, for clays from 10–15 to 40–50 Ω·m for water-bearing sands, from tens up to 300 Ω·m for water-bearing pebbles and boulders, over 300 Ω·m for dry boulder-pebble deposits.

In the very variable section, the true resistivity and thickness of each layer need to be calculated for each VES curve. In limited areas the layer thickness values are more stable than their resistivity. These values are crucial during interpretation, especially as is possible to distinguish the thick layer against a background of host layers among the great number of VES curves.

The variability of the geoelectric section, and sometimes the lack of clear electric discontinuities require special attention to qualitative constructions. In particular, ρ_a sections facilitate the mapping of high-resistive alluvial-proluvial and conductive clayey deposits, the detection of alluvial fans and inter-fan depressions. Near-vertical zones of decreased resistivities can indicate suggested fault zones (similar to decreased seismic velocities). Solutions can be facilitated by separating measurement data into regional and local components (as was done for gravity or magnetic survey findings), for example, by calculating the fourth order trend (regional component) and the difference between the measured ρ_a and its regional component (i.e., local component). In the section shown in Fig. 7.3a, the

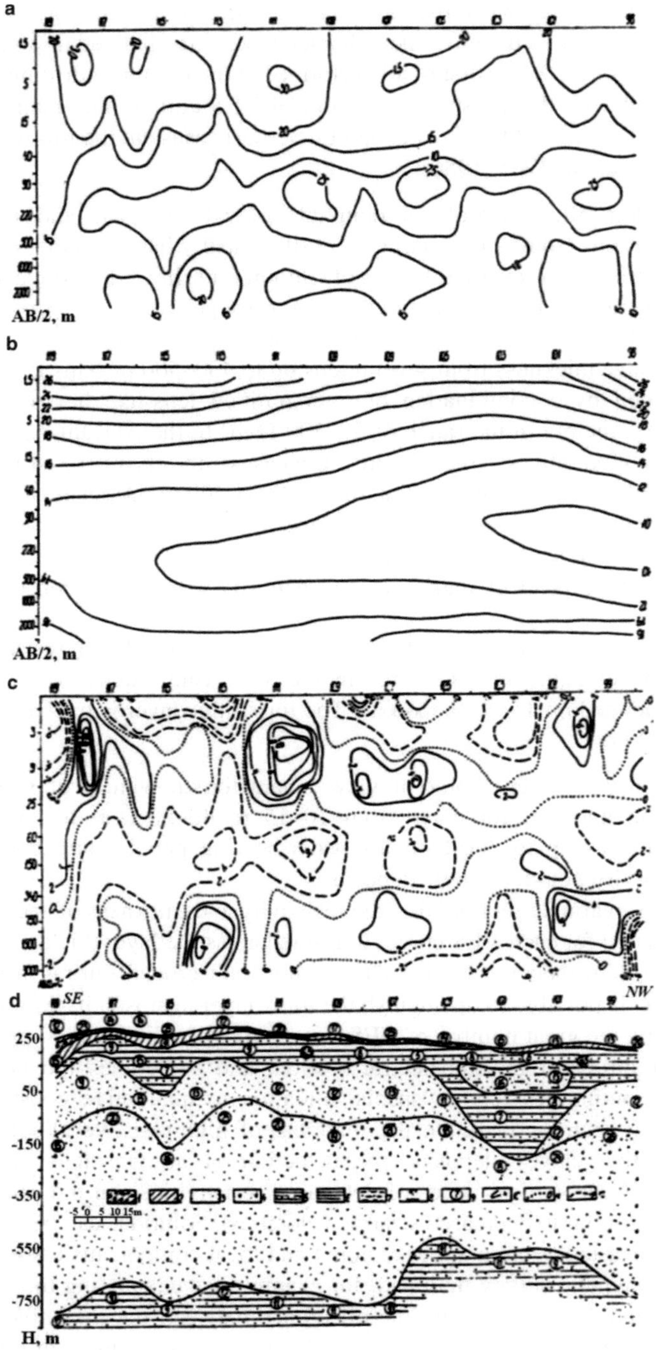

Fig. 7.3 Separation of ρ_a section into regional and local components. (**a**) ρ_a isolines, (**b**) ρ_a regional component isolines, (**c**) ρ_a local component isolines, (**d**) geological-geophysical section findings. (*1*) gravel pebbles with loam filler; (*2*) loams; (*3*) chiefly sand deposits; (*4*)

general sequence of layers (Fig. 7.3b) and local features of water control (Fig. 7.3c) emerge clearly.

Maps of isoOhm (equal resistivities) and maps of equal abscissas and ordinates of $(\rho_a)_{max}$ points are very important to characterize plane locations and the composition of water-bearing deposits. Resistivity maxima in the isoOhm map with a ρ_a range from 10 up to several hundred Ohm·m show sandy-pebble and sandy freshwater-bearing deposits, whereas zones of low resistivity (1–10 Ω·m) are usually related to inter-fan space.

7.1.3 Geophysical Specificities of Land Reclamation and Irrigation Studies

Irrigation and drainage cause active changes in water and salt in the upper part of a geological section (20–30 m) and the soil. Geophysical studies can provide uniform data on: (1) the detailed subdivision of aeration and water-saturation zones, (2) the depth of ground water and regional aquiclude, (3) the salinization of soil/ground and mineralization of ground water, (4) filtration properties of deposits in aeration zones (Khesin and Shakhnazaryan 1986).

The depth of a regional aquiclude can be found by conventional VES along relatively rare networks. Vertical electric soundings with induced polarization measurement (VES-IP) were applied to resolve these problems. Their AB/2 separation scale was denser for 1–500 m separations. The value of apparent polarizability (η_a) was measured by USSR equipment:

$$\eta_a = \frac{\Delta V_{IP}}{\Delta V} 100\%, \tag{7.2}$$

where ΔV and ΔV_{IP} are the voltage during and after charging, respectively.

To increase the reliability of geoelectrical sections and the correlation between hydrogeological and geophysical characteristics, parametric geophysical tests were performed on a series of reference wells (for example, on about 100 wells in the Samur-Absheron Massif). Cross-soundings were performed over wells with known geological sections and water-bearing horizon characteristics.

The groundwater level (GWL) is the most important factor in land reclamation. Assessment is based on the first break (first arrival) method of refraction survey (Table 7.1).

Fig. 7.3 (continued) gravel-sand deposits; (*5*) clayey deposits with sand bands; (*6*) sand deposits with clay bands; (7) potentially water-bearing sand deposits; (*8*) VES point and number; (*9*) ρ, Ohm·meters; (*10–12*) isolines of ρ_a, regional and local components: (*10*) positive, (*11*) zero, and (*12*) negative

Table 7.1 Correlation between GWL values according to drilling and seismic prospecting

Deposits	Number of measurements	Correlation coefficient
Clays	16	0.71
Clays with gravel	12	0.75
Loams	21	0.61
Sandy loams	12	0.71
Sands with gravel	22	0.75
Boulder-pebble deposits with sand and gravel	35	0.61

The difference between GWL data in the Samur-Absheron Massif (West Caspian onshore) measured in the wells and calculated from the results of refraction survey is, on average, 0.3–0.5 m. At times it attains1.0–1.5 m due to the local head high location of the capillary hem.

Solutions to complicated problems require a combination of geophysical methods or variants on them. For example, in the case of high mineralization of ground waters or salinization of the ground, sand and sandy-loam horizons are detected by very low resistivity, less than 3 Ω·m. Similar resistivities are characteristic of some clayey deposits. However, the integration of IP parameters and resistivity can be used to separate clayey deposits from sands with mineralized water.

Sharapanov et al. (1974) suggested the A^* parameter (relative polarizability) to achieve better detection of clay content. This parameter reflects both the amplitude and the time characteristics of the IP process:

$$A^* = \frac{\eta_{a1} - \eta_{a11}}{\rho_a}, \tag{7.3}$$

where η_{a1} and η_{a11} are the apparent polarizability in 1 s and 11 s after the current is interrupted, respectively.

The A^* parameter identified clay horizons on VES-IP curves by values of up to 15 and more arbitrary units (a.u.), whereas sand horizons with high mineralization waters were characterized by a few arbitrary units. The relative polarizability of aeration and saturation zones of arid areas in the Samur-Absheron Massif was studied in detail (Khesin and Shakhnazaryan 1986). The findings showed that polarizability depends only slightly on underground water mineralization but strongly on clay content (Tables 7.2 and 7.3).

It is obvious that the A^* parameter of terrigeneous sedimentary deposits depends on the clay particle content and (within small limits) the water mineralization in the sediments. The polarizability increases sharply with a decrease in the size of the particles and increases slightly with an increase in water mineralization in the pores and fractures. The inverse relation is observed for resistivity. Consequently, clay polarizability is higher than the polarizability of saltwater-bearing sands.

Table 7.2 Comparison of A^* values and underground water salinity in different sediments

Sediments	Number of measurements	A^* average value/ standard, (a.u.)	Salinity average value/ standard, (g/l)
Clays	24	12.27/10.30	16.81/15.44
	14	8.02/6.67	0.80/0.72
Loams	10	4.28/2.63	13.60/14.33
	36	3.67/2.47	0.39/0.04
Clay and sand interstratification	15	5.13/4.92	26.21/33.12
	18	4.13/2.19	0.39/0.04

Table 7.3 Relation between geological and electric characteristics

Sediments	Water in saturation zone	ρ_a average (range), Ohm·m	A^* average (range), a.u.
Clays	Fresh mineralized	18.1 (13.3–23.0)	8.02 (4.53–11.51)
	aeration zone	8.1 (5.0–11.2)	12.27 (8.15–16.39)
		4.0 (1.8–6.2)	6.64 (4.58–8.80)
Loams	Fresh mineralized	16.3 (13.6–19.0)	3.67 (2.90–4.60)
	aeration zone	6.9 (4.4–9.3)	4.28 (2.80–5.80)
		15.2 (13.2–17.2)	2.70 (2.30–3.10)
Sandy loams with gravel	Fresh mineralized	2.70 (2.30–3.10)	1.29 (1.01–1.57)
		17.3 (10.6–24.0)	4.69 (2.57–6.81)
Loams with gravel and sand bands	Aeration zone	49.8 (43.9–56.6)	0.79 (0.60–1.10)

On the basis of these data and the results of similar studies in other Caucasian regions, different nomographs for lithological differentiation, evaluation of mineralization and salinization were developed (Khesin and Shakhnazaryan 1986). As an example, Fig. 7.4 shows the possibility of a lithological division of an aeration zone.

Thus, using VES, IP and seismic data, numerous maps have been plotted of litho-facial varieties in the upper 1–5 m stratum, the salinization of the aeration zone, the mineralization and level of ground waters, and the spatial distribution of the earth's surface sandy-gravel horizon.

7.1.4 Methodology of Underground Water Geophysical Prospecting in Mountain Areas

Hydrogeophysical targets in the Caucasian and many other orogens are mainly hydrogeological massifs with numerous fracture-vein water-head systems. The

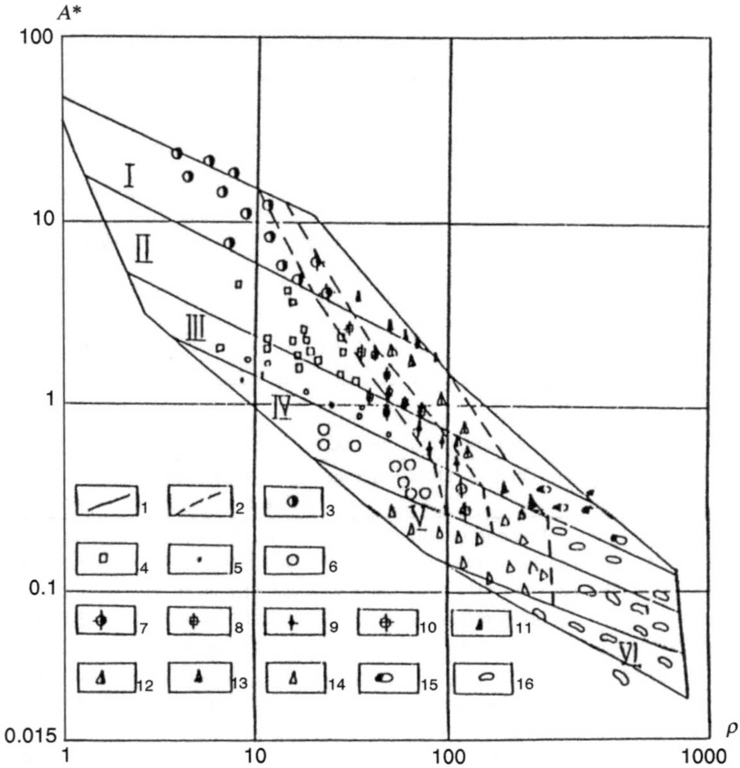

Fig. 7.4 Nomograph of a lithological division of aeration zone rocks according to their resistivity and polarizability. (*I*) clays; (*II*) loams; (*III*) sandy loams; (*IV*) sands; (*V*) pebbles gravels with filler sand; (*VI*) boulder-gravels with filler sand. (*1*) lithological boundaries; (*2*) boundaries of coarse rock intrusions, *initial data for graph construction*: (*3*) clays; (*4*) loams; (*5*) sandy loams; (*6*) sands; (*7*) clays with gravel inclusions; (*8*) loams with gravel inclusions; (*9*) sandy loams with gravel inclusions; (*10*) sands with gravel inclusion; (*11*) pebbles gravels with filler clay; (*12*) pebble gravels with filler loam; (*13*) pebble gravels with filler sandy loam; (*14*) pebble gravels with filler sand; (*15*) boulder gravels with filler sandy loams; (*16*) boulder gravels with filler sand

width of fault zones and their endogenic mineralization is usually tens of meters, sometimes, 200–300 m. Therefore, geophysical works were carried out in measurement steps of 50–500 m in general (Shakhnazaryan et al. 1986). Complicated patterns of anticipated targets, their relatively small dimensions, and the uneven topography require the integration of geophysical methods, the application of terrain correction techniques and the extraction of additional information from topography data, as is the case in mining geophysics. For example, fractured rocks can be illustrated by indenting topography that is emphasized by special processing (Khesin et al. 1996). The methodological specifics are different for prospecting in the Lesser Caucasus because of its relatively smooth topography and the presence of numerous roads.

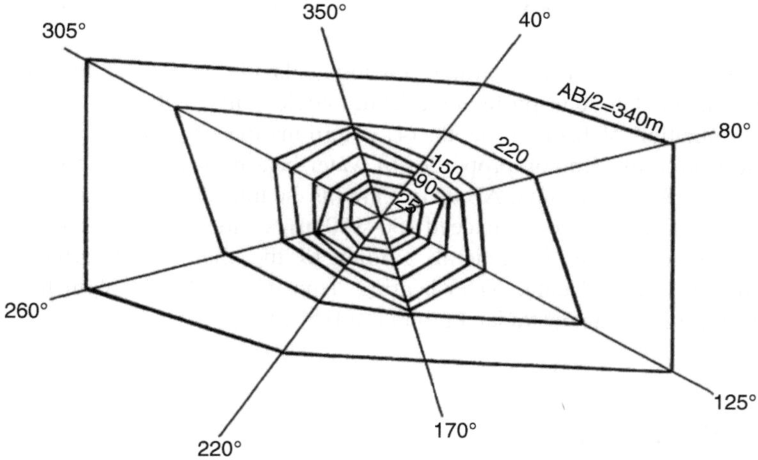

Fig. 7.5 Anisotropy ellipse according CVES data in the Shusha plateau (Lesser Caucasus)

7.1.4.1 Waters in Faulted and Fractured Zones

In underground waters in zones of disjunctive dislocations, exogenous and endogenous fracturing forms a united water-bearing complex. The main goals of geophysical studies are the following: (1) detection of faulted/fractured zones in bedrocks, (2) tracing of lithological boundaries, (3) characterization of fracture fillings, and (4) determination of the thickness and composition of the eluvial-dilivial cover. To solve these mapping problems electric profiling, VES, VES-IP and magnetic prospecting can be integrated (Shakhnazaryan et al. 1986; Khesin and Shakhnazaryan 1986).

Disjunctive dislocations are reflected by high gradients of the magnetic field or linear zones of decreased field in the boundaries between areas of different field patterns. The strike of fault zones can be captured by circular VES (CVES) data that coincide with the major axis of an anisotropy ellipse (Fig. 7.5).

Two-separation symmetrical profiling (AB = 680 m, A′B′ = 220 m) is the main tool for determining the dip of fault zones in area surveys. The ρ_a ratio for great/small separations over fault zones is less or equal to 1, whereas this ratio is usually greater than 1. Combined profiling is more effective for this problem, but is applied only in detailed work as it is expensive and cumbersome.

Decreased ρ_a values are characteristic of fractured zones. For example, fractured water-bearing limestone in the Mardakert district of Lesser Caucasus has a $\rho_a = 15–30\ \Omega\cdot m$, whereas the middle ρ_a level is 150–250 Ω·m. Such zones are well detected in isoOhms maps. To find water-bearing and clayey zones of decreased resistivity in detailed studies, VES-IP is widely used (Shakhnazaryan et al. 1986; Khesin and Shakhnazaryan 1986): the A^* parameter of clayey zones is usually higher than water-bearing fractured rocks. CVES-IP data are also useful for the detection of fractured zones with clayey filler: the major axis of the η_a anisotropy ellipse is perpendicular to the general fracturing direction; i.e. the major axes of ρ_a and η_a are mutually perpendicular (Melkanovitsky et al. 1982).

7.1.4.2 Waters of River Valleys and Infra-Mountain Hollows (Basins)

These geophysical studies have the following goals: (1) spatial and depth mapping of alluvium and deluvium-proluvium, (2) lithological differentiation of the section, (3) determination of bedrock depth, (4) determination of groundwater level, (5) characterization of filtration properties of water-bearing rocks. To approach these problems, VES with maximal AB up to 1 km (as the thicknesses of alluvial deposits in the Lesser Caucasus do not exceed 100–150 m) is widely used. Their processing and interpretation are, in general, similar to the methods described above. According to VES-IP data, water-bearing deposits can be distinguished by an increased η_a (2–4% against a background of 0.5–1.0%).

7.1.4.3 Waters of Lava Covers

The water-bearing Quaternary lava complex in the Kelbadjar plateau of the Lesser Caucasus has thicknesses ranging from several meters up to 120 m and is represented by five floods of intensive fractured, bubbled andesites-basalts. Dissection of the lava complex, revealing freshwater-bearing and impermeable layers were performed using VES, CVES, VES-IP, symmetrical and combined profiling and magnetic prospecting (Shakhnazaryan et al. 1986; Khesin and Shakhnazaryan 1986). These complicated geological-geophysical environments require sliding (moving) averaging of magnetic and electric profiling results with a window width of nine measurement steps. The main complication in interpretation of varied VES curves is determining the resistivity of the intermediate layers (lava floods). Thus, the, quantitative interpretation of VES curves was carried out in three stages, where these characteristic resistivities were obtained on the basis of regioning (Minasyan 1979).

According to work by Ogilvy and Bogoslovsky (1979) in the Greater Caucasus, the set of mobile methods of hydrogeophysics in mountain areas can also rely on water spring resistivimetry which is similar to ore prospecting (Khesin 1969).

7.2 Typical Geophysical Solutions to Hydrogeological Problems

7.2.1 Discovering Fresh Water in the Northwestern Foothills of the Greater Caucasus: The study of Pebble Collectors

The first significant hydrogeophysical studies were performed using VES to enhance the Baku water supply with high-quality underground waters from pebble aquifers of the Samur-Gusarchay interfluve in the Kusar-Divitchi sinclinorium (Khesin and Shakhnazaryan 1965). Aside from drilling, numerous electric soundings were carried out with AB separations up to 4 km along a 4 × 5 km network in an area of about 1,000 km^2.

The interfluve was separated into four geoelectric zones. The Samur and Gusarchay zones are related to the alluvial fans of Samur and Gusarchay rivers, the Hudat zone is linked to the depression between them, and the Seaside zone is associated with the seaside plain, where marine Upper Pliocene-Quaternary deposits prevail. Freshwater-bearing pebbles and sands have resistivities from several tens up to 300 Ω·m; higher resistivities are characteristic of dry pebbles and sands. The transition from continental to marine sediments is accompanied by decreasing resistivities.

The location of high resistivity pebbles is well contoured in isoOhms maps plotted for different separations of VES current electrodes (Fig. 7.6).

The maximum ρ_a in the map for AB/2 = 200 m is reflected in the relatively thin pebble cover in the limits of the modern river-bed and the alluvial fan of the Samur river. The Samur maximum displaces southward in the map for AB/2 = 500 m. It reflects thick pebble accumulations of an ancient (buried) alluvial fan within the central part of the sub-isometric synclinal that was detected by seismic and electric prospecting in the Mesozoic and younger strata. Several maxima and three minima of ρ_a were detected in the Samur zone and are shown in the map for AB/2 = 1,000 m. These maxima are evidently related to the most ancient fan deposits, whereas the minima are coincide with the Imamkulikend and Shirvanovka highs in northeastern Azerbaijan and the periphery of the Gilyar high in southeastern Dagestan revealed by seismic prospecting. The location of the Gusar maximum in all three maps is almost constant; thus, the alluvial fan of the Gusarchay River is extremely thick.

The thickness of high-resistive pebble aquifers and the depth of aquicludes are determined by quantitative interpretation of VES curves with an accuracy of about 20% (according to their comparison with drilling data). The geoelectrical sections and results of qualitative interpretation served to construct a schematic map where the location of geoelectric zones and pebble deposits and characteristics of underlying conductive clayey aquiclude are depicted (Fig. 7.7).

On the basis of this map and other VES data, the volume of required drilling was significantly reduced and the network of exploitation wells was optimized for the water supply of such a large industrial city as the capital of Azerbaijan. The costs of electric prospecting did not exceed 15% of the savings in the cost of the wells.

7.2.2 *Discovering Fresh Water in the Kura Depression: Aquifer Characteristics and their Relation to Deep Structure*

The key characteristics of water-bearing Upper Pliocene recent deposits, their thickness and lithofacial composition were obtained on the basis of a combination of VES, other geophysical methods and drilling. During two field seasons an area of

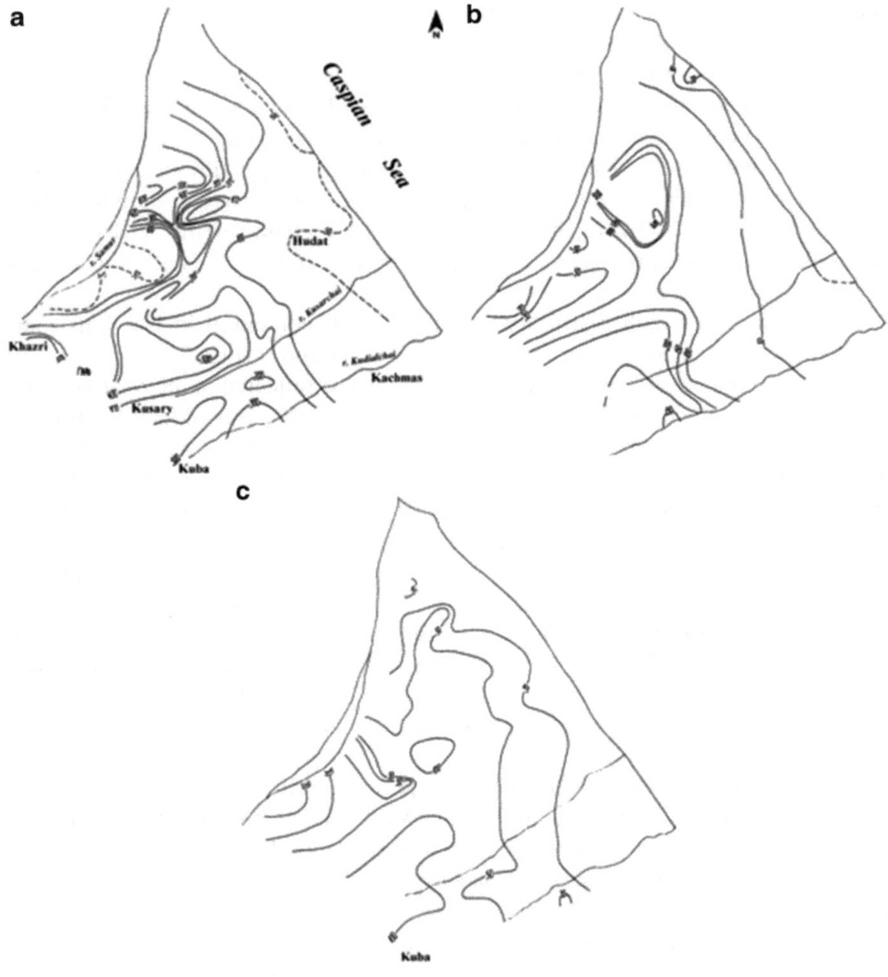

Fig. 7.6 Maps of equal resistivities (isoOhms) of the Samur-Gusarchay interfluve for AB/2 values: (**a**) 200 m, (**b**) 500 m, and (**c**) 1,000 m

5,400 km^2 was studied by 590 VES with ABs up to 10 km. Sandy-pebble collectors saturated by fresh water were identified on maps of isoOhms and in the upper part of geoelectrical sections on the basis of their high resistivity. Regularities in their distribution were revealed by comparison with dip structure features. A characteristic section is shown in Fig. 7.8.

The uppermost layer in Fig. 7.8 is composed of high-resistive Quaternary clayey-sand and pebble deposits, in particular deposits of alluvial fans of the Geranchay and Qyurokchay rivers. In the second layer of Upper Pliocene sandy-clay deposits (Ahchagyl and Absheron units), ρ values increase westward from 4.3 to 7 $\Omega \cdot m$ as a function of the increase in sand components of the layer in this

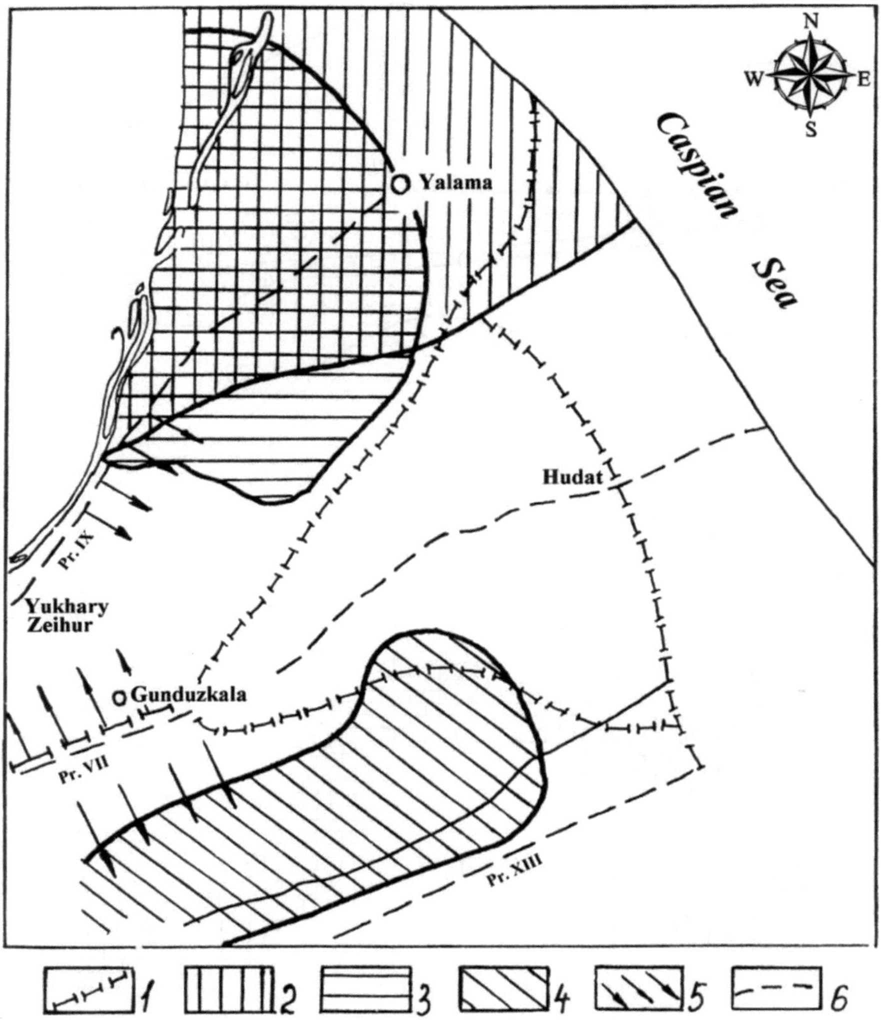

Fig. 7.7 Map of geoelectric zones and gravel mass locations in the Samur-Gusarchay interfluve. (*1*) boundaries of geoelectric zones ((*I*) Samur, (*II*) Hudat, (*III*) Kusarchai and (*IV*) Primorsk); (*2*) modern debris cone of the Samur river; (*3*) ancient buried debris cone of the Samur river; (*4*) debris cone of the Kusarchai river; (*5*) direction of argillaceous waterproof layer dips; (*6*) lines of geoelectric sections

direction. The underlying thick conductive layer is represented by a clayey Maykop unit (Oligocene-Upper Miocene). The lower high resistivity layer is represented by "foraminifer" deposits. The known Mesozoic Borsunly high is clearly reflected in structure of this layer, as well as in gravity field (Fig. 7.8). It is evident that the thickness of the sandy-pebble accumulations increases in synclinals and decreases

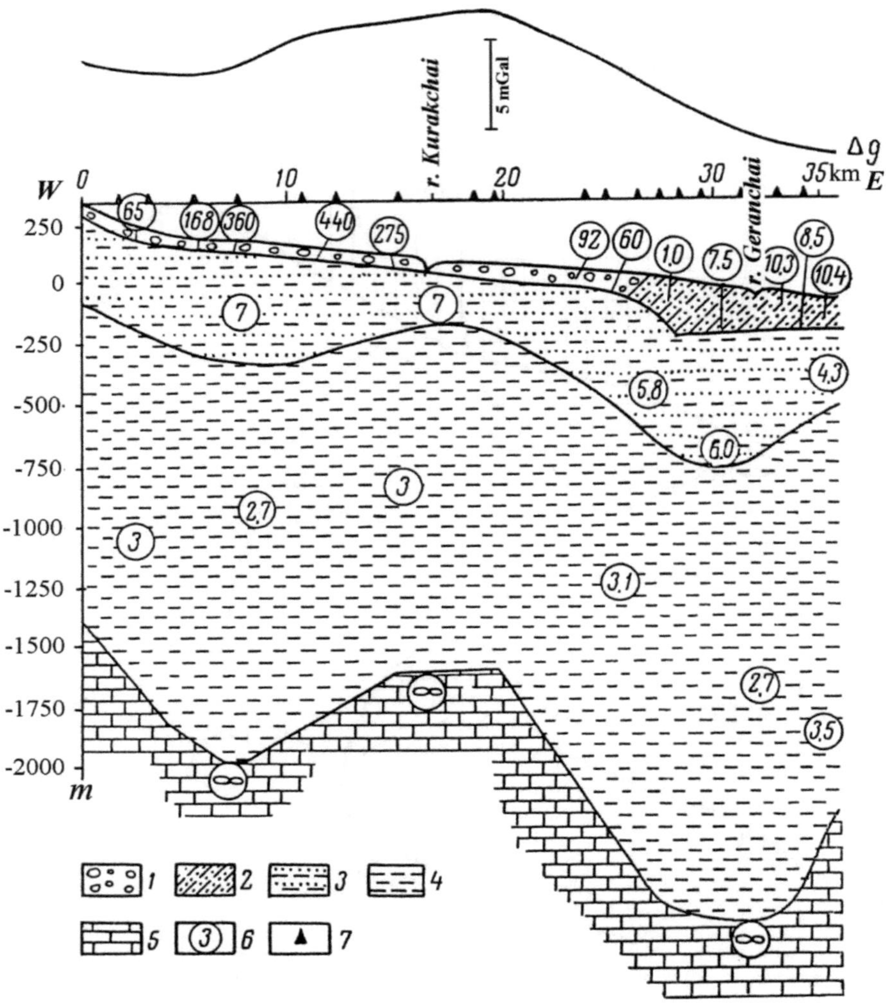

Fig. 7.8 Geoelectric section according to VES data in the Kazakh-Ganja Massif. (*1*) boulder-pebble deposits; (*2*) clay-sandy deposits of buried debris cone; (*3*) sandy-clay deposits of Ahchagyl and Absheron; (*4*) clay deposits of the Maikop series; (*5*) foraminiferal layer; (*6*) values of real electric resistivity, Ohm·m; (*7*) VES points

in high arches. Other geoelectric sections also showed a similar relation between dip structure and recent sedimentation with the greatest thickness of sandy-pebbles in depressions that separate neighboring anticlines. This regularity is frequent in the Kura Depression (Fig. 7.9).

Alluvial fans of recent rivers are clearly reflected by high-resistivity areas, for example, for the Indjachay, Terterchay, Hachinchay, Karkarchay and Araks rivers (Fig. 7.9). The Indjachay alluvial fan coincides with the synclinal structure between

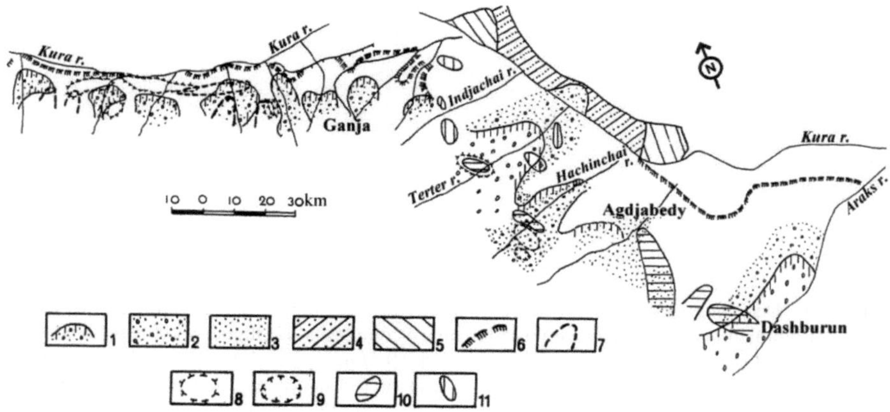

Fig. 7.9 Comparison of alluvial-fan deposits according to VES data and deep structure specificities according to geological and gravity data. (*1*) areas of maximal accumulation of mainly boulder-pebble deposits; (*2*) areas of extent of gravel-pebble deposits with sand inclusions; (*3*) mainly sandy deposits with pebble filler; (*4*) clayey-sandy deposits of PaleoKura; (*5*) clayey-sandy deposits of PaleoKura saturated by mineralized water; (*6*) areas of decreasing pebble components and their substitution by sands, sandy loams and loams; (*7*) Mesozoic structural projections (protrusions); (*8*) gravity maxima related to Mesozoic highs; (*9*) gravity minima related to maximal Cenozoic sedimentary accumulations and immersion of underlying strata; (*10*) Mesozoic highs; (*11*) Cenozoic highs

the Naftalan (in the north) and the Gyzylgadjily (in the south) highs. The Terterchay alluvial-fan deposits bypass the Mirbashir high in the north-west and are bounded by the Shirvan and Barda aniclines in the south-east. Pebbles of Hachinchay abut into the Gelludjy high in the north-east; eastward of this high, water-bearing Absheron horizons are absent. The Karkarchay alluvial-fan deposits are limited by the Agdjabedy high in the east. Increased resistivities on the left shore of the Kura river are indicative of a PaleoKura zone with a width of 8–10 km saturated by fresh water. Decreased resistivities (ρ_a = 3–6 Ω·m) differentiate the PaleoKura deposits, with mineralized waters to the north-west and south-east of this zone.

Thus, important hydrogeological information was obtained by electric prospecting in a very short time. It provided an optimal choice for explored and exploited wells and reduced expenses due to unnecessary drilling that would have exceeded the cost of electric prospecting 50-fold.

7.2.3 Searching of Thermal Waters

Thermal prospecting is a direct method for thermal water detection in the vicinities of thermal springs and shallow sources. However, ground thermal prospecting and infrared air surveys are ineffective in deep thermal water regions, because of the structural-lithological and hydrogeological chargeability of the thick cover

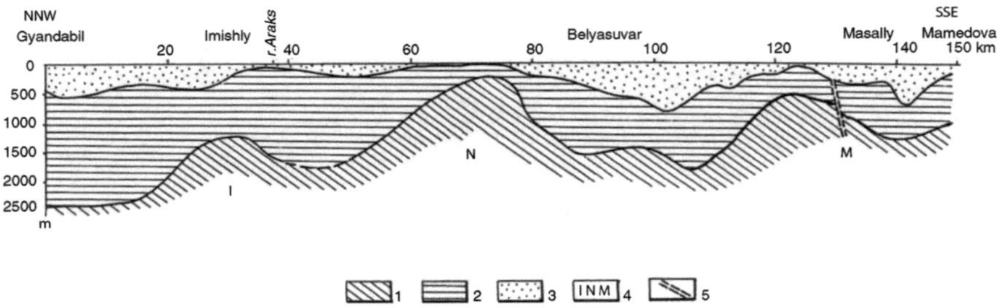

Fig. 7.10 Detecting thermal water-bearing structures along the eastern foothills of the Talysh Mts. (*1*) sedimentary and volcanogenic-sedimentary complexes of Oligocene-Miocene and older composing a high-resistivity reference horizon (15–60 Ω·m); (*2*) mainly clay marine deposits of Ahchagyl and Sarmat with a resistivity of 1–2 Ω·m; (*3*) sandy-clay and pebbles Quaternary deposits (mainly continental) with a resistivity of 3–30 Ω·m; (*4*) highs according to electric prospecting data: *I* Imishly (Mil), *N* Narimanov, *M* Massalin, (*5*) faults according to the electric prospecting data analysis

completely masks the pattern of the deep thermal field; one example is in Georgia, where the water temperature exceeds 100° at depths of 3–5 km in large areas (Molchanov et al. 1985). Nevertheless, in some cases it is possible to detect thermal water in a water-conductive channel at a depth of more than 100 m.

Because of the increased resistivity of warm deep layers, a magnetotelluric method can be used for their mapping (e.g., Kaufman and Keller 1981). However, decreased conductivity can also be caused by rock graphitization or pyritization, clayey rock masses or other sources. Therefore, indirect electric prospecting of thermal water is more widely exploited. The main tool for the study of thermal-water structural controls is the VES method. The most favorable structures likely to reveal thermal waters are zones of highs and faults (Melkanovitsky 1975). This type of study was performed at the northwestern immersion of the Talysh anticlinorium at a scale of 1:200,000 by VES with current electrode separation up to 10 km (Fig. 7.10).

Three geoelectric layers were clearly detected in this NNW-SSE geoelectric section. The upper layer is represented mainly by alluvial-proluvial deposits that are not infrequently saturated by fresh water. Its thickness is maximal in alluvial fans of the rivers flowing down from the Lesser Caucasus or the Talysh. The second layer of constant low resistivity is the regional aquiclude. The lower layer is usually reflected by the right branch of VES curve that rises under a 45° angle. On the basis of such clear differentiation, several buried highs were detected by electric prospecting (e.g., Fig. 7.10). Their locations were verified by the results of seismic, gravity and sometimes magnetic prospecting, as well as drilling data.

Disjunctive dislocations were detected by the sharp differences in geoelectric discontinuities. The construction of correlation schemas between different profiles, where resistivity graphs for different AB were plotted, was very useful for fault tracing (Khesin and Shakhnazaryan 1986). Comparison of the fault zone location in different separations of current electrodes helped determine the dip of the fault zone. As a result, the characteristics of thermal-water conductors were identified. For example, the transverse fault that limits the Masally high in SE reaches the known Arkevan springs of thermal waters in the SW. The Azerbaijan Geological Authority drilled 2.5–3.0 km wells at the node of fault zones near the town of Masally and other zones revealed by the VES survey with utilization of seismic and gravity data. These wells produce thermal waters with substantial outputs.

7.2.4 Detecting Fresh Water in Mountainous Areas

The greatest volume of integrated hydrogeophysical works (electric sounding and profiling, magnetic prospecting) has been carried out in the Askeran, Martuni and Shusha districts of the Lesser Caucasus, where an area of 1,700 km^2 was studied at a scale of 1:50,000 for a period of 3 years. The data (Khesin and Shakhnazaryan 1986) were used to plot numerous graphs and maps of graphs of $\rho_{a,}$ η_a and ΔZ, $\rho_{a,}$ and η_a sections, CVES ellipses, geoelectrical and geological-geophysical sections, isoOhms and isodynams maps, as well as structural maps on the bottom of alluvial deposits and roof of potentially water-bearing fractured limestone. The greatest thickness of alluvial and alluvial-proluvial freshwater-bearing deposits (150–180) m was found in the confluence of the Badara and Karkarchay rivers. Fractured Cretaceous limestone with potential fracture-layer fresh waters was contoured westward and southwestward of the village of Martuni. Recommended drilling revealed fracture and fracture-vein waters in fractured and fault zones detected by high gradients of magnetic field, which varied in terms of geoelectric sections and other geophysical features. A good example is the application of combined electric profiling, magnetic prospecting and VES in the area of the Karabachos silk factory (Fig. 7.11).

In this figure, the ΔZ graph depicts nonmagnetic limestone and relatively magnetic tufaceous sandstones. Fault zones are reflected by ΔZ local minima. These zones are clearly indicated by the typical "conductive crossing" of ρ_a graphs (e.g., Khesin 1969). Their location is constant in large and small electrode separations for a suggested fault near the 10th picket; thus, the fault is vertical. Conductive crossing location changes in different separations for the 5th picket show the dip of the suggested fault. An exploration well was recommended at the first fault. It revealed fresh water that provided the water supply to the largest factory of this type in the region.

A similar experience took place in an area near the village of Magavuz which has rugged topography, where symmetrical profiling with different AB separation was substituted for combined profiling. According to the VES data, water-impermeable compact tufaceous Albian rocks overlap the fractured water-bearing Callovian-Oxfordian limestone. In the dip fault zone detected by high ΔZ gradient, increased η_a (up to 3%) and decreased ρ_a values (for great separations) indicated the coordinates of a recommended well, which revealed pressure-head, fresh water at a depth of 150 m. It drifted with an output of 15 l/s.

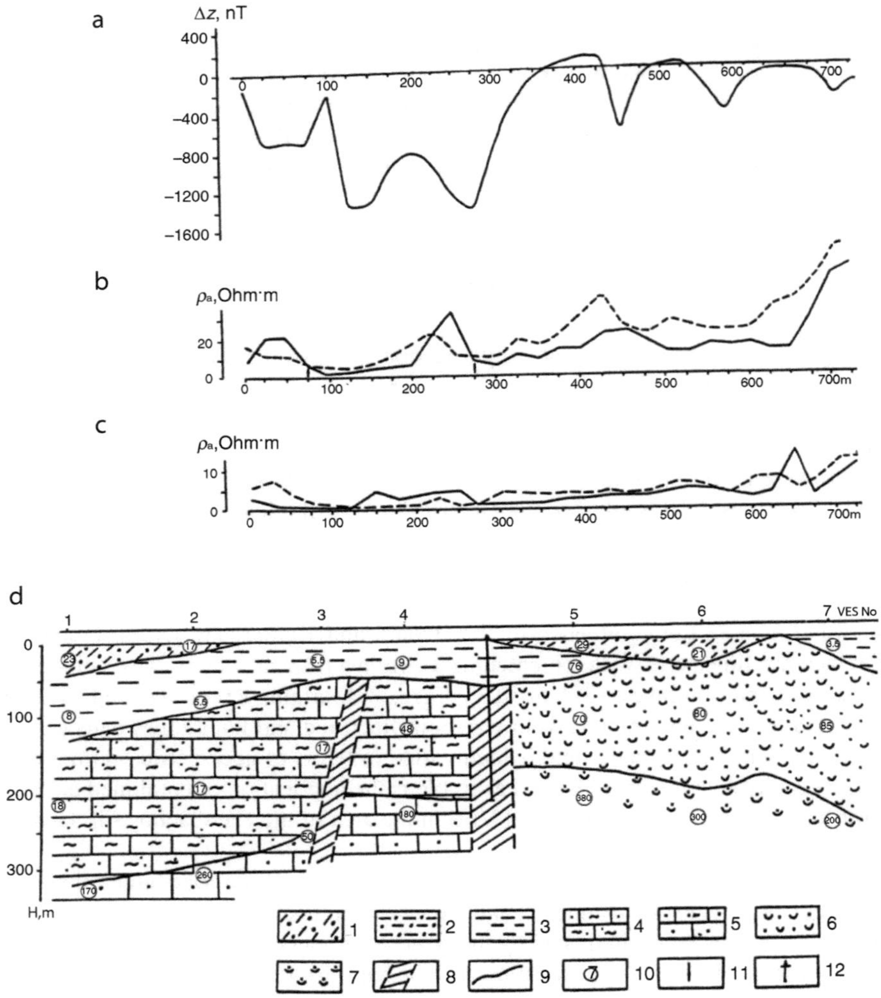

Fig. 7.11 The aquifer detection at the Karabachos silk factory. (**a**): magnetic ΔZ curve, (**b**) and (**c**) ρ_a curves at AB = 500 m and A'B' = 200 m, respectively, (**d**) geoelectric section according to VES data. (*1*) loamy sand and loam with rare inclusion of gravel and pebble; (*2*) clays with rare inclusion of gravel and pebble; (*3*) clays; (*4*) arenaceous limestone strongly fractured; (*5*) arenaceous limestone; (*6*) tuffaceous limestone; (*7*) high density tuffaceous limestone; (*8*) fracture zone; (*9*) lithological contact; (*10*) values of electric resistivity (Ohm·m); (*11*) direct intersection of resistivity curves; (*12*) borehole drilled according to geophysical data

7.3 Geophysical Investigations of the Caucasian Lakes

In the Greater and Lesser Caucasus there are hundreds and hundreds of mountainous lakes. Many of these are very small (less than 1 km^2), but there are larger lakes. Magnetic and electric (electromagnetic) methods are the most effective geophysical means of studying these lakes and their surroundings.

The largest lake of the Caucasus is Lake Sevan (1,360 km^2) located in the Sevan-Akera tectonic zone (see Fig. 2.4) of the Lesser Caucasus. E. G. Guloyan and R. S. Minasyan successfully applied vertical electric sounding (VES) combined with an airborne magnetic survey of the area around Lake Sevan from the south-west (Melkanovitsky 1984). All the VES curves clearly detected a reference horizon of lowered resistivity. Determination of it depth served to study the surface relief of water-resisting clay deposits. Unfortunately, analysis of VES data does not always make it possible to subdivide basalts and metamorphic complexes. The most interesting in terms of searching for underground water are zones of lowered relief of under-lava associations. On the spur of ridges the depressions may correspond to ancient river valleys along which move groundwater. The study of the relief of under-lava associations led to the conclusion that groundwater was often discharged directly into Lake Sevan. Several promising areas were selected on the basis of geophysical data to carry out prospecting hydrogeological works. Out of the 40 boreholes drilled according to geophysical data, 35 boreholes had abundant groundwater at a depth of 50–200 m. The most valuable areas are confined to zones of minimal resistance of lava associations within shallow gullies (Melkanovitsky 1984).

The most recent application of remote sensing and GIS technologies at Lake Sevan is for shore management (e.g., Agyemang et al. 2011).

Lake Sevan is a large pull-apart basin formed by two large NW strike-slip faults and a system of oblique extension faults and occurs in very complex tectonic pattern of basaltic bodies (Rebai et al. 1993; Karakhanian et al. 2001). The well-studied Sea of Galilee (Lake Kinneret) is characterized by very similar tectonic-geological features (e.g., Eppelbaum et al. 2007). The Israeli experience suggests that a combination of magnetic, paleomagnetic and K-Ar data along with seismic survey results can yield predictive data (Eppelbaum et al. 2004a, 2007).

Numerous Caucasian glacier lakes dammed by glaciers and moraine ridges are potentially hazardous (Petrakov et al. 2007). The Bashkara glacier lake (about 2,550 m over the m.s.l.) located in headwaters of the Adylsu River in the Kabardino-Balkaria Republic of Russia (Northern Caucasus) is a typical representative of such lakes. This lake was investigated by a geophysical set involving Ground Penetration Radar (GPR) and resistivity methods (Lekomte et al. 2010). The GPR measurements were successful with a penetration depth down to 70 m on icy ground, but the resistivity method only managed the first 4–6 m of the ground. Nevertheless, mutual analysis of these methods helped localize small water channels, temperate and compacted parts of the ice, and other important geological peculiarities (Lekomte et al. 2010). It should be noted that the recently developed wavelet methodology of integration of raw GPR and resistivity data (Eppelbaum et al. 2011b) should be valuable here.

The Mingechavir water reservoir is the largest artificial lake in the Caucasian region (15.7 km^3 of fresh water). Given the importance of this reservoir, the lake and surrounding areas were carefully studied tectonically and geophysically from a regional point of view (e.g., Gadjiev et al. 1989; Ismail-Zadeh and Khesin 1989b; Kerimov 1996; Gasanov 2001; Alizadeh 2005; Khain and Alizadeh 2005).

Detailed electric prospecting was carried out for the first time on the right shore of the Mingachevir reservoir by Salamov et al. (2010) to identify multi-directional fractures caused by tectonic movements. It was suggested (and later confirmed) that the recognized fractures are the edges of future landslides. Monitoring by electric observations indicated that the changes in the physical parameters of the ground are the harbingers of rockslide activation.

The analysis of variations of apparent electric resistivity of rocks were carried out in the area of Chirkey water reservoir (Dagestan, Northern Caucasus) for the period after the reservoir was filled (Idarmachev and Arefiev 2009). Dipole electric sounding was used with positioning of potential dipoles around the water reservoir 5.2–11.3 km apart. The data obtained for the period of observation (1976–1988) showed that the filling of the water reservoir affected the environment for lengthy periods of time.

Finally, integrated Remote Operated Vehicles (ROV) geophysical surveys at low altitudes (e.g., Eppelbaum and Mishne 2011) are a valuable tool for tectonic-structural investigations of lakes and their surroundings.

Chapter 8
Environmental and Near-Surface Geophysics

8.1 Investigations of Mud Volcanoes

8.1.1 Geological, Geothermal and Seismic Specificities

Mud volcanoes are widespread in the world both on land and in marine basins, in collision and transtensional settings (e.g., Kholodov 2002; Limonov 2004). Their presence is often an indicator of deep-seated hydrocarbon accumulations. At the same time, mud volcanism represents great environmental hazard that must be taken into account in the design of oil-and-gas pipelines and other constructions. The main conditions for mud volcano formation are a thick sedimentary cover (several kilometers) and plastic clayey members with an anomalously high formation of pore pressure and the presence of thermal water (Pilchin 1985; Limonov 2004). Nowadays, more than 900 terrestrial and 800 offshore mud volcanoes are known or presumed to exist (Dimitrov 2002). More than a quarter of all the known mud volcanoes are concentrated within the Caucasus (e.g., Kadirov et al. 2005a) and most (more than 220) (Kholodov 2002) are located within the "Abich triangle" (Abich 1863) near Baku (Fig. 8.1). Mud volcanoes are always confined to longitudinal faults or to the intersection nodes of longitudinal and transverse faults (Pilchin 1985). In general, pre-existing deep faults are the main controlling factors. Many mud volcanoes exist in the Black Sea and Taman Peninsula (northwestern Caucasus) as well as mid valley in the Yori River near the Georgia-Azerbaijan border (Fig. 8.1).

Edifices over 200 m high exist in Azerbaijan, with large eruptions in some cases producing flames of a similar scale. In Azerbaijan, eruptions are driven from a deep mud reservoir which is connected to the surface even during dormant periods, when seeping water still comes from the depths. Seeps have temperatures up to 2–3°C above the ambient temperature (Planke et al. 2003). The heat flow, gravity field and terrain relief of some mud volcanoes are shown in Fig. 8.2.

Seismic wave propagation in mud volcanoes zones is very complicated. In most cases, seismic prospecting has only provided some general data on the roots of mud

L.V. Eppelbaum and B.E. Khesin, *Geophysical Studies in the Caucasus*, Lecture Notes in Earth System Sciences, DOI 10.1007/978-3-540-76619-3_8,

Fig. 8.1 Distribution of mud volcanoes (*black* triangles) in the Eastern Caucasus and Bouguer gravity data (Kadirov and Mukhtarov 2004). Isoanomals (in milliGals): *solid lines* denote negative values, *dashed lines* denote positive values

volcanoes that are usually associated with crustal zones of deep-seated Mesozoic or Paleogene-Miocene highs (uplifts), where feeder channels originate from depths as low as 10 km (Rakhmanov 1987).

8.1.2 Gravity Prospecting

The first gravity studies (Arkhangelsky and Fedynsky 1932; Fedynsky 1937) showed that the location of mud volcanoes in southeastern Azerbaijan almost completely coincides with gravity depressions; i.e., a region of thick sedimentary cover of low density (see Fig. 2.7)

In the Near-Kura area, a chain of mud volcanoes is parallel to the eastern margin of the Talysh-Vandam maximum (Fedynsky 1937), where a near-vertical deep fault was recorded by Δg upward continuation data (e.g., Khesin et al. 1982). Notably, the modern gravity map in Bouguer reduction (e.g., Fig. 8.1) reflects the same regularities.

According to earlier pendulum surveys over mud volcanoes offshore (the islands of the Baku Archipelago) and the Caspian coastline, the relative gravity excess reached 15–20 mGal, which led researchers to hypothesize (Kovalevsky 1940) there were local dense rocks under the Bulla, Los, Svinoi, and Oblivnoi marine

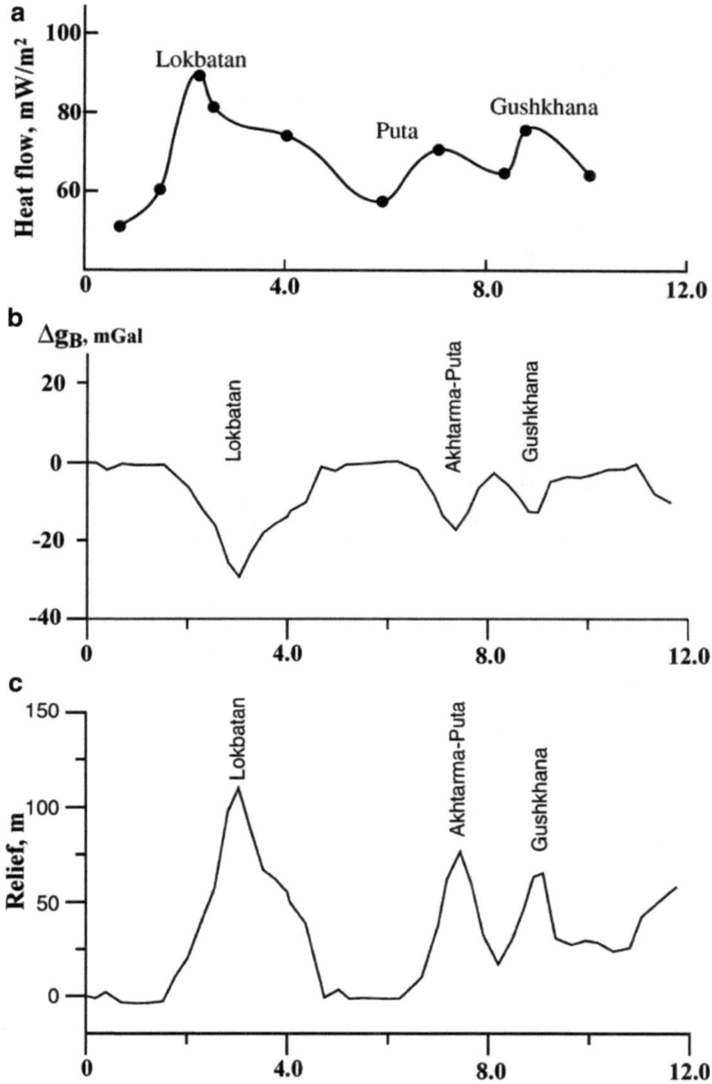

Fig. 8.2 Changes of heat flow (**a**), Bouguer gravity (**b**), and topography (**c**) along the profile crossing the Lokbatan – Akhtarma-Puta – Gushkhana mud volcanoes (Compiled from Kadirov and Mukhtarov 2004; Kadirov et al. 2005, with modifications)

volcanoes. This was verified by marine gravity survey data collected by the Southern Branch of the VNIIGeofizika over the Bulla-Sea area, where a residual gravity increase of 0.1–1.4 mGal was observed (Khesin et al. 1983, 1996).

By contrast ground gravity field studies show that mud volcanoes are characterized by local negative anomalies (Kadirov and Mukhtarov 2004; Feyzullayev et al. 2005a; Kadirov et al. 2005). It is likely that the clear correlation between gravity (Fig. 8.2b)

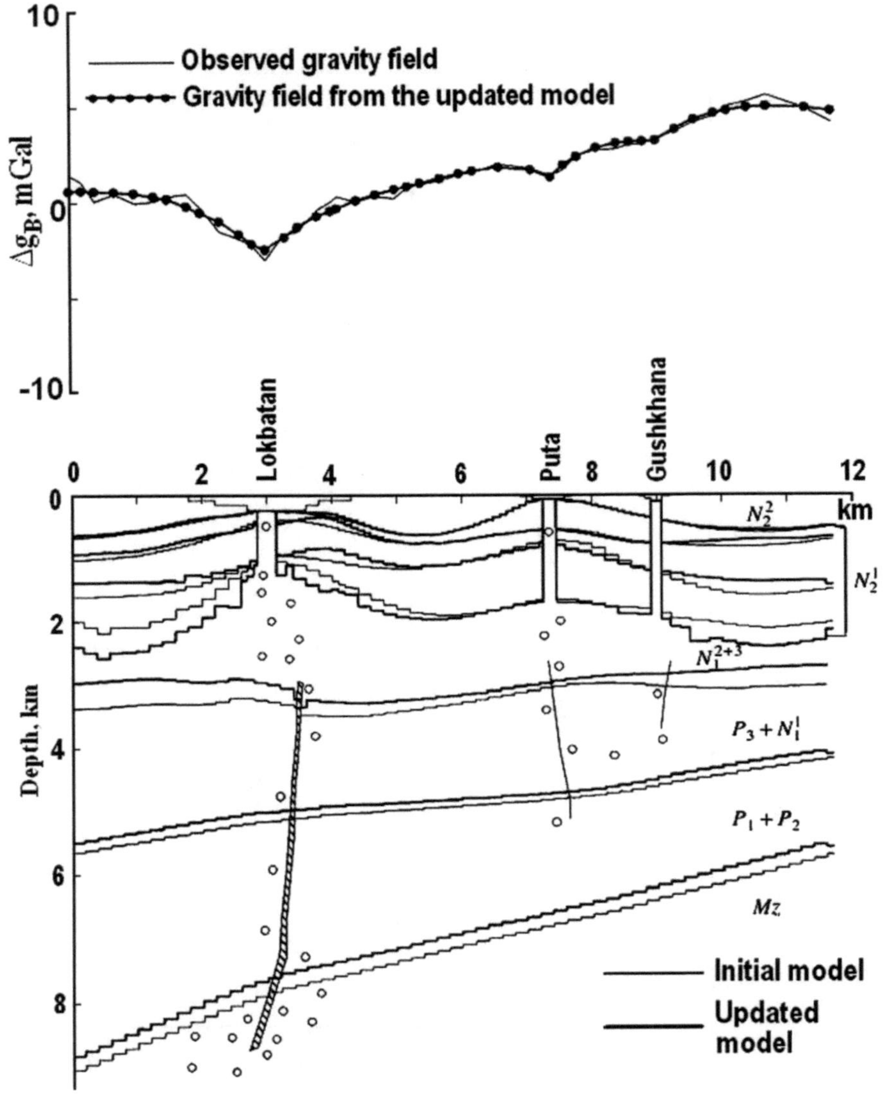

Fig. 8.3 Gravity model of the Lokbatan – Akhtarma-Puta – Gushkhana profile (Kadirov and Mukhtarov 2004; Kadirov et al. 2005, with minor modifications). Density contrasts (in g/cm^3) across the seven boundaries are (from shallow to deep): 0.01 at floor of Lower Ahchagyl (N_2^2); 0.04, 0.08 and −0.2 at apparent seismic horizons in the productive series (N_2^1); 0.25 at contact between the upper-middle miocene $(N_2^{2+3})$and lower miocene-oligocene $(P_3 + N_1^1)$ series; 0.15 at contact between the Oligocene and Eocene-Paleocene (P_1 + P_2) series, and 0.3 at the Mesozoic (*Mz*) surface. Density deficit for volcano cone is 0.3 g/cm^3

and topography (Fig. 8.2c) is related to insufficient Bouguer reduction (e.g., Khesin 1976; Khesin et al. 2001). Nevertheless, the ***PGM*** for the Lokbatan–Gushkhana profile was consistent with the observed data (Fig. 8.3).

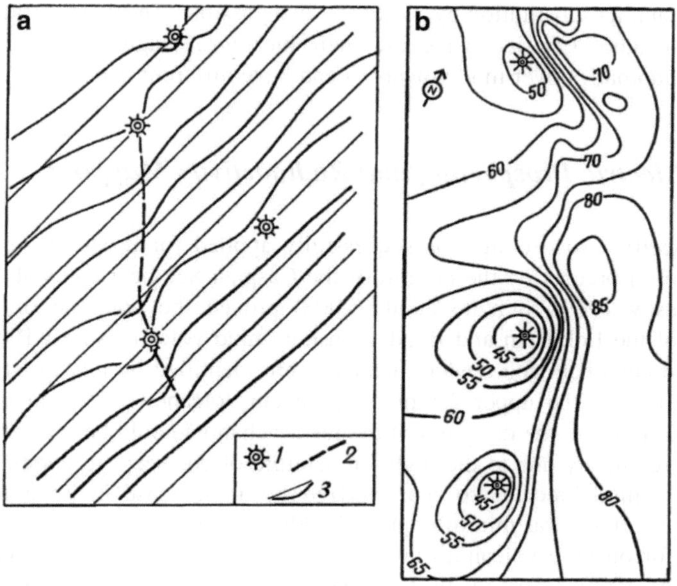

Fig. 8.4 Sites of ΔT map (Compiled from Dzabayev 1969): (**a**) graphs, (**b**) isodynams (in nanoTesla). (*1*) mud volcano, (*2*) fault zone, (*3*) ΔT graph

These differences in offshore and onshore gravity characteristics may reflect different relations between the mud volcano and host rocks: the density of volcanic rocks exceeds the density of the surrounding very young sediments in Caspian Sea, whereas the opposite density contrasts are found for onshore volcanoes.

8.1.3 Magnetic Prospecting

The largest volume of magnetic data on mud volcanoes was obtained in the South Caspian Basin and the adjacent territory, where Dzabayev (1969) performed 1:200,000 aeromagnetic surveys by fluxgate magnetometer. The increased magnetic susceptibility (up to several hundred 10^{-5} SI units) of several Neogene terrigenic strata within a thick sedimentary cover (Ismail-Zadeh et al. 1983b) facilitated successful magnetic prospecting. Ferromagnetic minerals are partly altered or chaotically oriented in fault zones. The latter are therefore detected by zones of decreased ΔT field down to several tens of nanoTesla. Isometric minima of higher amplitude indicate mud volcanoes (Fig. 8.4).

In general, mud volcanoes can be identified by magnetic maxima as well. Logachev and Zakharov (1979) described the results of V. I. Bagin and L. M. Malumyan on the productive series of the Absheron Peninsula who found magnetite-siderite associations in oil-bearing and mud volcano rocks unlike other rocks. This magnetic association was formed from the reduction of finely dispersed iron oxides and hydroxides under the influence of hydrocarbons. The reduction of magnetite and its substitution by pyrite are also possible. The near-surface burning of mud volcano deposits can cause their increased magnetization. Therefore, magnetic anomalies over mud volcanoes can have different patterns.

8.1.4 Electric Prospecting and Radioactivity Mapping

Marine electric prospecting was successfully applied for resistivity mapping of mud (chaotic) breccia on the bottom of the Caspian Sea near the Makarov Bank (e.g., Shatsov 1969). Scholte et al. (2003) proved the potential of resistivity mapping of the Lokbatan and Bozdag onshore mud volcanoes near Baku, where liquid mud chambers and feeder channels were identified by low resistivity. It is important that in the upper 2.5 m of Lokbatan volcano, exactly at the loci of surface combustion, the dry parts of the feeder channel could be detected by their increased resistivity due to the high temperature of gas seepage combustion.

Offshore mud breccia are characterized by an increase in integral gamma-radioactivity of several microRoentgen/h (Shatsov 1969). Similar excesses were observed for onshore volcanoes (Aliyev et al. 2001). The minimal level of mud breccia radioactivity depends on the area, presumably reflecting the radioactivity level in the mud volcano roots; for example, the increased radioactivity of Majkop clays. In marine volcanoes (Duvanny, Bulla, Garasu and Khara-Zire) radioactivity readings start from 10 to 10.5 μR/h; in volcanoes in the Near-Kura area (Bahar, Ayranteken, Byandovan, Hamamdag and Agzibir) this ranges from 11 to 12.5 μR/h, and in the Absheron volcanoes (Keyreki, Bozdag-Gobi, Sarynja, Lokbatan and Otman-Bozdag) the values are from 13.5to18 μR/h. Relatively high radioactivity is usually typical of the central part of the volcano, near the crater (Fig. 8.5).

Radon measurements on mud volcanoes in the Taman region indicate very strong changes in the concentration of radon in the soil of the volcano from day to day although the concentration depends on where the measurements were made (Tsvetkova et al. 2004). Near-active gryphon concentrations arbitrarily change from place to place. It was hypothesized that a dissolution of radon takes place by expulsion of gases; namely-hydrogen and methane, near gryphones. This proved to be the actual results of a hydrogen survey of mud volcanoes.

The radioactivity of mud breccia depends mainly (up to 80%) on the uranium (radium) content and, to a lesser extent, on the potassium content. In general, radioactivity decreases from older to younger deposits with an increase of potassium content and towards a regional subsidence of sedimentary deposits underlying the mud volcano. Thus, knoll radioactivity on island volcanoes is lower than that on land.

8.1.5 Relation Between Earthquakes and Mud Volcano Eruptions

Abich (1863) is thought to have been the first (on the basis of his comprehensive geological investigations in Azerbaijan and the Caucasus) to suggest that earthquakes

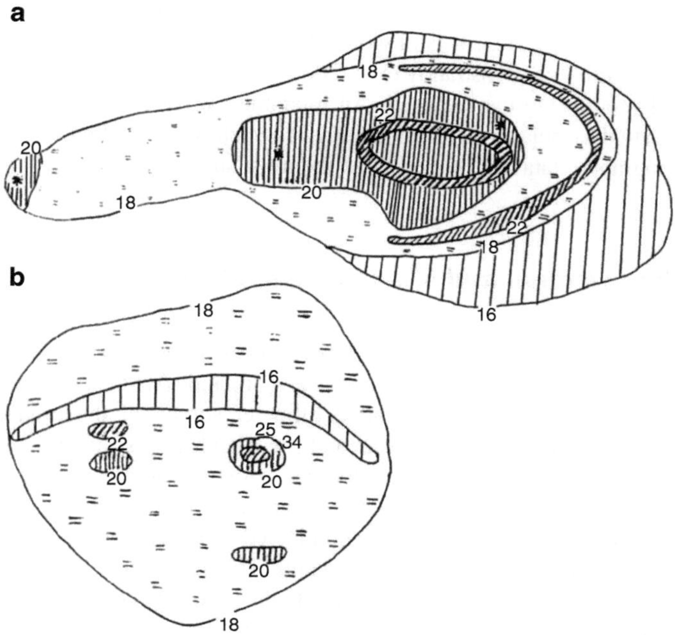

Fig. 8.5 Natural radioactivity fields over mud volcanoes; (**a**) Lokbatan, (**b**) Bozdag-Gobi (After Aliyev et al. 2001, with technical modification)

could trigger large mud volcano eruptions (Seibold and Seibold 2006). This was also found in Italy (Martinelli and Dadomo 2005), Japan (Chigira and Tanaka 1997), Turkmenistan (Guliyev and Feizullayev 1995) and other regions of the world.

Mellors et al. (2007) investigated the correlation between earthquakes and large mud volcano eruptions in a few regions (including Azerbaijan). The authors compared the occurrence of historical regional earthquakes ($M > 5$) with the occurrence of Azerbaijan mud volcano eruptions and found that the number of same-day earthquake/eruption pairs was significantly higher than expected if the eruptions and earthquakes were independent Poisson processes. Mellors et al. (2007) detected, for instance, a reliable correlation between the surrounding earthquakes and eruptions in mud volcanoes in Lokbatan, Keyreki and Shikzairli (central Azerbaijan).

Digits in the figure show values of radioactivity, μR/h

8.2 Engineering Geophysics

8.2.1 Monitoring of Oil-and-Gas Pipelines

Geophysical monitoring of oil-and-gas pipelines (primarily in Baku-Tbilisi-Ceyhan) (Balat 2006) is one of the most important problems in Caucasian engineering geophysics (Rabinowitz et al. 2004; Hetland and Gochitashvili 2004; Babazade et al. 2008; Tanircan et al. 2011). It is obvious that for this purpose ground (e.g., Modin 2010) and satellite (e.g., Kostianoy et al. 2008) observations as well as underground geophysical monitoring can be used. An integrated wavelet approach (Eppelbaum et al. 2011a) that associates different geophysical methods including satellite imaging as well topography data has proved to be the most successful.

8.2.2 Investigation and Monitoring of Dams

The Inguri hydroelectric power station (IHEPS) is the largest station in the Caucasus and is located on the border between Georgia and Abkhazia. Balavadze and Abashidze (1973) carried out a detailed tiltmeter survey of the modern tectonic movements of the Earth's crust in the area of the IHEPS dam structure. An integrated study of compressional and acoustic velocities, resistivity and pressure was successfully combined with the seismicity of this region and rock tension at the IHEPS by Savich et al. (1983). Bratus et al. (2005) conducted geoelectrical investigations at the Inguri and Zhinval (Southern Ossetia) dams. The main goal of this study was detection of voids and determination of the thickness of the concrete.

A comparative large number of studies have been devoted to seismic zonation and monitoring of the Caucasian dams. For instance, Agamirzoev et al. (1986) carried out a seismic micro-zonation of the vicinity of the Enikend dam located in central Azerbaijan at the junction of the Lesser Caucasus and the Kura Depression; Marchuk et al. (2001) investigated the risk of seismic activity in the area of the Sulak dam in Dagestan (Northern Caucasus); Sargsyan (2009) investigated reservoir-triggered seismicity in several large dams in Armenia; Peinke et al. (2006) and Matcharashvili et al. (2010) examined dynamic processes around the IHEPS associated with periodic changes in the water level.

8.2.3 Geophysical and Structural-Geological Analysis

A comprehensive interpretation of engineering refraction seismic data along seven profiles was carried out in the vicinity of a health resort called Red Field Valley in the Caucasus Mountain (Gylyjov and Piip 2002). A homogeneous function method made it possible to completely automate the processing and interpretation of the data. This yielded a detailed structural interpretation showing the layers, faults, rockslides, and zones of disturbed rocks and determined the structure of the bedrock to a depth of 50 m (Fig. 8.6).

Piip and Orlova (2006) employed seismic refraction data to study the structure of the Dallackau (Republic of Northern Ossetia-Allaniya, Northern Caucasus) landslide. Application of the homogeneous function method provided a more detailed resolution than conventional methods.

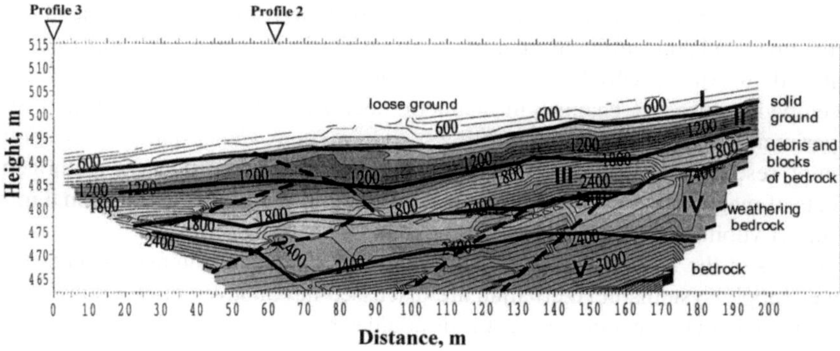

Fig. 8.6 Engineering refraction seismic section profile along profile 7 in the Red Field Valley (western slope of the Greater Caucasus) (After Gylyjov and Piip 2002, with minor modifications)

Morgounov and Zdorov (2007) investigated the effects of electromagnetic emissions as a precursor of landslides and avalanches in the Northern Caucasus. The recorded anomalous effects were explained in terms of microfracturing mechanisms in zones of steeply rising strains. During the process of microfracturing, the volume under stress is a source of acoustic and electromagnetic emissions due to the excitation of mechanoelectric transformers through mechanisms of relaxation of the mosaic electric charge, originating on the sides of opening cracks and other dislocation processes.

Radiometric and acoustic mapping has been successfully applied to select the area for the construction of a radio-TV transmitting station in Baku (Aliyev and Zolotovtskaya 2005). The building site measuring 0.04 km^2 was located on Absheron limestone. Based on gamma-field analysis, a dangerous fault zone in the immediate vicinity of the area was identified. In addition acoustic mapping indicated that there was heightened soil fracturing in this area. As a result, the planned building site was moved several hundred meters from the initial location.

8.2.4 Investigation of Geophysical Field Time Variations

8.2.4.1 Effects of Magnetic Field Variations and Human Health

Studies have confirmed the relationship geomagnetic activity and human health (e.g., Chizhevsky 1930, 1963; Rapoport et al. 2006). Babayev and Allahverdiyeva (2007) described the effect of variations of geomagnetic activity (GMA) on human health. Their experiments, as well as statistical and biomedical analyses confirmed the influence of changes in GMA on the brain and on human emotions.

Berishvili et al. (2008) noted that in some regions of Georgia the change in horizontal gradients of the total magnetic field exceeds by several fold the maximum value of the magnetic field during magnetic storms. The influence of these magnetic effects on human health calls for further investigation.

A reliable correlation between the Wolf number and solar geomagnetic activity on the one hand and cardiovascular mortality in Tbilisi between 1980 and 1992 was analyzed in Amiranashvili et al. (2002). Diagnostic magnetometers have been developed to measure changes in magnetic declination (e.g., Zverev et al. 1997). Experimental data from medical clinics and in some Caucasian resorts should prompt researchers to further explore the relationship between the pathogenic effects of magnetic storms, geoecological problems, and magneto-variational monitoring (Lyubimov et al. 1993).

Studies of the influence of geomagnetic storms of various intensities on electric power supply systems in Azerbaijan are discussed in (Babayev et al. 2006). It was concluded that large geomagnetic storms could seriously damage the electric power systems and geophysicists studying magnetic events should work closely with experts in power engineering.

8.2.4.2 Estimating the Impact of Electromagnetic Field Variations on Human Health

The influence of native electromagnetic radiation on living organisms was first studied by Chizhevsky (1930).

Khashiev and Shulga (1985), two researchers from the North-Caucasus Scientific Center, examined the influence of electromagnetic fields of extremely high frequencies on the central nervous system and the relationship to human depression. Khandokhov (2004) examined the influence of alternative electromagnetic fields of various frequencies on living organisms.

The sun is the main source of electromagnetic radiation on the human body. Many villages are located at heights of more than 2,000 m in the Caucasus, where the influence of solar radiation is especially strong. However, it is precisely these villages that hold longevity records (42% of the all individuals living to be 100 and over reside in the Caucasus). As frequently happens, there is doubtless a combined set of influences which need to be untangled for correct interpretation.

8.2.4.3 Analysis of Ancient Temperatures

Analysis of the climates of the past has attracted the attention of many experts working in the field of borehole temperature measurements who use this information to reconstruct ancient temperatures (e.g., Huang et al. 2000; Bodri and Cermak 2005; Eppelbaum et al. 2006b). In Azerbaijan, these investigations were employed in two boreholes – Garadzha-16 and Gemigaya-14 located in the Kura Depression (Mukhtarov et al. 2010). These boreholes were selected mainly because of the high accuracy and small steps of temperature observations in the boreholes. It was found that in the Kura Depression in the seventeenth and eighteenth centuries the temperature decreased 1°C, and in the second half of nineteenth century there was warming of 1–2°C. Undoubtedly, such investigations should be carried out in other regions of Azerbaijan and the Caucasus.

8.3 Archaeogeophysics

Geophysical methods (according to our experience and personal communications with Azerbaijanian, Armenian, Georgian and Russian geophysicists) were comparatively widely applied in archaeological sites in the Caucasus. However the number of publications is unjustifiably small. Thus the greater part of these archaeogeophysical investigations are still being held in archives as unpublished works. A. Shakhnazaryan (Azerbaijanian Geological Association) successfully employed a Vertical Electric Sounding (VES) method to study the foundation of the famous Maiden Tower ("Gyz Galasy") in the center of Baku (personal communication). VES helped define the position of the foundation and even determine that it was reconstructed over the centuries. Experimental dowsing (details of this procedure can be found for instance in van Leusen (1998)) by two geophysicists from the YuzhVNIIGeofizika on the tower also suggested there was a previously unknown underground cave under this town (personal communication).

In 1963, D. Tsitsishvili and G. Tabagua made various modifications to a resistivity survey at an ancient Roman site dating back to the first to sixth centuries AD in the area of Pizunda on the Black Sea coast of the Caucasus (Frantov and Pinkevich 1966). The findings revealed the remains of the ancient walls.

Detailed electric prospecting in the ancient city of Armaztsikhe-Bagineti (Eastern Georgia) revealed four anomalous areas. Their relevance to archaeological targets was confirmed by control trenches measuring 17 m on two terraces at the site. At the boundary between quaternary sediments and alluvial clays four significant archaeological sites with a total area of 1,170 m^2 were discovered (Apakidze et al. 2001).

The advanced geophysical field methodology for complex environments tested at archaeological sites in Israel (e.g., Eppelbaum et al. 2001, 2006, 2010; Eppelbaum 2011a) can also be employed in Caucasian archaeological sites.

8.3.1 Northern Caucasus

Shraibman et al. (1988) used electric and magnetic methods to localize burial grounds in the Northern Caucasus. In the first stage, which as aimed to detect burial boundaries and divide the area into separate zones, the authors used a VES method. They identified zones of thick clay and loam deposits and the zones which, most likely contained catacomb burials. In the second stage, micromagnetic exploration of the areas was used to delineate individual catacombs. These geophysical

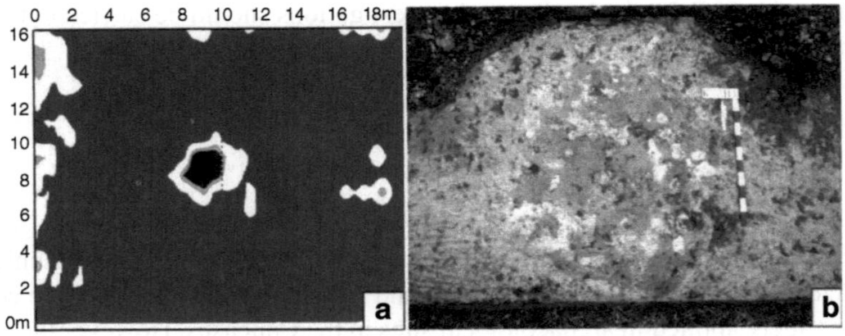

Fig. 8.7 Results of GPR investigations in the Sarmat era settlement in Novopavlovsk town (Northern Caucasus), (**a**): Horizontal cross-section of radar imaging, (**b**): burial site after cleanup (After Merkulov 2005, with minor modification)

methods reduce the amount of effort and time needed to carry out conventional archaeological excavations.

Fassbinder et al. (2007) carried out an integrated geophysical survey in the Bronze Age settlements of the Northern Caucasus (south of the town of Kislovodsk). Initially, satellite images, different types of aerial photos and topographical data were analyzed. Then a total magnetic field survey was made of several archaeological sites (e.g., Kabadinka 2), where many archaeological features were revealed. The largest area was Kabadinka 2 – where a grid of 200 $\times$ 240 m was employed. As a whole, the authors (Fassbinder et al. 2007) considered that the integrated survey was very successful and recommended further use of this methodology. A similar integrated study that combined satellite and aerial images and precise magnetic investigation in the Kislovodsk Basin of the Northern Caucasus was conducted by Reinhold et al. (2007). The continuation of these investigations (comprehensive magnetic survey with satellite imaging) revealed previously unseen and unknown structures beneath a large area of this plateau (Fassbinder et al. 2010). The association of these findings with selective archaeological excavations should yield further evidence that can be used for dating to the Bronze Age, and may also explain the purpose of the settlements – dubbed the "Stonehenge of the Causasus" and the rings of trenches in more detail.

An electric resistivity study was carried out in the "Mountain Echo" district (Kislovodsk) to search for Khazar remains of the seventh to ninth centuries (Modin 2010). The findings indicated two cultural layers with signatures of destruction between them (this destruction reflects the end of Khazar epoch).

According to some Russian Internet sites (e.g., site of the Inst. of Archaeology of the Caucasus (town of Nalchik)), GPR surveys have been extensively used for imaging ancient burial mounds and other ancient remains in the Northern Caucasus. Merkulov (2005) reported the successful application of GPR LOZA in a few field sites in the Northern Caucasus. Figure 8.7 presents a GPR examination of the Sarmat era settlement in the town of Novopavlovsk. In the anomaly identified in

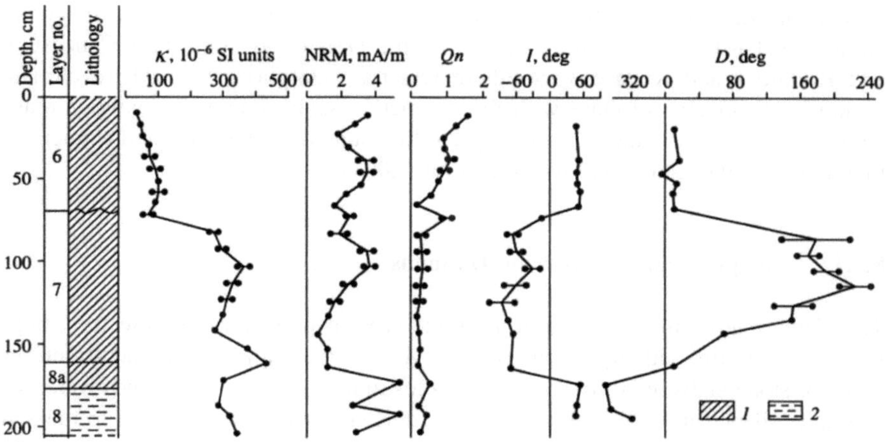

Fig. 8.8 Magnetic and paleomagnetic characteristics of the Paleolithic site deposit in the Matuzka Cave (northern Caucasus) (After Pospelova et al. 2006). (*1*) loam, (*2*) clayey loam

the central part of the horizontal section (Fig. 8.7a) archaeological excavations discovered remnants of wooden structures and ovens of the Russian stronghold of the last quarter of the eighteenth century (Fig. 8.7b).

8.3.1.1 Archaeopaleomagnetic Investigations

Precise paleomagnetic examinations were carried out in the Matuzka cave (western segment of the Greater Caucasus) (Pospelova et al. 2006). The typical magnetic and paleomagnetic characteristics in the Matuzka cave are presented in Fig. 8.8. The magnetic susceptibility of the directional samples varies in the range of 60–475 $\times$ 10^{-6} SI units. Interestingly, layer 6 is characterized by low κ values and a high Koenigsberger factor *Qn*. A record of the Matuzka excursion was discovered in the lowest cultural layer 7, where a human stone industry was revealed. The field direction of the excursion time was completely reversed. The excursion occurred in a period of a lower intensity of the geomagnetic field in a warm and humid climate. As a result of the VGP (virtual geomagnetic pole) position and climatic conditions, the Matuzka record can be identified with the ~130-ka Blake excursion.

Magnetic studies of the rock sequences were carried out in the Mezmaiskaya cave at its unique multilayer Paleolithic site (Pospelova et al. 2011). The magnetic properties were analyzed for 17 layers dating from over 73 ka ago to recent times. The rocks of layer 1X (Early Upper Paleolithic, ~38 ka) were found to have the highest magnetic susceptibility – up to 2,500 $\times$ 10^{-6} SI, which is related to the intense activity of Homo sapiens.

The question of paleoclimatic reconstruction by observation of scalar magnetic parameters has been solved for sediments of the Treugolnaya (Northern Caucasus), and Kudaro-1 and Kudaro-3 (Southern Ossetia, Georgia) caves, where climate variations precisely correlate with the distinctive scalar magnetic features: magnetic susceptibility κ, intensity of anhysteretic magnetic remanence and isothermal magnetic remanence (Pospelova et al. 1996, 2001).

8.3.1.2 Geophysical Effects from Dolmens

The Caucasian dolmens represent a unique type of prehistoric architecture, built with precisely dressed large stone blocks. Most of them are rectangular structures made of stone slabs or cut in rocks with holes in their facade. The stones were, for example, shaped into 90° angles, to be used as corners or were curved to make a circle. The monuments date from the end of the fourth millennium to the beginning of the second millennium B.C. These dolmens cover the Western Caucasus on both sides of the mountain ridge, over an area of more than 12,000 km^2 (Markovin 1978). Which people erected the dolmens remains a mystery (Valganov 2004). Clearly the dolmens are the most enigmatic ancient remains of the Caucasus. Boichenko et al. (2000) examined several less damaged dolmens located in the village of Shapsugskaya in the Krasnodar region. Radiometric investigations did not show any anomalous distribution. Some anomalies were detected in the observed magnetic field pattern. However, these anomalies could have been caused by the complex geometric form of the dolmens, and not their anomalous properties. Magnetic investigations of complex (and visible) objects should as a general rule be accompanied by 3D magnetic field modeling (e.g., Eppelbaum et al. 2000). These investigations involving several geophysical methods must be continued on dolmens situated in different regions of the Caucasus.

8.3.2 The Lesser Caucasus

A detailed paleomagnetic survey was carried out on the site of the earliest human occupation in Europe – the Dmanisi archaeological site in Southern Georgia (northern slope of the Lesser Caucasus) (Goguitchaichvili and Pares 2000). The Dmanisi site has yielded human remains and a lithic industry associated with the Late Pliocene – Early Pleistocene fauna. The site is composed of volcanogenic sediments overlying basaltic lava flows. The lithostratigraphic sequence comprises two basic depositional units: Unit A, overlying the basalt flows, and Unit B on top. A paleomagnetic and rock-magnetic study was carried out on 106 specimens from Units A and B and the uppermost basalt flow. The lava and Unit A yielded normal polarities, while reversed polarities and anomalous directions were observed in Unit B; the latter was probably due to overlapping of a secondary and a primary reversed polarity component. The lower part of the section shows a clear correlation with the Olduvai subchron, and the upper levels could be as young as 1.07 Ma. As human remains were found both in units with normal and reversed polarity, different non-contemporaneous human occupations might have been possible. This investigation was later itemized and defined (Calvo-Rathert et al. 2008).

A detailed resistivity survey of the basalt lava surface was conducted in the Dmanisi site to determine the maximal depth of occurrence of osseous wild animal and hominoid material (Dzhahutashvili 2006).

A new paleomagnetic study has recently been carried at the Pliocene site of Kvabebi (Sighnaghi region of Eastern Georgia, in the area around Kvabebi Mountain) to shed light on the origins of the early human occupation of Eurasia (Agusti et al. 2009), as evidenced by the early Pleistocene site of Dmanisi. The results indicate that this site should be situated in a reverse interval, which was identified as chron 2An.1r. The age of this site is therefore close to 3.07 Ma, coeval to the Hadar beds of the Afar Depression.

Electric Resistivity Tomography (ERT) imaging has been used to estimate the thickness of sediments in the Prehistoric Azokh Cave in the south-eastern segment of the Lesser Caucasus (Nagorny Karabakh, Azerbaijan) (Fernández-Jalvo et al. 2010).

Magnetic gradiometric surveys have been successfully carried out to identify remains of the Late Bronze Age fortress settlement in northwestern Armenia (Lindsay et al. 2010). The magnetic survey of basaltic ancient remains permitted not only to define the location of known archaeological features, but also to discover previously unknown ancient objects.

8.3.3 Taman and Kuban Regions

GPR imaging has been successfully applied in the Taman Peninsula both in land and shallow water. Archaeological investigations have confirmed the effectiveness of GPR analysis in dozens of investigated sites. The most significant GPR investigations were carried out in the Taman Gulf where numerous archaeological targets occur under a water layer of 0.5–1.0 m at a depth of 2–2.5 m under the sea bottom. The near-bottom layer is mainly composed of sand, shell rock and silt. Many ancient settlements are located in the littoral coastal zone at a distance of 100–300 m from the coast. These ancient targets were first detected only by chance for instance after strong storm wash-out. The development of a marine modification of GPR method initiated regular geophysical examinations of this zone (according to data from the Geoph. Dept. of *VNISMI*, Moscow).

The VES method, accompanied by investigation of samples from near-surface boreholes was used in the Semibratnee city-site (ancient town of Labris) in the Kuban region (Vnukov et al. 2008). The analysis of VES profiles revealed the bottom of a high resistivity cultural layer overlaid by more conductive bedrock. As a result, two unfortified settled areas were found stretching to the west and to the south of the fortified part of the settlement.

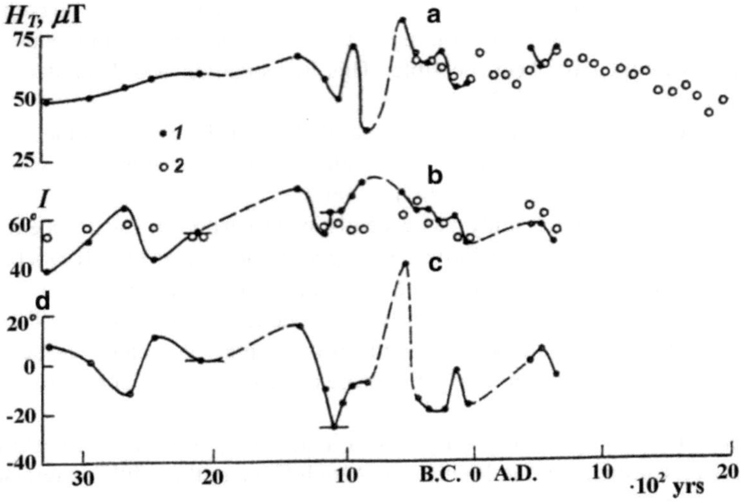

Fig. 8.9 Variations of components of the archaeogeomagnetic field in Georgia: (**a**): intensity (in microTesla), (**b**): inclination, (**c**) declination (Burlatskaya and Chelidze 1987). (*1*) data from Georgia, (*2*) world data

Magnetic survey at this site was carried out in 2006 in the southern, most elevated area, where recent excavations have taken place and on the western field (in 2007) (Smekalova et al. 2008). On the magnetic map of the western field there are two long positive anomalies (in average 40 nT in the amplitude), which come together to form a 90º angle. There are two very strong positive magnetic anomalies and also two interruptions in the long anomalies in the western and south-eastern parts of these linear structures. When the magnetic map was overlaid on the aerial photo, it became clear that the two long anomalies corresponded very precisely to the south-western border of the southern rectangular part of the site (Smekalova et al. 2008). The source of the anomaly is probably a city wall that was built of clay, which was subjected to fire afterwards either occasionally or deliberately to harden the clay "body" of the wall.

8.3.4 *Transcaucasus*

Numerous archaeomagnetic investigations have been carried out in the Transcaucasian (unfortunately, most remain unpublished or appeared in journals that remain hard to access).

Figure 8.9 displays time variations of magnetic field intensity, inclination and declination in Georgia for a period of 4,000 years (Burlatskaya and Chelidze 1987). The geomagnetic field intensity distribution (Fig. 8.9a) indicates an increase in the values from the third millennium to the beginning of second millennium B.C., and a slow decline of intensity from the end to the first millennium B.C. to the present time. These regularities are compounded by superimposed fluctuations with periods of several hundred years. This investigation was continued in (Burlatskaya and Chelidze 1990) where geomagnetic field variations in Georgia over last fifteenth centuries were analyzed.

Burlatskaya and Chernykh (1989) studied the geomagnetic field strength in Azerbaijan over the last 2,200 years, and Burlatskaya et al. (1991) investigated the geomagnetic field intensity in the Caucasus in the last 6,000 years.

Undoubtedly, all these paleomagnetic investigations must implement magnetic data analysis in the Caucasian archaeological sites and resolve various engineering problems through magnetic and micromagnetic surveys.

Recent archaeomagnetic investigations in the Transcaucasian include a detailed magnetic examination of three prehistoric sites: Cherula, Kotias-Klde and Dzudzuana (Chiatura region of Imeretia province, Georgia) (Itkis et al. 2011). In all three sites the targets were prehistoric caves. Magnetic surveys at these sites and magnetic susceptibility measurements of the soil were performed using a 1×1 m grid. The soil within such caves is usually affected by a number of natural and anthropogenic processes. The enhancement of the cave fill due to processes associated with human habitation (e.g., repeated heating and accumulation of organic debris) provided favorable conditions for the conversion of the iron oxide found within the soil into a highly ferromagnetic form (Dalan and Banerjee 1998). The results at the prehistoric cave sites testify to the high efficiency of the method for revealing these targets and providing a detailed characterization of their specific features (Itkis et al. 2011).

8.4 Environmental Geophysics

8.4.1 Landslide Geophysical Monitoring

Landslide geophysical monitoring is one of the most crucial problems in environmental geophysics. Bogoslovsky and Ogilvi (1977) carried out an effective combined self-potential (SP) and temperature investigation of a flowing landslide near the town of Adler (Black Sea coast of the Caucasus). The SP measurements reflect the details of groundwater flow and the degree of water saturation of the landslide body (Fig. 8.10). The position of the landslide body is characterized by distinct negative SP anomalies. The equipotential contours generally delineate the landslide boundaries and the minimum gradients are in the direction of groundwater flow.

The negative anomaly at the head of the landslide is associated with water infiltration through the fractures located near the wall of detachment. Temperature observations of the landslide body show increased values reaching +31°C (Fig. 8.10). These anomalies are associated with heating of the upper soil strata in

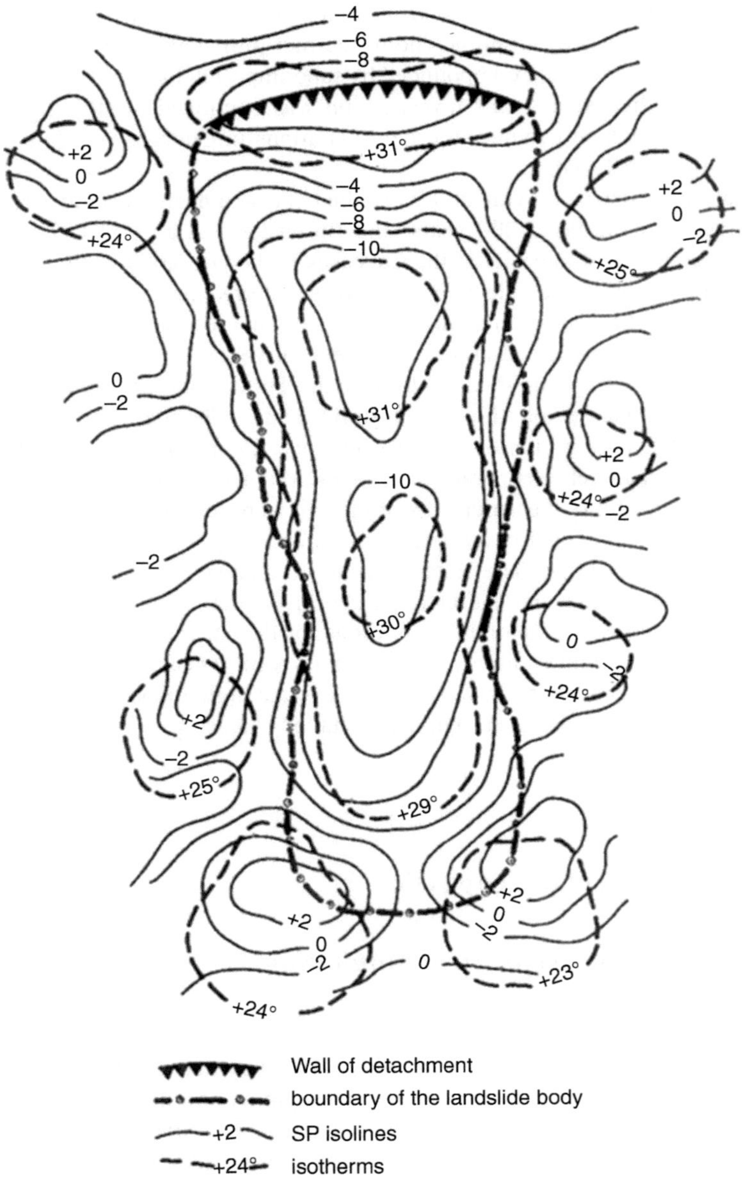

Fig. 8.10 The results of SP and near-surface temperature measurements on a flowing landslide

those places where the groundwater level is at a greater depth. At the sites of seepage outflow (in the peripheral parts of the landslide and near its toe) the temperature drops to +23°C.

Electrical sounding curves obtained at an interval of one to two months in fixed points on the banks of one of the large water reservoirs in the Caucasus situated in fissured igneous rock clearly indicated a change in hydrogeological conditions (Ogilvy and Bogoslovsky 1979). It was associated with a gradual filling of pores and fissures with water from the reservoir. The authors detected a clear displacement of maxima of the ρ_a curves towards higher values with time.

8.4.2 Study of Radioactive Parameters

The gamma-background was monitored in the underground laboratory in Tbilisi (Georgia) at the depth of 50 m (Nevinsky 1987). For a more precise determination of background characteristics the detector (crystal NaI(Tl)) was placed into a special metallic absorber. The whole energy range of gamma-ray (0.3–5.0) Mev was divided into intervals of 1.7–2.0, 2.5–3.4 and 3.4–5.0 Mev. These intervals correspond to gamma-radiation of radon, thoron and cosmic rays, respectively.

A gamma-survey at the Absheron Peninsula (Azerbaijan) revealed a few spots of heightened radioactivity and a few districts with extremely high radioactivity up to 3,000 μR/h that obviously constitutes a danger for people working at these sites (Aliyev and Zolotovtskaya 1996). These districts are located near old oil-fields and around iodine factories. Unfortunately, contamination of some new (previously "clean") areas increases very quickly. For instance, gamma-radiation on Gum isle (Caspian Sea) rose from 4–5 to 600–700 μR/h in 5 years (Zolotovitskaya 2003).

Feyzullayev et al. (2005b) carried out a detailed analysis of Oligocene clay radioactivity in the Greater Caucasus. They found that Oligocene clay integral radioactivity ranges from 29 to32 μR/h with an average value of 31.4 μR/h. Such values are known to be typical for clays.

The Kura-Araks river Basin (within Armenia) was tested for water radioactivity and radionuclide composition (Nalbandyan and Saghatelyan 2008). Collating the obtained monthly data on 40 K and ^{226}Ra for the period revealed high concentrations of 40 K versus ^{226}Ra.

Radon (Rn) measurements were conducted along the geological faults and the vicinities of petroleum storehouses near the Black Sea coast of the Caucasus (Tsvetkova et al. 2004). At the time of the measurements at the borders of faults there was an increased concentration of radon; namely, one peak approximately in the middle of a fault could be observed. The combined application of radon and hydrogen surveys in the vicinity of petroleum storehouses showed the presence of masked petroleum pollution not detectable from the surface.

It should be pointed out that radon measurements in the subsurface are influenced by a number of factors (Finkelstein et al. 1998; Eppelbaum and Finkelstein 1998):

$$F_{sum} = F_1 + F_2 + F_3 + F_4 + F_5 + F_6 + F_7 + F_8, \quad (8.1)$$

where F_1 is the function of Rn emanation from the soil (including the influence of the external temperature); F_2 is the daily variation of Rn due to vertical warm convection arising shortly after sunrise; F_3 is the variation of Rn due to rain and Rn dissolved in underground water; F_4 is the variation of Rn due to fluctuations in atmospheric pressure; F_5 is the variation of Rn due to air movement (direction and speed of wind or ventilation); F_6 is the variation of Rn related to deep geodynamics, especially near earthquake epicenters; F_7 is the variation of Rn due to strong geomagnetic storms in periods of solar chromo spherical bursts; F_8 is the variation of Rn caused by the settling of aerosols following dust storms, which can transfer a number of radioactive elements and may lead to a sharp increase in radon.

The best ways of eliminating different kinds of noise, including shallow noise occurring down several meters in caves and more deeply in underground tunnels are described in Finkelstein et al. (1998, 2006).

8.4.3 Revealing Ring Structures

The methodology involved in delineating ring structures (RS) was presented in Sect. 3.7.3.2. Here we consider several examples of different kinds of delineation. First, the geophysical signatures of mud volcanoes discussed in Sect. 8.1 are typical examples of ring structures (see the clay diapirs box in the chart in Fig. 3.34).

8.4.3.1 Identification of a Meteorite Crater

As was noted in Eppelbaum (2007b), one of the most important types of RS are meteorite craters (aside from the largest impacts, the number of small meteorite signatures is considerably larger than people usually assume). To identify impact structures of various sizes, different geophysical methods can be applied.

The first meteoritic crater found in the Caucasus – Arkhys – is located in the valley of the Large Zelenchuk River (Northern Caucasus) (Hryanina and Vityasev 1999). This crater is a depression with a diameter of 2 km and the visible depth from the foot to the top of the socle rim is about 250 m. The geophysical signature of the crater is a negative gravity anomaly in the Bouguer reduction of 3–4 mGal. Statistically, there are dozens of these significant craters (with diameter > 500 m) in the Caucasus. To locate them, different procedures for RS delineation against the significant noise background should be applied.

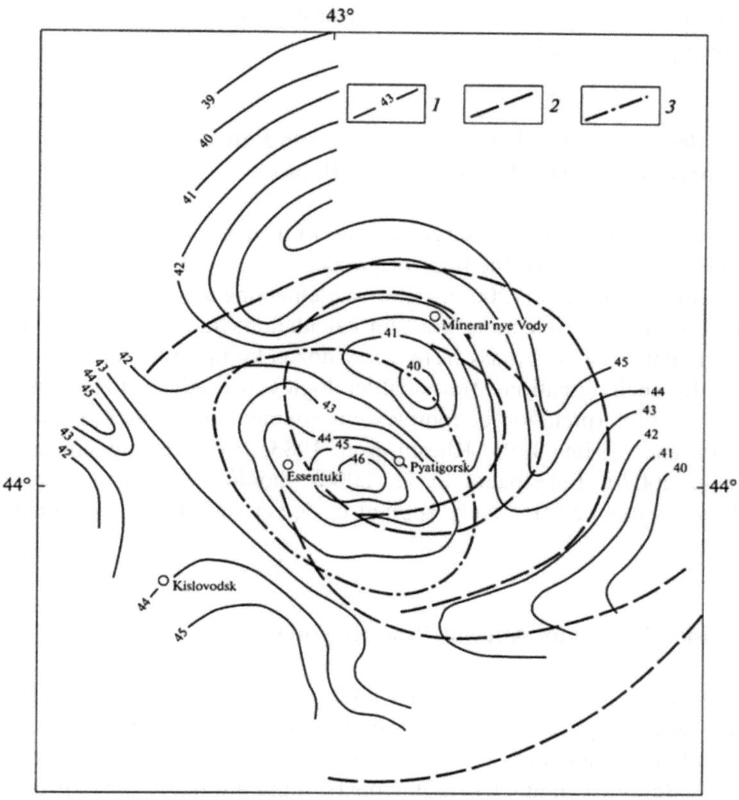

Fig. 8.11 Location of the Moho discontinuity and circular faults in the Pyatigorsk volcanic center (northern Caucasus) (After Masurenkov et al. 2009b). (*1*) depth to the Moho discontinuity, km, (*2*) circular faults at the earth's surface, (*3*) axis of circular reclamation which delimits the topmost dome fan pressure

8.4.3.2 Identification of the Relationship Between Surface and Deep Circular Structures

On the basis of comprehensive geophysical-geochemical investigations a deep ring structure was identified in the area of the Caucasian Mineral Waters (Garetovskaya et al. 1986; Bashorin et al. 2005). Recent petrological and geochemical studies linked this deep RS with the position of subsurface faults (Fig. 8.11) (Masurenkov et al. 2009b). The subsurface RS reflects the position of the Pyatigorsk volcanic center, which is associated to the dynamic vertically stretched Earth's crust. This may be the vortex-like structure described by Aleinikov et al. (2001). Another map by Masurenkov et al. (2009a) for the same region shows the distribution of heat flux of carbonaceous mineral water as a superposition of ring structures with different orders. The correlation between the subsurface and deep geological-geophysical signatures could clarify the role of structural heterogeneity (Masurenkov et al.

2009a, 2009b) and may be used both for seismological prognosis and searching for economic deposits (e.g., Khain 2000).

8.4.3.3 Identification of Ring Anomalies from Potentially Dangerous Metallic Objects Occurring at a Depth

A very precise marine magnetic survey was performed to localize potentially dangerous (metallic) objects along the trace of projected oil-and-gas pipeline in the North Caspian (Brusilovsky et al. 2009). After implementing a number of interpretation procedures (frequency selection, gradient calculation, transformations, etc.), more than 90 anomalies (with a mass from a few dozen kg up to 50 tons) were reliably detected. In such situations the methodologies for advanced magnetic field processing and interpretation in complex environments presented in Chap. 4 (e.g., Eppelbaum and Mishne 1995; Khesin et al. 1983, 1996; Eppelbaum et al. 2001; Eppelbaum 2007b; Eppelbaum et al. 2008) should be applied. Clearly such methodologies can be effectively applied to detect such targets anywhere in the world.

8.4.3.4 Identificationof Ancient Ring Remains

Several ancient ring targets (including prehistoric ditches) were identified in the Northern Caucasus by the use of geographic information systems (GIS), magnetic surveys and GPR data analysis (e.g., Belinskiy 2005; Merkulov 2005; Fassbinder et al. 2010; Belinskiy et al. 2011). Obviously, more are likely to be discovered.

Chapter 9
Investigation of Seismic Activity

The Caucasus is one of the most active segments of the Alpine-Himalayan seismic belt (Khain 2000). The Caucasian region is characterized by intensive deformation and seismicity that accommodates the continental shortening between the Eurasian and Arabian plates, which are converging at a rate of about 30 mm/year (De Mets et al. 1990; Jackson 1992). The Caucasus is considered a key area for seismic hazard assessment for the following main reasons (Balassanian et al. 1999): (1) the active tectonics and seismicity rate of the whole area, (2) availability of abundant multi-disciplinary data and a long established tradition of hazard assessment, (3) the unique opportunity to test different methodologies in one test area.

There are several thousand publications on Caucasian seismic activity. Among the most significant are Borisov (1982), Gorshkov (1984); Jackson (1992), Vinnik et al. (1992); Rebai et al. (1993); Khain and Lobkovsky (1994), Ismail-Zadeh (1996), Kerimov (1996), Balassanian et al. (1997), Giardini and Balassanian (1997), Chernobay and Gabsatarova (1999), Martirosyan et al. (1999), McClusky et al. (2000), Gasanov (2001), Rogozhin (2002), Akhmedbeili and Gasanov (2004), Alizadeh (2005), Bashorin et al. (2005), Ismail-Zadeh (2005a), Ulomov et al. (2007), Babazade (2010), Babayev et al. (2010).

9.1 Earthquakes in the Caucasus: A Short Historical Review

Ancient earthquakes in the Caucasus have been reported in hundreds of publications. Nikonov (1982) described a very powerful earthquake in the Eastern Caucasus that took place on January 14, 1668. The author noted that it was the strongest detected earthquake in the last five centuries with an estimated magnitude of $M = 8 \pm 0.5$. At the same time Gasanov (2001) described a catastrophic earthquake at $M \approx 9$ in the Goygol Lake area (Lesser Caucasus) in 1139 (generally speaking this earthquake triggered the formation of this famous lake). The Dagestan earthquake of 1830 with $M = 6 \pm 0.9$ was one of the strongest in the Northern Caucasus (New Catalog 1977). Numerous strong earthquakes ($M \geq 6$) were

L.V. Eppelbaum and B.E. Khesin, *Geophysical Studies in the Caucasus*, Lecture Notes in Earth System Sciences, DOI 10.1007/978-3-540-76619-3_9,

registered in the Shamakha area (close to the southern slope of the Greater Caucasus) in 1192, 1667, 1669, 1828, 1859, 1868, 1872 and 1902. Earthquakes with a magnitude $M \geq 6$ were also detected in the Mashtaga area (1842), and the Caspian Sea (957, 1812, 1842, 1852, 1911, 1935, 1961, 1963, 1986, 1989 and 2000) (New Catalog 1977; Gasanov 2001; Kovachev et al. 2009). Four catastrophic seismic events were detected in the last 5,500 years in the Elbrus region of the Northern Caucasus (Bogatikov et al. 2003).

It should be noted that one of the first seismological stations in the world – "Baky" – was founded by the famous Nobel businessmen brothers in Baku in 1903 after the catastrophic Shamakha 1902 earthquake.

Evaluations of the risk of a seismic hazard in the Caucasus region, based on cluster analysis of the geological, geophysical, and seismological data (Reisner and Ioganson 1993) shows that the Elbrus area (North Caucasus) has a potential earthquake source (PES) with a predicted maximum earthquake magnitude of 7.2 (Rogozhin 2002). Vikulin (2003) reported that the strong earthquake cyclicity for the Caucasus region has a period of 90–140 years.

On the basis of the Sadovsky et al. (1983) research for Caucasian earthquakes, the following simple formula was derived (Babazade 2010) to calculate the magnitude M:

$$M = \frac{\lg V - 2}{1.6},$$

where V is the volume of the earthquake focus in cm^3. However, it is clear that calculating the value of V is not a trivial problem.

9.2 Studying Petrophysical Properties in Seismogenic Regions

As emphasized above, the analysis of the physical properties of rocks plays an important role in all types of geophysical investigations. Analysis of seismological data is no exception (e.g., Kerimov 1996).

A comprehensive examination of petrophysical properties was carried out for the seismogenic crustal blocks of the southeastern Greater Caucasus (Safarov 2006). The author analyzed experimental *P-T* data on compressional wave velocities and densities of sedimentary carbonate rocks of the Shamakha-Ismailly seismogenic blocks (southern slope of the southeastern Greater Caucasus). The petrophysical models of the crust of the region defined the geometric boundaries of the blocks including their lower edges.

Smirnova (1968) studied the influence of several earthquakes on the Gudermes oil field in the Northeastern Caucasus. One of the wells under study, located at 10–15 km from the epicenters of two earthquakes occurring on March 23, 1950 ($M = 3.5$ and 4.5) showed an increase in oil production of more than 30% on the day the earthquakes took place. Following two other earthquakes, occurring in August 1955 ($M = 4.5$ and 4.2), many neighboring wells showed immediate or delayed changes in oil production. Similar effects were recorded in many other oil boreholes in the Caucasus and around it. Beresnev and Johnson (1994) believe that this effect is caused mainly by the effect of elastic-wave excitation on saturated porous media.

Megakhed et al. (1985) analyzed the thermal characteristics of the effusive rocks from the Lesser Caucasus. These investigations led to some specific conclusions about heat transport in the rocks, and help account for some temperature precursors of dangerous geodynamic events (see for instance, Sect. 9.3.7).

Abdullaev et al. (2011) studied variations in apparent resistivity at the Dmitrovskoe oil-and-gas field situated on the western coast of the Caspian Sea in a seismic region of Dagestan (Northern Caucasus). These authors detected an increase of apparent resistivity before large earthquakes ($M = 6.8–7.4$) that was apparently caused by gas emissions from the field along fractured rock zones whose permeability is affected by varying tectonic stresses in the precursory zone of a large earthquake.

It is well-known that paleomagnetic studies have a direct relationship to tectonic processes and dangerous geodynamic events. Among the numerous investigations in this field are the works of Asanidze and Pecherskiy (1979), Asanidze et al. (1980), Van der Voo and Channel (1980), Ismail-Zadeh (1983), Bazhenov et al. (1996), Camps et al. (1996), Bazhenov and Burtman (2002), Goguitchaichvili et al. (2000, 2001), Ismail-Zadeh et al. (2005b), Issayeva and Khalafli (2006), Khalafli (2006), Shcherbakova et al. (2009).

A detailed review of petrophysical investigations in Azerbaijan (including seismogenic regions) can be found in Alizadeh (2005).

9.3 Modern Geodynamic Events and Geophysical Detection Methods

In the Greater Caucasus the main seismoactive zone are located in its southern foothills. The catastrophic Shamakha earthquake of 1902 was centered in the eastern part and the Racha earthquake of 1991 (the Racha–Lechkhumy Fault Zone is shown in Fig. 2.4) occurred in the eastern part of this zone. In the eastern part of the Greater Caucasus, seismic activity has been observed in its northern slope, from the Dagestan wedge to the Tersk-Sunzhensk range of the fore-Caucasus and the submarine Caspian margins of Dagestan and Azerbaijan (Khain and Lobkovsky 1994). Some authors believe that a relict Benioff zone dipping under the Greater Caucasus is present here and testifies to the collision between the Transcaucasian microplates and Eurasia (e.g., Khain and Lobkovsky 1994). Analysis of the Racha earthquake – the largest instrumentally recorded earthquake in the Greater Caucasus (Triep et al. 1995) – confirmed the earlier theory of the thrusting of the Transcaucasian Massif under the Greater Caucasus (Khain 2007).

One of the most powerful modern earthquakes in the Caucasus was the Spitak (Armenia) earthquake of December 7, 1988 (Cisternas et al. 1989). The main seismoactive zone was the Pambak-Sevan fault – the eastern extension of the North Anatolian fault. A fault formed during the Spitak earthquake (thrust with a strike-slip component) and the epicentral area had an east–west orientation. This seismicity was restricted to the intersection of the Pambak-Sevan fault and the diagonal Nalband fault zone. This whole seismically active area lies within the Transcaucasus transverse uplift caused by pressure from the Arabian Plate (Khain and Lobkovsky 1994).

Thus, the geophysical determination of earthquake signatures in both the Greater and Lesser Caucasus is a difficult problem since very complex geological media and a superposition of faults with different orientations highly complicates the identification of these signals.

It is beyond the scope of this book to present all the geophysical methods for short-term earthquake prediction in the Caucasus. However, the several geophysical methods briefly described below should provide a useful background for readers.

9.3.1 Gravity Temporary Tideless Variations

Gravity precursors of earthquakes associated with crustal deformations at a depth and other effects are well known (e.g., Walsh and Rice 1979; Lambert and Bower 1991).

Gravity field variations were measured in the towns of Ambrolauri and Oni (Georgia) within the epicentral zone of the Caucasus earthquake of April 29, 1991 ($M = 6.8$) (Skovorodkin et al. 1994). Twenty days prior to the aftershock of June 15, 1991 ($M = 6.3$), gravity field variations reached -50 μGal. On the basis of gravity field modeling it was suggested that the recorded gravity effect was due to rock density changes within the aftershock area as the result of closure-opening of fissures (Skovorodkin et al. 1994).

An interesting phenomenon was detected before the Agdash earthquake of 1999 in Azerbaijan (Kadirov et al. 2005b). At the gravity points of the geodynamic polygon of the southern slope of the Greater Caucasus (area Ismailly – Agdash) negative gravity variations of 170 μGal were detected from 1984 to 1999. After the 1999 Agdash earthquake, the gravity field stabilized back to the 1984 level.

Measurements of second derivatives of the gravity potential W_{xy} and W_{yz} were carried out over the course of 1981 at the Sheki geophysical station (southern slope of the Greater Caucasus). Five to six days before a series of geodynamic events in Ismailly and Shamakha regions in November 1981 these second derivatives manifested anomalous behavior. The differences in distribution of these effects before the Ismailly and Shamakha earthquakes suggested that mechanisms underlying these events were different (Kadirov et al. 2005b).

Since 2002 continuous measurement of tideless variations of gravity have been carried out in the "Binagadi" station in located in the Absheron Peninsula 24 km from Baku (Khalilov 2008). The measurements are captured simultaneously by four high-precision quartz gravimeters. As a result of these measurements gravity variations were detected preceding strong earthquakes whose epicenters were extremely remote (from two thousand to tens of thousands of km) from the station. Changes in tideless gravity variations have been recorded before strong earthquakes in Indonesia, Pakistan, Japan, Taiwan, India, the Philippines and Iran. Statistical data indicate that in 85% of the cases, the gravity signals appeared on average 8–15 days before powerful earthquakes (Khain and Khalilov 2006).

9.3.2 *Temporary Magnetic Variations Associated with Geodynamic Events*

Short and long-term magnetic precursors are discussed abundantly in the geophysical literature (e.g., Pudovkin et al. 1973; Gokhberg et al. 1979; Eppelbaum and Finkelstein 1998; Alperovich and Zheludev 1999; Gaffet et al. 2003).

Anomalous geomagnetic variations (AVG) have different characteristics and their tectonomagnetic effects may only be seen after removing all the noise and incident factors (Fig. 9.1). As can be seen from this diagram, this is not a simple problem.

The most important feature of these magnetic measurements is the amplitude of the dynamic band of the signal – more than 200 dB. Another peculiarity is the apparent error of magnetic measurements when a device with a reduced relative error of about 0.1% is used for the measurements of values about 0.0001% of its total scale. Thus, in such conditions the selection of parameters characterizing measurement accuracy is crucial.

Differential investigations of geomagnetic effects in Ambrolauri and Oni (Georgia) indicated a seismomagnetic effect of 7 nT during the aftershock period (Skovorodkin et al. 1994).

Figure 9.2 shows the negative tectonomagnetic effect 1.5 days before the Salyany earthquake ($M = 5$–5.3) January 09, 1998. Such a clear magnetic effect could be recorded because of the proximity of the observation station to the earthquake epicenter. However analysis of the differential magnetic function is preferable (Finkelstein and Maslatsov 1984; Eppelbaum and Finkelstein 1998).

Differential satellite magnetometry may also be used for this purpose, in that it can separate internal from external magnetic effects (Kharitonov et al. 2005).

In 2007, the H, D, and Z components of the magnetic field along 44 registered earthquakes with a magnitude >5 were analyzed in the North Caucasian Observatory. In 32 cases typical quasi-periodic electromagnetic disturbances (phase and amplitude-time peculiarities) were identified (Sobisevich and Sobisevich 2010). Quasi-harmonic disturbances of the magnetic field with a period of 100–400 s and a

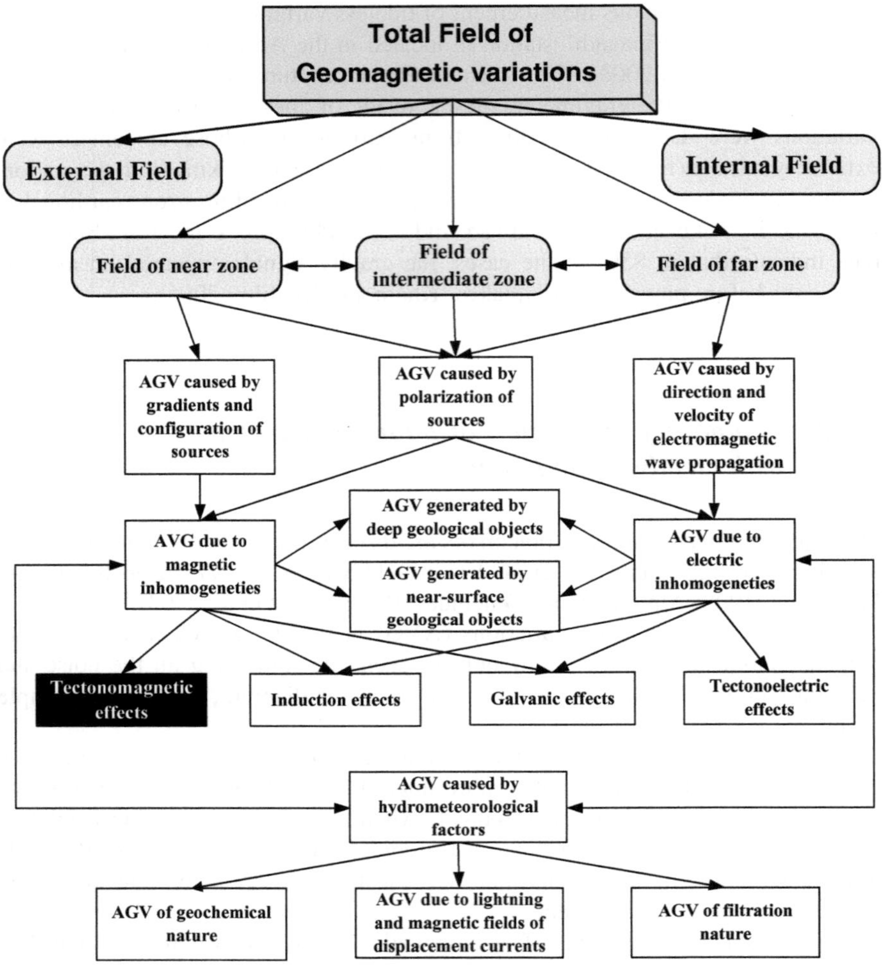

Fig. 9.1 Types of anomalous magnetic variations and their genesis (Finkelstein et al. 2012)

duration of 10–12 min delineate geomagnetic variations from the regional background of the Earth. These anomalous quasi-harmonic variations are observed generally 2–4 h before a strong seismic event. The appearance of anomalous variations before strong earthquakes is obviously caused by electrodynamic processes in the zone of an upcoming geodynamic event. Figure 9.3 illustrates geomagnetic field variations (components H and Z) and tilt changes (north–south, east–west) just prior to and during a strong earthquake in South Sumatra 12.09.2007 ($M = 8.2$, depth of focus – 33 km). About 3 h before the earthquake, magnetic variometers recorded quasi-harmonic ULF electromagnetic variations with a period of 150–250 s (Fig. 9.3). The evolution of the seismic process showed that this period was not constant. Rather it increased and reached a value of 450 s (Sobisevich and Sobisevich 2010).

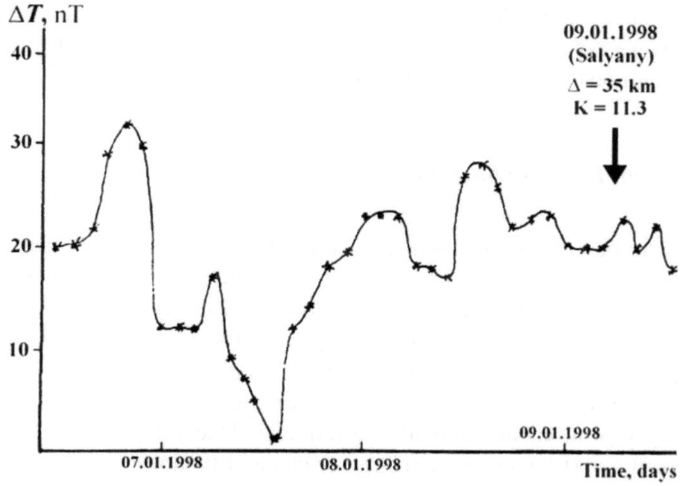

Fig. 9.2 Seismomagnetic effect before the Salyany earthquake January 09, 1998 (After Rzayev 2005, with small modification)

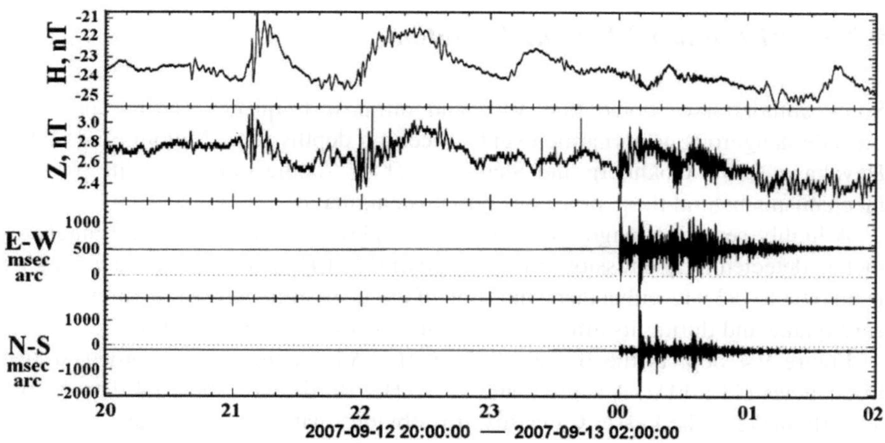

Fig. 9.3 Magnetic and tiltmeter variations recorded at the Baksan geophysical observatory (Northern Caucasus) before the Sumatra September 12, 2007 earthquake (After Sobisevich and Sobisevich (2010), with slight modifications)

9.3.3 Electric Field Potential Gradient

The effects of disturbances of the electric field potential before earthquakes has been analyzed in many publications (e.g., Gufeld and Shuleikin (1988); Liperovsky et al. 2008). Kachakhidze et al. (2009) examined the anomalous disturbances of the electric field potential gradient (EFPG) as possible precursors of earthquakes in the Caucasus. After eliminating seasonal and other variations, so-called "clear" EFPG anomalies from (−148.9 V/m) to 188.5 V/m were obtained. These "clear" EFPG anomalies were associated with 29 of the 41 earthquakes under study. The EFPG anomalies manifest themselves in 11-day precursors.

9.3.4 Laser Interferometer

A laser interferometer–deformograph (LID) with a 75 m long arm was developed by Moscow State University to record and study lithospheric deformations over a wide frequency range (Milyukov et al. 2005). It is located near Mount Elbrus, at the Baksan Ravine and assesses a range of geophysical and geodynamical effects. The Baksan apparatus is a Michelson two-beam unequal-arm laser interferometer operating in the regime of separate beams. Among other geodynamic effects, the LID reliably registered the Sumatra earthquake (December 26, 2004) at (a) LF channel (band $f < 0.1$ Hz); and (b) Tremor channel (band 0.1–10 Hz). Integration of the laser interferometer measurements with the GPS/GLONASS network defines specific geodynamical characteristics with greater precision (Milyukov et al. 2010).

9.3.5 VLF and ULF Time Variations

Some authors have noted that VLF and ultralow frequency (ULF) anomalies precede dangerous geodynamic events at certain depths (e.g., Meloni et al. 2004; Hayakawa 2007; Gokhberg and Shalimov 2008). At the same time the physical mechanisms behind these processes remains enigmatic.

A highly sensitive magnetovariational complex using magnetostatic magnetometers detected ULF emissions in the range of 0.1–1 Hz with an amplitude ranging from 0.03 to 0.2 nT, with a mean duration of about 30 min, before the Spitak (1988) earthquake and during its aftershock activity (Kopytenko et al. 1993).

Figure 9.4 depicts the parcel-like negative VLF anomalies recorded at two frequencies (11.9 kHz, Australia, and 15.1 kHz, Bordeaux, France) 4 days before an earthquake in the Lesser Caucasus. It is obvious that such a simultaneous, sharp drop in VLF field intensity at two frequencies from VLF-transmitters located in different regions of the world cannot be fortuitous.

The recent publication by Sobisevich et al. (2009) indicates that researchers at the North Caucasian geophysical observatory have discovered typical ULF configurations preceding strong Caucasian geodynamic events.

9.3.6 Radon Precursors

Radon (Rn) gas precursors have been studied in several hundred international publications. The most complete analysis of noise appearing in radon precursors

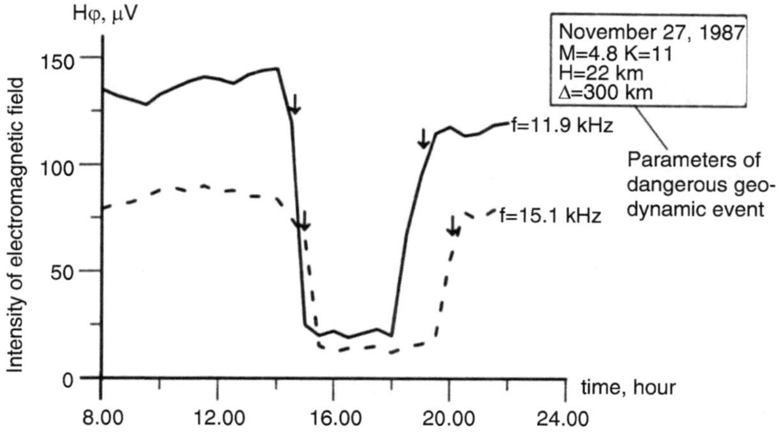

Fig. 9.4 Investigations of *VLF*-variations as a possible earthquake precursor (results of observations at the Mardakert district, Lesser Caucasus, November 23, 1987) (Eppelbaum and Finkelstein 1998)

was presented in Finkelstein et al. (1998) and Eppelbaum and Finkelstein (1998) – see Eq. 8.1 in Chap. 8. Finkelstein et al. (2006) proposed a new physical-environmental model that relates radon concentrations in underground environments to air temperature variations at the Earth's surface. The applicability of this model was tested by probing for radon indicators of geodynamic processes in two underground tunnels in central and southern Israel.

Radon variations were detected in 1987–1989 in the underground laboratory in Tbilisi before several large-scale earthquakes in the Caucasus (Nevinsky and Tsvetkova 1989). An extensive study using solid state nuclear track detectors (SSNTD) along the Northern Caucasus was carried out by Nevinsky and Tsvetkova (2005). An earthquake on September 28, 2002 was predicted by a variation in radon concentration that took place 3 days before the earthquake

A survey of soil gas radon concentrations was carried out at three sites in the seismic area of the Araks Basin in Armenia from 1996 to 1999 (Kharatian et al. 2002). Higher radon concentrations were found inside the tectonic micro-blocks during local seismic activity.

A number of earthquake prediction observation sites have been set up in the Causasus since 1990 in the mining galleries of Krasnodar and Stavropol regions (Tsvetkova 2001). These authors noted that the results demonstrate the applicability of radon monitoring for predicting large earthquakes in the Caucasus.

9.3.7 Temperature Precursors

A relationship between the temperature anomalous variations in observed in boreholes and danger geodynamic activity is known (e.g., Mogi et al. 1989; Singh and Dey 2003; Ouyang et al. 2009).

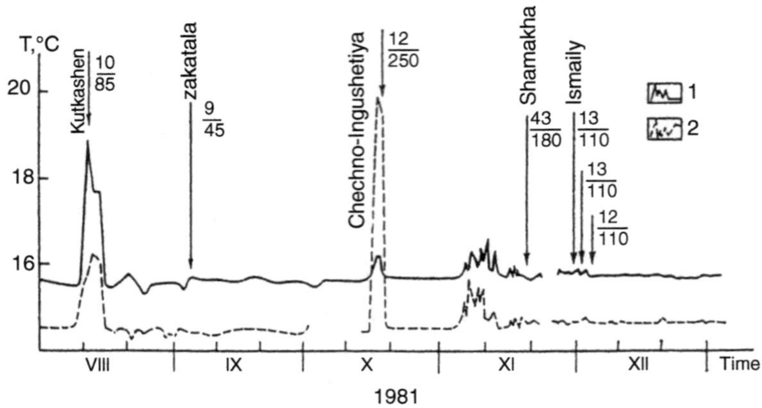

Fig. 9.5 Temperature variations in the Biladzhik well (Azerbaijan) and their comparison with the geodynamic events in the Caucasus. (*1*) and (*2*) are the temperatures observed at a depth of 335 and 110 m, respectively. The *arrows* indicate the time of the earthquake. The numbers in the numerator and denominator show the energy class of the earthquake and the epicentral distance in km between the earthquake and the well, respectively (Lyubimova et al. 1982)

Temperature monitoring was carried out in the Biladzhik well (Sheki region of Azerbaijan, southern part of the Greater Caucasus) to correlate temperature variations with the earthquakes recorded in the region (Lyubimova et al. 1982). Monitoring was conducted at a depth of 110 and 335 m (Fig. 9.5). This figure shows that in general the temperature graphs are synchronous at both depths, but in the two most striking cases the distribution of the anomalous curves goes in opposite directions. Before the Kutkashen event, the temperature anomaly observed at a depth of 335 m exceeded the temperature anomaly at 110 m (which seems logical). However, before the Checheno-Ingushetiya event the temperature anomaly observed at 110 m exceeded the 335 m anomaly tenfold. This phenomenon is difficult to explain from a physical-tectonical point of view. Undoubtedly, such investigations must be continued in various regions of the Caucasus.

9.4 Long-Term Seismicity Prognosis

Earthquake risk assessment combines the likelihood and the consequences of a set of earthquake scenarios for a given period of time (Beer and Ismail-Zadeh 2003). The risk can be estimated as the probability of harmful consequences or expected losses (injuries to people, damage to property, decline in economic activity, etc.) resulting from interactions between seismic hazards (*H*), vulnerability (*V*), and exposed values (*E*). Conventionally, earthquake risk (*R*) is expressed by the convolution of these three parameters (Babayev et al. 2010):

$$R = H \otimes V \otimes E. \tag{9.1}$$

Interestingly this structurally Eq. 9.1 is similar to Eq. 3.88 for optimal integration of geophysical-geological tools (Eppelbaum et al. 2003b).

Ismail-Zadeh (1996) suggested a nonconventional way to study earthquake migrations in the Caspian Sea. He presented mathematically formulated physical-geological processes that cause the formation and evolution of sedimentary basins. Ismail-Zadeh et al. (2005a) developed a methodology of 3D modeling of earthquake prognosis on the basis of a combined analysis of seismic, heat flow, and gravity observations.

9.5 Algorithms for Geodynamic Event Prediction

9.5.1 Thermoelastic Characteristics and Their Relationship to Earthquakes

On the basis of 3D geothermal and thermoelastic models of the Caucasus and adjacent areas (Alexidze et al. 1993, 1995) several zones characterized by essential anomalies in the lateral gradients of the vertical component of thermoelastic displacements were discovered (Gugunava et al. 2006). The Spitak (Armenia 1988) and Racha-Java (Georgia 1991) earthquakes were significantly correlated with zones of lateral gradients. Gugunava et al. (2006) suggested the following mechanism of earthquake genesis. Initially energy accumulates as a result of thermoelastic stresses, subduction, or transform motions. Then the tidal effect together with the approach of the "plastic" wave of accumulated energy leads to rock failure along definite zones, the formation of deep faults, and therefore seismicity. Obviously, this hypothesis needs to be evaluated and elaborated in the future.

9.5.2 Intraplate Seismicity Studies

Intraplate seismicity studies have shown that the average statistical distances δ_M (km) or δ_K(km) between neighboring epicenters of the closest pairs of seismic sources with dimensions (lengths) L_M (km) or L_K(km) are located along lineament structures and are characterized by magnitude M or energy class K (Ulomov et al. 2007):

$$L_M = 10^{(0.6M-2.5)}, \; L_K = 10^{(0.333K-3.832)}, \tag{9.2}$$

where $K = \log E = 1.8M + 4.0$, and

$$\delta_M = 10^{(0.6M-1.94)}, \delta_K = 10^{(0.333K-3.272)}, \quad (9.3)$$

where E is the energy of deformation.

The value $\delta/L \approx 3.63$does not depend on the magnitude and reflects the self-similarity in the hierarchy of dimensions of interacting geoblocks and their earthquake sources. Formulas (9.2) corresponds to this expression with a clearer physical meaning:

$$L = \sqrt[3]{2E/\mu\varepsilon^2}, \quad (9.4)$$

which simply relates the energy of deformation E (J) to the crust volume L^3 (m^3) under critical conditions of its fracture, when the strain field ε of the medium is represented only by the shear component μ. The generally accepted average values are $\mu = 3.0 \cdot 10^{10}$ Pa and $\varepsilon = 1.45 \cdot 10^{-4}$. This approach was applied to the Iran-Caucasus-Anatolia region and the following formula was suggested (Ulomov et al. 2007): $\delta_M = 10^{(0.54M-1.92)}$.

9.5.3 Areal Autocorrelation Analysis

High reliability can be attained in geophysical field regioning and can reveal regional structures, if the method of areal autocorrelation analysis is used (Khesin et al. 1996). However, the results of this method are highly dependent on the field description interval and on the size of an applied apparent graticule. Furthermore this analysis is accompanied by a considerable loss of area on the map edges and only enables the derivation of general results, which are rare and of no specific value.

9.5.4 Geophysical Field Complexity as Factor of Seismicity Prognosis

Relief complexity, which as described above, governs the spatial distribution of topographic corrections, can be estimated by a map of the specific sinuosity of height isolines. Alexeyev and Khesin (1971) suggested applying the specific sinuosity of isolines (*SSI*) to characterize the complexity of a geophysical field (see Sect. 4.2.1).

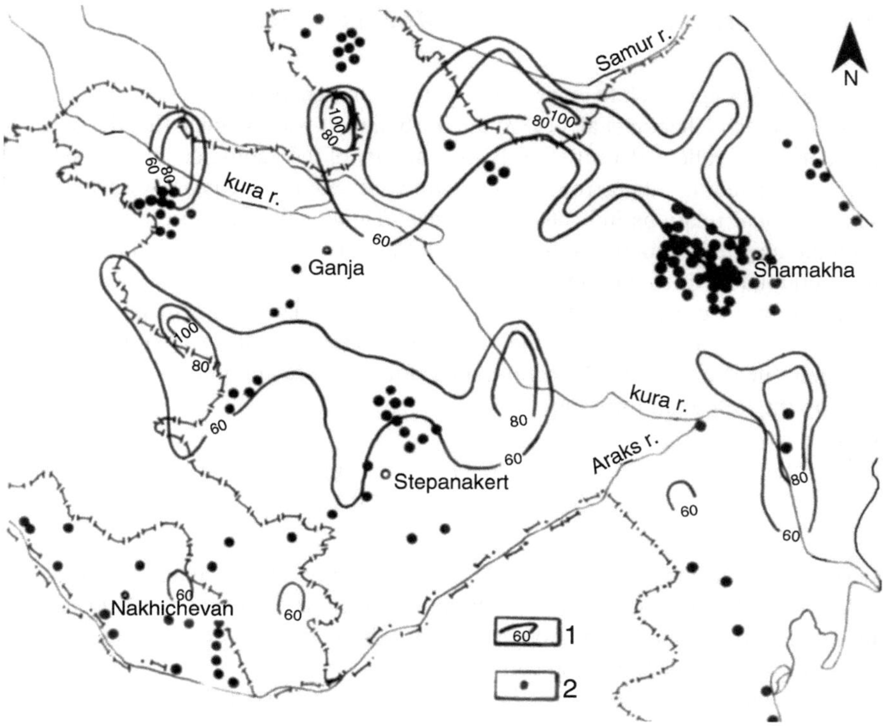

Fig. 9.6 Average lengths of disjunctive dislocations in sliding cells for the Eastern Caucasus (Khesin and Metaxas 1974). (*1*) density isolines of disjunctive dislocations determined on the basis of magnetic and gravimetric prospecting data (km/1,000 km^2); (*2*) earthquakes of magnitude of 7 and over from macroseismic evidence

The *SSI* is measured in km/km^2, and the value obtained refers to the center of this cell. The map not only characterizes the complexity of terrain relief and determines the requirements for eliminating its effect, but also allows for geological interpretation. The *SSI* distribution (Fig. 4.20) reflects certain geological features of the region. Interestingly, this chart correlates rather well with the map of seismic activity (Riznichenko et al. 1983). The latter can contribute to the formation of the terrain relief and is also caused by some relief-forming factors.

The average lengths of disjunctive dislocations indicated by magnetic and gravimetric data on 25×25 km^2 sites are presented in Fig. 9.6.

To determine the general features of geophysical fields (and the height field, as well) it is often advisable to calculate tertiary indicators; i.e. *SSI* by formula (9.5), and the length of dislocations emerging from the results of interpretation in a sliding window (Khesin and Metaxas 1974).

There is a consensus that the relative dislocation density increases in zones of main disjunctive dislocations. These zones are of prime importance as ore and oil-and-gas controls (Khesin 1991), and because they tend to be the foci of earthquakes (see Figs. 4.42 and 9.6).

9.5.5 Correlations Between the Magnetic Field and Seismicity

An interesting approach to correlations between magnetic field and seismicity was applied by Riznichenko et al. (1983) in Azerbaijan. It was assumed that the regional magnetic field reflected a sufficiently long history of the geological evolution of the region (including dangerous geodynamic events). Analysis of magnetic-seismological relations indicated that zones of high seismicity are not associated with the magnetic anomalies themselves but rather with zones of high gradients of magnetic field, which are indicative of the presence of large faults between blocks of the Earth's crust, and magnetic field behavior. Therefore, maps of seismic activity were compared with the maps of modules of horizontal gradients of magnetic field intensity. Magnetic field gradients were computed on a 5 × 5 km grid by the following formula:

$$\left|\frac{\Delta U}{\Delta l}\right| = \sqrt{\left(\frac{\Delta U_x}{\Delta x}\right)^2 + \left(\frac{\Delta U_y}{\Delta y}\right)^2}, \tag{9.5}$$

where ΔU_x is the mean increment of U along the x axis; ΔU_x is calculated on parallel sides of a cell with l dimensions in a sliding rectangular "apparent graticule"; ΔU_y is the corresponding increment along the y axis; $\left|\frac{\Delta U}{\Delta l}\right|$ is the mean value of the gradient modulus on the transformation cell (*grad U*).

A comparison of these maps is shown in Fig. 9.7. Zone I includes the high-seismicity areas of the Greater Caucasus, its SE submersion to the side of the Lower Kura Depression as well as the Saatly-Kurdamir outcrop of the consolidated crust. Zone II is revealed in the Lesser Caucasus. Zone III represents the Talysh mountains and their NE submersion.

The correlation field (Fig. 9.8) contains a concentration of points representing zone I. This cluster also includes three points from zone III which located near the regression of A_T on (grad u) for zone I. Several points from zone II are close to the same line. These points correspond to areas of the highest seismicity and the greatest magnetic gradients coinciding in the plane. Obviously, these points represent the general regularity between the values investigated in other areas masked by different kinds of noise.

9.5.6 Revealing Hidden Intersections of Linear Structures

Identifying the nodes of linear structures is a crucial problem, since such nodes indicate the location of economic ore, oil-and-gas and underground water deposits of various types. This problem is also a high priority in earthquake prediction (Keilis-Borok and Levshin 1984). Therefore, a method was developed for direct localization of structural intersections by digital field models (Khesin et al. 1996).

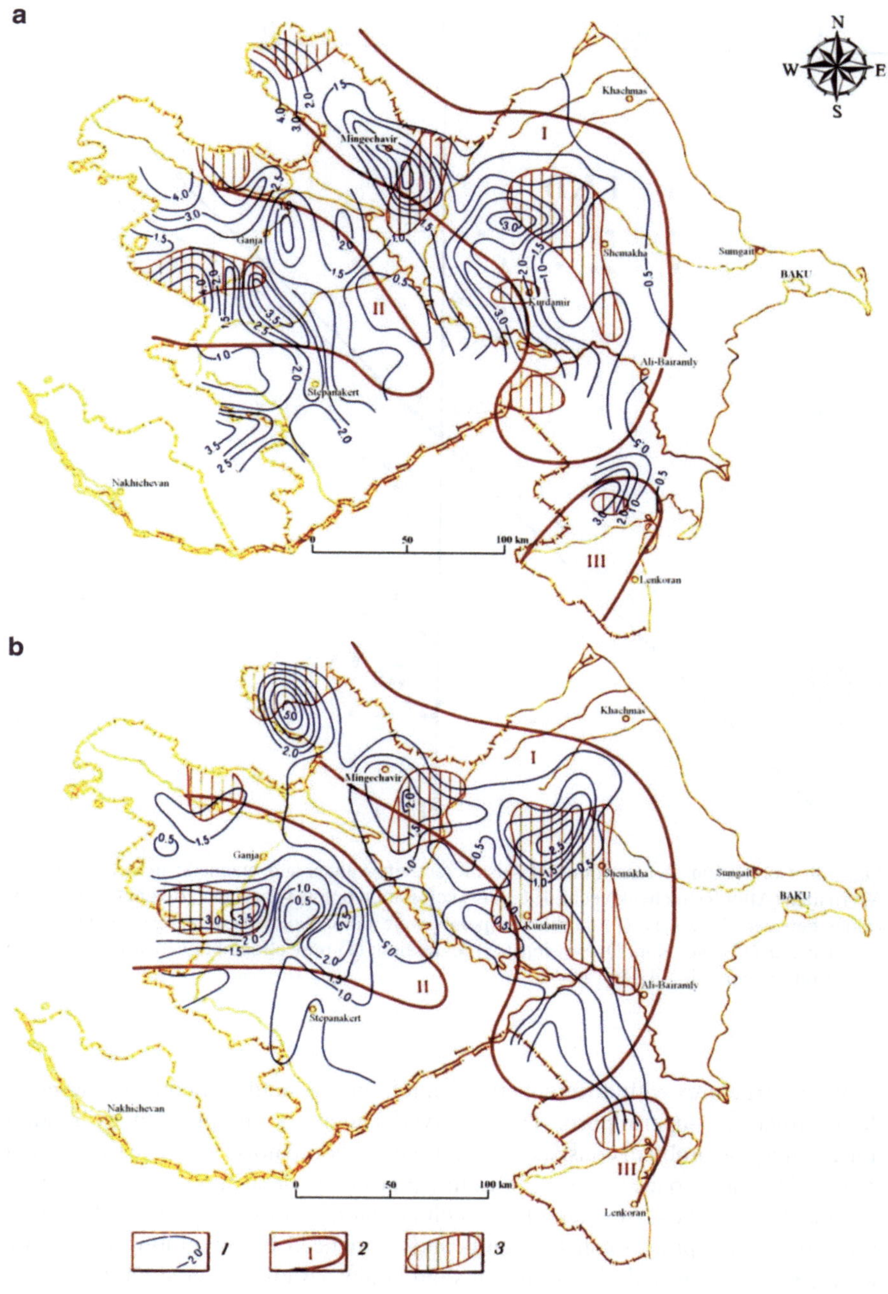

Fig. 9.7 Comparison of horizontal gradients of magnetic field maps with elements of seismic activity (After Riznichenko et al. 1983). (**a**) constant level of detail, (**b**) constant accuracy. (*1*) isolines of horizontal gradients in milliErsted/km; (*2*) contours and numbers of zones of heightened seismic activity; (*3*) areas of heightened ($A > 0.2$) seismic activity

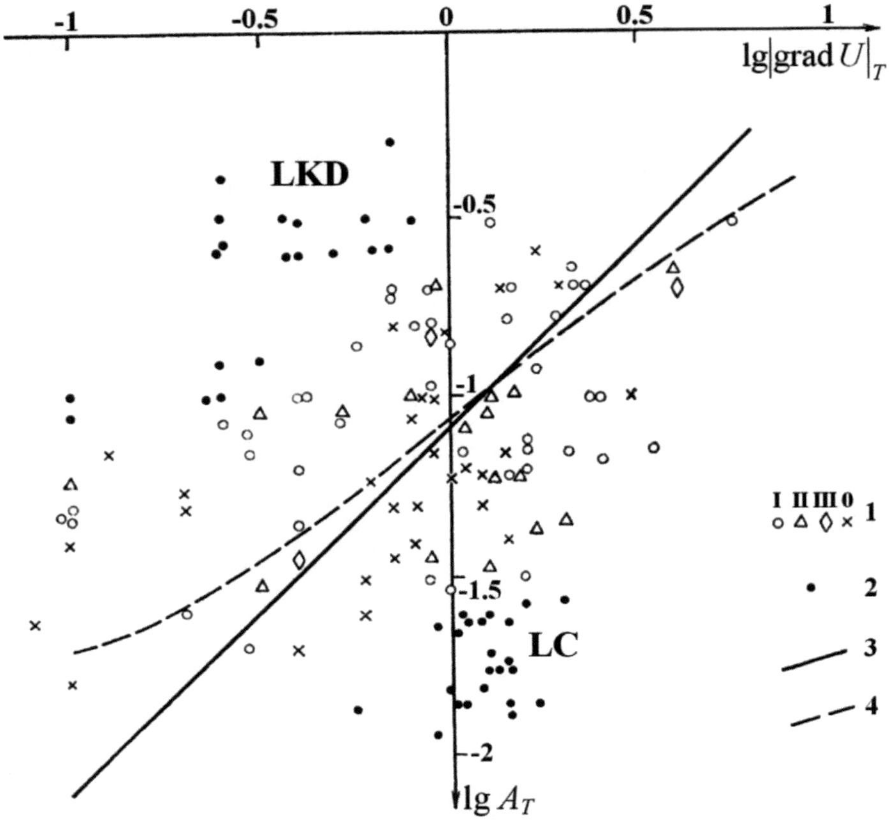

Fig. 9.8 Correlation of logarithms of seismic activity values and magnetic field gradients for Azerbaijan (After Riznichenko et al. 1983). (*1*) correlation points for zones *I*, *II*, and *III*, 0 displays values between the zones; (2) correlation points were not calculated by regression (*LC* Lesser Caucasus, *LKD* Lower Kura Depression); (*3*) regression; (*4*) projection in this logarithmic system of rectilinear regression A_T on $(\text{grad } u)_T$

When structures with different strikes intersect, the field components caused by these structures superimpose. The observed pattern is complicated by random noise. Thus, the problem consists of suppressing random noise and revealing isoline strikes. The latter, once removed from the intersection area, approximate the strike of anomalous bodies. An example of such an interpretation is shown in Fig. 3.32. Note that this approach can be significantly enhanced by integrating it with the advanced theory of fault generation in solid bodies (Aleinikov et al. 2000).

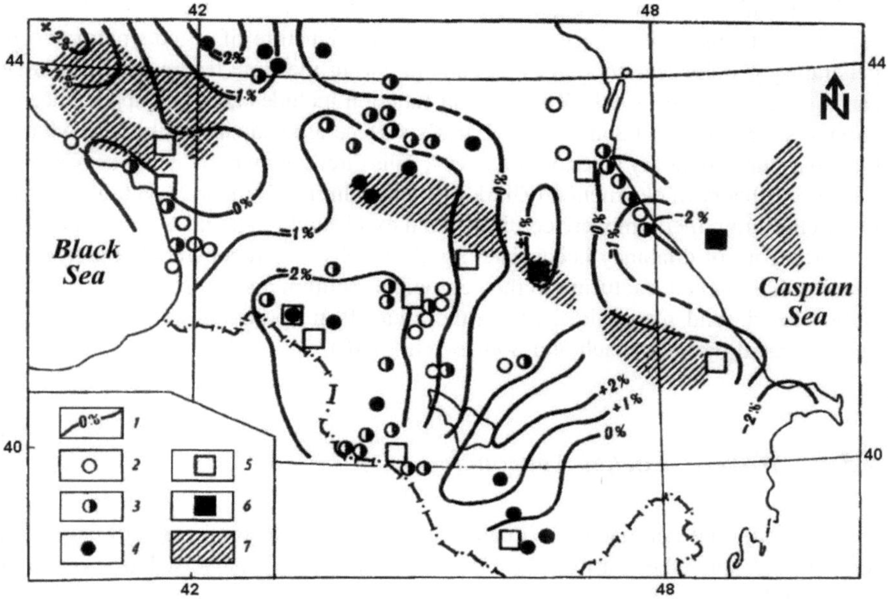

Fig. 9.9 Comparison of horizontal velocity variations in the *upper* mantle with thermal flow data, epicenters of earthquakes and gravity isostatic anomalies (After Vinnik et al. (1978)). (*1*) isolines of variations of compressional wave velocity in *upper* mantle (in% from average velocity), values of thermal flow: (*2*) <0.8 unit; (*3*) >0.8 and <1.6 unit; (*4*) >1.6 unit; (*5*) epicenters of large earthquakes in the Earth's crust; (*6*) epicenters of strong earthquakes; (*7*) significant gravity isostatic anomalies

9.5.7 Pattern Recognition of Regional Seismicity

Studies of earthquake seismology focused for years on the inner structure of the Earth by assessing compression (*P*), shear and other wave types. In the last few decades, an effective method involving converted waves of earthquakes has been developed. Analysis of remote earthquakes are used to estimate the structure of very deep (several hundred kilometers) strata that are related to the upper mantle. Informative data from such studies of natural seismicity fields were obtained in the Caucasus. For example, Egorkina et al. (1976, 1978) successfully applied a seismological apparatus called "Zemlya" (Earth) to study the thickness of the Earth's crust and deep faults in Armenia. Vinnik et al. (1978) used the waves of remote earthquakes to calculate the velocity characteristics of the upper mantle and compared this to gravity and heat flow data (Fig. 9.9).

An approach based on pattern recognition of distributed regional seismicity was developed by Keilis-Borok (1990). This innovative approach includes clustering of events and changes in aftershock statistics (Molchan et al. 1990). The algorithms were formulated prior to the destructive Spitak (1988) earthquake (Turcotte 1995).

A general space-time function of the density of number of faults in the area of earthquake focus was suggested as a quantitative signature of a geodynamic event generating geophysical anomalies-precursors during the time leading up to a large earthquake (Alekseev 1993). This general function includes zones of appearance of anomalous geophysical zones at the Earth's surface associated mainly with what is known as the "boundary layer of dilatancy". This direction in particular was based on the kinetic theory of destruction (Zhurkov 1968). Zhurkov's theories on fault generation in the strain medium were confirmed and explicated in Aleinikov et al. (2000).

The theory of dilatancy is consistent with the physical-mechanical properties of zones of heightened fracturing in the geological medium and the structure of quasi-static electric and magnetic fields in the area of an upcoming seismic event, as demonstrated by Sobisevich and Sobisevich (2010).

9.5.8 *Wavelet Packet Approach*

The conventional approach to calculating gravity, geomagnetic, electromagnetic, temperature and other earthquake precursors is Fourier analysis. However, it is poorly suited for problems concerned with the detection of low-magnitude short-lived events, especially in the absence of *a priori* information on the duration and frequency content of the perturbation (Alperovich and Zheludev 1999). By contrast, Wavelet Analysis (WA) and its extensions, such as wavelet packet analysis, for example, are extremely useful tools for such cases (e.g., Averbuch et al. 2010).

Wavelet analysis provides a rich library of available waveforms, and fast, computationally efficient procedures of representation of signals and selection of pertinent waveforms. The basic assumption justifying the application of wavelet analysis is that the essential structure of a signal does not consist of a large number of waveforms. The best way to reveal this structure is to represent the signal by a set of basic elements containing waveforms consistent with the signal. Large coefficients are thus attributed to a few basic waveforms, whereas small coefficients are predicted for noise and structures that are inconsistent with the basic waveforms (Averbuch et al. 2010; Eppelbaum et al. 2011a).

The wavelet approach can be applied to derive enhanced (e.g., coherence portraits) and composite images of geophysical indicators to identify earthquake signatures. The methodology, based on matching with wavelet packet dictionaries, can be used to extract the target signals even from data that are highly contaminated with noise.

The recently developed technique of diffusion clustering combined with the abovementioned wavelet methods can be utilized to integrate geophysical data and detect existing signals caused by dangerous geodynamic events at a depth. The goal is to detect the geophysical signatures of upcoming earthquakes in a noisy area with a minimal number of false alarms and miss-hits. This is done through analysis of specific physical parameters (which vary for different regions). Robust algorithms presented in Averbuch et al. (2010) could be effectively used for this purpose. The geophysical signals are characterized by the distribution of their energies among blocks of wavelet packet coefficients. To derive the geophysical precursors, the best discriminant basis method followed by dimensionality reduction via diffusion maps is used. The decision is made by combining the results from linear discriminant analysis, and classification and regression trees. This approach can be applied to geophysical fields and geochemical indicators. The outputs from different sources are combined and their dimensionality is reduced via diffusion maps. Subsequently the combined data are embedded into a low (3–6)-dimensional space where data related to earthquake precursors are well separated from the different kinds of noise. This low-dimensional set of the data representatives can be used as a reference set for the classification of newly arriving data.

WA was successfully applied to earthquake prediction in Japan (e.g., Alperovich and Zheludev 1999; Zheludev and Alperovich 1999). Clearly the enhanced modification of WA packet analysis may be applied for the detection both single and integrated geophysical precursors in the Caucasian region.

9.5.9 Earthquakes as a Strongly Nonlinear Event

There is growing recognition of the importance of nonlinear phenomena in many branches of geophysics. It was shown that such phenomena as earthquakes (e.g., Keilis-Borok 1990; Voisin 2002) and their reflection in geophysical fields as well as the propagation of geophysical fields in non-uniform geological media is a nonlinear process (e.g., Aleinikov et al. 2001). However, it is still common practice to solve complex geophysical problems by removing visible nonlinear effects and reducing the problem to a linear one. Such a linearization can eliminate useful effects and even radically change the notion of targets (Eppelbaum and Kardashov 1998). Therefore, many important geophysical problems cannot be solved without special nonlinear procedures. Analysis of such physical phenomena as thermal and gravity effects in the Earth shows that the nonlinear component may play a significant role in the process.

By contrast, the problem of influence of *strongly* nonlinear sources on transitional dynamics has scarcely been investigated in geophysics (Kardashov et al. 2000; Kardashov and Eppelbaum 2008).

The process of thermal wave propagation in the Earth can be used as a possible geodynamic precursor. In general, the model of thermal wave propagation can be described by the following evolution equation:

$$\frac{\partial u}{\partial t} = \Delta u + f(u), \tag{9.6}$$

where $\Delta = \frac{\partial^2}{\partial x^2} + \frac{\partial^2}{\partial y^2} + \frac{\partial^2}{\partial z^2}$ is the Laplace operator, and $f(u)$ is a nonlinear function of a thermal source. Let us assume the following form of the thermal source term:

$$f(u) = \alpha + \varepsilon g(u), \tag{9.7}$$

where α and ε are some parameters, and $g(u)$ is the nonlinear function which describes disturbances of the thermal field. Function $g(u)$ can be presented both as a conventional deterministic function and a random distribution function. If $\varepsilon = 0$, we obtain a linear equation corresponding to the conventional problem with a linear source term: $\frac{\partial u}{\partial t} = \Delta u + \alpha u$. The case of $\varepsilon \neq 0$ corresponds to a situation where the normal thermal field is disturbed by anomalous sources.

Under specific conditions a nonlinear disturbance can cause the appearance of a shock wave even for extremely small values of the parameter ε (Eppelbaum and Kardashov 1998). This effect can be used to identify the target. On the other hand, Eq. 9.6 can take transient solutions under special conditions such as localized or switch (kink) waves. Note that the properties of both localized and shock waves are determined by the characteristics of the source function $f(u)$. Consequently, this makes it possible to estimate certain parameters of the medium. Here this is limited to a nonlinearity of the type $g(u) = u^3$, but other types of nonlinearity have been analyzed as well. The temperature source term of the form $f(u) = \alpha u + \varepsilon u^3$ provides a solution to the traveling wave (Eppelbaum and Kardashov 2001):

$$u(t, x, y, z) = \psi\left(k_x x + k_y y + k_z z + vt\right), \tag{9.8}$$

where v is the wave propagation speed and $\vec{k} = \left(k_x, k_y, k_z\right)$ is the unit vector in the direction of wave propagation. Self-similar solutions of this kind were considered in detail in Barenblatt (1996). Substitution of Eqs. 9.8 for 9.6 yields:

$$v\psi'(s) = \psi''(s) + \alpha\psi(s) + \varepsilon\psi(s), \tag{9.9}$$

where $s = k_x x + k_y y + k_z z + vt$ since $k_x^2 + k_y^2 + k_z^2 = 1$. It was shown in Kardashov et al. (2000) that transitive waves (both switch and localized waves) corresponding to the source function of type Eq. 9.7 are the steady-state solutions ($v = 0$) of Eq. 9.8. Equation (9.6) may be generalized by including the effect of nonlinear diffusion. Such models are widely used, for instance, in descriptions of thermal propagation processes in media with nonlinear coefficients of thermal conductivity. Finally, we consider the following generalization of Eq. 9.6:

$$\left\{\begin{aligned} \frac{\partial u}{\partial t} &= \sum_{i=1}^{3} \frac{\partial}{\partial x_i}\left[k_i\left(\frac{\partial u}{\partial x_i}\right)\frac{\partial u}{\partial x_i}\right] + f(u), \\ \frac{\partial u}{\partial t} &= \sum_{i=1}^{3} \frac{\partial}{\partial x_i}\left[\bar{k}_i(u)\frac{\partial u}{\partial x_i}\right] + f(u). \end{aligned}\right\}. \tag{9.10}$$

This is a novel approach for the identification of nonequilibrium potential sources by using the corresponding transitional wave analysis. Transitional waves deterministically characterize the structural parameters of the nonlinear sources. Parameters of the transitional waves characterize the structure and the main characteristics of the geophysical (thermal) source function. Models with nonlinear diffusion and sources can be applied to the analysis of various geophysical events associated with nonlinear effects (first of all, earthquakes). The transitional waves obtained using these models serve to determine and estimate not only the parameters of the sources but also the nonlinear diffusion.

It is obvious that Sects. 9.3.7 and 9.5.10 should be associated in future work.

Conclusion

Over many dozens of years (and primarily in the late Soviet period from 1960 to 1990), the Caucasus was a key geophysical polygon for different geophysical methods and types of analysis.

The goal of this book was to provide an overview of Caucasian geophysics and present the main works of the authors. It is very difficult to produce a mid-length book of this nature and thus many important geophysical findings could not be included. We hope that other researchers will continue this series of books on Caucasian geophysics.

A number of chapters were devoted to specific methodologies used in the Caucasus. These include gravitational and magnetic fields, anomalies of temperature field, fields of VLF transmitters, resistivity, self-potential, and induced polarization, all of which have been applied in complex Caucasian environments (rugged relief, considerable heterogeneity of media, oblique polarization and unknown levels of the normal field). In addition, methods of forecasting dangerous geodynamic events in the Caucasus were also presented.

We also focused on interpretation. These include: (1) elimination of noise (primarily the topographic effect) typical of mountainous regions, (2) classification of targets and summarizing their characteristics, estimating the possibilities of various sets and the role of natural geophysical fields, (3) development of various information methods for integrated interpretation using a set of geophysical fields and the interpretation of complex individual fields such as methods based on gathering useful information and the elimination of superfluous information to identify important hidden objects, (4) the development of interpretation methods to solve inverse problems under conditions of the rugged terrain relief, arbitrary polarization of objects and unknown levels of the normal field, (5) development of interactive integrated selection on the basis of an effective algorithm for solving a direct 3-D problem of gravimetric and magnetic prospecting, (6) estimation of prospecting reliability on the basis of an information-statistical approach, (7) development of sequential interpretation procedures which is reflected in the structure of this book.

L.V. Eppelbaum and B.E. Khesin, *Geophysical Studies in the Caucasus*,
Lecture Notes in Earth System Sciences, DOI 10.1007/978-3-540-76619-3,

This interpreting system and its components make it possible to solve geological problems of various types at all stages of geophysical prospecting. Often one and the same problem is solved in contiguous stages, whereas during a single stage several problems can be solved. This is conditioned by the specific type of investigations carried out in the Caucasus and in other mountainous regions that impede repeated surveys of areas that are extremely hard to access.

The Caucasian mountain conditions primarily affect large-scale prospecting, where the stochastic nature of the region and its geophysical fields have the greatest impact. Therefore, the key interpretational tools are statistical techniques for processing the results and revealing certain features of the geological structure from the geophysical field which manifest as a combination of noise and the target signal. This can be extended to the interpretation of geophysical data integration, which involves singling out an object in a given class through the accumulation of information from various fields.

When conducting regional and detailed surveying, the major focus is the deterministic variants of interpretation. These are oriented to the investigation of marker interfaces and localized anomalous bodies. Naturally, deterministic and probabilistic methods do not exclude but rather complement each other. Combinations of these methods are applied in all regions.

The interpretation system and its components have been successfully tested on a complicated model and in real situations in the Caucasus and the Kura Depression. The system has been used to study deep structures, ore, oil-and-gas and underground water controls. Moreover, it helped reveal promising areas and new deposits, and resolve other geological problems including those associated with long-term seismic activity prediction. In addition, various components of the system (some of which were tested in Israel and the Eastern Mediterranean) were successfully applied to the study of near-surface inhomogeneties, such as engineering geology and archaeology targets. Therefore, the techniques involved and the interpretation system on the whole can be widely applied to investigate a number of regions and areas with complex structures and a range of challenges.

A brief review of geophysical methods used to search for ore deposits, underground water, hydrocarbon prospecting, environmental and archaeological geophysics in Caucasus, as well an examination of earthquake precursors was also presented.

In conclusion, authors of the book wish to express their thanks to the thousands and thousands of geophysicists and geologists who have worked in the Caucasus during many decades, and who continue to work in this region to this day. Geophysical studies in the Caucasus have a promising future ahead. . . .

References

Abakelia MS (1937) A problem of Kyurdamir gravity ridge in Transcaucasian. Azerbaijan Petroleum Industry (Azerbaijdzhanskoe Neftyanoe Khozyaistvo), No. 9:6–10 (in Russian)
Abdullaeyev DI, Bagirov AA (1996) Mine-borehole geophysics. In: Kerimov KM (ed) Geophysical Investigations in Azerbaijan Sharg-Garb, Baku 272–274 (in Russian)
Abdullayev NR (2000) Seismic stratigraphy of the upper pliocene and quaternary deposits in the South Caspian Basin. J Petrol Sci Eng 28:207–226
Abdullayev RN, Azizbekov ShA, Kashkai MA, Kerimov GI, Mustafabeily MA, Sitkousky IN, Shikhalibeily ESh (1962) Metallogeny of Azerbaijan. Acad. of Sci. of Azerbaijan, Baku (in Russian)
Abdullaev Sh-SO, Idarmachev ShG, Aliev MM, Aliev IA, Magomedov AG (2011) Variations in apparent resistivity associated with geodynamic processes in the area of an oil-and-gas field. J Volcanol Seismol 5(3):209–215
Abich GV (1863) New Islands on the Caspian Sea and the cognition of Mud Volcanoes of the Caspian Region. Mem Acad Sci Peterbourg, Ser VIII, 6(5)
Adler JL (1942) Simplification of tidal corrections for gravity meter surveys. Geophysics 7:35–44
Affleck J (Special ed) (1964) Special issue on magnetic methods. Geophys 29(4)
Agamirzoev RA, Kuliev FT, Korobanov VV, Panakhi BM, Alieva CT (1986) Seismic microzonation of territory of the Enikend hydrosystem. Izv AN Azerb SSR, Ser: Earth Sci (1): 79–83 (in Russian)
Agusti J, Vekua A, Oms O, Lordkipanidze D, Bukhsianidze M, Kiladze G, Rook L (2009) The pliocene-pleistocene succession of Kvabebi (Georgia) and the background to the early human occupation of Southern Caucasus. Quat Sci Rev 28:3275–3280
Agyemang T, Heblinski J, Schmeider K, Sajadyan H, Vardanyan L (2011) Accuracy assessment of supervised classification of submersed macrophytes: the case of the Gavaraget region of Lake Sevan, Armenia. Hydrobiologia 661(1):85–96
Aiken CL (1982) The variable Bouguer reduction datum, its relation to the prediction of gravity anomalies from topography, and computation of residual Bouguer gravity anomalies. In: Proceedings of the 52nd annual international meeting, society. Exploration Geophysics, Expanded Abstracts, pp 247–250
Akhmedbeili FS, Gasanov AG (2004) Tectonic types of seismic sources in Azerbaijan. Elm, Baku (in Russian)
Aki K, Richards PG (1980) Quantitative seismology: theory and methods. W.H. Freeman, San Francisco
Akselrod SM, Danevich VI, Sadykhov DM (1991) Nuclear-magnetic investigations of borehole sections in Azerbaijan. Azerb. State, Baku (in Russian)

L.V. Eppelbaum and B.E. Khesin, *Geophysical Studies in the Caucasus*,
Lecture Notes in Earth System Sciences, DOI 10.1007/978-3-540-76619-3,

Akselrod SM, Putkaradze LA (1979) Integrated interpretation of logging with due account of variations in physical properties of the object under study. Review of VIEMS, Ser. Regional, exploration, and borehole geophysics, Moscow (in Russian)

Aleinikov AL, Belikov VT, Eppelbaum LV (2001) Some physical foundations of geodynamics (in Russian, contents and summary in English). Kedem Printing-House, Tel Aviv

Aleinikov AL, Belikov VT, Eppelbaum LV, Nemzorov NI (2000) Mountainous rock destruction and metamorphic processes in the earth: a view from classical physics. Sci Israel (3): 65–87

Alekseev AS (1993) A multidisciplinary mathematical model of combined foreshock for earthquake prediction research. J Earthquake Pred Res 2(2):137–150

Alexeyev VV (1971) Application of the self-potential method under mountainous terrain relief. Prospecting and protection of entrails (*Razvedka i Okhrana Nedr*), (9): 38–43 (in Russian)

Alexeyev VV (1976) Development and improvement of the magnetic anomalies interpretation methods under conditions of inclined magnetization and sloping relief (a specific example of Belokan-Zakatala ore district, southern slope of the Greater Caucasus). Ph.D. thesis, Mining Institute, Dnepropetrovsk (in Russian)

Alexeyev VV, Khesin BE (1971) Interpretation of magnetic survey data in orogens of the South of the USSR. Review of VIEMS, Ser. Regional, exploration, and borehole geophysics, Moscow (in Russian)

Alexeyev VV, Khesin BE (1987) Processing and interpretation of gravity data in mountainous regions. Appl Geophys (Prikladnaya Geofizika) (117): 90–100 (in Russian)

Alexeyev VV, Khesin BE (1989) Grouping of gravity measurement points as a mean of suppressing near-surface noise effects and determining the effective density. Deposited by VINITI, USSR Academy of Sciences (5721-B85) (in Russian)

Alexeyev VV, Khesin BE, Eppelbaum LV (1996) Geophysical fields observed on different heights: common interpretation technique. In: Extended Abstracts of the Jakarta-96 international geophysics conference, pp 104–108

Alexidze MA, Gugunava GG, Kiria DK, Chelidze TL (1993) A three-dimensional stationary model of the thermal and thermoelastic fields of the Caucasus. Tectonophysics 227 (1–4):191–203

Alexidze MA, Gugunava GG, Kiria DK, Chelidze TL (1995) A three-dimensional stationary model of the thermal and thermoelastic fields of the Caucasus. Erratum. Tectonophysics 243 (3–4):297–298

Aliev FS, Askerov FS (2002) Groundwater protection in the Republic of Azerbaijan related to the production and transportation of oil. In: Israfilov RG, Howard KWF (eds) Current problems of hydrogeology in urban areas, urban agglomerates and industrial centres. Kluwer Acad. Publ. NATO Science Series: IV: Earth and Environmental Sciences, vol 8. pp 301–315

Aliyev ChS (1994) Radioactive fields of depression regions of Azerbaijan. D.Sci. thesis, Baku (in Russian)

Aliyev ChS, Zolotovitskaya TA (1996) Results of radiometric investigations. In: Kerimov KM (ed) Geophysical Investigations in Azerbaijan Sharg-Garb, Baku, pp 386–389

Aliyev ChS, Zolotovitskaya TA (2005) Comparison of gamma-field with geological structure, geophysical fields, oil&gas bearing, and seismicity. In: Alizade AkA (ed) Geology of Azerbaijan, vol V. Physics of the Earth, Nafta-Press, pp 292–312

Aliyev ChS, Fezullaeyv AA, Zolotovitskaya TA, Aliyeva EA, Aliyeva CK, Aliyev KA (2001) Radioactive fields of mud volcanoes of Azerbaijan. Geophys N Azerbaijan 3:25–32

Alizadeh AkA (ed) (2005) Geology of Azerbaijan, vol V, Physics of the Earth. Nafta-Press, Baku (in Russian)

Al-Lazki AI, Sandvol E, Seber D, Barazangi M, Turkellii N, Mohamad R (2004) P_n tomographic imaging of mantle lid velocity and anisotropy at the junction of the Arabian, Eurasian and African plates. Geophys J Int 158:1024–1040

Allen MB, Vincent SJ, Alsop JI, Ismail-Zadeh A, Flecker R (2003) Late cenozoic deformation in the South Caspian region: effects of a rigid basement block within a collision zone. Tectonophysics 366:223–239

Alperovich LS, Zheludev VA (1999) Long-period geomagnetic precursors of the Loma-Prieta earthquake discovered by wavelet method. In: Hayakawa M (ed) Atmospheric and ionospheric electromagnetic phenomena associated with earthquakes. TerraScientic, Tokyo, pp 123–136

Alpin LM, Dayev DS, Karinsky AD (1985) Theory of fields employed in geophysical prospecting. Nedra, Moscow (in Russian)

Amiranashvili AG, Cornelissen G, Amiranashvili V, Gheonjian L, Chikhladze VA, Gogua RA, Matiashvili TG, Paatashvili T, Kopytenko YuA, Siegelová J, Dušek J, Halberg F (2002) Circannual and circadecennian changes in mortality from cardiovascular causes in Tbilisi, Republic of Georgia (1980–1992). Scripta Medica (Brno) 75(5):255–260

Amiraslanov TC (1987) Interpretation of gravitational magnetic anomalies for complex constructed regions on the basis of geophysical field integrated analysis. D. Sci. thesis, Baku (in Russian)

Amiraslanov TC (1990) Geological interpretation of gravity and magnetic anomalies in the central part of the Middle Kura Depression. Azerbaijan Oil Industry (Azerbaidzhanskoe Neftyanoe Khozyaistvo) (7): 13–17 (in Russian)

Andreyev BA, Klushin IG (1962) Geological interpretation of gravity anomalies. Gostoptekhizdat, Leningrad (in Russian)

Andruschuk VL (1968) Geology of the USSR, vol IX, Northern Caucasus. Nedra, Moscow (in Russian)

Apakidze A, Tabagua G, Chanturishvili L, Chelidze T, Kipiani G, Giunashvili G, Jakhutashvili M (2001) Results of archaeogeophysical research at the Armaztsikhe-Bagineti ancient city (eastern Georgia). Bull Georgian Acad Sci (Tbilisi) 164(2):284–286 (in Russian)

Appel K, Haken W (1989) Every planar map is four colorable, vol 98, Contemporary mathematics. Amer. Math Soc, Providence

Arkhangelsky AD (1933) Geology and gravimetry. ONTI, Baku (in Russian)

Arkhangelsky AD, Fedynsky VV (1932) Geological results of gravity prospecting in eastern Azerbaijan. Bull MOIP (3–4) (in Russian)

Aronov VI (1976) Computer processing of gravitational anomalies under arbitrary relief of observation surface. Nedra, Moscow (in Russian)

Artemenko VI, Malovitsky YaP (1977) Results of near-bottom temperature survey at oil & gas deposit Bakhar. Geol Oil Gas (Geologiya Nefti i Gaza) (4): 50–54 (in Russian)

Artem'yev MYe, Balavadze BK (1973) Isostasy of the Caucasus. Geotectonics (Geotektonika) (6): 20–33 (in Russian)

Artem'yev MYe, Golland VE, Niauri GA (1985) New data on the isostasy of the Caucasus. Izv Earth Phys 21(2):85–93

Artyushkov EV (2007) Formation of the superdeep South Caspian basin: subsidence driven by phase change in continental crust. Russ Geol Geophys 48:1002–1014

Arutyunyan GB (1986) Application of mine modification of the method of electric correlation in polymetallic deposits in Armenia. In: Brodovoi VV, Badalyan SV Gamoyan VB (eds) Methodology, technics and results of geophysical prospecting of ore deposits. Publication of the Academy Sciences of Armenia, Yerevan, pp 78–82

Asanidze BZ, Pecherskiy DM (1979) Paleomagnetic investigations of the Jurassic rocks of Georgia and Northern Caucasus. Izv Earth Phys 10(10):77–92

Asanidze BZ, Pecherskiy DM, Adamiya ShA (1980) Results of paleomagnetic investigations of the paleozoic rocks of the Caucasus. Izv Earth Phys 16(9):90–109

Aslanyan AT (1970) Geology of the USSR, Armenia, vol XLIII. Nedra, Moscow (in Russian)

Auzin AK (1977) Electrical prospecting. Nedra, Moscow (in Russian)

Avdevich MM, Fokin AF (1978) Electrical modeling of potential geophysical fields. Nedra, Leningrad (in Russian)

Averbuch A, Zheludev V, Neittaanm P, Koren J (2010) Block based deconvolution algorithm using spline wavelet packets. J Math Imaging Vis 38(3):197–225

Azizbekov ShA (1972) Geology of the USSR, Azerbaijan, vol XLVII. Nedra, Moscow (in Russian)

Babayev ES, Allahverdiyeva AA (2007) Effects of geomagnetic activity variations on the physiological and psychological state of functionally healthy humans: some results of Azerbaijanian studies. Adv Space Res 40:1941–1951

Babayev ES, Hashimov AM, Yusifbeyli NA, Rasulov ZG, Asgarov AB (2006) Geomagnetic storm risks to electric power distribution and supply systems at mid-latitude locations and their vulnerability from space weather. In: Proceedings of the 3rd international TPE conference, Sect: Technical and physical problems in power engineering, Gazi University, Ankara, pp 1097–1104

Babayev G, Ismail-Zadeh A, Le Mouël J-L (2010) Scenario-based earthquake hazard and risk assessment for Baku (Azerbaijan). Nat Hazards Earth Syst Sci 10:2697–2712

Babazade OB (2010) Static and kinematic foundations of seismic geodynamics of earthquake's focus zones and space-time prognosis. D.Sci. thesis, Moscow, Russ. Acad. of Sci. (in Russian)

Babazade OB, Babazade NO, Griesser N Romanov B (2008) Dynamic prediction and earthquake monitoring – one of possible approaches to decrease the risk for oil and gas pipeline. In: Transaction of the 14th world conference on earthquake engineering, Beijing, China, Paper 14_07-0078, pp 1–10

Baba-Zadeh VM, Makhmudov AM, Ramazanov VG (1990) Copper-porphyry and molybdenum-porphyry deposits. Azerneshr, Baku (in Russian)

Baba-Zadeh VM, Makhmudov AM, Ramazanov VG (2002) Mineral resource base of Azerbaijan: a gaze into the XXI century. In: Tvalchrelidze AG, Morizot G (eds) Mineral resource base of the southern Caucasus and systems for its management in the XXI century. Proceedings of the NATO advanced research workshop, Series IY: Earth and Environmental Sciences, 17. Kluwer, Dordrecht, pp 105–124

Badalyan SV (1968) Possibilities and peculiarities of application induced polarization method in mines. Problems Appl Geophys (Voprosy Razvedochnoi Geofiziki), 7 (in Russian)

Bagin VI, Malumyan LM (1976) Iron-bearing minerals in oil saturated sedimentary rocks of productive series of Azerbaijan. Izv Ser Phys Earth (4): 73–79

Bagirov AA, Bektashi AP, Malakhova VA (1996) Copper-porphyry deposits. In: Kerimov KM (ed) Geophysical Investigations in Azerbaijan, Sharg-Garb, Baku pp 286–288

Bagirov E, Nadirov R, Lerche I (1998) Hydrocarbon evolution for a north–south section of the South Caspian Basin. Mar Petrol Geol 14(7/8):773–854

Balassanian S, Nazaretian S, Avanessian A, Arakelian A, Igumnov V, Badalian M, Martirossian A, Ambartsumian V, Tovmassian A (1997) The new seismic zonation map for the territory of Armenia. Nat Hazards 15:231–249

Balassanian S, Ashirov T, Chelidze T, Gassanov A, Kondorskaya N, Molchan G, Pustovitenko B, Trifonov V, Ulomov V, Giardini D, Erdik M, Ghafory-Ashtiany M, Grünthal G, Mayer-Rosa D, Schenk V, Stucchi M (1999) Seismic hazard assessment for the Caucasus test area. Annali Di Geofisica 42(6):1139–1151

Balat M (2006) The case of Baku-Tbilisi-Ceyhan oil pipeline system: a review. Energ Sources Part B 1:117–126

Balavadze BK (1939) Results of a geological interpretation of gravity measurements on the Akhaltsikhi coalbed. Proc Geophys Inst Georgian Acad Sci (Tbilisi), IV: 97–108 (in Russian)

Balavadze BK (1944) The results of electrometric investigations of Dashkesan cobalt deposits. Bull Georgian Acad Sci (Tbilisi) 5(10):365–374 (in Russian)

Balavadze BK, Abashidze VG (1973) Tiltmeter survey of the modern tectonic movements of earth's crust in the area of the dam Inguri hydroelectric power station building. Dokl Georgian Acad Sci Tbilisi 71(1):597–600 (in Russian)

Balavadze BK, Shengelaya GSh, Mindeli PSh (1979) Gravity model of earth crust of the Caucasus and Caspian Sea. In: Gravity models of earth's crust and upper mantle. Naukova Dumka, Kiev, pp 149–158 (in Russian)

Baranov V (1957) A new method for interpretation of aeromagnetic maps: pseudo-gravimetric anomalies. Geophysics 22(2):359–383

Barenblatt GI (1996) Scaling, self-similarity, and intermediate asymptotics. Cambridge University Press, Cambridge, UK

Barr R, Jones DL, Rodger CD (2000) ELF and VLF radio waves. J Atmos Solar-Terrestrial Phys 62:1689–1718

Baryshnikova IA, Krasnov AI, Akhmetzyanova GP (2011) Airborne electric prospecting for searching uranium. Russ Geophys J 49–50:96–110 (in Russian)

Bashirov AE, Eppelbaum LV, Mishne LR (1992) Improving Eötvös corrections by wide-band noise Kalman filtering. Geophys J Inter 108(1):193–197

Bashorin VN, Laverova NI, Pronin AP (2005) Seismically active fluid magmatic systems of northern Caucasus. Institute of the Physics of the Earth, Russian Academic Science, Moscow (in Russian)

Batoroyev KB (1974) Cybernetics and analogy method. Vysshaya Shkola, Moscow (in Russian)

Batsevich L (1881) Geological description of the Absheron Peninsula. Mater Geol Caucasus III:38 (in Russian)

Bazhenov ML, Burtman VS (2002) Eocene paleomagnetism of the Caucasus (southwest Georgia): oroclinal bending in the Arabian syntaxis. Tectonophysics 344:247–259

Bazhenov ML, Burtman VS, Levashova NL (1996) Lower and middle Jurassic paleomagnetic results from the south lesser Caucasus and the evolution of the Mesozoic Tethys ocean. Earth Planet Sci Lett 141:79–89

Beer T, Ismail-Zadeh A (eds) (2003) Risk science and sustainability. Kluwer, Dordrecht

Belinskiy A (2005) Remote sensing archaeological monuments in Stavropol region (Russia). In: Transaction of the 6th international conference on Archaeological prospection, Rome, pp 384–386

Belinskiy A, Fassbinder JWE, Reingold S (2011) Caucasian ring ditches – the easternmost Prehistoric rondels in Europe? In: Drahor MG, Berge MA (eds) Transaction of the 9th international conference on Archaeological Prospection Izmir, pp 106–107

Beresnev IA, Johnson PA (1994) Elastic-wave stimulation of oil production: a review of methods and results. Geophysics 59(6):1000–1017

Berezhnaya LT, Telepin MA (1966) Reduction of the observed potential field to the common level and a problem of its further transformation. Appl Geophys (Prikladnaya Geofizika) 16:38–46 (in Russian)

Berezkin VM (1967) Calculation of terrain relief and intermediate layer effects in gravity prospecting. Nedra, Moscow (in Russian)

Berezkin VM (1988) Total gradient method in geophysical prospecting. Nedra, Moscow (in Russian)

Berezkin VM, Kirichek MA, Kunarev AA (1978) Application of geophysical survey methods for direct oil and gas prospecting. Nedra, Moscow (in Russian)

Berishvili G, Gabisonia I, Tarkhnishvili A, Kereselidze Z, Lominadze J, Lomouri M, Mebaghishvili N, Tabaghua G, Kartvelishvili G, Chkhitunidze M (2008) Research of some characteristics of Guria magnetic anomaly. Bull Georgian Natl Acad Sci 2(2):49–52

Berkovitch AL, Eppelbaum LV (2009) From seismic inversion to applied optics: a novel approach to design of complex systems. Sci Israel 2:12–26

Bhattacharyya BK, Chan KC (1977) Reduction of magnetic and gravity data on an arbitrary surface acquired in a region of high topographic relief. Geophysics 42(7):1411–1430

Binkin IG, Pystulnikov LM, Odintzov NI, Kambarly CE, Oifa VYa, Pererva VM (1986) Results of experimental investigations of zones facies replacements of Maikop oil&gas perspective horizons in the West-Kuban depression. In: Aleksin AG (ed) Prognosis of geological section and searching complex-buried traps. Nauka, Moscow, pp 117–126 (in Russian)

Blakely RJ (1995) Potential theory in gravity and magnetic applications. Cambridge University Press, Cambridge

Bodri L, Cermak V (2005) Borehole temperatures, climate change and the pre-observational surface air temperature mean: allowance for hydraulic conditions. Global Planet Change 45:265–276

Bogatikov OA, Rogozhin EA, Gurbanov AG, Marakhanov AV, Spiridonov AV, Shevchenko AV, Burkanov EE (2003) Ancient earthquakes and volcanic eruptions in the Elbrus region. Dokl Russ Acad of Sci 390(4):524–528

Bogoslovsky VA, Ogilvi AA (1977) Geophysical methods for the investigation of landslides. Geophysics 42(3):562–571
Bogush IA, Burtzev AA, Sendetsky II, Yakushev VV (2004) Gold-löllingite mineralization in the Northern Caucasus (the Bolshaya Laba river area). Proc Russ Mineral Soc, Part CXXXIII (3) (in Russian)
Boichenko AP, Korobova EG, Ananskih AC (2000) Study of dolmens of village Shapsugskaya by some physical methods. Unpublished report of the Kuban State University, Krasnodar (in Russian)
Bongard MM (1967) Problem of recognition. Nauka, Moscow (in Russian)
Borisov BA (1982) Strong earthquakes in the Eastern Caucasus: interpretation of historical data and analysis of geological situation. Izv Earth Physics 18(9):726–737
Borisovich VT, Eppelbaum VM (eds) (1988) Optimization of ore exploration in mountainous regions. Nedra, Moscow (in Russian)
Borisovich VT, Eppelbaum LV and Ginzburg SN (1988) Optimization of geophysical investigations in mountainous regions on the basis of integration of geophysical and mining-drilling works. In: Borisovich VT, Eppelbaum VM (eds) Optimization of ore exploration in mountainous regions, Nedra, Moscow, pp 123–133 (in Russian)
Borodaevskaya MB, Krivtsov AT, Shirai EP (1977) Types of pyrite-bearing provinces and methods for predicting pyrite mineralization. Review of VIEMS, Ser. Geology and Methods for Searching Useful Minerals, Moscow (in Russian)
Borovko NN (1971) Statistical analysis of spatial geological regularities. Nedra, Leningrad (in Russian)
Borovko NN (1972) Applicability and effectiveness of various recognition schemes for typical geological problems. Methods of Explor Geophys (Metody Razvedochnoi Geofiziki) 15:5–14 (in Russian)
Borovko NN (1979) Optimization of geophysical investigations in ore deposit prospecting. Nedra, Leningrad (in Russian)
Borovko NN, Mishin LT, Latikainen VI (1969) Techniques of quantitative field description for predicting stanniferous deposits on a scale of 1:50,000. Methodological Directions, VIRG, Leningrad (in Russian)
Bratus A, del Negro E, Yabar DN (2005) Geoelectrical investigation of dams in Georgia – GIS. In: Transaction of the Balkan geophysics society, Section: Environmental, Engineering. and Archaeogeophysics, Bucharest, pp 28–31
Brillouin LN (1962) Science and information theory, 2nd edn. Academic, New York
Brodovoi VV (1989) Searching and prospecting of useful deposits: copper. In: Brodovoi VV (ed) Borehole and mining geophysics, vol II. Nedra, Moscow, pp 190–208 (in Russian)
Brodovoi VV, Ginzburg SN, Haldeyev OD (1989) Searching and prospecting of lead and zinc. In: Brodovoi VV (ed) Borehole and mining ore geophysics, Nedra, Moscow, vol II. pp 297–310
Brodskaya SYu, Gogua RA (1988) Formation conditions of Adjarian intrusives based on petromagnetic data. Izv AN USSR, Earth Phys 24(4):294–300
Brodsky MA, Strakhov VN (1987) On the solution of the inverse potential problem for polyhedrons with polynomial densities. Dokl AN SSSR 293(2):336–339
Brunet M-F, Korotaev MV, Ershov AV, Nikishin AM (2003) The South Caspian Basin: a review of its evolution from subsidence modeling. Sediment Geol 156:119–148
Brusilovsky YuV, Ivanenko AN, Filin AM (2009) High-precision differential geomagnetic survey in regions perspective on hydrocarbons deposits (North Caspian). In: Transaction of the XVIII international conference on marine geology, Moscow, pp 20–22
Bugaets AN, Dudenko LN (1976) Mathematical methods for predicting economic mineral deposits. Nedra, Leningrad (in Russian)
Bukhnikashvili AV, Kebuladze VV, Tabagua GG, Dzhashi GG, Gugunava GE, Tatishvili OV, Gogua RA (1974) Geophysical exploration of Adzar group of copper-polymetallic deposits. Metsniereba, Tbilisi (in Russian)
Bulakh EG, Markova MN, Timoshenko VI, Boiko PD (1983) Mathematical support for automated system of gravity anomalies interpretation (Minimization Method). Naukova Dumka, Kiev (in Russian)

Bulakh EG, Rzhanitsyn VA, Markova MN (1976) Application of minimization method for solution of structural geology problems using gravity data. Naukova Dumka, Kiev (in Russian)

Buryakovsky L, Chilingar GV, Aminzadeh F (2001) Petroleum geology of the south Caspian Basin. Gulf Professional Publishing, Woburn, USA

Burlatskaya SP, Chelidze ZA (1987) Geomagnetic field variations in Georgia from the third millennium B.C. to the first millennium A.D. Izv Earth Phys 23(9):774–778

Burlatskaya SP, Chelidze ZA (1990) Geomagnetic field variations in Georgia during the last 15 centuries BC. Izv Earth Physics 26(7):602–609

Burlatskaya SP, Chernykh IYe (1989) Change in geomagnetic field strength in Azerbaijan during last 2200 years determined from archaeomagnetic data. Izv Earth Physics 25(7):594–597

Burlatskaya SP, Vydrin AI, Chernykh IYe (1991) Global and regional geomagnetic field variations in the Caucasus. Izv Earth Physics 27(10):856–867

Camps P, Ruffet G, Shcherbakov VP, Shcherbakova VV, Prevot M, Moussine-Pouchkine A, Sholpo L, Goguitchaichvili A, Asanidze B (1996) Paleomagnetic and geochronological study of a geomagnetic field reversal or excursion recorded in pliocene volcanic rocks from Georgia (Lesser Caucasus). Phys Earth Planet In 96:41–59

Calvo-Rathert M, Goguitchaichvili A, Sologashvili D, Villalan JJ, Bogalo MF, Carrancho A, Maissuradze G (2008) New paleomagnetic data from the hominine bearing Dmanisi paleo-anthropologic site (southern Georgia, Caucasus). Quat Res 69:91–96

Carslaw HS, Yaeger JC (1959) Conduction of heat in solids, 2nd edn. Oxford University Press, Oxford, UK

Chapin DA (1996a) The theory of the Bouguer gravity anomaly: a tutorial. Leading Edge 15 (5):361–363

Chapin DA (1996b) A deterministic approach toward isostatic gravity residuals – a case study from South America. Geophysics 61(4):1022–1033

Chekalyuk EB, Fedortsev IM, Osadchy VG (1974) Geothermal field survey. Naukova Dumka, Kiev (in Russian)

Cheremensky GA (1972) Geothermics. Nedra, Leningrad (in Russian)

Chernobay IP, Gabsatarova IP (1999) Source classification in the Northern Caucasus. Phys Earth Planet In 113:183–201

Chernov AA (2008) Taman Peninsula research with the use of gravitational field anomalies. In: Proceedings of the Caspian & Black Sea Geosciences Conf., Baku, pp 1–5

Chigira M, Tanaka K (1997) Structural features and the history of mud volcano in southern Hokkaido, northern Japan. J Geol Soc Japan 103:781–793

Chizhevsky AL (1930) Epidemic catastrophes and periodic activity of the Sun. Moscow (in Russian)

Chizhevsky AL (1963) The sun and we. Moscow (in Russian)

Cisternas A, Philip H, Bousquet JC, Cara M, Deschamps A, Dorbath L, Dorbath C, Haessler H, Jimenez E, Nercessian A, Rivera L, Romanowicz B, Gvishiani A, Shebalin NV, Aptekman I, Arefiev S, Borisov BA, Gorshkov A, Graizer V, Lander A, Pletnev KA, Rogozhin I, Tatevossian R (1989) The Spitak (Armenia) earthquake of 7 December 1988: field observations, seismology and tectonics. Nature 339:675–679

Dalan RA, Banerjee SK (1998) Solving archaeological problems using techniques of soil magnetism. Geoarchaeology 13(1):1–36

De Mets C, Gordon RG, Argus DF, Stein S (1990) Current plate motions. Geophys J Int 101:425–478

Demidovich OA (1969) Singling out weak geophysical anomalies by a statistical method. Nedra, Moscow (in Russian)

Demura GV, Nikitin AA, Tarkhov AG (1974) Classification of geological objects by the data obtained using a set of geophysical methods based on the principle of self-training. Izv VUZov Ser Geol Prospect (2): 133–142 (in Russian)

Dell ZE, Franklin SV (2009) The buffon-laplace needle problem in three dimensions. J Stat Mech Theory Exp 1–11. doi:10.1088/1742-5468/2009/09/P09010

Detkova NV, Shopin YuG (1969) Estimation of correlation between the elastic wave velocity and the density for magmatic and metamorphic rocks in the ore-bearing regions of the Altai and lesser Caucasus. Problems Appl Geophys (Voprosy Razvedochnoi Geofiziki) 9:160–163 (in Russian)

Diaconescu CC, Kieckhefer RM, Knapp JH (2001) Geophysical evidence for gas hydrates in the deep water of the South Caspian Basin, Azerbaijan. Mar Petrol Geol 18:209–221

Dimitrov LI (2002) Mud volcanoes – the most important pathway for degassing deeply buried sediments. Earth-Sci Rev 59:49–76

Dmitriyev VI (ed) (1982) Computing mathematics and computer methods in exploration geophysics. Geophysicist's manual. Nedra, Moscow (in Russian)

Dmitriyev VI, Baryshnikova IA, Zakharov EV (1977) Anomalous electromagnetic fields of the buried beds. Nedra, Leningrad (in Russian)

Dmitriyeva TA, Kaner VV, Skornyakov SM, Semenov VG (1985) Theoretical analysis of temperature fields in near-surface horizons for 2-D and 1-D ore body models. Deposited by VINITI, USSR Acad Sci (5191-85) pp 1–12 (in Russian)

Dobrin MB (1976) Introduction to geophysical prospecting, 3rd edn. McGraw-Hill, New York, 1976

Dobrynina ZK, Dmitriyeva TA, Pletnev AA, Semenov VG, Skornyakov SM (1985) Shaping of seasonal thermal anomalies over ore bodies. Izv VUZov Ser Geol Prospect (8): 98–101 (in Russian)

Donovan D, Forgey RL, Roberts AA (1979) Aeromagnetic detection of diagenetic magnetite over oil fields. AAPG Bull 63:245–248

Duckworth K, Calvert HT (1995) An examination of the relationship between time-domain integral chargeability and the cole-cole impedance model. Geophysics 60(4):1249–1252

Duda RO, Hart PE (1973) Pattern classification and scene analysis. Wiley, New York

Dukhovsky AA, Ilayev MG, Kronidov II (1970) Geophysical investigations / methodological instructions for geological survey on a 1:50,000 scale. Nedra, Leningrad (in Russian)

Dyadkov PG (1985) On the linearity of the geomagnetic variation secondary effects in anomalous fields by modulus measurements. Geol Geophys (Geologiya i Geofizika) (4): 129–136 (in Russian)

Dzabayev AA (1969) Principles of the searching and study of oil-and-gas bearing structures by aeromagnetic method (South Caspian Basin). Statistika, Ashkhabad (in Russian)

Dzhahutashvili MN (2006) Effectiveness of geophysical methods in delineation of archaeological features in Georgia. Ph.D. Thesis, Tbilisi State University, Tbilisi

Egan SS, Mosar J, Brunet M-F, Kangarly T (2009) Subsidence and uplift mechanisms within the South Caspian Basin: insights from the onshore and offshore Azerbaijan region. Geol Soc Lond Spec Publ 312:219–240

Egorkina GV, Sokolova IA, Egorova LM (1976) Deep structure of ophiolite belts in Armenia. Sov Geol 3:127–133 (in Russian)

Egorkina GV, Egorova LM, Sokolova IA et al (1978) New data on structure and age of the basement in Armenia. Applied Geophysics (*Prikladnaya Geofizika*), 92, Nedra, Moscow, pp 93–104 (in Russian)

Eliseyeva TV (2007) Mathematical modeling of temperature and potential fields in piecewise media. Ph.D. Thesis, Penza University (in Russian)

Eppelbaum LV (1984a) Multimodel approach to the study of geophysical targets. Deposited by VINITI, USSR Academy of Sciences, No.7842-87 pp 1–10 (in Russian)

Eppelbaum LV (1984b) A correlation method for eliminating topographic effect in the *VLF*-method. Deposited by VINITI, USSR Academy of Sciences, No.6492-84, pp 1–14 (in Russian)

Eppelbaum LV (1989) The development of methods for processing and interpretation of natural geophysical fields in prospecting for pyrite ores under mountainous conditions. Ph.D. Thesis, Institute of Geophysics (Georgian Acad. of Sciences), Tbilisi (in Russian)

Eppelbaum LV (1991) Examples of terrain corrections in the VLF-method in the Caucasian region, USSR. Geoexploration 28:67–75

Eppelbaum LV (1999a) Quantitative interpretation of resistivity anomalies using advanced methods developed in magnetic prospecting. In: Transaction of the XXIV general assembly of the Europe geophysics society, Strasburg, 1, No. 1, p 166

Eppelbaum LV (1999b) Near-surface thermal prospecting in petroleum geology. In: Applied geothermics for petroleum engineers, Elsevier, pp 274–301

Eppelbaum LV (2006) Methodology of 3-D combined modelling of magnetic and gravity fields in the eastern Mediterranean. In: Proceedings of the 3rd EUG meeting, vol. 8, 02057, pp 1–4.

Eppelbaum LV (2007a) Revealing of subterranean karst using modern analysis of potential and quasi-potential fields. In: Proceedings of the 2007 SAGEEP conference, Denver, USA, pp 797–810

Eppelbaum LV (2007b) Localization of ring structures in earth's environments. J Soc Slovakian Acad Sci Spec Issue: Arch Prosp XLI:145–148

Eppelbaum LV (2010a) Archaeological geophysics in Israel: past, present and future. Adv Geosci 24:45–68

Eppelbaum LV (2010b) An advanced methodology for remote operation vehicle magnetic survey to delineate buried targets. In: Transaction of the 20th general meeting of the international mineralogical association, CH30G: Archaeometry (general session): Composition, technology and provenance of archaeological artifacts, Budapest, p 103

Eppelbaum LV (2011a) Study of magnetic anomalies over archaeological targets in urban conditions. Phys Chem Earth 36(16):1318–1330

Eppelbaum LV (2011b) Boris E. Khesin, Memorials. The leading Edge, (2), p 232

Eppelbaum LV (2012) Geophysical investigations in the Caucasus (1925–2012): initial, basic and modern stages. In: Proceedings of the EAGE conference, geophysical research abstracts, EGU2012-1341, pp 1–3

Eppelbaum LV, Alperovich L, Zheludev V, Pechersky A (2011a) Application of informational and wavelet approaches for integrated processing of geophysical data in complex environments. In: Proceedings of the 2011 SAGEEP conference, vol 24. Charleston, pp 24–60

Eppelbaum LV, Alperovich L, Zheludev V, Pechersky A (2011b) Integration of different geophysical fields for development of reliable 3D models of subsurface in transport infrastructure areas. FP7 ISTIMES EuroScience Report, pp 1–125

Eppelbaum L, Ben-Avraham Z, Itkis S (2003a) Ancient roman remains in Israel provide a challenge for physical-archaeological modeling techniques. First Break 21(2):51–61

Eppelbaum L, Ben-Avraham Z, Katz Y (2004a) Integrated analysis of magnetic, paleomagnetic and K-Ar data in a tectonic complex region: an example from the Sea of Galilee. Geophys Res Lett 31(19):L19602

Eppelbaum LV, Ben-Avraham Z, Katz YI (2007) Structure of the Sea of Galilee and Kinarot Valley derived from combined geological-geophysical analysis. First Break 25(1):21–28

Eppelbaum L, Eppelbaum V, Ben-Avraham Z (2003b) Formalization and estimation of integrated geological investigations: informational approach. Geoinformatics 14(3):233–240

Eppelbaum LV, Ezersky MG, Al-Zoubi AS, Goldshmidt VI, Legchenko A (2008) Study of the factors affecting the karst volume assessment in the Dead Sea sinkhole problem using microgravity field analysis and 3D modeling. Adv Geo Sci 19:97–115

Eppelbaum LV, Finkelstein MI (1998) Radon emanation, magnetic and VLF temporary variations: removing components not associated with dynamic processes. Collection of selected papers of the XXVI general assembly of the European seismological commission (Tel Aviv, Israel), pp 122–126

Eppelbaum LV, Gasanova EYu (1990) Quantitative interpretation of anomalies in the long-wire method. In: Transaction of the conference of young scientists on problems of geology and geophysics (in Russian), USSR, Baku, p 28

Eppelbaum LV, Itkis SE, Khesin BE (2000) Optimization of magnetic investigations in the archaeological sites in Israel. In: Special issue of Prospezioni Archeologiche "Filtering, Modeling and Interpretation of Geophysical Fields at Archaeological Objects", pp 65–92

Eppelbaum LV, Kardashov VR (1998) Nonlinear geothermal processes in the Earth crust and transition waves. In: Transaction of the international conference "The Earth's Thermal Field and Related Research Methods". Moscow, Russia, 82–85 (in Russian)

Eppelbaum LV, Kardashov VR (2001) Analysis of strongly nonlinear processes in geophysics. In: Moresi L, Müller D (eds) Proceedings of the Chapman conference on exploration geodynamics. Dunsborough, pp 43–44

Eppelbaum L, Katz Y (2011) Tectonic-geophysical mapping of Israel and eastern Mediterranean: implication for hydrocarbon prospecting. Positioning 2(1):36–54

Eppelbaum LV, Khesin BE (1988) Physical-geological models for pyrite deposits of the filizchay and lesser-Caucasian types. In: Transactions of all-union meeting "Multifactor ore deposit models as the basis for developing effective methods of search, evaluation and prospecting", Tbilisi, pp 126–127 (in Russian)

Eppelbaum LV, Khesin BE (1992) *VLF*-method: elimination of noise and quantitative interpretation. Transactions of regional symposium on electromagnetic compatibility "1992 – From a unified region to a unified world", Section "LF to ULF electromagnetics and the earth", 5.2.1, Tel Aviv, pp 1–6

Eppelbaum LV, Khesin BE (2002) Some common aspects of magnetic, induced polarization and self-potential anomalies interpretation: implication for ore target localization. Collection of selected papers of the IV international symposium on problems of eastern Mediterranean geology, pp 279–293

Eppelbaum LV, Khesin BE (2004) Advanced 3-D modelling of gravity field unmasks reserves of a pyrite-polymetallic deposit: a case study from the greater Caucasus. First Break 22(11):53–56

Eppelbaum LV, Khesin BE (2011) Development of 3-D gravity-magnetic models of earth's crust of Azerbaijan and adjacent areas: a generalized review. Positioning 2(2):84–102

Eppelbaum L, Khesin B, Ginzburg A, Ben-Avraham Z (1992) Quantitative interpretation of magnetic anomalies and preliminary 3-D modeling of gravity and magnetic fields. In: Transaction of the conference of Israel geology society, Annual Meeting, Ashqelon, p 35

Eppelbaum LV, Khesin BE, Itkis SE (2001) Prompt magnetic investigations of archaeological remains in areas of infrastructure development: Israeli experience. Archaeol Prospect 8 (3):163–185

Eppelbaum LV, Khesin BE, Itkis SE (2006a) Some peculiarities of geophysical investigations at archaeological sites in Israel. Russ Archaeol 1:59–70 (in Russian)

Eppelbaum LV, Khesin BE, Itkis SE (2010) Archaeological geophysics in arid environments: examples from Israel. J Arid Environ 74(7):849–860

Eppelbaum LV, Khesin BE, Itkis SE Ben-Avraham Z (2004b) Advanced analysis of self-potential data in ore deposits and archaeological sites. In: Proceedings of the 10th European meeting of environmental and engineering geophysics, Utrecht, pp 1–4

Eppelbaum LV, Kutasov IM, Barak G (2006b) Ground surface temperature histories inferred from 15 boreholes temperature profiles: comparison of two approaches. Earth Sci Res J 10(1):25–34

Eppelbaum LV, Livshits Ya, Flexer A, Ben-Avraham Z (1998) Integrated geological-geophysical analysis of ring structures phenomenon in the eastern Mediterranean Transaction of the conference of Israel geological society, annual meeting, Mizpe-Ramon, pp 23–24

Eppelbaum LV, Mishne LR (1988) Some theoretical aspects of optimum filtration of geophysical signals with dependent noises. Deposited by VINITI, USSR Academy of Sciences, No. 6034-B88 (in Russian)

Eppelbaum LV, Mishne AR (1995) High-precision magnetic survey: elimination of secondary time variations. In: Transaction of the conference of the geology society of America, Rocky Mountain, 27, No.4, p 10

Eppelbaum LV, Mishne AR (2011) Unmanned Airborne magnetic and VLF investigations: effective geophysical methodology of the near future. Positioning 2(3):112–133

Eppelbaum LV, Pilchin AN (1998) Western Caspian Sea Basin and Dead Sea Basin: comparison of some physical-geological characteristics. In: Transaction of the III international conference on the eastern Mediterranean, Cyprus, Aug 1998, p 20

Eppelbaum LV, Pilchin AN (2006) Methodology of curie discontinuity map development for regions with low thermal characteristics: an example from Israel. Earth Planet Sci Lett 243 (3–4):536–551

Eppelbaum L, Yakubov Ya (2010) Method for comprehensive computing of water flows geodynamics in the Dead Sea basin. In: Transaction of the XXXV EGS meeting, Zurich, 24, pp 1–4

Ernstson K, Scherer V (1986) Self-potential variations with time and their relation to hydrogeologic and meteorological parameters. Geophysics 51(10):1967–1977

Erofeyev LYa, Avteneyev GK (1971) On statistical relationship between the terrain relief and magnetic field on gold deposits. In: Problems of geography of Kuzbass and Gornyi Altai (Voprosy geografii Kuzbassa i Gornogo Altaya), No.4, 270–275 (in Russian)

Erofeyev LYa, Vakhromeyev GS, Zinchenko, VS, Nomokonova GG (2009) Physics of rocks. Tomsk Polytechnical University (in Russian)

Fassbinder J, Korobov DS, Reinhold S (2007) Magnetometrie auf neu entdeckten früheisenzeitlichen Siedlungslandschaften bei Kislovodsk im Nordkaukasus. Denkmalpflege Informationen (München) 137:58–59

Fassbinder J, Reinhold S, Belinskiy A (2010) Bronze age settlements and new discovered ring ditches of the Maykop culture in the North-Caucasus (Russia). In: Proceedings of the conference "Recent Work in Archaeological geophysics", The Geological Society, Burlington House, London, pp 8–11

Fedynsky VV (1937) Gravity anomalies of Azerbaijan. AzONTI, Baku – Moscow (in Russian)

Fernández-Jalvo Y, King T, Andrews P, Yepiskoposyan L, Moloney N, Murray J, Domínguez-Alonso P, Asryan L, Ditchfield P, van der Made J, Torres T, Sevilla P, Nieto Díaz M, Cáceres I, Allué E, Monfort MMD, Martín TS (2010) The Azokh cave complex: middle pleistocene to Holocene human occupation in the Caucasus. J Hum Evol 58:103–109

Feyzullayev AA, Kadirov FA, Aliyev CS (2005a) Mud volcano model resulting from geophysical and geochemical research. In: Martinelli G, Panahi B (eds) Mud volcanoes, geodynamics and seismicity, pp 251–262

Feyzullayev AA, Kheirov MB, Aliyev ChS, Abbasova SB, Aliyev KA (2005b) Comparative characteristics of composition and radioactivity of oligocene clays in the greater Caucasus and Talysh. Lithol Miner Resour 40(4):376–385

Fidanyan FM (1986) Integrated gravi-magnetic investigation of Bazum iron-ore deposit. In: Brodovoi VV, Badalyan SV, Gamoyan VB (eds) Methodology, technics and results of geophysical prospecting of ore deposits. Publ. of the Acad. Sci. of Armenia, Yerevan, pp 121–127 (in Russian)

Finkelstein M, Brenner S, Eppelbaum L, Ne'eman E (1998) Identification of anomalous radon concentration due to geodynamic processes by elimination of ***Rn*** variations caused by other factors. Geophys J Int 133(2):407–412

Finkelstein M, Eppelbaum L (1997) Classification of the disturbing objects using interpretation of low-intensive temporary magnetic variations. In: Transaction of the conference of geological society of America. Salt Lake City, 29, No.6, p 326

Finkelstein M, Eppelbaum L, Price C (2006) Analysis of temperature influences on the amplitude-frequency of Rn gas concentration. J Environ Radioact 86(2):251–270

Finkelstein MI, Maslatsov GI (1984) Precise component geomagnetic study of variations in the search for manifestation of tectonomagnetic effects. In: Experimental studies of the geomagnetic field. IZMIRAN, USSR Acad. of Sci., Moscow, 152–160 (in Russian).

Finkelstein M, Price C, Eppelbaum L (2012) Is the geodynamic process preparation of strong earthquake reflected in a magnetic field? Submitted to J Geophys Eng

Fitterman DV (1984) Calculation of self-potential anomalies near vertical contacts. Geophysics 44 (2):195–205

Fokin AF (ed) (1971) Method of transient processes in the search for sulphide ore deposits. Nedra, Leningrad (in Russian)

Fotiadi EE (ed) (1970) Geology and mathematics; problems of diagnosis and identification in geology, geochemistry and geophysics. Nauka, Novosibirsk (in Russian)

Frantov GS, Pinkevich AA (1966) Geophysics in archaeology. Nedra, Leningrad (in Russian)

Freund F (2011) Pre-earthquake signals: underlying physical processes. J Asian Earth Sci 41 (4–5):383–400

Gadirov VG (2009) Results of gravity and magnetic prospecting application for prognosis of oil-and-gas deposits in the Kura Depression of Azerbaijan. Geophys (Geofizika) (2): 51–56 (in Russian)

Gadjiev RM (1965) Deep geological structure of Azerbaijan. Azerneshr, Baku (in Russian)

Gadjiev TG, Karkoshkin AI, Khesin BE, Alexeyev VV, Potapova EI, Salekhli TM (1984) Petrodensity characteristics of geological associations in Azerbaijan. Azerneshr, Baku (in Russian)

Gadjiev TG, Karkoshkin AI, Khesin BE, Alexeyev VV, Potapova EI, Salekhli TM (1985) Petrodensity map of the Azerbaijan SSR. Scale 1:500,000. Map Printing Factory, Leningrad

Gadjiev TG, Nechaev YuV, Potapova EI, Sattarova VM (1989) The map of deep structure of the Caucasus according to cosmic data. Map Printing Factory of the Ministry of Geology of the USSR, Baku

Gaffet S, Guglielmi Y, Virieux J, Waysand G, Chwala A, Stolz R, Emblanch C, Auguste M, Boyer D, Cavaillou A (2003) Simultaneous seismic and magnetic measurements in the low-noise underground laboratory (LSBB) of Rustrel, France, during the 2001 January 26 Indian earthquake. Geophys J Int 155:981–990

Gamkrelidze PD (ed) (1964) Geology of the USSR, vol X, Georgia. Nedra, Moscow (in Russian)

Gamoyan VB (1986) Sources of stray currents in mines. In: Brodovoi VV, Badalyan SV, Gamoyan VB (eds) Methodology, technics and results of geophysical prospecting of ore deposits. Publ. of the Acad. Sci. of Armenia, Yerevan, pp 47–53 (in Russian)

Gamoyan VB, Unusyan FS, Chilingaryan AZ (1986) Self-potential fields of sulphide deposits of Armenia and their physical-geological models. In: Brodovoi VV, Badalyan SV Gamoyan VB (eds) Methodology, technics and results of geophysical prospecting of ore deposits. Publ. of the Acad. Sci. of Armenia, Yerevan, pp 54–60 (in Russian)

Garetovskaya IV, Krasnopevtseva GV, Sizov FV et al (1986) Investigation of deep structure *of the North Caucasian* seismic zone with the help of seismic and gravity methods (Caucasian Mineral Waters and Elbrus region). In: The main problems of seismotectonics, Nauka, Moscow, pp 105–119 (in Russian)

Gasanov AG (2001) Deep structure and seismisity of Azerbaijan in relation to oil-and-gas prognosis. Elm, Baku (in Russian)

Gazaryan GO (1965) Experince of VLF method application in underground. Results of induced method application in a chrome deposit. Methodology and technics of prospection, No. 49 (in Russian)

Gazaryan GO (1968) Experince of VLF method application in underground. P Explor Geophys 7:169–170 (in Russian)

Gezin AE (1964) Application of induced polarization method in ore areas of the northern Caucasus. In: Employment of induced polarization method for searching ore deposits. Nedra, Moscow (in Russian)

Giardini D, Balassanian S (eds) (1997) Historical and prehistorical earthquakes in the Caucasus. In: Proceedings of the NATO advanced research workshop, Yerevan, Armenia (1996). Series: NATO Science Partnership, vol 28. Springer Heidelberg/NY

Ginzburg SN (1982) VLF-method for searching and prospecting of pyrite-polymetallic deposits. Prospecting and protection of entrails (Razvedka i Okhrana Nedr), No.9, pp 39–44 (In Russian)

Ginzburg S et al (1974) Integrated geophysical investigations in ore deposits of in Georgia (southern slope of the Greater Caucasus). Unpublished Report of *TzNIGRI* (Central Scient. Inst. of Non-Ferrous and Precious Metals), Moscow (in Russian)

Ginzburg S et al (1981) Underground geophysical investigations in ore deposits of Belokan-Zakatala ore field. Unpublished Report of *TzNIGRI* (Central Scient. Inst. of Non-Ferrous and Precious Metals), Moscow (in Russian)

Ginzburg SN, Vitkova ID, Istratov VA, Maslennikov AL (1986) Peculiarities of performing combined geophysical investigations by searching copper-pyrite deposits in mountainous conditions. In: Brodovoi VV, Badalyan SV, Gamoyan VB (eds) Methodology, technics and results of ceophysical prospecting of ore deposits. Publ. of the Acad. Sci. of Armenia, pp 183–188 (in Russian)

Gladky KV (1967) Gravity and magnetic prospecting. Nedra, Moscow (in Russian)

Goguitchaichvili A, Pares JM (2000) A recognition palaeomagnetic study of volcanic and sedimentary rocks from Dmanissi (Caucasus): implications for the oldest human occupation in Europe. Earth Planet Sci 331:183–186

Goguitchaichvili A, Calvo M, Sologashvili D, Alva L, Urrutia J (2000) Palaeomagnetism of Georgian plio-quaternary volcanic provinces (Southern Caucasus): a pilot study. Earth Planet Sci 331:683–690

Goguitchaichvili A, Camps P, Urrutia-Fucugauchi J (2001) On the features of the geodynamo following reversals or excursions: by absolute geomagnetic paleointensity data. Phys Earth Planet In 124:81–93

Gokhberg MB, Morgunov AV, Aronov EL (1979) On the high-frequency radiation by seismic activity. Dokl AN USSR 248(5):1077–1081

Gokhberg MB, Shalimov SL (2008) Influence of earthquakes and explosions to ionosphere. Nauka, Moscow (in Russian)

Goldman M (1990) Non-conventional method in geoelectric prospecting. Ellis Horwood, Chichester

Goldshmidt VI, Kuzmin YuI, Shabaldin VN (1981) An automated system of gravity data interpretation. In: Theory and methods for interpretation of gravity and magnetic fields. Naukova Dumka, Kiev, pp 58–70 (in Russian)

Golizdra GYa (1977) The main methods for computer-aided solution of direct problems of gravity prospecting. Review of VIEMS, Ser. Region., Applied and Borehole Geoph., Moscow (in Russian)

Golizdra GYa (1980) Starting the problem of combined interpretation of gravitational fields and seismic observations. Izv Phys Earth 7:95–100

Golubyatnikov DV (1916) Detailed geological map of the Absheron Peninsula. Bibi-Eibat, Part II. Geothermal observations in Bibi-Eibat and Surakhany. In: Transactions of the Geological Committee, vol 141, II, Sankt-Petersburg. (in Russian)

Golovin IV, Suprunenko EI (1971) Some techniques of statistical analysis of geophysical data in regional prognosis. Methods Explor Geophys (Metody Razvedochnoi Geofiziki) 12:53–57 (in Russian)

Goltzman FM (1971) Statistical models of interpretation. Nauka, Moscow (in Russian)

Gongadze SA (2006) Anomalous gravity field in Georgia and some aspects of its geological-geophysical interpretation. Ph. D. Thesis, M. Nodia Geophysical Institute, Tbilisi (in Russian)

Gorbatikov AV, Stepanova MYu (2008) Statistical characteristics and stationary properties of low-frequency seismic signals. Izv Phys Earth 1:57–68

Gorbatikov AV, Sobisevich AL, Ovsyuchenko AN (2008a) Model of deep structure of Akhtyr flexure-disjunctive zone and Shugo mud volcano. Dokl Russ Acad Sci 421(5):670–674

Gorbatikov AV, Stepanova MYu, Korablev GE (2008b) Microseismic field affected by local geological heterogeneities and microseismic sounding of the medium. Izv Phys Earth 7:66–84

Gordeyev SG (1970) On the problem of influence of terrain relief by VLF field measurements. In: Proceedings. of TzNIGRI (Central Scientific Institution of Non-Ferrous and Precious Metals), Moscow, 89, pp 188–195 (in Russian)

Gordeyev SG, Sedelnikov ES (1974) On the interpretation of the results of *VLF* method on the basis of the conductive bed simulation. In: Proceedings of TzNIGRI (Central Scientific Institution of Non-Ferrous and Precious Metals), Moscow, 116, pp 88–99 (in Russian)

Gordeyev SG, Sedelnikov ES, Tarkhov AG (1981) Electric prospecting by the radio comparing and bearing method. Nedra, Moscow (in Russian)

Gordin VM (1974) Methods for eliminating terrain relief effects under highly accurate gravimetric observations. Review of VIEMS, Ser. Region., Applied and Borehole Geophysics, Moscow (in Russian)

Gorodnitsky AM, Zonenshain LP, Mirlin EG (1978) Reconstructions of the continents in phanerozoic. Nauka, Moscow (in Russian)

Gorshkov GP (1984) Regional seismotectonics of the southern part of the USSR. Nauka, Moscow

Gorshkov AI, Niauri GA (1984) Reflection of fault tectonics of the Caucasus in zones of horizontal gravity anomaly gradients. Izv Phys Earth 20(9):699–704

Gorzhevsky DI, Kurbanov NK, Filatov EI et al (1987) Methodological principles of prediction and prospecting for lead and zinc deposits. Nedra, Moscow (in Russian)

Granath JW, Soofi KA, Baganz OW, Bagirov E (2007) Gravity modeling and its implications to the tectonics of the South Caspian Basin. In: Yilmaz PO, Isaksen GH (eds) Oil and gas of the greater Caspian Area. AAPG Studies in Geology, 55, pp 43–46

Green T, Abullayev N, Hossack J, Riley G, Roberts AM (2009) Sedimentation and subsidence in the south Caspian Basin, Azerbaijan. In: Brunet M-F, Wilmsen M, Granath JW (eds) South Caspian to central Iran Basins, vol 312, Special Publications. The Geological Society, London, pp 241–260

Griffin WR (1949) Residual gravity in theory and practice. Geophysics 14(1):39–56

Guberman ShA (1987) Non-formal analysis of data in geology and geophysics. Nedra, Moscow (in Russian)

Gudmundsson A, Nilsen K (2006) Ring-faults in composite volcanoes: structures, models and stress fields associated with their formation. Geological Society, Special Publications (London), 269, pp 83–108

Gufeld IP, Shuleikin VJ (1988) Variations of atmospheric field during earthquakes preparing. Izv Phys Solid Earth 2:12–16

Gugunava GE (1981) Relation between the some geophysical fields and Caucasus deep structure. Mezniereba, Tbilisi (in Russian)

Gugunava GE (1988) Structure of Caucasus inferred from the analysis of some geophysical fields. D. Sci. Thesis, Institute of Geophysics of the Georgian Academy of Science – Institute of Geology of the Azerbaijanian Academy of Sci. (in Russian)

Gugunava G, Kiria J, Gogashvili J, Kiknadze D (2006) On genesis of Shamkhori, Spitak and Racha-Java earthquakes. J Georgian Geoph Soc Phys Solid Earth 11:63–68

Guliyev IS, Feizullayev AA (1995) All about mud volcanoes. Institute of Geology, Azerbaijan Academic of Science, Baku, pp 1–52

Gurbanov AG, Sobisevich AL, Sobisevich LE, Nechaev YuN, Arbuzkin VN, Prutskii NI, Trofimenko EA, Grekov II (2004) Activity of elbrus volcano (North Caucasus). Russ J Earth Sci 6(4):1–13

Gurevich B, Keydar S, Landa E (2002) Multifocusing imaging over an irregular topography. Geophysics 67(2):639–643

Gurevich YuM, Kormiltsev VV, Ulitin RV (1975) On the similarity of induced polarization curves for alternating and direct currents. Izv Phys Earth (1): 114–116

Guseinov AN, Shirinov FA (1985) In: Mamedov AV (ed) Map of geological regioning for oil of the Azerbaijan SSR. Scale 1:500,000. Printing Map Factory, Leningrad

Guseinov AN (1988) Paleomagnesim of the Jurassic and cretaceous deposits of the eastern part of the lesser Caucasus. Ph.D. Thesis, Inst. of Geology, Baku, Azerbaijan (in Russian)

Gylyjov RM, Piip VB (2002) Automatic processing and interpretation of engineering seismic data in Mountain Caucasus area. In: Transaction of the EAGE 64th conference & exhibition, Florence, pp 1–4

Hänl H, Maue AW, Westpfahl K (1961) Theorie der Beugung. Springer, Berlin

Hayakawa M (2007) VLF/LF radio sounding of ionospheric perturbations associated with earthquakes. Sensors 7:1141–1158

Heinrichs WE, Ludwig CS (1966) Results of an induced-polarization survey employing both time and frequency domain techniques. Mining Geophys Soc Explor Geophysicists 1:317–320

Hetland J, Gochitashvili T (eds) (2004) Security of natural gas supply through transit countries. NATO Science Series, II – Mathematics, Physics and Chemistry, vol 149. Springer Heidelberg/NY

Hryanina LP, Vityasev AV (1999) Arches – the first meteorite crater in the Caucasus. In: Proceedings of the XXX lunar and planetary conference, Houston, USA No. 1183

Huang S, Pollack HN, Sheen PY (2000) Temperature trends over past five centuries reconstructed from borehole temperatures. Nature 403:756–758

Hughes DS, Pondrom WS (1947) Computation of vertical magnetic anomalies from total magnetic field measurements. Trans Amer Geophys Union 28(2):193

Idarmachev ShG, Arefiev SS (2009) Results of dipole electric sounding in the area of Chirkey water reservoir after its filling. Izv Phys Solid Earth 45(9):75–85

Inan S, Yalcin MN, Guliev IS, Kuliev K, Feizullayev AA (1997) Deep petroleum occurrences in the lower Kura depression, South Caspian Basin, Azerbaijan: an organic geochemical and basin modeling study. Mar Petroleum Geol 14(7/8):731–762

Ismail-Zadeh AT (1996) Migration of seismic activity in the Caspian Sea. Comput Seism Geodyn 3:125–129

Ismail-Zadeh AT (ed) (2005). In: Proceedings of the international workshop on recent geodynamics, Georisk and Sustainable development in the Black Sea to Caspian Sea Region (Baku, 2005), American Institute of Physics, Melville

Ismail-Zadeh A, Mueller B, Schubert G (2005a) Three-dimensional modeling of present-day tectonic stress beneath the earthquake-prone southeastern Carpathians based on integrated analysis of seismic, heat flow, and gravity observations. Phys Earth Planet Inter 149:81–98

Ismail-Zadeh A, Tackley P (2010) Computational methods for geodynamics. Cambridge University Press, Cambridge, UK

Ismail-Zadeh TA (1983) Paleomagnetic investigations of Meso-Cenozoic of Azerbaijan. D.Sci. Thesis, Institute of the Physics of the Earth, Moscow (in Russian)

Ismail-Zadeh TA, Gadjiev TG, Khesin BE, Karkoshkin AI, Alexeyev VV, Metaxa KhP (1983a) Petromagnetic map of the Azerbaijan SSR, Scale 1:500,000. Printing Map Factory, Leningrad

Ismail-Zadeh TA, Gadjiev TG, Khesin BE, Karkoshkin AI, Alexeyev VV, Potapova EI (1983b) Petromagnetic characteristics of Azerbaijan. Elm, Baku (in Russian)

Ismail-Zadeh TA, Issayeva MI, Khalafli AA, Novruzov ZA, Sadygova TD, Mamedova NP (2005b) Solving tectonical problems by paleomagnetic methods. In: Alizadeh AkA (ed) Geology of Azerbaijan, vol 4.(Tectonics), Nafta-Press, pp 139–152 (in Russian)

Ismail-Zadeh TA, Khesin BE (eds), Alexeyev VV, Gadjiev TG, Karkoshkin AI, Khesin BE (1989a) Gravity and magnetic anomalies map of Azerbaijan SSR. Scale 1:500,000, Printing Map Factory, Leningrad

Ismail-Zadeh TA, Khesin BE (eds), Alexeyev VV, Gadjiev TG, Karkoshkin AI, Khesin BE (1989b) Gravity and magnetic anomalies of Azerbaijan and their geological interpretation. Printing Map Factory, Leningrad (in Russian)

Issayeva MI, Khalafli AA (2006) Paleomagnetic Studies of the Caucasian segment of the Alpine Fold Belt. In: Ismail-Zadeh A (ed) Proceedings of the international workshop on recent geodynamics, georisk and sustainable development in the Black Sea to Caspian Sea region (Baku, 2005), American Institute of Physics, Melville, pp 132–135

Itkis S, Matskevich Z, Mesheveliani T (2011) Pre-excavation studies of Prehistoric caves in Georgia. In: Proceedings of the near-surface EAGE meeting, Leicester, P44, pp 1–5

Jackson J (1992) Partioning of strike-slip and convergent motion between Eurasia and Arabia in eastern Turkey and Caucasus. J Geophys Res 97(B9, 12):471–12479

Jackson J, Priestley K, Allen M, Berberian M (2002) Active tectonics of the South Caspian Basin. Geophys J Int 148:214–245

Johnson IM (1984) Spectral induced polarization parameters as determined through time-domain measurements. Geophysics 49(11):1993–2003

Kachakhidze N, Kachakhidze M, Kereselidze Z, Ramishvili G (2009) Specific variations of the atmospheric electric field potential gradient as a possible precursor of Caucasus earthquakes. Nat Hazards Earth Syst Sci 9:1121–1126

Kadirov FA (2000) Application of the Hartley transform for interpretation of gravity anomalies in the Shamakhy–Gobustan and Absheron oil- and gas-bearing regions, Azerbaijan. J Appl Geophys 45:49–61

Kadirov FA, Lerche I, Guliyev IS, Kadyrov AG, Fezullayev AA, Mukhtarov ASh (2005a) Deep structure model and dynamics of mud volcanoes, southwest Absheron Peninsula (Azerbaijan). Energy Explor Exploit 23(5):307–332

Kadirov FA, Mukhtarov ASh (2004) Geophysical fields, deep structure, and dynamics of the Lokbatan Mud Volcano. Izv Phys Solid Earth 40:327–333

Kadirov FA, Rzayev AG, Tagiyev TSh (2005b) Investigation of gravity precursors of earthquakes. In: Alizadeh AkA (ed) Geology of Azerbaijan, vol 5. Physics of the Earth, pp 35–40

Kappelmayer O, Hänel R (1974) Geothermics with special reference to application. Borntraeger

Karaev NA, Rabinovich GYa (2000) Ore seismic prospecting. Geoinformmark, Moscow (in Russian)

Karakhanian A, Tozalakyan P, Grillot JC, Philip H, Melkonyan D, Paronyan P, Arakelyan S (2001) Tectonic impact on the Lake Sevan environment (Armenia). Environ Geol 40 (3):279–288

Karatayev GI, Pashkevich IK (1986) Integrated analysis of geological and geophysical data. Naukova Dumka, Kiev (in Russian)

Kardashov VR, Eppelbaum LV (2008). Mathematical models of strongly nonlinear geophysical phenomena. In: Transaction of the 5th Europe conference of mathematicians, Amsterdam, p 2

Kardashov VR, Eppelbaum LV, Vasilyev OV (2000) The role of nonlinear source terms in geophysics. Geophys Res Lett 27(14):2069–2073

Karkoshkin AI (1972) Characteristics of ore-control factors by petrophysical data on example of the Kedabek ore field. Geol Ore Deposits 14(4):43–53 (in Russian)

Karkoshkin AI (1979) Petrophysical characteristics of the Somkhet-Agdam zone (in connection with ore prognosis from the geophysical data). Ph.D. Thesis, Leningrad State University (in Russian)

Karkoshkin AI, Alexeyev VV, Khesin BE (1974) On the similarity in palaeomagnetic characteristics of middle Jurassic deposits in the Lesser Caucasus and some other areas in the USSR. Geol Geophys 7:62–69 (in Russian)

Katz B, Richards D, Long D, Lawrence W (2000) A new look at the components of the petroleum system of the South Caspian Basin. J Petrol Sci Eng 28:161–182

Kaufman AA (1992) Geophysical field theory and method, vol 1, Gravitational, Electric and Magnetic Fields. Academic, San Diego

Kaufman AA, Keller GV (1981) The magnetotelluric method. Elsevier, Amsterdam

Kekelia SA, Kekelia MA, Kuloshvili SI, Sadradze NG, Gagnidze NE, Yaroshevich VZ, Asatiani GG, Doebrich JL, Goldfarb RJ, Marsh EE (2008) Gold deposits and occurrences of the greater Caucasus, Georgia Republic: their genesis and prospecting criteria. Ore Geol Rev 34(3): 369–386

Kekelia SA, Kekelia MA, Otkhmezuri Z, Özgür N, Moon C (2004) Ore-forming system in volcanogenic-sedimentary sequences by the example of base metal deposits of the Caucasus and East Pontic metallotectonics. Miner Res Expl Bul 129:1–16

Keilis-Borok VI (1990) The lithosphere of the earth as a nonlinear system with implications for earthquake prediction. Rev Geophys 28:19–34

Keilis-Borok VI, Levshin AL (eds) (1984) Mathematical modeling and interpretation of geophysical data. Computational Seismology, Issue 16, Nauka, Moscow (in Russian)

Kerimov KM (ed) (1996) Geophysical Investigations in Azerbaijan. Sharg-Garb, Baku (in Russian)

Kerimov KM, Pilchin AN, Gadzhiev TG, Buachidze GY (1989) Geothermal map of the Caucasus, Scale 1:1,000,000 Baku, Cartographic Plant No. 11 (in Russian)

Khain VE (1984) Regional geotectonics, The Alpine Mediterranean Belt. Nedra, Moscow (in Russian)

Khain VE (1991–1993) Geology of Northern Eurasia (ex-USSR). Part 2 (1993). Berlin. Borntraeger. In: Milanovsky EE (ed) Geology of the USSR, Part 3 (1991). Moscow University Press (in Russian)

Khain VE (1995) Main problems of the modern geology. Nauka, Moscow (in Russian)

Khain VE (2000) Tectonics of continents and oceans. Nauchyi Mir, Moscow (in Russian)

Khain VE (2007) Mesozoic-Cenozoic accretionary complexes of the greater Caucasus. Dokl Earth Sci 413A(3):376–379

Khain VE, Alizadeh AkA (eds) (2005) Geology of AzerbaijaTectonics, vol IV. Nafta-Press, Baku (in Russian)

Khain VE, Khalilov EN (2006) Tideless variations of gravity before strong distant earthquakes. Science without Borders (ICSD/IAS H&E, Innsbruck), 2, pp 319–339

Khain VE, Koronovsky NV (1997) Caucasus. In: Moores EM, Fairbridge RW (eds) Encyclopedia of European and Asian regional geology, Chapman and Hall, London, pp 127–136

Khain VE, Levin LE, Polyakova ID (2005) Petroleum potential of deep Oligocene–Miocene basins in southern Russia. Dokl Earth Sci 404(7):979–981

Khain VE, Lobkovsky LI (1994) Relict seismicity in the Alpine belt of Eurasia: mode of occurrence. Geotectonics (Geotektonka) 28(3):192–198 (in Russian)

Khain VE, Ryabukhin AG (2002) Russian geology and the plate tectonics revolution. Geol Soc Special Publ Lond 192:185–198

Khalafli AA (2006) Paleomagnetism and some problems of the shear deformations of the Lesser Caucasus. Takhsil, Baku

Khalfin LA (1958) Information theory of geophysical interpretation. Dokl AN USSR 122 (6):1007–1010 (in Russian)

Khalilov AA (1996) Chovdar barite deposit. In: Kerimov KM (ed) Geophysical Investigations in Azerbaijan, Sharg-Garb, Baku, pp 292–293 (in Russian)

Khalilov EN (2008) Forecasting of earthquakes: the reasons of failures and the new philosophy. Sci Without Borders (ICSD/IAS H&E, Innsbruck) 3:300–315

Khandokhov TH (2004) Influence of alternative electromagnetic fields of various frequencies to living systems. Ph.D. Thesis, Nalchik University (in Russian)

Kharatian K, Travi Y, Igoumnov V (2002) Activit 'du gaz radon dans l'air du sol et sismicit 'locale: exemple du bassin de l'Arax (Arm'nie). Surf Geosci 334:179–185

Kharitonov AL, Fonarev GA, Gaidash SP, Eppelbaum LV, Kischa PV (2005) Application of differential satellite magnetometry for fields separation of external (Solar) and internal origin. In: Proceedings of the All-Russian conference on space experiment "Astrometry", Sankt-Petersburg, pp 335–340

Khashiev H-MA, Shulga EL (1985) On the some mechanisms of depressive influence of electromagnetic fields of very high frequencies to the central nervous system. In: Proceedings of the north-Caucasian center of high education, Ser: Natural Sciences, No. 2, pp 74–78

Khesin BE (1961) Application of geothermics in Azerbaijan and analogous regions for the study of deep structure and mineral prospecting. For the Tech Prog (Za Technicheskii Progress) 10:19–24 (in Russian)

Khesin BE (1962a) Application of geophysical methods for geological mapping on the scale of 1:10,000 and 1:5,000. Scient. Notes of the Azerb. State University, Ser. Geology-Geograph Science, No. 1, 37–41 (in Russian)

Khesin BE (1962b) Application of emanation survey in Azerbaijan for detail geological mapping. In: Experience of application of radioactive methods for prospecting of nonradioactive ores. Gosgeoltekhizdat, Moscow, pp 27–33 (in Russian)

Khesin BE (1965) A simple method of considering terrain relief effect on the magnetic survey results. Prospecting and Protection of Entrails (Razvedka i Okhrana Nedr), No.9, pp 30–38 (in Russian)

Khesin BE (1968) Geophysical characteristics of tectono-magmatic zones of Azerbaijan. Geotectonics (Geotektonika) (6): 61–69 (in Russian)

Khesin BE (1969) Mining geophysics in mountainous regions. Nedra, Moscow (in Russian)

Khesin BE (1971) Possibilities of geophysical investigations for the study of tectonics and magmatism during ore prognosis. Method Explor Geophys (Metody Razvedochnoi Geofiziki) (12): 100–108 (in Russian)

Khesin BE (1972) On the possibility of integrated interpretation of geological and geophysical data on the southern slope of the Greater Caucasus using the information theory. Izv AN Azerb SSR, Ser: Earth Sci (2): 141–148 (in Russian).

Khesin BE (1974) Informational technique for prognosis of geological objects in the absence of standards in the area under investigation. Method Explor Geophys (Metody Razvedochoi Geofiziki) (24): 98–102 (in Russian)

Khesin BE (1975a) Geophysical base of deep drilling in Azerbaijan for prognosis of endogenous ores. In: Geophysical investigations in Azerbaijan (3) Baku, pp 117–124 (in Russian)

Khesin BE (1975b) Statistical reduction of regional gravity anomalies of Azerbaijan. In: Geophysical investigations in Azerbaijan, 3, Baku, pp 125–130 (in Russian)
Khesin BE (1976) Prognosis and localization of hidden mineralization in mountainous regions on the basis of geophysical data. Nedra, Moscow (in Russian)
Khesin BE (1978a) Elimination of topographic effects in magnetic prospecting. Explor Geophys (Razvedochnaya Geofizika) (83): 102–111 (in Russian)
Khesin BE (1978b) The peculiarities of geophysical prospecting for ore deposits in the Alpine mountainous regions. Soviet Geol (12): 76–86 (in Russian)
Khesin BE (1981) Geophysical studies in the mountainous regions. D.Sci. Thesis, Moscow Geol. Prospecting Institute (in Russian)
Khesin BE (1987) Common aspects of prognosis of oil-and-gas and ore deposits using the methods of exploration geophysics. In: Prognosis of oil & gas deposits by methods of exploration geophysics. Baku, pp 91–98 (in Russian)
Khesin BE (1991) On the convergence of geophysical studies for the ore and oil-and-gas prognosis. Appl Geophys (Prikladnaya Geofizika) (125): 110–119 (in Russian)
Khesin B (1998) Effective magnetization of the Precambrian in Sinai and southern Israel: implication of new methods for ΔT field analysis. Isr J Earth Sci 47(1):47–60
Khesin B (2005) Use of geophysical methods for the solution of environmental problems in Israel. HAIT J Sci Eng 2(1–2):95–124
Khesin B (2007) Use of natural electric fields in mineral prospecting, geoecological and antiterrorism activity. Scientific issues of the "Scientists of the South" Association, No. 5, 5–14
Khesin BE, Alexeyev VV (1986) On statistical reduction and interpretation of the measured values of Δg in mountainous regions. In: Methods and results of geophysical prospecting of ore deposits. Yerevan, pp 100–105 (in Russian)
Khesin BE, Alexeyev VV (1987) A method for determination of rock magnetization in their natural occurrence. Author's Certificate No.1125578 USSR MKI 01 3/08
Khesin BE, Alexeyev VV, Eppelbaum LV (1988) Optimization of geophysical investigations in mountainous regions by increasing the effectiveness of interpretation. In: Borisovich VT, Eppelbaum VM (eds) Optimization of ore exploration in mountainous regions. Nedra, Moscow, pp 79–122 (in Russian)
Khesin BE, Alexeyev VV, Eppelbaum LV (1993a) Investigation of geophysical fields in pyrite deposits under mountainous conditions. JAppl Geophys 30:187–204
Khesin BE, Alexeyev VV, Eppelbaum LV (1993b) 3-D modeling of gravity and magnetic fields as a final stage of application of effective interpretation system of geophysical data under difficult geological conditions. Geoinformatics 4(3):177–188
Khesin BE, Alexeyev VV, Eppelbaum LV (1996) Interpretation of geophysical fields in complicated environments, Ser.: Modern Approaches in Geophysics. Kluwer, Dordrecht/Boston/London
Khesin BE, Alexeyev VV, Eppelbaum LV (1997) Rapid methods for interpretation of induced polarization anomalies. J Appl Geophy 37(2):117–130
Khesin BE, Alexeyev VV, Metaxa KhP (1983) Interpretation of magnetic anomalies in the conditions of oblique magnetization and rugged topography. Nedra, Moscow (in Russian)
Khesin BE, Eppelbaum LV (1986) Optimization of the set of methods in the system of geophysical prospecting of economic minerals. Izv AN Azerb SSR, Ser: Earth Sci(1): 89–93 (in Russian)
Khesin BE, Eppelbaum LV (1988) Methods for quantitative interpreting the results of near-surface thermal prospecting under complicated physical-geological conditions. Baku, IL AzNIINTI, No.109 (in Russian)
Khesin BE, Eppelbaum LV (1994) Near-surface thermal prospecting: review of processing and interpretation. Geophysics 59(5):744–752
Khesin BE, Eppelbaum LV (1997) The number of geophysical methods required for target classification: quantitative estimation. Geoinformatics 8(1):31–39

Khesin BE, Eppelbaum LV (2007) Development of 3-D gravity/magnetic models of Earth's crust in complicated regions of Azerbaijan. In: Proceedings of the 69th EAGE conference, P343, London, Great Britain, pp 1–5

Khesin B, Feinstein S, Eyal Y (2001) Identification of hidden geological features in Sinai using statistical reduction of gravity anomalies. J Africa Geosci Rev 8(3 & 4):365–378

Khesin BE, Metaxas ChP (1974) Investigation of magnetic and other fields to reveal faults and boundaries of tectonic blocks in Azerbaijan. In: The study of seismicity and deep structure of Azerbaijan. Elm, Baku, pp 92–96 (in Russian)

Khesin BE, Metaxas ChP (2000) Revealing of main faults within collision zone of Arabian/ Russian plates by geophysical features. In: Transaction of the 25th General Assembly of the European geophysical society, SE1, Nice

Khesin BE, Metaxas KhP, Alexeyev VV (1982) On geological nature of the Talysh-Vandam gravity maximum. Depon. VINITI AN USSR, No. 3680–82. p 29 (in Russian)

Khesin BE, Muradkhanov S (1962) Technique of geophysical survey of chromite ores in ophiolite belt (Azerb. SSR). For the Technical Progress (Za Tekhnicheskii Progress), No. 8, 20–24 (in Russian)

Khesin BE, Muradkhanov SA (1967) Integration of geophysical investigations for the revealing of pyrite deposits at the southern slope of the greater Caucasus. In: Fedynsky VV (ed) Techniques, technology and results of geophysical prospecting. Nedra, Moscow, pp 423–426 (in Russian)

Khesin BE, Sattarov I (1967) Efficiency of gamma-ray log in Dashkesan and Filizchay ore fields. Izv Acad Nauk Azerb SSR, Ser Nauki o Zemle(6): 43–46 (in Russian).

Khesin BE, Shakhnazaryan AL (1965) Hydrogeological investigations of Azerbaijan with the use of electric prospecting. VIEMS, Moscow (in Russian)

Khesin BE, Shakhnazaryan AL (1986) Ground geophysical investigations of submountain and mountain regions for water supply and reclamation problems solution in Azerbaijan. Book manuscript, Baku (in Russian)

Kholodov VN (2002) Mud volcanoes, their distribution regularities and genesis: communication 1. Mud volcanic provinces and morphology of mud volcanoes. Lithology Miner Resour 37 (3):197–209

Kholodov VN, Kiknadze ZR (1989) Massive sulfide deposits of the greater Caucasus. Nauka, Moscow

Khutorsky MD (1982) Heat flow in areas of structural geological inhomogeneities. Nauka, Moscow (in Russian)

Khutorsky MD, Salnikov VE, Fotogdinov RA (1983) Geothermal prospecting of ore deposits. In: Methodological and experimental principles of geothermometry. Nauka, Moscow, p 212–229 (in Russian)

Kilty KT (1984) On the origin and interpretation of self-potential anomalies. Geophys Prospect 32 (1):51–62

Klett TR, Schenk CJ, Charpentier RR, Brownfield ME, Pitman JK, Pollastro RM, Cook TA, Tennyson ME (2011) Assessment of undiscovered oil and gas resources of the Azov–Kuban Basin Province, Ukraine and Russia, 2010. *U.S. Department of the Interior*, Fact Sheet 2011–3052

Klichnikov VA (ed) (1970) Geophysical prospecting of ore deposits. Kazakhstan Geophysical Inst. Alma-Ata (in Russian)

Klushin IG, Tolstikhin IN (1961) Singling out linear tectonic dislocations in geophysical maps. Geol Geop (Geologiya i Geofizika) (6):98–103 (in Russian)

Knapp CC, Knapp JH, Connor JA (2004) Crustal-scale structure of the South Caspian Basin revealed by deep seismic reflection profiling. Mar Petrol Geol 21:1073–1081

Kobranova VN (1986) Petrophysics. Nedra, Moscow (in Russian)

Kogan AB (1975) The theory and methods of directed summation and frequency filtration and results of the geomagnetic field investigation of the Siberian platform. Nedra, Leningrad (in Russian)

Komarov VA (1980) Electric prospecting by the induced polarization method, 2nd edn. revised and supplemented. Nedra, Leningrad (in Russian)

Komarov VA, Pishpareva NA, Semenov MV, Khloponina LS (1965) Principles of the interpretation in the induced polarization method. Nedra, Leningrad (in Russian)

Kondakov NI (1975) Logic dictionary, 2nd edn. Nedra, Moscow (in Russian)

Kopf A, Deyhle A, Lavrushin VY, Polyak BG, Gieskes JM, Buachidze GI, Wallmann K, Eisenhauer A (2003) Isotopic evidence (He, B, C) for deep fluid and mud mobilization from mud volcanoes in the Caucasus continental collision zone. Int J Earth Sci 92:407–425

Kopytenko YuA, Matiashvili TG, Voronov PM, Kopytenko EA, Molchanov OA (1993) Detection of ultra-low-frequency emissions connected with the Spitak earthquake and its aftershock activity, based on geomagnetic pulsations data at Dusheti and Vardzia observatories. Phys Earth Planet In 77:85–95

Kostianoy AG, Ermakov PN, Soloviev DM (2008) Complex satellite monitoring of the Nord Stream gas pipeline construction. In: Transactions of the IEEE conference, No. 978-1-4244-2268:1–5

Kovachev SA, Kazmin VG, Kuzin IP, Lobkovsky LI (2009) Novel data about the mantle seicmicity of the Caspian region and their geological interpretation. Geotectonics (Geotektonika) (3):30–44 (in Russian)

Kovalevsky SA (1940) Mud volcanoes of South Caspian-side (Azerbaijan and Turkmenistan). Azgostoptekhizdat, Baku (in Russian)

Kozlovsky EA, Krivtsov AI (1988) Modeling of ore deposits: trends and targets. Sov Geol (8): 3–8 (in Russian)

Krasnopevtseva GV (1984) Deep structure of the seismically active Caucasus region. Nauka, Moscow (in Russian)

Krumbein WC, Graybill FA (1965) An introduction to statistical models in geology. McGraw-Hill Book, New York

Kulkov BN, Bakharev LV, Kagan GF (1991) Some mining-techmological aspects of underground registrating cosmic rays method in the combined exploitation of the Tyrnyauz deposit. Min J (Gornyi Zhurnal) (8):51–54 (in Russian)

Kunin NYa (1968) Problems of singling out the observed gravity field components for structural-tectonic regioning (as applied to the Southern Kazakhstan), In: Geophysical investigations in Kazakhstan. Kazakhstan Geophysical Inst. Alma-Ata, p 76–88 (in Russian)

LaFehr TR (1991) An exact solution for the gravity curvature (Bullard B) correction. Geophysics 56(8):1179–1184

Lachenbruch AN (1968) Rapid estimation of the topographic disturbance to superficial thermal gradients. Rev Geophys 6(3):365–400

Lakhtionov MO, Tarkhov AG (1967) Experience in thermal prospecting at pyrite deposits of the Urals. Izvestiya VUZov Ser Geol Prosp (5):87–94 (in Russian)

Lakhtionov MO, Tarkhov AG (1970) Geothermal investigations at ore deposits. Sov Geol (3):121–124 (in Russian)

Lambert A, Bower DR (1991) Constraints on the usefulness of gravimetry for detecting precursory crustal deformations. Tectonophysics 193:369–375

Landau LD, Lifshitz EM (1960) Electrodynamics of continuous media. Pergamon, Oxford

Lekomte I, Köllner F, Petrakov D, Chernomorets S, Shakhmina M, Hamran SE, Juliussen H, Kääb A (2010) Geophysics in Glacial-hazard initiation zones, Russian Caucasus. In: Proceedings of the Near-Surface EAGE Meeting, Zurich, A39, p 1–5

Leroy SAG, Marretb F, Giraltc S, Bulatov SA (2006) Natural and anthropogenic rapid changes in the Kara-Bogaz Gol over the last two centuries reconstructed from palynological analyses and a comparison to instrumental records. Quatern Int 150:52–70

Leschak HA, Lewis JE (1983) Geothermal prospecting with shallow-temperature surveys. Geophysics 48(7):975–996

Limonov AF (2004) Mud volcanoes. Soros' Educ J 8(1):63–69 (in Russian, abstract in English)

Lindsay I, Smith AT, Badalyan R (2010) Magnetic survey in the investigation of sociopolitical change at a Late Bronze Age fortress settlement in Northwestern Armenia. Archaeol Prospect 17:15–27

Lines LR, Schultz AK, Treitel S (1988) Cooperative inversion of geophysical data. Geophysics 53 (1):8–20

Liperovsky VA, Meister CV, Liperovskaya EV, Bogdanov VV (2008) On the generation of electric field and infrared radiation in aerosol clouds due to radon emanation in the atmosphere before earthquakes. Nat Hazards Earth Sci 8:1199–1205

Litvinov SYa (1941) Marine electric prospecting. Gostoptekhizdat, Moscow (in Russian)

Logachev AA, Zakharov VP (1979) Magnetic prospecting. Nedra, Leningrad (in Russian)

Lomtadze VV (1982) Software for geophysical data processing. Nedra, Leningrad (in Russian)

Lukavchenko PI (1951) Tables and nomographs for calculating terrain gravity corrections in survey with gravimeters. Gostoptekhizdat, Moscow (in Russian)

Lukavchenko PI (1961) On reduction and interpretation of gravity anomalies under conditions of severely rugged relief. Appl Geophys (Prikl Geofiz) 31:219–229 (in Russian)

Lustikh EN (1947) Geological meaning of the different methods for gravity anomalies calculation. In: Transactions of the Theoretical Geophysics Institute of the USSR Academy Science vol 3, Moscow, pp 3–45 (in Russian)

Lyubavin LM, Shaub YuB (1968) Integrated interpretation of airborne electromagnetic and aeromagnetic data using correlation functions. Prospect Protect Entrails (Razvedka i Okhrana Nedr) (6):40–45 (in Russian)

Lyubimov AA (1983) Rapid geological interpretation of gravity and magnetic data. Nedra, Moscow (in Russian)

Lyubimov GA, Lyubimov AA (1988) Computer-aided methods of gravity and magnetic interpretation. Nedra, Moscow (in Russian)

Lyubimov VV, Gurfinkel YuI, Oraevskii VN (1993) Experience of application of diagnostic magnetometers in city and resort conditions. IZMIRAN Publ (99): 1046, Moscow (in Russian)

Lyubimova EA (1968) Thermometry of the earth and the Moon. Nauka, Moscow (in Russian)

Lyubimova EA, Nikitina VN, Tomara GA (1976) Heat fields of the internal and marginal seas of the USSR. State of observation level and interpretation theory for 2-D inhomogeneities. Nauka, Moscow (in Russian)

Lyubimova EA, Mukhtarov ASh, Ismail-Zadeh TA (1985) Temperature variations in the "Biladzhik" drill hole (Azerbaijan) during regional seismic activity. Izv Phys Earth 21 (4):319–322

McClusky S, Balassanian S, Barka A, Demir C, Ergintav S, Georgiev I, Gurkan O, Hamburger M, Hurst K, Kastens K, Kekelidze G, King R, Kotzev V, Lenk O, Mahmoud S, Mishin A, Nadariya M, Ouzounis A, Paradissis D, Peter Y, Prilepin M, Reilinger R, Sanli I, Seeger H, Tealeb A, Toksöz MN, Veis G (2000) Global positioning system constraints on plate kinematics and dynamics in the eastern Mediterranean and Caucasus. J Geophys Res 105:5695–5719

Magakyan IG (1961) Ore deposits, 2nd edn. Yerevan, (in Russian)

Malakhova VA (1996) Iron-ore deposits of Dashkesan district. In: Kerimov KM (ed) Geophysical Investigations in Azerbaijan, Sharg-Garb, Baku, pp 292–293 (in Russian)

Malovitsky YaP, Kogan LI, Mistrukov YuM et al (1977) Marine geophysical investigations. Nedra, Moscow (in Russian)

Malyshev VP (1981) Probabilistic-deterministic design of experiment. Nauka, Alma-Ata (in Russian)

Mamedov AV, Guseinov AN, Shirinov FA (1985) Map of geological regioning for oil of the Azerbaijan SSR. Scale 1:500,000. Printing Map Factory, Leningrad

Mangino S, Priestley K (1998) The crustal structure of the southern Caspian region. Geophys J Int 133:630–648

Manley DM, Mohammed SF, Robinson ND, Thomas RW (2005) Structural interpretation of the deepwater Gunashli Field, facilitated by 4-C OBS seismic data. Leading Edge (9):922–926

Marchuk AN, Abakarov AR, Daniyalov MG, Asmanov OA, Mirzaliev MM, Levkovich RA (2001) Geodynamic effects on the dams in cascade of the Sulak hydraulic power systems. Hydrotechnical Constr 35(9):468–477

Markovin VI (1978) Dolmens of the western Caucasus. Nauka, Moscow (in Russian)

Markus MA (2002) The formation of massive sulfide ores in black shales of the Eastern Caucasus: evidence from the Kizil Dere orefield. Lithology Miner Resour 37(2):157–161

Martinelli G, Dadomo A (2005) Volcano monitoring and seismic events. In: Martinelli G, Panahi B (eds) Mud volcanoes, geodynamics and seismicity, vol 51, NATO Sci. Ser. Earth Environ. Springer, New York, pp 211–220

Martirosyan A, Balassanian S, Simonian R, Asatryan L (1999) Computation of probabilistic seismic hazard for the GHSAP test area 'Caucasus'. Nat Hazards 20:1–20

Masurenkov YuP, Sobisevich AL, Likhodeev DV, Shevchenko AV (2009a) Thermal anomalies of the Northern Caucasus. Dokl Earth Sci 428(5):667–670

Masurenkov YuP, Sobisevich AL, Laverova NI (2009b) Relation between structural and material circular motions in a volcanic center. Dokl Earth Sci 429(8):1359–1363

Matcharashvili T, Chelidze T, Abashidze T, Zhukova N, Meparidze E (2010) Changes in dynamics of seismic processes around Inguri high dam reservoir induced by periodic variation of water level. In: Proceedings of Intern. Conf. Synchronization and triggering: from fracture to earthquake processes, earth and planetary sciences, vol 1., Part 3, pp 273–286

Matusevich YuF, Priyma VI (1967) Techniques for the increase of depth penetration in seismic studies of northwestern Ciscaucasus by reflection method. In: Fedynsky VV (ed) Techniques, technology and results of geophysical prospecting. Nedra, Moscow, pp 125–132 (in Russian)

Megakhed A, Petrunun GI, Popov VG, Ladigin VM (1985) Thermal characteristics of effusive rocks from the lesser Caucasus. Izv Phys Earth 21(2):957–961

Mekhtiev ShF, Gadjiev TG, Kashkai MA, Akhmedov AM (1976) Geology of the USSR, vol XLVII, Azerbaijan Republic. Nedra, Moscow (in Russian)

Melikadze G, Chelidze T, Leveinen J (2005) Modeling of heavy metal contamination within an irrigated area. In: Baba A (ed) et al Groundwater and ecosystems. Springer Heidelberg/NY, p 243–253

Melkanovitsky IM (1975) Regional geophysical studies of hydrogeological features in Artesian Basins. Nedra, Moscow (in Russian)

Melkanovitsky IM (1984) Geophysical studies by regional hydrogeological investigations. Nedra, Moscow (in Russian)

Melkanovitsky IM, Ryapolova VA, Khordikainen MA (1982) Methodology of geophysical studies for fresh waters prospecting. Nedra, Moscow (in Russian)

Mellors R, Klib D, Aliyev A, Gasanov A, Yetirmishli G (2007) Correlations between earthquakes and large mud volcano eruptions. J Geophys Res 112(B04304):1–11

Meloni M, Di Mauro D, Lepidi S, Mele G, Palangio P (2004) Tectonomagnetic and VLF electromagnetic signals in Central Italy. Ann Geophys 47(1):29–37

Merkulov S (2005) Experiences with GPR LOZA and special software applications to locate archaeological objects in Northern Caucasus. In: Transaction of the 6th International Conference on Archaeological Prospection, Rome p 264–267

Mikhailov VO, Smolyaninova EI (2002) Numerical modeling of neotectonic movements and the state of stress in the North Caucasus foredeep. Tectonics 21(4):1–14. doi:10.1029/2002TC001379

Mikov BD (1963) Practical techniques for magnetic anomalies interpretation under conditions of complicated terrain relief. Institute of Geology and Geophysics, Novosibirsk (in Russian)

Milanovsky EE, Khain VE (1963) Geological structure of the Caucasus. Moscow State University, Moscow (in Russian)

Milyukov VK, Klyachko BS, Myasnikov AV, Striganov PS, Yanin AF, Vlasov AN (2005) A laser interferometer–deformograph for monitoring the crust movement. Instrum Exp Tech 48 (6):780–795

Milyukov V, Kopaev A, Zharov V, Mironov A, Myasnikov A, Kaufman M, Duev D (2010) Monitoring crustal deformations in the Northern Caucasus using a high precision long base laser strainmeter and the GPS/GLONASS network. J Geodyn 49:216–223

Minasyan RS (1979) Prospecting for subsurface water in the central volcanic highland of Armenia by geophysical methods. Geophys Prospect 27(4):898–814
Mironov VS (1980) A course in gravity prospecting, 2nd edn., revised and supplemented. Nedra, Leningrad (in Russian)
Modin IN (2010) Electric prospecting in technical and archaeological geophysics. D. Sci. thesis, Moscow State University (in Russian)
Mogi K, Mochizuki H, Kurokawa Y (1989) Temperature changes in an artesian spring at Usami in the Izu peninsula (Japan) and their relation to earthquakes. Tectonophysics 159:95–108
Molchan GM, Dmitrieva OE, Rotwain IM, Dewey J (1990) Statistical analysis of the results of earthquake prediction, based on bursts of altershocks. Phys Earth Planet Int 61:19–34
Molchanov AA, Sidorov VA, Kaltashev SA, Petrov VS, Sidorova IYa, Bil'kov MN (1985) Geophysical works for thermal water prospecting. Exploration Geophysics, Review of VIEMS, Moscow (in Russian)
Morgounov V, Zdorov A (2007) Electromagnetic emission precursors of landslides, avalanches and earthquakes. In: EGU Proceedings of the geophysical research abstracts, Vienna, vol. 9, 01357
Movshovich EB, Knepel MN, Cherkashin MS (1987) Formalization of geological data for mathematical processing. Nedra, Moscow (in Russian)
Mudretsova EA (1981) Gravimetric prospecting. Geophysicist's manual. Nedra, Moscow (in Russian)
Mukhtarov ASh, Kadirov FA, Mamedov VA (2010) Reconstruction of the surface temperature in the Kura depression (Azerbaijan) by the inversion of borehole data. Izv AN RAN Phys Solid Earth 46(6):524–528
Muradkhanov SA (1971) Interpretation of geophysical data from the wells of Filizchay deposit using statistical methods. Express-information of VIEMS, Ser. Region., Applied and Borehole Geoph., Moscow, No.88 (in Russian)
Muradkhanov SA, Magerramov MR (1996) Borehole logging. In: Kerimov KM (ed) Geophysical Investigations in Azerbaijan, Sharg-Garb, Baku, 275 (in Russian)
Murty BV, Haricharan P (1984) A simple approach toward interpretation *SP* anomaly due to 2-D sheet model of short dipole length. Geophys Res Bull 22(4):213–218
Mustafabeily M, Khesin BE, Muradkhanov SA, Alexeyev VV (1964) Searching of copper-polymetallic (lead-zinc) deposits on southern slope of the Greater Caucasus. Prospect Prot Entrails (Razvedka i Okhrana Nedr) (9):30–38 (in Russian)
Mustafayev GV (2001) Geodynamic situation during formation of Dashkeasn ore-magmatic system. Dokl Azerbaijan Acad Sci Ser Earths Sci (1):40–48 (in Russian)
Nabighian MN (1972) The analytic signal of two-dimensional magnetic bodies with polygonal cross-section: its properties and use for automated anomaly interpretation. Geophysics 37 (3):507–517
Nagatha T (1961) Rock magnetism. Maruzen, Tokyo
Nalbandyan A, Saghatelyan A (2008) Radiogeological monitoring to the south Caucasus' rivers as a tool for early warning system. In: Apikyan S et al (eds) Prevention, detection and response to nuclear and radiological threats. Springer, Dordrecht, pp 183–190
Napadensky GB, Sheremet OG (1976) Computer-aided integrated interpretation of gravity and magnetic fields. Review of VIEMS, Ser. Region., Applied and Borehole Geoph., Moscow, 1976 (in Russian)
Nasruyev NP, Andreyev LI, Eleseyenko NA (1975) In: Akhmedov GA et al (eds) Geophysical investigations in Azerbaijan, vol III., pp 87–91 (in Russian)
Neishtadt N, Eppelbaum L, Levitski A (2006) Application of seismo-electric phenomena in exploration geophysics: review of Russian and Israeli experience. Geophysics 71(2):B41–B53
Nemtsov LD (1967) High-precise gravity prospecting. Nedra, Moscow (in Russian)
Nettleton LL (1940) Geophysical prospecting for oil. McGraw-Hill, New York
Nevinsky IO (1987) Underground low background laboratory. GruzNIINTI Information, 87–74 (in Russian)

Nevinsky IO, Tsvetkova TV (1989) Gamma-background variations in underground low background setups. Atom Energy 72(6):622 (in Russian)

Nevinsky IO, Tsvetkova TV (2005) SSNTDs in the automatic detector of radon. Radiat Meas 39:115–119

New Catalog of Strong Earthquakes on the Territory of the USSR Since Ancient Time Till 1975, 1977. Nauka, Moscow (in Russian)

Nikonov AA (1982) A very powerful earthquake in the Greater Caucasus Mountains, January 14, 1668. Izv Earth Phys 18(9):713–725

Nikitin AA (1986) Theoretical principles of geophysical data processing. Nedra, Moscow (in Russian)

Nikitin AA (1993) Statistical processing of geophysical data. Series of Advanced Geophysics, Russian Experience, No. 22, Electromagnetic Research Centre, Moscow (in Russian)

Nikitin AA, Medovsky IG, Vlasova II (1971) Analysis of simple techniques for integrated interpretation of geophysical data. Explor Geophys (Pikladnaya Geofiz) (62): 167–174 (in Russian)

Nikitsky VE, Glebovsky YuS (eds) (1990) Magnetic prospecting. Geophysicist's manual. Nedra, Moscow (in Russian)

Nikolsky YuI (1971) On inapplicability of Bouguer reduction for the study of structure of Earth crust and upper mantle using gravimetric data. In: Methods of geophysical prospecting, Issue 12, Nedra, Leningrad, p 13–23 (in Russian)

Nikolsky YuI, Milai TA, Kogan LZ (1975) Geological-geophysical studies of the tectonics, magmatism, and metallogeny of the Caucasus. Nedra, Leningrad (in Russian)

Nodia MZ (1939) Magnetic field of the Caucasian isthmus. In: Proceedings of the Geophysics Institute of Georgian Academy of Sciences, Tbilisi (in Russian)

Novikova OV, Muradkhanov SA, Karkoshkin AI et al (1983) Unpublished report "Construction of geological-geophysical etalons for various types of ore objects for north-eastern slope of the lesser Caucasus (within territory of Azerbaijan)". Azerbaiian Geol. Assoc, Baku

Ogilvy AA, Bogoslovsky VA (1979) The possibilities of geophysical methods applied for investigating the impact of man of the geological medium. Geophys Prospect 27:775–789

Olsson O (1983) Computation of *VLF* response over half-plane and wedge models. Geophys Prospect 31:171–191

Orlov VKh (1984) Analysis and singling out the relation between Bouguer anomalies and the altitudes of relief. Ph.D. thesis, Leningrad State University (in Russian)

Ortenberg DL (1930) Essay of the Dashkesan iron ore deposit by the data of magnetic surveys of *1923–1924*. In: Transaction of the main geological prospecting authority of the VSNKh of the USSR, Issue 11, Gosgeolizdat, Moscow – Leningrad (in Russian)

Ostrovsky EYa (1980) Comparative analysis of results of geo-observations. Sov Geol (10): 105–114 (in Russian)

Ouyang Z, Zhang H, Fu Z, Gou B, Jiang W (2009) Abnormal phenomena recorded by several earthquake precursor observation instruments before the M_s 8.0 Wenchuan, Sichuan earthquake. Acta Geol Sin 83(4):834–844

Ovnatanov ST, Tamrazyan GP (1970) Thermal studies as a technique in subsurface structural investigations. Geothermics (2):1184–1190

Ovsepyan RV, Badalyan SV, Gamoyan VB, Arutyunyan GV, Gevorkyan MG, Garibyan ZV (1986) Results of underground electric prospecting of Alaverdy copper deposit (Armenian SSR) in exploration stage. In: Brodovoy VV, Badalyan SV, Gamoyan VB (eds) Techniques, technology and results of geophysical prospecting of ore deposits. Publ. House of the Acad. of Sci. of Armenian SSR, Yerevan, pp 258–265 (in Russian)

Paffenholtz KA (1959) Geological essay of the Caucasus. Acad. of Sci. of Armenian SSR, Yerevan (in Russian)

Parasnis DS (1966) Mining geophysics. Elsevier, Amsterdam

Parasnis DS (1986) Principles of applied geophysics, 4th edn. revised and supplemented. Chapman & Hall, London

Paul M (1935) Uber Messingew der erdbodentemperatur and salzodomen. Z Geophys 11 (7/8):388–392

Peinke J, Matcharashvili T, Chelidze T, Gogiashvili J, Nawroth A, Lursmanashvili O, Javakhishvili Z (2006) Influence of periodic variations in water level on regional seismic activity around a large reservoir: field data and laboratory model. Phys Earth Planet In 156:130–142

Peresada VP (1970) Automated pattern recognition. Energiya, Leningrad (in Russian)

Petrakov DA, Krylenko IV, Chernomorets SS, Tutubulina OV, Krylenko IN, Shakhmina MS (2007) Debris flow hazard of glacial lakes in the Central Caucasus. In: Chen C-l, Major JJ (eds) Debris-flow hazards mitigation: mechanics, prediction, and assessment. Millpress, Rotterdam, pp 703–714

Petrovsky AA, Skaryatin RI (1929) Ondometrical works of Institute of Applied Geophysics. In: Proceedings of the IAG, No. 5 (in Russian)

Pilchin AN (1983) Geothermal regime of the earth crust of Kura Depression and its influence on the pressure distribution, Ph. D. thesis, Geophysical Institute of Georgian Academy of Science (in Russian)

Pilchin AN (1985) On the genesis of mud volcanoes. Sov Geol (10):78–81 (in Russian)

Pilchin AN, Eppelbaum LV (1997) Determination of magnetized bodies lower edges by using geothermal data. Geophys J Int 128(1):167–174

Pilchin AN, Eppelbaum LV (2006) Iron and its unique role in the earth evolution. Monogr Mex Geophys Soc 1–67

Pilchin AN, Eppelbaum LV (2007) Stabilities of the iron oxides in the Earth and their role in the formation of rock magnetism. Acta Geofis 55(2):133–153

Pilchin AN, Khesin BE (1981) On possible nature of the magnitoactive bodies of bottom edges. Appl Geophys (Prikl Geofiz) 92:123–127 (in Russian)

Piip VB, Orlova NM (2006) Structure of Dallackau landslide on the refraction data with interpretation by Homogeneous Function method. In: Transaction of the EAGE 68th Conference Vienna, p 1–5

Pirajno F (2000) Ore deposits and mantle plumes. Kluwer Acad. Publ, Dordrecht/Boston/London

Pirzhalava ML, Lyakhov LL, Simoniya LV (1992) Geological-geophysical modeling of barite deposits of the Western Georgia. Izv VUZov, Geol Prospect (3):124–127 (in Russian)

Planke S, Svensen H, Hovland M, Banks DA, Jamtveit B (2003) Mud and fluid migration in active mud volcanoes in Azerbaijan. Geo-Mar Lett 23:258–268

Poley JP, Steveninck JV (1970) Delineation of shallow salt domes and surface faults by temperature measurements at a depth of approximately 2 meters. Geophys Prospect 18:666–700

Poltoratsky VV, Ginzburg SN (1989) Gravity prospecting. In: Brodovoi VV (ed) Borehole and mining geophysics, vol II. Nedra, Moscow, pp 190–209 (in Russian)

Polyakov AS (1969) Methodological manual on electric profiling. Nedra, Leningrad (in Russian)

Popkhadze N, Beridze T, Moritz R, Gugushvili V, Khutsishvili S (2009) Facies analysis of the volcano-sedimentary host rocks of the cretaceous madneuli massive sulphide deposit, Bolnisi District, Georgia. Bull Georgian Natl Acad Sci 3(2):103–108

Pospelova GA, Golovanova LV, Dronochev VB, Tselmovich VA (2011) Magnetic and mineralogical characteristics of rocks at the Mezmaiskaya cave Paleolithic site (Northern Caucasus). Izv Phys Solid Earth 47(7):641–651

Pospelova GA, Golovanova LV, Sharonova ZV, Semenov VV (2006) Paleomagnetic studies of Paleolithic site deposits in the Matuzka cave (Northern Caucasus). Izv Phys Earth 42(7): 585–597

Pospelova GA, Kapicka A, Lyubin VP, Sharonova ZV (2001) Application of scalar magnetic rock parameters to the reconstruction of climate coeval with the deposition of sediments in the Kudaro-1 and Kudaro-3 Caves, South Ossetia, Georgia. Izv Phys Solid Earth 37(10):844–854

Pospelova GA, Sharonova ZV, Mironov TV, Levkovskaya GM (1996) Paleomagnetic study and climatic record in sedimentary rocks of the Treugolnaya cave (Northern Caucasus). Izv Phys Solid Earth 32(9):754–765

Pototsky E, Khesin B (1971) Probabilistic-informational approach to the revealing of fracture zones in the Near-Caspian region. Azerb Oil Ind (5):15–16 (in Russian)

Pudovkin IM, Pogrebnikov MM, Pochtarov VI, Bekzhanov GG (1973) On the direct correlation between the geomagnetic variations and earthquakes. Dokl AN USSR 208(5):1074–1077

Quick DH (1974) The interpretation of gradient array chargeability anomalies. Geophys Prospect 22(4):736–746

Rabinowitz PD, Yusifov MZ, Arnoldi J, Hakim E (2004) Geology, oil and gas potential, pipelines, and geopolitics of the Caspian Sea region. Ocean Dev Int Law 35(1):19–40

Raiffa H (1968) Decision analysis. Introductory lectures on choices under uncertainty. Addison-Wesley, London

Rakhmanov RR (1987) Mud volcanoes and their implications for the oil and gas content of deep formations. Nedra, Moscow (in Russian)

Rapoport CI, Btreus TK, Kleimenova NG (2006) Geomagnetic pulsations and cardiac infarction. Therapeutic Archive (4):56

Rapoport SYa, Shapirovsky NI, Ganbarov YuG (1964) Marine seismic prospecting in Azerbaijan. In: Transaction of the Azerbaijan Science Institute for Oil Production (AZNII DN), Baku, p 11–25

Rebai S, Philip H, Dorbath L, Borisoff B, Haessler H, Gisternas A (1993) Active tectonics in the lesser Caucasus: coexistence of compressive and extensional structures. Tectonics 12(5): 1089–1114

Reford MS, Sumner IS (1964) Aeromagnetics. Geophysics 29(4):482–516

Reinhold S, Belinskij AB, Korobov DS, Fassbinder J, Borisov AV, Peters S (2007) Landscape archaeology in the north Caucasus mountains: initial findings of exploration of promontory landscape in Kislovodsk during the late Bronze Age and early Iron Age. Eurasia Antiqua 13:139–179

Rempel GG, Parshukov NP (1976) Experience in coordinated computer simulation of gravity anomalies and vertical electric sounding. Trudy SNIIGIMS 238:5–8 (in Russian)

Reisner GI, Ioganson LI (1993) The earthquake potential of Western Russia and other former USSR republics. Seismicity Seism Reg North Eurasia (1):186–195

Riznichenko YuV, Khesin BE, Metaxa KhP (1983) Relation between seismicity and magnetic field according to observations in Azerbaijan. Izv Ser Phys Solid Earth (1):3–14 (in Russian)

Rodionov PF, Sofronov NI (1935) On the possibility of thermometry application to prospecting sulphide deposits. Prob Sov Geol (Problemy Sovetskoi Geologii) (8):257–268 (in Russian)

Rogachev BV (1965) Handmanual for the method of charge with measurement of magnetic field. Nedra, Moscow

Rogozhin EA (2002) Modern geodynamics and potential earthquake sources in the Caucasus region. In: Modern mathematical and geological models of the natural environment, Institute of the Physics of the Earth, Moscow, p 244–254 (in Russian)

Roy A (1970) Gravity and magnetic interpretation on uneven topography by sin *x*/*x* method of continuation. Geoexploration 8(1):37–40

Ruderman EN, Golubkov VV (1985) Some algorithms for processing and interpretation the results of electromagnetic survey by Induction, VLF and IP methods, *Review of VIEMS*, Ser. *Exploration Geophysics* (*Razvedochnaya Geofizika*), Moscow (in Russian)

Ruppel C, McNutt M (1990) Regional compensation of the Greater Caucasus mountains based on an analysis of Bouguer gravity data. Earth Planet Sci Lett 98(3–4):360–379

Ryss YuS (1983) Geoelectrochemical prospecting methods. Nedra, Leningrad (in Russian)

Rzayev AG (2005) Seismoprognostic investigations. In: Alizadeh AkA (ed) Geology of Azerbaijan, vol 5, Physics of the Earth, Nafta-Press, p 23–34

Sadovsky MA, Pisarenko VF, Shteinberg VO (1983) On the dependence of earthquake energy from volume of seismic focus. Dokl Acad Sci USSR 271(3):598–602

Safarov IB (2006) Petrophysical properties of rocks in seismogenic crustal blocks of the south-eastern Greater Caucasus and their geological and geophysical interpretation. Izv Phys Solid Earth 42(3):260–270

Saintot A, Brunet M-F, Yakovlev F, Sebrier M, Stephenson R, Ershov A, Chalot-Prat F, Mccann T (2006) The Mesozoic-Cenozoic tectonic evolution of the Greater Caucasus. Geol Soc Lond Mem 32:277–289

Salamov AM, Gabibov FG, Bogomolova OA (2010) Investigation of landslide on the right coast of the Mingachevir reservoir by electrical prospecting methods. Bull Volgograd State Archit-Buil Univ (20):33–39 (in Russian)

Sargsyan LS (2009) Reservoir-triggered seismicity in Armenian large dams. J Seismol Earthquake Eng 11(3):153–157

Savich AI, Ilin MN, Ezersky MG, Kalinin NI (1983) Long-term geophysical observations on the Inguri dam rock foundation. Bull Int Assoc Eng Geol (26–27):315–319

Samykina E, Surkov A, Eppelbaum L, Semenov S (2005) Do old spoils contain large amounts of economic minerals? Miner Eng 18(6):643–645

Sandwell DT, Smith WHF (2009) Global marine gravity from retracked Geosat and ERS-1 altimetry: ridge segmentation versus spreading rate. J Geophys Res 114(B01411):1–18. doi:10.1029/2008JB006008

Seigel HO (1959) Mathematical formulation and type curves for induced polarization. Geophysics 24(3):547–565

Seibold I, Seibold E (2006) Hermann Wilhelm Abich im Kaukasus: Zum zweihundertsten Geburtstag. Int J Earth Sci 95:1087–1100

Semenov AS (1974) Electric prospecting by self-potential method, 3rd edn., revised and supplemented. Nedra, Leningrad (in Russian)

Shakhnazaryan AL, Agamaliev GN, Elenbogen AM (1986) Peculiarities of geophysical prospecting of fresh underground water in mountainous region of the Lesser Caucasus (Azerbaijan part). Explor Geophys (Razvedochnaya Geofiz) 103:89–97 (in Russian)

Shaknazaryan AL, Khesin BE (1967) Efficiency of electrical prospecting for hydrogeological studies in Azerbaijan. In: Fedynsky VV (ed) Techniques, technology and results of geophysical prospecting. Nedra, Moscow, pp 510–516 (in Russian)

Shalayev SV (1972) Geological interpretation of geophysical anomalies using linear programming. Nedra, Leningrad (in Russian)

Shannon C (1948) A mathematical theory of communication. Bell Syst Techn J 27(3–4):379–423, 623–656

Shapirovsky NI, Ganbarov YuG (1961) On the problem of seismic prospecting study of Mesozoic deposits in marine areas. Azer Oil Ind (Azerbaidzhanskoe Neft Khoz) (1):8–11

Sharapanov NN, Chernyak GYa, Baron VA (1974) Methodology of geophysical studies in hydrogeological surveys for land reclamation. Nedra, Moscow (in Russian)

Shatsov AN (1969) Marine radiometry. Nedra, Moscow (in Russian)

Shaub YuB (1963) Application of correlation analysis to geophysical data processing. Izv AN SSSR Ser Geophys (4):578–589 (in Russian)

Shcherbakova VV, Perrin M, Shcherbakov VP, Pavlov VE, Alvaz'yan A, Zhidkov GV (2009) Rock magnetic and paleointensity results from Mesozoic baked contacts of Armenia. Earth Planets Space 61:23–39

Shempelev AG, Nevsky LN (2008) The abyssal structure of Ossetian Depression and perspectives of its oil-and-gas presence. In: Transaction of the EAGE Conference, Saint-Petersburg, P042, p 1–5

Shempelev AG, P'yankov VYa, Lygin VA, Kuhmazov SU, Morozova AG (2005) Results of geophysical investigations along profile volcano Elbrus – Caucasian Mineral Waters. Regional geology and metallogeny. In: Transactions of the VSEGEI, Sankt-Petersburg, No. 25, p 178–185 (in Russian)

Shikhalibeily ES, Grigoriants BV (1980) Principal features of the crustal structure of the South-Caspian Basin and the conditions of its formation. Tectonophysics 69:113–121

Scholte KH, Hommels A, Van der Meer FD, Slob EC, Kroonenberg SB, Aliyeva E, Huseynov D, Guliev I (2003) Subsurface resistivity in combination with hyperspectral field and satellite data for mud volcano dynamics. In: IEEE Proceedings, Azerbaijan, p 3395–3397

Shraibman VI, Serkerov SA, Sidelnikova TA, Flerov VS (1988) The new elements of magnetic and electric prospecting in studying ground burials in the Northern Caucasus. Sov Archaeol (1):101–112
Shraibman VI, Zhdanov MS, Vitvitsky OV (1977) Correlation methods for transformation and interpretation of geophysical anomalies. Nedra, Moscow (in Russian)
Simmons G (1967) Interpretation of heat flow anomalies. Rev Geophys 5(2):42–52
Singh RP, Dey S (2003) Surface latent heat flux as an earthquake precursor. Nat Hazards Earth Syst Sci 3:749–755
Sirotkina TN, Nikolsky YuI (1971) Method of successive geological approximations by interpretation of geophysical fields in ore regions (on example of Armenia). Method Explor Geophys (Metody Razvedochnoi Geofiz) (12):36–44 (in Russian)
Skovorodkin YuP, Guseva TV, Sigalov BA, Butazov SV, Bezuglaya PI, Boikov PI, Mishin AV, Mamatsuev KA (1994) Geophysical observations during the Caucasus earthquake of April 29, 1991: Aftershock period. Izv Phys Earth 30(1):55–68
Smekalova TN, Voss O, Smekalov SL (2008) Magnetic surveying in archaeology: more than 10 years using the overhauser GSM-19 gradiometer. Publ. House of the Polytechnical University, St. Petersburg
Smirnov AA (1975) Introduction to theory of electromagnetic field. Nedra, Moscow (in Russian)
Smirnov VI (1978) Geology of useful minerals, 2nd edn. Nedra, Moscow (in Russian)
Smirnova MN (1968) Effect of earthquakes on the oil yield of the Gudermes field (Northeastern Caucasus). Izv Phys Solid Earth (12):71–76
Smith WHF, Sandwell DT (1997) Global seafloor topography from satellite altimetry and ship depth soundings. Science 277:1957–1962
Sobisevich AL, Gorbatikov AV, Ovsuchenko AN (2008) Deep structure of the Mt. Karabetov mud volcano. Dokl Earth Sci 422(7):1181–1185
Sobisevich AL, Kanonidi KKh, Sobisevich LE, Gridnev DG (2009) On the one class of electromagnetic disturbances arising before strong earthquakes. Seism Equip (Seismicheskie Prib) 45 (3):48–55 (in Russian)
Sobisevich LE, Sobisevich AL (2010) Dilatancy structures and electromagnetic disturbances of the ULF diapason on stages of preparation and evaluation of large seismic event. Bull Earth Sci Branch Russ Acad Sci 2, NZ6027, doi: 10.2205/2010NZ000045
Sobolev RN, Starostin VI (1993) Petrophysicai features of the ore-bearing Eldzhurtinsky massif (northern Caucasus). J Southe Asian Earth 8(1–4):605–609
Sokolov BA, Khain VE (1982) Presence of oil and gas in overthrusting margins of folded mountainous structures. Sov Geol (12):53–58 (in Russian)
Solovov AP (1985) Geochemical methods for searching useful minerals. Nedra, Moscow (in Russian)
Spichak VV (1999) Magnetotelluric fields in three-dimensional models of geoelectrics. Nauchnyi Mir, Moscow (in Russian)
Spichak VV, Borisova VP, Fainberg EB, Khalezov AA, Goidina AG (2007) Electromagnetic 3D tomography of the Elbrus volcanic center according to magnetotelluric and satellite data. J Volcanol Seismol 1(1):53–66
Starostenko VI, Kostyukevich AS, Kozlenko VG (1988) Comprehensive interpretation of seismometric and gravimetric data. I. Principles and methods. II. Application. Izv Phys Earth (English version) 24(4):371–282, 6:456–462
Stopnevich AD (1913) Value of geothermal gradient in the oil-bearing areas. In: Transactions of the Baku Branch of the Imperial Russian Technical Society, XXVII, Nos. 3–4 (in Russian)
Strakhov VN (1960) Experience in magnetic anomaly interpretation by plotting of ΔZ isolines in the vertical plane. Appl Geophys (Prikl Geofiz) 27:116–130 (in Russian)
Strakhov VN (1976) Main ideas and methods for deriving information from gravitational and magnetic observations. Nedra, Moscow (in Russian)
Strakhov VN, Lapina MI, Efimov AB (1986) Solving direct problems of gravimetry and magnetometry on the basis of novel analytical concepts for field elements from standard approximating bodies, 1–2. Izv Ser Phys Earth (6):55–69, (7):66–78

Sudzhadinov RA, Kosmodemyansky VV (1986) Mapping for oil and gas structures at large depth, In: Ismail-Zadeh TA (ed) Combined interpretation of geological-geophysical materials for investigating the structure and prospecting for oil and gas at large depths, "YuzhVNII-Geofizika", p 94–103 (in Russian)

Suggestions to Authors of the Reports of the United States Geological Survey (1978) US Government Publishing Office, Washington

Sundararajan N, Al-Garni MA, Ramabrahmam G, Srinivas Y (2007) A real spectral analysis of the deformation of a homogeneous electric field over a thin bed – A Hartley transform approach. Geophys Prospect 55:901–910

Surkov A, Samykina E, Eppelbaum L, Semenov S (2008) The main reason for mineral loss in gravity dressing. Open Miner Process J (1):37–44

Suryaninov EYu, Vishnevsky PV, Vasserman VA (1983) Investigation of geothermal anomaly behavior over geological objects with contrasting thermal conductivity. Izv Phys Solid Earth (12):96–102

Svanidze GG, Tzomaya VSh (eds) (1988) Water reserves of the transcaucasian. Gidrometeoizdat, Leningrad (in Russian)

Svetov BS (1992) Information theory basis of geophysics. Electromagnetic Research Centre, Moscow

Tafeyev YuP, Sokolov KP (1981) Geological interpretation of magnetic anomalies. Nedra, Leningrad (in Russian)

Talwani M (1998) Errors in the total Bouguer reduction. Geophysics 63(4):1125–1130

Tan O, Taymaz T (2006) Active tectonics of the Caucasus: earthquake source mechanisms and rupture histories obtained from inversion of teleseismic body waveforms. Geol Soc Am Spec Pap 409:531–578

Tanircan G, Tsereteli N, Garaveliev E, Siyahi B, Varazanashvili O, Mammadli T, Yethirmishli Q, Cehlidze T, Axundov A, Safak E (2011) Seismic hazard assessment for Southern Caucasus-Eastern Turkey Energy Corridor. In: Proceedings of the EGU Conference, Vienna, Geophysical Research Abstracts, vol 13, No. 1744

Tarkhov AG (ed) (1980) Electrical prospecting. Geophysicist's manual. Nedra, Moscow (in Russian)

Telford WM, Geldart LP, Sheriff RE (1990) Applied geophysics. Cambridge University Press, Cambridge

Terekhova RV (1977) Determination of the magnetic anomaly nature by daily geomagnetic variations. In: Geology, rock magnetism and paleomagnetism of the Southern Urals. Ufa, Ural Scientific Center, p 85–95 (in Russian)

Tesmull FA, Crossley DI (1981) Inversion of *VLF*-data for simple lateral inhomogeneities, Geol Surv Canada (81–15):79–86

Thomas R (1998) An update on the Four-Color Theorem. Not Am Math Soc 45(7):848–859

Tikhonov AN, Bulanzhe YuD (1945) On the gravity field averaging. Izv Ser Geophys (2) (in Russian)

Tikhonov AN, Goncharsky AV (eds) (1987) Ill-posed problems in the natural sciences. Mir Publ, Moscow (in Russian)

Tikhonov AN, Samarsky AA (1963) Equations of mathematical physics. Pergamon, Oxford

Triep EG, Abers GA, Lerner-Lam AL, Mishatkin V, Zakharchenko N, Starovoit O (1995) Active thrust front of the Greater Caucasus: the April 29, 1991, Racha earthquake sequence and its tectonic implications. J Geophys Res 100(B3):4011–4033

Tsvetkova T, Monnin M, Nevinsky I, Perelygin V (2001) Research on variation of radon and gamma-background as a prediction of earthquakes in the Caucasus. Radiat Meas 33:1–5

Tsvetkova T, Nevinsky I, Nevinsky V, Paniyshkina A (2004) Experience of working with SSNTDs in Caucasus: a survey. Radiat Meas 38:263–269

Tzimelzon IO (1959) A problem of geological interpretation of Talysh-Vandam gravity maximum. Geol Oil-and-Gas (3):55–65 (in Russian)

Tzimelzon IO (1965) Earth's crust deep structure and tectonics of Azerbaijan by geophysical data. Sov Geol (4):103–111 (in Russian)

Tzimelzon IO (1970) Connection between the sedimentary deposits of Azerbaijan and Earth's crust deep structure. Geotectonics (Geotektonika) (5):69–81 (in Russian)
Turcotte DL (1995) Fractals and chaos in geology and geophysics. Cambridge University Press, Cambridge
Tvalchrelidze AG (2002) Mineral resource base of Georgia in the XXI century. In: Tvalchrelidze AG, Morizot G (eds) Mineral resource base of the Southern Caucasus and systems for its management in the XXI century. Proceedings of the NATO advanced research workshop, Series IY: Earth and Environmental Sciences, vol 17, Kluwer Academic Dordrecht/Boston/London, p 19–70
Tvalchrelidze GA (ed) (1976) Tectonics and metallogeny of the Caucasus. Metsniereba, Tbilisi (in Russian)
Tvalchrelidze GA (1978) On types of pyrite deposits and provinces. Izv Ser Geol (10):5–16 (in Russian)
Ulomov VI, Danilova TI, Medvedeva NS, Polyakova TP, Shumilina LS (2007) Assessment of seismic hazard in the North Caucasus. Izv Phys Solid Earth 43(7):559–572
Vakhromeyev GS, Davydenko AYu (1987) Modeling in applied geophysics. Nedra, Moscow (in Russian)
Valganov SV (2004) Dolmens of the Caucasus. Reconstruction of worship. Agency Business-Press, Moscow (in Russian)
van den Bouwhuysen JNA (1934) The thermocouple proves useful on a geophysical survey. Eng Min J 135:342–344
Van der Voo R, Channel JET (1980) Paleomagnetism in orogenic belts. Rev Geophys Space Phys 18:455–481
van Leusen M (1998) Dowsing and archaeology. Archaeol Prospect 5:123–138
Varlamov AS, Filatov VG (1983) Determination of the density of rocks and geological objects. Nedra, Moscow (in Russian)
Ventsel ES (1969) The probability theory, 3rd edn. revised. Nauka, Moscow (in Russian)
Veselov KE (1986) Gravimetric survey. Nedra, Moscow (in Russian)
Vikulin AV (2003) Physics of wave seismic process. Petropavlovsk State University. Petropavlovsk (in Russian)
Vinnik LP, Godzikowskaya AA, Pataraya YeI, Sikharulidze DI, Bagramyan AKh (1978) Velocity anomalies of upper mantle in the Caucasus. Izv Phys Earth (7):22–31 (in Russian)
Vinnik LP, Makeyeva LI, Milev A, Usenko AYu (1992) Global patterns of azimuthal anisotropy and deformations in the continental mantle. Geophys J Int 111:433–447
Vinogradov PA (1964) The experience of application of low-frequency inductive methods for copper-sulphide ore prospecting in the Northern Caucasus. In: Transaction of Conference on inductive methods in mining geophysics. Nedra, Moscow (in Russian)
Vincent SJ, Davies CE, Richards K, Aliyeva E (2010) Contrasting Pliocene fluvial depositional systems within the rapidly subsiding South Caspian Basin; a case study of the palaeo-Volga and palaeo-Kura river systems in the Surakhany Suite, Upper Productive Series, onshore Azerbaijan. Mar Petrol Geol 27:2079–2106
Vnukov SYu, Porotov AV, Pushkarev PYu, Kelterbaum D (2008) Archaeological and palaeogeographical studies of the Semibratnee city-site. Antiques Bosphorus (12):127–147
Voisin S (2002) Dynamic triggering of earthquakes: The nonlinear slip-dependent friction case. J Geophys Res 107, B12, 2356, doi: 10.1029/2001JB001121, 10_1-10_11
Vysokoostrovskaya EB, Zelenetsky DS (1968) On quantitative evaluation of the prospects of a territory when searching for ore mineral deposits. Sov Geol (8):58–71 (in Russian)
Walker D (2008) Recent developments in low frequency spectral analysis of passive seismic data. First Break 26(2):69–77
Walsh JB, Rice JR (1979) Local gravity change resulting from deformation. J Geophys Res 84:165–170
Yaglom AM, Yaglom IM (1973) The probability and Information, 2nd edn., revised and supplemented. Nauka, Moscow (in Russian)
Yakubovsky YuV, Lyakhov LL (1956) Electric prospecting. Gosgeolizdat, Moscow (in Russian)
Yanovsky BM (1978) Earth's magnetism. Leningrad State Univ. Publ.. Leningrad (in Russian)

Zaborousky AI (1963) Electric prospecting. Gostoptekhizdat, Moscow (in Russian)
Zadeh LA (1983) The role of fuzzy logic in the management of uncertainty in expert systems. Fuzzy Set Syst 11:199–227
Zaitseva IV, Nagiev VI, Kurbanov NK et al (1988) Detailed prospecting of Katekh pyrite-polymetallic deposit. vols 1 & 2, "*Azerbaijangeologiya*" *Association* (in Russian)
Zakariadze GS, Dilek Y, Adamia SA, Oberhänsli RE, Karpenko SF, Bazylev BA, Solov'eva N (2007) Geochemistry and geochronology of the Neoproterozoic Pan-African Transcaucasian Massif (Republic of Georgia) and implications for island arc evolution of the late Precambrian Arabian–Nubian Shield. Gondwana Res 11:92–108
Zeinalov GA (2000) Importance of remote-sensing data in structural geologic analysis of oil- and gas-bearing regions of Azerbaijan. Nat Resour Res 9(4):307–313
Zemtsov EE (1967) On possibility of prognosis of oil-and-gas-bearing structures by seismic prospecting. In: Fedynsky VV (ed) Techniques, technology and results of geophysical prospecting. Nedra, Moscow, pp 264–271 (in Russian)
Zhdanov MS (1984) The analogues of the Cauchy type integral for the theory of geophysical fields. Nauka, Moscow (in Russian)
Zhdanov MS, Keller GV (1994) The geoelectrical methods in geophysical exploration. Elsevier, Amsterdam
Zheludev V, Alperovich L (1999) Wavelet analysis as a tool for revealing geomagnetic precursors of an earthquake. *Seismo-Electromagnetics*, *TERRAPUB*, 255, Tokyo
Zhurkov CN (1968) Kinetic conception of solid body durability. Bull Acad Sci USSR (3):46–52
Zolotovitskaya TA (2003) Mechanisms of formation of radionuclide contamination at the oil fields of Azerbaijan. In: Birsen N, Kadyrzhanov KK (eds) Environmental protection against radioactive pollution. Kluwer Acad. Publisher NATO Science Series, Earth & Environmental Sci., vol 33, 74–77
Zonenshain LP, Le Pichon X (1986) Deep basins of the Black Sea and Caspian Sea as remnants of Mesozoic back-arc basins. Tectonophysics 123:181–211
Zonge KL, Sauck WA, Sumner JS (1972) Comparison of time, frequency, and phase measurements in induced polarization. Geophys Prospect 20(3):626–648
Zor E (2008) Tomographic evidence of slab detachment beneath eastern Turkey and the Caucasus. Geophys J Int 175(3):1273–1282
Zorin YuA, Lysak SV (1972) Quantitative interpretation of geothermal anomalies. Izv Ser Phys Earth (9):68–73
Zverev AS, Kiriyakov VH, Lyubomov VV (1997) Registration of magnetic activity. Instrum Tech of Exp (1):168 (in Russian)
Zverev VP, Dolnikov VA, Khutorsky MD, Fotogdinov RA (1982) Thermal effect and sulphide oxidation rate in the Katekh deposit (as applied to the southern slope of the Greater Caucasus). Dokl USSR Acad Sci 265(4):960–962

Index

L.V. Eppelbaum and B.E. Khesin, *Geophysical Studies in the Caucasus*, Lecture Notes in Earth System Sciences, DOI 10.1007/978-3-540-76619-3,

T

MIX
Papier aus verantwortungsvollen Quellen
Paper from responsible sources
FSC® C105338

If you have any concerns about our products,
you can contact us on
ProductSafety@springernature.com

In case Publisher is established outside the EU,
the EU authorized representative is:
Springer Nature Customer Service Center GmbH
Europaplatz 3, 69115 Heidelberg, Germany

Printed by Libri Plureos GmbH
in Hamburg, Germany